Springer Series in Materials Science

Volume 231

More information about this series at http://www.springer.com/series/856

The Springer Series in Materials Science covers the complete spectrum of materials physics, including fundamental principles, physical properties, materials theory and design. Recognizing the increasing importance of materials science in future device technologies, the book titles in this series reflect the state-of-the-art in understanding and controlling the structure and properties of all important classes of materials.

Arcady Zhukov
Editor

Novel Functional Magnetic Materials

Fundamentals and Applications

Springer

Editor
Arcady Zhukov
UPV/EHU
Basque Country University
San Sebastian, Spain

ISSN 0933-033X ISSN 2196-2812 (electronic)
Springer Series in Materials Science
ISBN 978-3-319-26104-1 ISBN 978-3-319-26106-5 (eBook)
DOI 10.1007/978-3-319-26106-5

Library of Congress Control Number: 2015959945

Printed on acid-free paper

This Springer imprint is published by Springer Nature
The registered company is Springer International Publishing AG Switzerland

Preface

This book on functional magnetic materials was inspired by the rapidly growing interest in research on magnetism and magnetic materials spurred by the discovery of novel magnetic materials, including nanomaterials, and recent progress in the development of functional materials with improved magnetic and magneto-transport properties for use in a wide variety of applications. The expert chapters from leading researchers from around the world cover a wide range of experimental and theoretical work highlighting the following main topics:

- Soft magnetic materials and sensor applications
- Magnetocaloric materials and magnetic refrigeration
- Magnetic shape alloys and related applications
- Amorphous and nanocrystalline magnetic materials and applications
- Hard magnetic materials
- Magnetic semiconductors
- Composites

The aim throughout is to provide the most up-to-date information on recent developments in magnetic materials with an eye toward industry and advanced technologies.

The technology industry requires cost-effective materials with reduced dimensionality and desirable magnetic properties such as enhanced magnetic softness, giant magnetic field sensitivity, large magnetocaloric effect, and large shape memory effect for use in magnetic sensors, microelectronics, security, and energy-efficient magnetic refrigerators. In particular, the miniaturization of modern magnetoelectronic devices tends to stimulate rapid development of nanoscaled magnetic materials. This, in turn, has led to the development of novel magnetic materials in the form of ribbons, wires, microwires, and multilayered thin films, which have attracted significant attention from the scientific community.

In the area of magnetic materials, the discovery of the so-called giant magnetoimpedance effect in these materials makes them very attractive for a wide range of high-performance sensor applications ranging from use in electric surveillance to biomedicine. In another research area, the development of novel magnetocaloric materials for advanced magnetic refrigeration technology has also generated growing interest among scientists. The majority of magnetic refrigeration needs economical, environmentally friendly materials that possess high cooling efficiencies (i.e., large magnetocaloric effect over a wide temperature range). To increase the heat exchange rate, the surface-to-volume ratio must be enhanced. Therefore progress in the development of magnetocaloric materials with low dimensionality, like ribbons, films, or wires, has drawn significant industry attention. Other magnetic materials such as hard magnets, tuneable composites, and magnetic semiconductors are also demanded by industry.

In all of these cases, a comprehensive understanding of the processing-structure-property relationship in the fabricated materials is of critical importance. Consequently, great efforts have been and continue to be focused on systematic theoretical and experimental studies with the overall goal of advancing our current knowledge of the origins of material properties in relation to some special arrangements at the nanometric scale and, relatedly, to the prediction of novel, unusual macroscopic properties based on nano- and microstructure. These efforts are a common theme throughout the chapters in this book.

I hope this publication will stimulate further interest in magnetic materials research. Last but not least, I would like to acknowledge all the contributing authors for their invaluable time, great contributions, and assistance with this book. Without such efforts we would not be able to accomplish and bring this special volume to the interested readers.

San Sebastian, Spain Arcady Zhukov

The original version of the book was revised because Arcady Zhukov's name was misspelled. An erratum explaining this can be found at DOI 10.1007/978-3-319-26106-5_11

Contents

Contributors

N. Ali Department of Physics, Southern Illinois University Carbondale, Carbondale, IL, USA

W. Anwand Helmholtz-Zentrum Dresden-Rossendorf, Dresden, Germany

J.M. Barandiaran BC Materials and University of the Basque Country (UPV/EHU), Bilbao, Spain

R. Caballero-Flores Department of Physics, University of Oviedo, Oviedo, Spain

V.A. Chernenko BC Materials & University of the Basque Country (UPV/EHU), Bilbao, Spain

Ikerbasque, Basque Foundation for Science, Bilbao, Spain

I. Dubenko Department of Physics, Southern Illinois University Carbondale, Carbondale, IL, USA

Rhimou ElKammouni Instituto de Ciencia de Materiales de Madrid, Madrid, Spain

Laboratory of Magnetic Sensoric URFU, Ekaterinburg, Russia

L. Escoda Girona University, Girona, Spain

E. Gan'shina Faculty of Physics, M.V. Lomonosov Moscow State University, Moscow, Russia

J. González Department of Materials Physics, University of the Basque Country, San Sebastián, Spain

L. González-Legarreta Department of Physics, University of Oviedo, Oviedo, Spain

A. Granovsky Faculty of Physics, M.V. Lomonosov Moscow State University, Moscow, Russia

B. Hernando Department of Physics, University of Oviedo, Oviedo, Spain

Mihail Ipatov Department of Materials Physics, University of the Basque Country, San Sebastián, Spain

Ludek Kraus Institute of Physics, Academy of Sciences of the Czech Republic, Prague, Czech Republic

Galina V. Kurlyandskaya Laboratory of Magnetic Sensors, Ural Federal University, UrFU, Ekaterinburg, Russia

Departamento de Electricidad y Electrónica, Universidad del País Vasco, UPV/EHU, Bilbao, Spain

V.A. L'vov Taras Shevchenko National University, Kyiv, Ukraine

Institute of Magnetism, Kyiv, Ukraine

E. Lähderanta Lappeenranta University of Technology, Lappeenranta, Finland

Dmitri V. Louzguine-Luzgin WPI Advanced Institute for Materials Research, Tohoku University, Sendai, Japan

Yang Luo Advanced Composites Centre for Innovation and Science, Department of Aerospace Engineering, University of Bristol, Bristol, UK

Julia Lyubina Experimental Solid State Group, Blackett Laboratory, Department of Physics, Imperial College London, London, UK

A. Orlov Federal State Research and Design Institute of Rare Metal Industry, Moscow, Russia

Hua-Xin Peng Institute for Composites Science Innovation (InCSI), School of Materials Science and Engineering, Zhejiang University, Hangzhou, China

N. Perov Faculty of Physics, Lomonosov Moscow State University, Moscow, Russia

K. Potzger Helmholtz-Zentrum Dresden-Rossendorf, Dresden, Germany

V.M. Prida Department of Physics, University of Oviedo, Oviedo, Spain

V. Prudnikov Faculty of Physics, M.V. Lomonosov Moscow State University, Moscow, Russia

Faxiang Qin Institute for Composites Science Innovation (InCSI), School of Materials Science and Engineering, Zhejiang University, Hangzhou, China

Valeria Rodionova Immanuel Kant Baltic Federal University, Kaliningrad, Russia

W.O. Rosa Centro Brasileiro de Pesquisas Físicas, Urca, Rio de Janeiro, Brazil

Fabrizio Scarpa Advanced Composites Centre for Innovation and Science, Department of Aerospace Engineering, University of Bristol, Bristol, UK

A.S. Semisalova Faculty of Physics, Lomonosov Moscow State University, Moscow, Russia

Lappeenranta University of Technology, Lappeenranta, Finland

Helmholtz-Zentrum Dresden-Rossendorf, Dresden, Germany

R. Skomski Department of Physics and Astronomy and Nebraska Center for Materials and Nanoscience, University of Nebraska, Lincoln, NE, USA

A. Smekhova Faculty of Physics, Lomonosov Moscow State University, Moscow, Russia

Fakultät für Physik, Experimentalphysik, Universität Duisburg-Essen, Duisburg, Germany

S. Stadler Department of Physics and Astronomy, Louisiana State University, Baton Rouge, LA, USA

J.J. Suñol Girona University, Girona, Spain

Manuel Vázquez Instituto de Ciencia de Materiales de Madrid, Madrid, Spain

Arcady Zhukov UPV/EHU, Basque Country University, San Sebastian, Spain

Valentina Zhukova Faculty of Chemistry, Basque Country University, San Sebastian, Spain

Chapter 1
Magnetic Shape Memory Materials with Improved Functional Properties: Scientific Aspects

V.A. L'vov, V.A. Chernenko, and J.M. Barandiaran

1.1 Introduction

The ferromagnetic shape memory alloys (SMAs) attract considerable attention of researchers mainly because they proved to be easy deformable by a moderate magnetic field. The record-breaking magnetic field-induced strain (MFIS), $\varepsilon \approx 12\ \%$, was induced recently in the ferromagnetic Heusler-type Ni–Mn–Ga alloy by the external magnetic field $H \approx 10$ kOe [1].

Historically, as early as in 1996, the nearly 0.2 % strain was induced by the magnetic field application to the Ni–Mn–Ga single crystal [2]. As it was noticed by the authors of the pioneer work [2], the observed value of MFIS was too large, to be attributed to the magnetostriction which is inherent to all magnetically ordered solids. In this connection, the idea about the strong influence of the moderate magnetic field on the microstructure of internally twinned alloy via the *magnetically induced mechanical stress* (magnetostress) was put forward [2–4]. Furthermore, it was shown that the magnetostress is caused by the magnetoelastic interaction [5, 6].

The most studied ferromagnetic SMAs belong to the Heusler Ni_2MnGa compound and its off-stoichiometric analogues. Basically, these alloys exhibit the

V.A. L'vov
Taras Shevchenko National University, Kyiv 01601, Ukraine

Institute of Magnetism, 36-b, Vernadsky Street, Kyiv 03142, Ukraine
e-mail: victor.a.lvov@gmail.com

V.A. Chernenko (✉)
BC Materials & University of the Basque Country (UPV/EHU), 48080 Bilbao, Spain

Ikerbasque, Basque Foundation for Science, Bilbao 48013, Spain
e-mail: volodymyr.chernenko@ehu.eus

J.M. Barandiaran
BC Materials & University of the Basque Country (UPV/EHU), 48080 Bilbao, Spain
e-mail: manub@we.lc.ehu.es

A. Zhukov (ed.), *Novel Functional Magnetic Materials*, Springer Series in Materials Science 231, DOI 10.1007/978-3-319-26106-5_1

cubic–tetragonal and cubic–orthorhombic martensitic transformations (MTs), in some cases accompanied by the lattice modulations and slight monoclinic distortion. MFIS is induced by the magnetic field-activated twin boundary motion in the internally twinned martensitic tetragonal or orthorhombic phases. MFIS of about -5 % and -6 % was observed in Refs. [4, 7] and [8], respectively, under the external magnetic field of about 5 kOe applied to the tetragonal modulated martensite with $c/a < 1$ (a, c are the lattice parameters of tetragonal crystal lattice). The maximal deformation of this martensite takes place when the magnetic field $\mathbf{H}||[001]||z$ transforms the tetragonal lattice with the fourfold symmetry axis parallel to a crystallographic direction $[010]||y$ into the lattice with the fourfold symmetry axis directed along $[001]||z$ (y, z are the coordinate axes). In practice, this twin variant conversion (*martensite reorientation*) goes through the intermediate twinned state of the crystal, which consists of the alternating y- and z-domains of tetragonal lattice. MFIS measured in Refs. [4, 7, 8] is close to the "theoretical maximum," $\varepsilon_m = c/a - 1$, which corresponds to the case of complete martensite reorientation. Moreover, the similarity between the strain–field and stress–strain curves was observed. A comparison of the strain field and stress–strain dependencies showed that a giant MFIS arises when the magnetostress exceeds the threshold value $\sigma_{\text{th}} \sim 1$ MPa and approaches theoretical maximum at the magnetostress value equal to about 2 MPa [4, 7, 8]. An elevation of the achieved MFIS value from 0.2 to 6 % stimulated further intense studies of the ferromagnetic SMAs causing sharp increase of the number of publications concerning the magnetically induced deformation (see, e. g., the review article [9]). The mentioned before record-breaking MFIS (≈ 12 %) was obtained by the magnetic field application to a single variant of the non-modulated martensite with $c/a = 1.152$ [1].

A majority of the recent studies on ferromagnetic SMAs focus at the technological advances of these multifunctional materials, while the physical understanding of the phenomena, they exhibit, is not enough elaborated and/or explained. A special emphasis in the current literature is put upon the following physical effects important for applications: (1) the low threshold value of magnetostress, which "triggers" the twin reorientation, (2) an anhysteretic deformational behavior under the cyclic mechanical load and/or magnetic field, (3) a stability of the deformational behavior with respect to the alloy aging and thermomechanical cycling, and (4) the magnetovolume effects and their role in the heat evolution/absorption in the course of martensitic transformation. A rigorous consideration of these effects goes beyond the results obtained for the ferromagnetic SMAs and should involve also the ones available for the ordinary SMAs. In the present chapter, we briefly review the state of the art of these important effects including their analysis in the framework of a Landau theory as an efficient and predictive tool capturing their underlying physics at the phenomenological level.

The chapter is organized in the four main sections, according to the aforementioned sequence of problems: (1)–(4). Considering a big amount of literature related to these problems, we restrict ourselves by a citation of some pioneering works and selected articles, which provide a conceptual comprehension of phenomena under the questions. More references can be found in the selected articles.

1.2 Giant Magnetically Induced Deformation of Ferromagnetic SMA

1.2.1 Magnetostress as the Origin of Martensite Reorientation

The magnetostress induced in the Ni–Mn–Ga single crystals by a field aligned with one of the $\langle 100 \rangle$ crystallographic directions was considered in the early works [3–8]. In particular, a dependence of the magnetostress on the magnetic field was determined in Ref. [6] from the experimental stress–strain curves measured in the presence of an external magnetic field. According to Ref. [6], the compressive axial load and external magnetic field were applied perpendicularly to each other along the $\langle 100 \rangle$ crystallographic directions. The stress–strain curves showing an occurrence of the martensite reorientation in zero magnetic field and stress-induced MT in the stationary magnetic field were obtained. The stress-induced MTs were observed under the constant field values, H_n, where $n = 1, 2...15$. Three of the stress–strain curves taken at the presence of magnetic field are shown in Fig. 1.1a. The magnetostress induced by the constant field can be estimated from the stress values corresponding to a middle point of the stress–strain curve obtained in the presence of this field and a middle point of the zero-field curve, as shown in Fig. 1.1a. The values of magnetostress induced by the fields $H_7 = 4$ kOe and $H_{13} = 10$ kOe are presented by the two-side arrows. More precise procedure of a determination of the magnetostress was also described in Ref. [6]. The determined magnetostress values are shown in Fig. 1.1b by the open circles; the line connecting these circles is a guide for the eye.

As it is seen from Fig. 1.1a, the stress values of about 0.7 and 1.8 MPa are needed for the start and finish of martensite reorientation, respectively, in zero magnetic field. Figure 1.1b shows that these stress values can be induced by the external magnetic fields of 2.5 and 4.5 kOe, respectively. These fields are close in value to those directly determined in Refs. [7, 8] from the strain–field curves.

Figure 1.2 illustrates a correspondence between the magnetostress evolution determined from the stress–strain curves and magnetization behavior. The field dependence of magnetostress occurred close to a quadratic function of the magnetization ζM^2, where ξ is constant. (For the experimental details concerning the magnetization measurement, see Ref. [6].) It should be noticed that some theoretical models stated that magnetostress, $\sigma(H)$, linearly depends on the magnetic field in the low field range (see, e. g., [8, 10]). The data shown in Fig. 1.2 disproves this statement and points to the quadratic function $\sigma(H)$ in the fields $H \leq 4$ kOe,

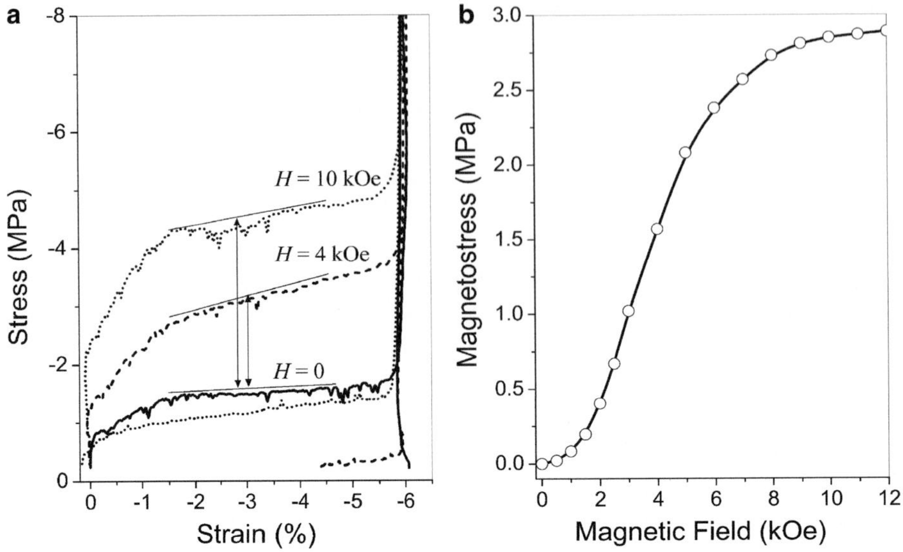

Fig. 1.1 Stress–strain loops measured for Ni–Mn–Ga single crystal in zero magnetic field and under constant magnetic field applied perpendicularly to the compressive load, (**a**). Magnetostress as a function of applied magnetic field, (**b**). See Ref. [6]

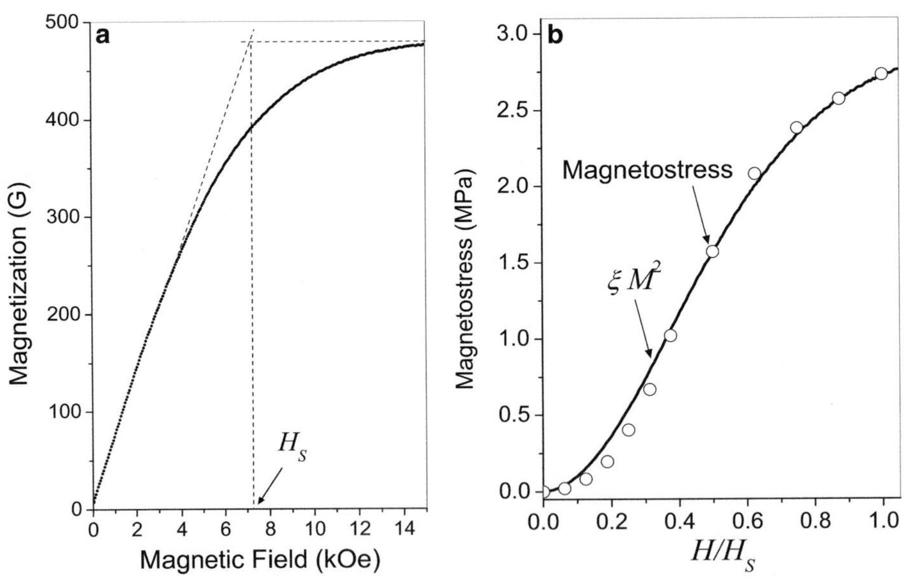

Fig. 1.2 Magnetization curve measured for the Ni–Mn–Ga single crystal, (**a**). The magnetostress (*circles*) and square of magnetization (*line*) as a function of the reduced magnetic field (**b**). See Ref. [6]

because in this field range, the magnetization linearly depends on the magnetic field.

As it became clear from the very beginning of the studies of giant MFIS, the magnetic field-induced martensite reorientation is observable only if the magnetic field direction is close to the $\langle 100 \rangle$ directions in the Ni–Mn–Ga single crystalline specimen and the temperature of the specimen is close enough to the MT temperature. The temperature dependence of magnetostress and limits of MFIS were studied in Ref. [11]; the contributions of different factors to this dependence were analyzed in Ref. [12]. A generalization of the magnetostress conception to the other directions of magnetic field was further given in Ref. [13].

1.2.2 Magnetoelastic Coupling as the Origin of Magnetostress

The determination of the temperature range and field directions being favorable for the observation of large MFIS is a practically important task, which needs a special consideration. For the task solution let us consider the cubic crystal lattice undergoing the cubic–tetragonal MT. This MT is accompanied by the appearance of the large (about several percents) shear strains and comparatively small volume change (≤ 0.5 %). As so, for z-variant of a tetragonal phase: $\varepsilon_{xx} = \varepsilon_{yy} \approx -\varepsilon_{zz}/2$, and the MT strain, ε_{M}, is expressed through the lattice parameters a_0 for cubic and $a,\ c$ for tetragonal phases as

$$\varepsilon_{\mathrm{M}} = \varepsilon_{zz} = (c - a)/a_0 \approx -(2/3)(1 - c/a) \tag{1.1}$$

The nondiagonal components of strain tensor are equal to zero in both the austenitic and martensitic phases and can be disregarded hence.

The cubic–tetragonal MT is traditionally described in terms of the variables

$$u_1 = \left(\varepsilon_{xx} + \varepsilon_{yy} + \varepsilon_{zz}\right)/3, \tag{1.2}$$

and

$$u_2 = \sqrt{3}\left(\varepsilon_{xx} - \varepsilon_{yy}\right),\ \ u_3 = 2\varepsilon_{zz} - \varepsilon_{yy} - \varepsilon_{xx}, \tag{1.3}$$

which express the uniform compression/extension and shear straining of the crystal, respectively. (Here the notations from Refs. [5, 6] are used).

The Gibbs potential of ferromagnetic SMA is expressed through the elastic (F_e), magnetic (F_m), and magnetoelastic (F_{me}) parts of the Helmholtz free energy as

$$G = F_e(u_\alpha) + F_m(m_i) + F_{me}(u_\alpha, m_i) + 3Pu_1 - \frac{1}{6}(\sigma_2 u_2 + \sigma_3 u_3) \tag{1.4}$$

where

$$\begin{aligned} \sigma_2 &= \sqrt{3}(\sigma_{xx} - \sigma_{yy}), \\ \sigma_3 &= 2\sigma_{zz} + \sigma_{yy} + \sigma_{xx}, \end{aligned} \tag{1.5}$$

$\alpha = 1,\ 2,\ 3,\ i = x,\ y\ z$, and m_i are the components of unit magnetic vector $\mathbf{m}$, which is related to magnetization vector $\mathbf{M}$ as $\mathbf{m} = \mathbf{M}/M$.

The expressions for the elastic and magnetic energies will be presented in this chapter when they are necessary. The magnetoelastic free energy is expressed as

$$\begin{aligned} F_{me} = &-\delta_{ex} M^2(T) u_1 \\ &- \delta M^2(T)\left[\sqrt{3}u_2\left(m_x^2 - m_y^2\right) + \left(2m_z^2 - m_y^2 - m_x^2\right)u_3\right], \end{aligned} \tag{1.6}$$

where the magnetoelastic constants δ_{ex} and δ describe an influence of the deformation on the energies of spin–exchange and spin–orbit interactions [9]. The strong inequality $\delta_{ex} >> \delta$ is fulfilled if the Curie temperature of ferromagnetic solid is of the order of room temperature or higher.

For the sake of definiteness, let us consider the Ni–Mn–Ga alloy undergoing MT to the tetragonal martensitic phase with $c/a < 1$. In this case MT goes in the ferromagnetic state because its temperature, T_M, is lower than the Curie temperature, T_C.

For z-variant of the martensitic state $u_2 = 0$, $u_3 = 3\varepsilon_M < 0$, and therefore,

$$F_{me} = -3\delta\varepsilon_M M^2(T)\left[2m_z^2(H) - m_y^2(H) - m_x^2(H)\right]. \tag{1.7}$$

The MT strain, ε_M, is related to the lattice parameters of tetragonal phase (see Eq. (1.1)). Comparing the energies of the states with $\mathbf{m}||\mathbf{z}$ and $\mathbf{m}\perp\mathbf{z}$, one can see that MT results in the appearance of the magnetocrystalline anisotropy energy:

$$F_{me} = -K_u m_z^2(H), \tag{1.8}$$

with magnetocrystalline anisotropy parameter:

$$K_u \equiv 9\delta\varepsilon_M M^2(T) = -6\delta M^2(T)(1 - c/a), \tag{1.9}$$

which linearly depends on the "tetragonality" of crystal lattice, $1 - c/a$. The magnetoelastic constant is negative for Ni–Mn–Ga alloys; therefore, the case $\delta < 0$ should be considered. In this case $K_u > 0$ for the martensitic phase with $c/a < 1$ and the easy-axis magnetic state with magnetic vector $\mathbf{m}||\mathbf{z}||\mathbf{c}$

corresponds to the minimum of anisotropy energy. This state is stable in the absence of magnetic field and in the magnetic field applied along the easy axis **c**.

The magnetoelastic energy, Eq. (1.6), can be presented in the form

$$F_{me} = 3P^{(me)}u_1 - \sum_{i,k} \sigma_{ik}^{(me)} \varepsilon_{ik}, \tag{1.10}$$

where $P^{(me)}$ and $\sigma_{ik}^{(me)}$ may be interpreted as the magnetoelastic pressure and uniaxial compressive stress, respectively. Both pressure and stress arise in the course of the ferromagnetic ordering of SMA and result in the spontaneous magnetostriction. The ordinary magnetostriction is present in both the cubic and tetragonal phases because MT occurs in a ferromagnetic state. Equations (1.3), (1.6), and (1.10) show that

$$\begin{aligned} P^{(me)} &= -\delta_{ex} M^2(T)/3, \\ \sigma_{zz}^{(me)}(H,T) &\equiv 2\delta M^2(T)\left[2m_z^2(H) - m_y^2(H) - m_x^2(H)\right]. \end{aligned} \tag{1.11}$$

The relationship $\varepsilon_{xx} = \varepsilon_{yy} \approx -\varepsilon_{zz}/2$, which corresponds to the volume conservation during the spontaneous deformation process, is accepted for the second term at the right side of Eq. (1.10), because in this equation the shear strains, ε_{ik}, and volume change, $\Delta V/V = 3u_1$, are separated.

For z-variant of the tetragonal lattice, we have $m_z(0) = 1$, $m_y(0) = m_x(0) = 0$, and therefore, the diagonal component of magnetoelastic stress, $\sigma_{zz}^{(me)}$, is negative (see Eq. (1.11)). This stress causes a magnetostrictive contraction of the crystal lattice in z-direction and hence stabilizes z-variant of the martensitic phase with $c/a < 1$. It may be shown in the same way that the stress components $\sigma_{yy}^{(me)}$ and $\sigma_{xx}^{(me)}$ stabilize y- and x-variants of martensite with $\mathbf{m}(0)||\mathbf{y}$ and $\mathbf{m}(0)||\mathbf{x}$, respectively.

Let the increasing magnetic field, H_y, be applied to the martensitic structure formed by the alternating z- and y-variants of martensite. This field does not change the direction of magnetic vector in y-variants but does rotate the magnetic vectors in z-variants from z- to y-direction. The field induces the positive temperature-dependent stress in z-variants of martensite. This stress reaches the maximal value

$$\sigma_{zz}(T) \equiv \sigma_{zz}^{(me)}(H_S, T) - \sigma_{zz}^{(me)}(0,T) = -6\delta M^2(T), \tag{1.12}$$

when the magnetic state with $m_z = m_x = 0$, $m_y = 1$ is established in z-variants.

The positive stress $\sigma_{zz}(T)$ expands the crystal lattice in z-direction, and so, the magnetic field applied in y-direction destabilizes z-variant of martensite with $c/a < 1$. In the case of Ni–Mn–Ga alloys, the z- and y-variants of martensite form the twin structure with the highly mobile twin boundaries. The field-induced destabilization of z-components of twins results in the twin boundary motion, which, in turn, leads to a decrease of the volume fraction of these components in the twin structure and the appropriate increase of the volume fraction of y-components of twins. This is how the martensite reorientation process proceeds.

Equations (1.9) and (1.12) result in the relationship

$$\sigma_{zz}(T) = \frac{K_u(T)}{1 - c(T)/a(T)}, \quad (1.13)$$

which is commonly used for the rough estimation of magnetostress (see Refs. [14, 15] and references therein).

As it was mentioned above, the magnetostress was considered in Refs. [3–8] only for the magnetic field applied in the high symmetric directions of the crystal lattice. The theoretical analysis of the dependence of magnetostress on the field direction was carried out later on [13]. The influence of the external magnetic field **H** on the twin structure formed by the neighboring martensite variants (twin components) with $\mathbf{c}||[001]||\mathbf{z}$ and $\mathbf{c}||[010]||\mathbf{y}$ was considered. It was taken into account that the martensite reorientation is caused by the difference of stress components:

$$\sigma_{zz}(\mathbf{H}) - \sigma_{yy}(\mathbf{H}) \equiv \sigma(\mathbf{H}), \quad (1.14)$$

where $\sigma_{ii}(\mathbf{H}) = \sigma_{ii}^{(me)}(\mathbf{H}) - \sigma_{ii}^{(me)}(0)$. The stress values in the twin components, $\sigma^{[001]}(\mathbf{H}, T)$ and $\sigma^{[010]}(\mathbf{H}, T)$, were calculated using Eq. (1.6) for magnetoelastic energy. It was shown that these values depend on the magnetic field direction as

$$\begin{aligned} \sigma^{[001]}(\mathbf{H}) &= 6\delta M^2(T)\left\{\cos\left[2\varphi^{[001]}(\mathbf{H})\right] - \cos\left[2\varphi(0)^{[001]}\right]\right\}, \\ \sigma^{[010]}(\mathbf{H}) &= 6\delta M^2(T)\left\{\cos\left[2\varphi^{[010]}(\mathbf{H})\right] - \cos\left[2\varphi(0)^{[010]}\right]\right\}, \end{aligned} \quad (1.15)$$

where $\varphi^{[001]}(\mathbf{H})$ and $\varphi^{[010]}(\mathbf{H})$ are the angles between the [001] direction and the magnetic vectors of the martensite variants with $\mathbf{c}||[001]$ and $\mathbf{c}||[010]$, respectively [13]. It was concluded that the magnetic field application is equivalent to the mechanical loading of the twinned crystal by the equivalent stress:

$$\sigma_{\mathrm{eq}}(\mathbf{H}, T) = \frac{1}{2}\left[\sigma^{[001]}(\mathbf{H}, T) + \sigma^{[010]}(\mathbf{H}, T)\right]. \quad (1.16)$$

It should be emphasized that the *equivalent stress* is just what has been referred to as the *magnetostress* and determined above from the experimental stress–strain curves taken in the external magnetic field (Fig. 1.1). The factor 1/2 was introduced in Eq. (1.16) since the magnetostress $\sigma^{[001]}(\mathbf{H}, T)$ vanishes when the magnetic field is aligned with [001] direction (because $\varphi^{[001]}(\mathbf{H}) = \varphi^{[001]}(0) = 0$) and $\sigma^{[010]}(\mathbf{H}, T)$ exists only in one of twin components, while the mechanical load stresses both components of a twin [6, 13].

The dimensionless magnetoelastic constant $\delta = -23$ evaluated from the magnetostriction of cubic phase provides correct experimental values of the magnetic anisotropy of tetragonal phase and magnetostriction of cubic phase of the Ni–Mn–Ga alloys (see [9, 12] and references therein). The saturation magnetization of these alloys is about of 500 G at room temperature. In this case the equivalent stress that is induced by the saturation magnetic field applied in y-direction to z-variant of martensite is about 3.4 MPa. This theoretical value is rather close to the saturation value of magnetostress ($\approx$ 2.9 MPa) determined from the experimental stress–strain loops (see Fig. 1.1b). The value $\sigma_{eq} = 2.9$ MPa results from Eqs. (1.12)–(1.16) for magnetization value equal to 460 G at room temperature.

1.2.3 Limits of MFIS Observability

The threshold stress initiating the martensite reorientation is about of 10 MPa for the majority of SMAs. For this reason, the observation of a large (about few percents) MFIS occurred possible only in some ferromagnetic SMAs, after the elaboration of the complicate modes of production and pretreatment of the samples. It was shown, moreover, that MFIS is observable only if the temperature of ferromagnetic SMA (the Ni–Mn–Ga alloys were used for experiments) lies in a certain temperature interval below MT temperature [11]. It happened that the threshold stress, σ_{th}, which starts the martensite reorientation, increases with a decreasing of temperature quicker than the magnetostress, σ_{eq}. For this reason, the lower limit of MFIS observation can be found from a condition $\sigma_{eq} < \sigma_{th}$. It was demonstrated, however, that the width of the temperature interval of MFIS existence (~100 K) is large enough for the practical applications [11].

The magnetic field orientation along with one of the $\langle 100 \rangle$ crystallographic directions is the most suitable for the MFIS observation. In practice, a mutual orientation of the magnetic field vector and crystal axes deviates from the optimal one, especially, in the films and small structural elements of technical devises. Therefore, the orientation dependence of magnetostress and angular limits of MFIS observation deserve an analysis as provided below.

The direction of magnetic vector of a twin component (martensite variant) in the increasing magnetic field is controlled by the condition $\partial F/\partial\varphi = 0$ for the free energy:

$$F = \pm K_u m_z^2 - \mathbf{mH}, \tag{1.17}$$

where $K_u > 0$ and the sign "+" applies to the y-component and the sign "−" to the z-component of twin showing that $\mathbf{m}(0) \perp \mathbf{z}$ in y-component and $\mathbf{m}(0) || \mathbf{z}$ in z-component. The equation for determination of the magnetic vector direction is

$$\sin 2\varphi \pm \frac{2H}{H_A(T)} \sin(\theta - \varphi) = 0, \tag{1.18}$$

where θ is the angle between the magnetic field and z-axis and $H_A(T) = 2K_u(T)/M(T)$ is the magnetic anisotropy field.

The angular dependencies of the equivalent stress computed for the representative Ni–Mn–Ga alloy (that is using the values $M = 500$ G at room temperature and $\delta = -23$) are presented in Fig. 1.3 for the different values of external magnetic field. The equivalent stress, $\sigma_{eq}(\theta)$, is equal to zero at $\theta = 45^\circ$, because the neighboring twin components have the same magnetoelastic energy when magnetic field is parallel to the [011] direction. The absolute value of $\sigma_{eq}(\theta)$ function reaches maximum when the magnetic field vector approaches the [001] or [010] direction. The negative stress may induce the transformation of y-variant into z-variant ($y \rightarrow z$ martensite reorientation), and the positive one may initiate the $z \rightarrow y$ martensite reorientation. The angular dependencies of the equivalent stress show that the stronger is magnetic field, the wider is angular range, which is admissible for the observation of martensite reorientation and giant MFIS.

Let the stress value, which is needed for the start of $y \rightarrow z$ martensite reorientation, is equal to -2 MPa. This value is close to the ones measured in the course of stress–strain tests performed, e.g., in Refs. [6, 16]. In this case the field-induced martensite reorientation process can start if the deviation of the magnetic field vector from [001] or [010] direction is less than 10°, 22.5°, or 26° for magnetic fields H_A, $2H_A$, or $4H_A$, respectively (the limiting θ values are marked by arrows in Fig. 1.3). The field $H = 0.6H_A$ cannot induce the martensite reorientation. It should be noticed that the decrease of the absolute value of equivalent stress $|\sigma_{eq}(0)|$ results in the drastic decrease of the angular limits of an observability of the martensite reorientation. If, for example, $\sigma_{eq}(0) = -2.5$ MPa, the limiting values of the angle are 2.5°, 12°, and 16° for the magnetic fields H_A, $2H_A$, and $4H_A$, respectively [17].

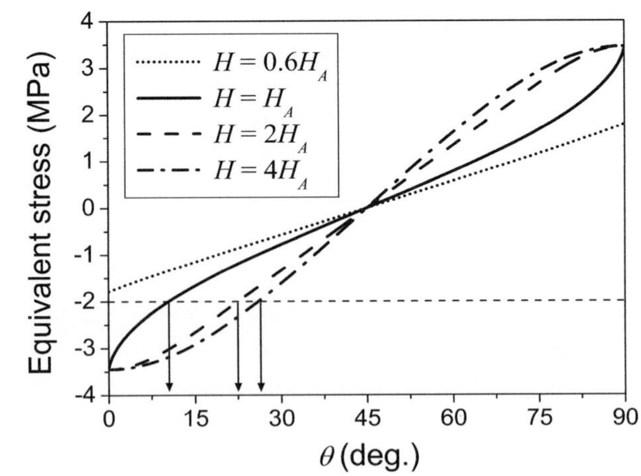

Fig. 1.3 The angular dependencies of equivalent stress computed for the different values of external magnetic field. The *arrows* show the theoretical angular limits of MFIS observation in the inclined (with respect to [001] direction) magnetic field. See also Ref. [13]

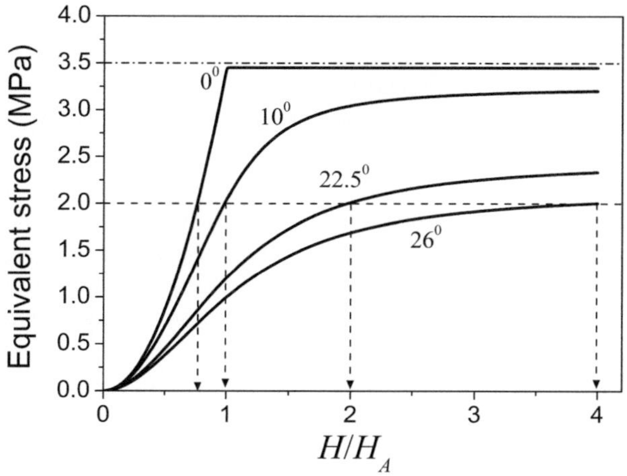

Fig. 1.4 The equivalent stress versus magnetic field computed for the different angles between the magnetic field vector and [001] crystallographic direction. The *arrows* show the field values needed to start the martensite reorientation in the case if the threshold stress value is equal to −2 MPa (*dashed line*). The *dash-dotted line* shows that the magnetic field cannot start the martensite reorientation if the threshold stress value is equal to −3.5 MPa

The dependencies of the equivalent stress on the magnetic field are shown in Fig. 1.4 for the different field directions. The arrows point to the fields corresponding to the start of $y \rightarrow z$ martensite reorientation for the case when a threshold stress value is equal to −2 MPa. It is seen that the threshold field values, which start the martensite reorientation, are equal to H_A, $2H_A$, or $4H_A$, if the angle between the magnetic field vector and [001] direction is equal to 10°, 22.5°, or 26°, respectively. Moreover, the figure shows that the observation of the martensite reorientation becomes practically impossible if the angle θ exceeds some critical value, in the given case, 26°. Indeed, an increase of the angle from 22.5° to 26° results in the doubling of the threshold field value. Taking into account that the magnetic anisotropy fields of about 8 kOe were measured for the Ni–Mn–Ga alloys exhibiting giant MFIS, we can see from Fig. 1.4 that the magnetic field should exceed 32 kOe to start the twin reorientation at $\theta = 26°$. It is seen, furthermore, that the elevation of the absolute value of threshold stress to 3 MPa results in a decrease of the angular limit of martensite reorientation to the value of about 10°. If the absolute value of threshold stress is equal to 3.5 MPa, the martensite reorientation cannot be initiated by the magnetic field application.

1.2.4 Ferromagnetic SMAs with Improved Functional Properties

An improvement of the functional properties of a ferromagnetic SMA presumes a reduction of the mechanical stress and magnetic field values needed for maximal deformation of the alloy, on one hand, and the improvement of the fatigue properties of a specimen subjected to cyclic stressing, on the other hand.

As it was argued above, MFIS appears when the magnetostress induced by the increasing magnetic field reaches the threshold value needed for the start of martensite reorientation process. The field corresponding to the threshold stress exceeds 1 kOe in most cases. This value is rather large, and therefore, it is desirable to reduce the threshold magnetic field, whereby enabling a design of the compact magnetomechanical actuators and sensors. For that, generally speaking, two possibilities exist: first, a reduction of the so-called twinning stress that is the mechanical stress, which causes a complete (or at least considerable) martensite reorientation, and, second, the creation of elevated magnetostress by the reduced magnetic field.

The impressive reduction of the twinning stress has been already achieved (see [18] and references therein). In particular, a 4 % of MFIS was induced by the stress of -0.5 MPa in the course of compressive stress–strain tests [18]. The tests were performed for the Ni–Mn–Ga single crystalline bar ($1 \times 2 \times 12\ \text{mm}^3$) initially being in the twinned state. It was also reported, that in the single-variant martensitic state of the bar, the stress of about -1.2 MPa is needed to trigger the twinning process.

The possibility of creation of the elevated magnetostress by a reduced magnetic field can be considered as follows. To create a large stress by a low field, the magnetic vector must be easily reoriented from [001] to [010] direction. It is possible if the magnetocrystalline anisotropy constant K_u is small. At first sight, the decrease of K_u certainly leads to the magnetostress reduction, according to Eq. (1.13). On the other hand, K_u value is proportional to the "tetragonality" of crystal lattice, $1 - c/a$, and can be diminished by decreasing this value. As Eqs. (1.9) and (1.13) show, the magnetic anisotropy constant (and, therefore, magnetic anisotropy field) can be diminished in this case without a noticeable change of the magnetostress value. This conclusion also follows from Eq. (1.12), which shows that the magnetostress, basically, depends on the two values, namely, saturation magnetization and magnetoelastic constant. This theoretical result is obvious from the point of view of general physics. One can say that the larger is magnetization value, the higher is the magnetic energy of a ferromagnetic SMA. The larger is magnetoelastic constant, the more pronounced is the influence of magnetic energy variation under magnetic field on the crystal lattice, and magnetostress characterizes this influence.

Two other possibilities of the MFIS observation in the low magnetic field were predicted some time ago. First of them follows from a competition between the magnetocrystalline anisotropy energy and the magnetostatic anisotropy energy, which is large in the magnetic films/platelets and rod-shaped samples [19]. The second possibility arises in the case of a strongly nonlinear dependence of the magnetic anisotropy constant on the "tetragonality" of crystal lattice. This possibility may be expected in the case of large $1 - c/a$ values (for more details see Ref. [17]). The experimental direct check of these possibilities has not been carried out yet, but experimental data showing that the increase of “tetragonality” results in the deviation of the “K_u versus $1 - c/a$” dependence from linearity were published recently [20].

Lastly, the studies of functional stability of the ferromagnetic Ni–Mn–Ga alloys should be mentioned. The methods and results of these studies are described in Ref. [21] and references therein. In the representative work, Ref. [21], ten single crystalline bars ($1 \times 2.5 \times 20$ mm^3) having edges parallel to the $\langle 100 \rangle$ crystallographic directions were prepared for the stress–strain tests. The periodic stress–strain cycles with the strain amplitude of 3 % and different frequencies were performed for five bars, while the tests with the strain amplitude of 2 % and frequency of 250 Hz were carried out for the rest five bars. The tests with the strain amplitude of 2 % are especially interesting, because more than 7×10^7 cycles were performed for each of the five bars and only one of them was broken (after 1.6×10^8 cycles). Two billion cycles were performed for one of the bars without its breaking. The majority of bars withstood a multiple cycling, but the noticeable changes in the dynamic stiffness of bars and strain/stress hysteresis were observed. So, the obtained experimental data show that the fatigue properties of Ni–Mn–Ga ferromagnetic SMAs in the martensitic state are rather good, although their further improvement is needed.

1.3 Large Anhysteretic Deformation

1.3.1 Problem Statement

An anomalously low hysteresis of the large deformation of Fe_3Pt and Ni–Fe(Co)–Ga ferromagnetic SMAs was observed recently [22, 23]. The deformation values of about -5 and 10 % were achieved for these compounds, respectively, in the course of compressive and tensile stress–strain cycles exhibiting a near-zero hysteresis. The realization of a large anhysteretic deformation of different (not only ferromagnetic) SMAs has a long-standing interest in both solid-state physics and materials science (see Refs. [24–28] and references therein). This interest is caused mainly by the commercial use of SMAs as the thermomechanical actuators whose actuation speed is limited by the width of thermal hysteresis of deformation whereby the SMA actuators exhibiting a low thermal hysteresis of deformation can be driven at a high actuation frequency. Moreover, the lowering of hysteresis results in a reduction of the fatigue effects accompanying cyclic deformation of SMA [26].

In most cases a large reversible deformation of SMA arises due to the reversible MT; that is why the temperature- and stress-induced MTs exhibiting extremely small hysteresis attract a great attention of experimentalists and theorists in the last 10 years. In particular, the possibility of anhysteretic MTs in the ternary and quaternary Ti–Ni-based alloys was searched intensively in view of their advanced engineering and biomedical applications. It has been shown that the thermal hysteresis characteristics obtained for these alloys crucially depend on the crystallographic features of MTs [24–26, 28].

In this section we discuss the physical factors facilitating the achievement of a giant anhysteretic deformation of SMAs.

1.3.2 Crystallographic Aspect of Anhysteretic Deformation

For a systematic search of SMAs with a minimal thermal hysteresis of MT, the working hypothesis was put forward by James and Zhang [24]. According to this hypothesis, the main physical feature providing the anhysteretic character of MT is the ideal compatibility of the crystal lattices of parent and product phases. Two physical conditions of the ideal compatibility were formulated as (1) zero volume change during MT and (2) existence of the crystallographic planes, which are common for the crystal lattices of austenitic and martensitic phases. When MT starts, these planes become the undistorted interfaces between the spatial domains of martensite and austenite. As so, the physical meaning of formulated in Ref. [24] conditions is very simple: they facilitate a nucleation of product phase inside the matrix of parent phase and minimize the energy of interfaces.

For the presence of undistorted plane, a very special relationship between the values of lattice parameters of austenitic and martensitic phases should be fulfilled. By this reason, the undistorted planes are absent in the majority of SMAs, and the interfaces separating the spatial domains of austenitic and martensitic phases are the thin layers, which are periodically strained in such a way that the average strain value tends to be zero. These interfaces are parallel to certain crystallographic planes, which are referred to as the invariant planes [29]. The lattice parameters of the austenitic and martensitic phases of Ti–Ni–X (X = Cu, Pd, Pt, Au) and Ti–Ni–Cu–Pd alloys may be tuned by a variation of concentration of the element X to obtain the almost undistorted interfaces [25, 26]. This fact enabled a purposeful search for Ti–Ni-based alloys with the near-zero thermal hysteresis and improved functional stability [25, 26]. As a result of this search, $Ti_{50}Ni_{41}Pd_9$, $Ti_{50}Ni_{39}Pd_{11}$, and $Ti_{50}Ni_{37}Au_{13}$ alloys with the lattice parameters providing the existence of undistorted interfaces were designed. These alloys, compared to the binary TiNi single crystal, showed ten times reduced thermal hysteresis of MT. The minimum width of thermal hysteresis (~7 K) was observed for $Ti_{50}Ni_{37}Au_{13}$ alloy [25].

Further advance in the observation of anhysteretic MTs was achieved as a result of experiments with the quaternary Ti–Ni–Cu–Pd alloys whereby around those materials, the alloys with the negligibly small (within the accuracy of measurement) thermal hysteresis were found [26].

The development of Ti–Ni-based alloys with near-zero thermal hysteresis of MT leaded to a conclusion that the presence of undistorted interfaces is a key condition for the minimum hysteresis width, while the volume effect of MT is a secondary factor. Moreover, after studying of the large number of Ti–Ni–X alloys, it became clear that some alloys exhibiting a small volume change during MT appeared in the group of alloys with near-zero hysteresis width, while some others showed the largest hysteresis among the all alloys studied in Ref. [25]. On the one hand, this is a surprising finding in view of the physically obvious influence of the volume change on the nucleation of resultant phase inside the parent phase. On the other hand, the determination of abovementioned key condition for a narrowing of hysteresis width simplifies a search for the alloys with near-zero hysteresis width, if they allow a fine-tuning of the lattice parameters.

Three remarks are to the point. First, the determination of a small volume effect of MT from the changes in lattice parameters needs a very precise measurement of these parameters in both phases. Second, the alloys with the almost undistorted interfaces show a reduced transformation heat. Third, the more precise lattice parameters are tuned to obtain the undistorted interfaces, the less pronounced are the jumps of physical properties (in particular, electric resistance [26]), indicating a weakening of the first-order character of MT. These remarks point to the probable relationship between the two scientific problems, namely, the reduction of hysteresis width and occurrence of the quasi-second-order MTs.

1.3.3 Quasi-Second-Order MTs and Giant Quasi-Linear Deformation of SMAs

According to the fundamentals of thermodynamics, the transformational behavior of SMAs is controlled by the principle of minimum of Gibbs potential, which involves the elastic energy. The Landau theory is widely used for the description of MTs in the single crystals of both ordinary and ferromagnetic SMAs. This theory starts from the series expansion for the elastic energy in terms of the order parameter of MT. In accordance with the basic principle of the theory, the Landau expansion for the free energy must be invariant with respect to the symmetry group of the high-symmetry phase, in the case of SMAs, the symmetry group of cubic crystal lattice. It means that all invariant combinations of the order parameter of MT must be included in the Landau expansion for the elastic energy [30]. It is especially important that this expansion includes the term, which is cubic in order parameter of MT, because the well-known Lifshitz criterion states that this term makes impossible the second-order phase transition (see [31] and references therein).

Numerous experiments show that MTs normally are the first-order phase transitions, in agreement with the Lifshitz criterion. However, the quasi-second-order MTs are observed in Ti–Ni-based and Fe_3Pt alloys [20, 32]. It means that the coefficients of the third-order terms in the Landau expansions for the elastic energies of some SMAs can be abnormally small under conditions of experiment [31]. These conditions may concern the temperature interval of MT, hydrostatic compression, mode of preparation and magnetic state of the alloy, defect concentration, etc. [33].

The quasi-second-order B2 → R transformation (which is the cubic–rhombohedral phase transition) with the thermal hysteresis of about 2 K was observed long ago [32]. The rhombohedral R-phase appears in Ti–Ni–Fe and in aged or thermomechanically treated Ni-rich Ti–Ni alloys prior to the formation and stabilization of the monoclinic phase [32]. It means that the appearance of this phase is conditioned by the different physical, chemical, and technological factors. However, new possibilities of the observation of B2 → R transformation in the bulk Ti–Ni-based systems and films were found recently [27]. These alloy systems are promising for the applications because of the narrow (of about 3 K) temperature interval of B2 → R transformation, which shows the features of the quasi-second-order MT. Furthermore, a smooth variation of the electric resistance of $Ti_{50}Ni_{34}Cu_{11.5}Pd_{4.5}$ thin film in the course of an anhysteretic MT from cubic austenite to orthorhombic martensite was observed.

Commercially important Ti–Ni-based alloys exhibit the cubic–orthorhombic, cubic–monoclinic, and cubic–rhombohedral MTs in the paramagnetic state. Physically more interesting is the behavior of Fe_3Pt alloy exhibiting quasi-second-order MT in the ferromagnetic state.

The character of MT in Fe_3Pt strongly depends on the degree of atomic order, S. An "ordinary" first-order MT is observed in the partly ordered alloy (with $S = 0.57$, $T_M = 145$ K), while the highly ordered alloys exhibit the quasi-second-order MT from a cubic austenite to tetragonal martensite with the lattice parameter ratio $c/a < 1$ (for $S = 0.75$, $T_M = 85$ K) or $c/a > 1$ (for $S = 0.88$, $T_M = 60$ K) [20]. The atomic ordering of Fe_3Pt alloy leads to an increase of the Curie temperature and decrease of the MT temperature [20]. It means that the ordering stabilizes both the ferromagnetic state and austenitic phase in the alloy. Note that the same feature is inherent to the $Ni_{45}Mn_{36.7}In_{13.3}Co_5$ metamagnetic SMA undergoing the phase transformation from ferromagnetic austenite to nonmagnetic martensite [34].

Martensitic transformation of the partially ordered Fe_3Pt alloy ($S = 0.57$) is a non-thermoelastic because of the large MT strain, $(c - a_0)/a_0 \sim -13\ \%$.

The lattice parameters of the martensitic phase in the alloy with $S = 0.75$ smoothly vary with temperature pointing to the thermoelastic quasi-second-order character of MT. A low-temperature limit of the MT strain is about -4%. The magnetic field $H \approx 13$ kOe applied along the [001] crystallographic direction at the temperature of 4.2 K induces the large partly reversible deformation of $-1.2\ \%$, indicating field-induced changes in the martensite microstructure. It is noteworthy that the magnetically induced deformation is approximately equal to the one third of the MT strain [20].

MT of the alloy with $S = 0.88$ also has a quasi-second-order character, but it is characterized by a small MT strain, $(c - a_0)/a_0 \sim 0.5\ \%$. The magnetic field $H \approx 13$ kOe applied along the [001] crystallographic direction at the temperature of 4.2 K induces the partly reversible deformation, which is close in value to the MT strain [20].

Properties of Fe_3Pt with $S = 0.75$ are especially interesting not only because of a large magnetically induced deformation. Large ($\sim 4.5\ \%$) quasi-linear anhysteretic deformation of this alloy was induced at 90 K in the course of a stress–strain cycle with the maximum stress of 200 MPa [22]. It is important that the stress–strain loop with the maximum stress of 500 MPa indicated the hysteretic deformation being typical for the usual stress-induced MT [22]. The stress-induced MT was started at the stress and strain values of about 280 MPa and 7 %, respectively. It has been concluded, therefore, that the elastic strain of 4.5 % was achieved due to the very pronounced softening of the shear elastic modulus $C'(T)$ in the vicinity of MT temperature; the other physical factors extending the elastic limit of deformation have been also discussed [22].

The thermoelastic cubic–tetragonal MTs, which are observed in the ordered Fe_3Pt alloys, can be described by a minimization of free energy. Assuming that MT obeys a volume conservation principle, the tetragonal distortion of cubic lattice in the course of MT can be described by the strain tensor component, ε_{zz}, because in this case $\varepsilon_{zz} \approx -2\varepsilon_{xx} = -2\varepsilon_{yy}$. The Landau expansion for the free energy is

$$F = \frac{1}{2} c_2(T) u^2 + \frac{1}{3} a_4 u^3 + \frac{1}{4} b_4 u^4, \tag{1.19}$$

where $u = 3\varepsilon_{zz}$ and the coefficients $c_2(T)$, a_4, and b_4 of the second-, third-, and fourth-order terms are denoted as in Refs. [23, 31, 33, 35]. The variable u is very relevant, because it is related to the lattice parameters of cubic and tetragonal phases as $u = 2(c/a - 1) = 3(c - a_0)/a_0$ and the estimation $u \sim 0.1$ is valid, as a rule, for the tetragonal phases of ferromagnetic SMAs.

The equilibrium value of tetragonal distortion in the tetragonal phase, u_0, satisfies the condition $\partial F/\partial u = 0$, which results in the expression

$$u_0 = -(a_4/2b_4)\left[1 + \sqrt{1 - c_2(T)/c_t}\right], \tag{1.20}$$

where $c_t = a_4^2/4b_4 > 0$. The coefficient $c_2(T)$ is a decreasing function of temperature. MT starts when this function reaches the positive value of c_t and finishes when this function changes sign. Therefore, the MT start and the MT finish temperatures satisfy the conditions $c_2(T_{MS}) = c_t$ and $c_2(T_{MF}) = 0$. The smaller is a_4 value, the narrower is the temperature interval of MT. In the limiting case of $a_4 \rightarrow 0$, the width of MT temperature interval tends to zero and the cubic–tetragonal MT becomes the second-order phase transition. It is in agreement with the Lifshitz criterion because in this limiting case the cubic term disappears from Landau expansion, Eq. (1.20).

Using Eq. (1.20) and conditions for the martensite start and martensite finish temperatures, the coefficients of Landau expansion for the free energy can be related to the lattice parameters measured at MT finish temperature as

$$a_4 = \frac{2c_2(T_{MS})}{1 - c(T_{MF})/a(T_{MF})},$$
$$b_4 = \frac{c_2(T_{MS})}{[1 - c(T_{MF})/a(T_{MF})]^2}. \tag{1.21}$$

Equation (1.21) shows that the coefficient b_4 is positive which is in accordance with the essentials of Landau theory. The coefficient of the third-order term of Landau expansion, a_4, is positive for Fe_3Pt alloys with $S = 0.57$ and $S = 0.75$, because for these alloys $c/a < 1$. For the alloy with $S = 0.88$, this coefficient is negative and small in the absolute value. For this alloy, $c/a > 1$ and $|1 - c(T_{MF})/a(T_{MF})| < 10^{-2}$ in the whole temperature range of martensitic phase (as follows from lattice parameters measurements [20]). It means that a_4 is a decreasing function of the parameter S, which changes sign at its certain critical value, S_c. In view of the very small lattice distortion in the tetragonal phase of the alloy with $S = 0.88$, the S_c must be slightly smaller than 0.88.

1.3.4 Postcritical Deformational Behavior of Shape Memory Alloy

It was shown that the low thermal hysteresis and narrow stress–strain loops can be observed in SMAs if the transformation path in the stress–temperature plane approaches the critical point in the stress–temperature phase diagram, and moreover, a special postcritical deformation behavior of an alloy can be also observed [23, 36]. The postcritical behavior implies a large nonlinear deformation, up to 10 %, which goes on in the absence of phase transition. To explain an essence of this behavior, let us plot a stress–temperature phase diagram of SMA loaded in [001] direction aligned with z-axis of coordinate system and c-edges of the unit cells of cubic lattice. The stability domains of cubic and tetragonal phases in a stress–temperature plane are bounded by the lability lines, which can be plotted using the minimum conditions for the Gibbs potential.

$$G = \frac{1}{2}c_2(T)u^2 + \frac{1}{3}a_4u^3 + \frac{1}{4}b_4u^4 - \frac{1}{3}\sigma u, \tag{1.22}$$

where $\sigma = \sigma_{zz}$ is the stress tensor component. These conditions are

$$\frac{\partial G}{\partial u} = 0, \quad \frac{\partial^2 G}{\partial u^2} > 0 \tag{1.23}$$

The thermodynamic phases become unstable when the inequality in Eq. (1.23) fails, and therefore, the equations of lability lines can be derived from the equalities

$$c_2(T)u + a_4u^2 + b_4u^3 - \sigma/3 = 0, \tag{1.24}$$

and

$$c_2(T) + 2a_4u + 3b_4u^2 = 0. \tag{1.25}$$

Equation (1.24) shows that in the limiting case of a linear elasticity, $\sigma = 9c_2\varepsilon_{zz} \equiv E\varepsilon_{zz}$. Young's modulus E is related to the bulk, B, and shear, C', modules as $1/E = 1/B + 1/3C'$, and therefore, the approximate relationship

$$E \approx 3C' \approx 9c_2 \tag{1.26}$$

is valid in the cubic phase if $C' \ll B$. Equations (1.21) and (1.26) enable an estimation of all coefficients of the Landau expansion for Gibbs potential from the experimental values of elastic modules and lattice parameters measured for the austenitic and martensitic phases, respectively.

The function $c_2(T)$ can be excluded from Eq. (1.24) using Eq. (1.25), and in such a way the expression

$$\sigma = -3u^2(a_4 + 2b_4u) \tag{1.27}$$

can be easily obtained. The solutions of Eq. (1.25) are

$$u_\pm = -\frac{a_4}{2b_4}(1 \pm R), \tag{1.28}$$

where

$$R \equiv R(T) = \sqrt{1 - 3c_2(T)/4c_t} \tag{1.29}$$

is a function of temperature and parameter c_t has been already introduced in Eq. (1.20). Substituting the values $u_\pm$ into Eq. (1.27), one can find the following equations for the lability lines:

$$\sigma = -\frac{a_4^3}{9b_4^2}(1 - R)^2(1 + 2R) \equiv \sigma_1 \quad \sigma = -\frac{a_4^3}{9b_4^2}(1 + R)^2(1 - 2R) \equiv \sigma_2. \tag{1.30}$$

The martensitic phase is stable if $|\sigma| > |\sigma_2|$, the austenitic one is stable if $|\sigma| < |\sigma_1|$, and in the stress range $|\sigma_2| < |\sigma| < |\sigma_1|$, the mixed austenite–martensite state can be observed.

As it is seen from Eq. (1.30), the lability lines cross each other at $R = 0$ that is when:

$$3c_2(T) = 4c_t = a_4^2/b_4. \tag{1.31}$$

The cross point of the lability lines is referred to as a critical point at the stress–temperature phase diagram. The coordinates of this point are $\sigma*$ and $T*$. The critical stress value can be easily found from Eqs. (1.23), (1.28), and (1.30):

$$\sigma* = -\frac{8}{27}C'(T_{MS})[1 - c(T_{MF})/a(T_{MF})]. \tag{1.32}$$

Equations (1.26) and (1.31) result in the relationship

$$C'(T*) = 4C'(T_{MS})/3, \tag{1.33}$$

which can be used for the estimation of critical temperature if the experimental temperature dependence of the shear modulus is available.

According to Eqs. (1.29) and (1.30), the view of the stress–temperature phase diagram of a cubic–tetragonal MT is specified by the temperature dependence of the coefficient of second-order term in the Landau expansion for Gibbs potential of the crystal. This is true only for the ideal crystal being in the paramagnetic state. As we will see later on, the presence of crystal defects and ferromagnetic ordering leads to the appearance of temperature dependence of the parameter a_4, which is involved in the equations for the lability lines. However, if the physical state of defect subsystem weakly depends on temperature in the temperature interval of MT, the coefficient a_4 can be considered constant.

As it was mentioned above, the function $c_2(T)$ is equal to c_t at the MT start and to zero at the MT finish temperature. The Landau theory presumes a linear character of this function, and therefore, the expression

$$c_2(T) = c_t(T - T_{MF})/\Delta T, \tag{1.34}$$

where $\Delta T = T_{MS} - T_{MF}$, is assumed to be valid in the temperature interval of MT. Extrapolating the function, Eq. (1.34), to the temperatures $T > T_{MS}$ and taking into account Eq. (1.31), one can obtain the relationship $T* - T_{MS} = \Delta T/3$. It will be shown below that the critical point of the phase diagram ($T*, \sigma*$) is, in the same time, the endpoint of the phase transition line (similar to the well-known critical point "liquid–gas"). In this case the stress-induced MT can be observed only in a very narrow temperature interval, $\Delta T/3$, and the width of this interval is less than 10 K for the most of SMAs. This conclusion contradicts to the experiments since, usually, the temperature interval of an observation of the stress-induced MTs exceeds considerably the difference between the MT start and MT finish values. As so, the function

$$c_2(T) = c_t + \varsigma(T - T_{MS}) \tag{1.35}$$

with $\varsigma < c_t/\Delta T$ has to be used for the computations at the temperatures $T > T_{MS}$.

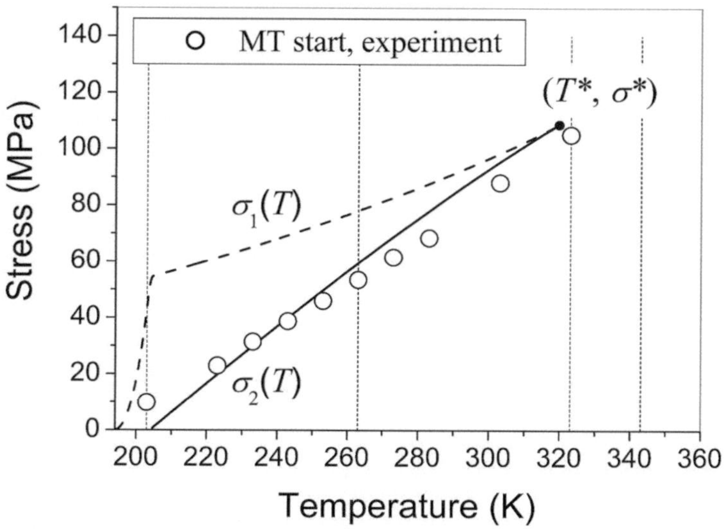

Fig. 1.5 Theoretical phase diagram of the stress-induced phase transitions in SMA. The *solid* and *dashed lines* correspond to the stabilization of martensitic phase and destabilization of austenitic phase, respectively. The *thin vertical lines* correspond to the isothermal stress–strain cycles. Experimental stress values corresponding to the start of forward stress-induced MT in $Ni_{50}Fe_{19}Co_4Ga_{27}$ (at %) alloy are shown by *open circles*. See also Ref. [23]

To the best of our knowledge, the first experiments discovering the postcritical deformational behavior were carried out recently for the $Ni_{50}Fe_{19}Co_4Ga_{27}$ (at %) alloy [23], so it is instructive to plot the stress–temperature phase diagram using the physical values measured for this alloy: $T_{MS} = 204$ K, $1 - c(T_{MF})/a(T_{MF}) = 0.22$, and $E(T_{MS}) = 5$ GPa. A substitution of these values into Eqs. (1.23), (1.28), and (1.32) gives the critical stress value $\sigma^* = 109$ MPa. Using Eqs. (1.23), (1.28), and (1.12), one can evaluate the coefficients c_2, a_4, and b_4 and plot the phase diagram shown in Fig. 1.5. The diagram was obtained using the value $\varsigma = 1.5$ MPa/K, which sets the critical temperature $T^* = 328$ K, in accordance with Eqs. (1.28), (1.33), and (1.17).

Figure 1.5 shows the lability lines of the austenitic ($\sigma_1(T)$ function) and martensitic ($\sigma_2(T)$ function) phases; the thin vertical lines correspond to the stress–strain cycles performed at the fixed temperatures. Theoretically, the maximal hysteresis of stress-induced MT takes place if the forward MT occurs in the whole experimental specimen at $\sigma = \sigma_1(T)$ and reverse MT happens at $\sigma = \sigma_2(T)$. In this case the horizontal plateaus are present at the stress–strain curves depicting the forward and reverse MTs. The isothermal curves computed from Eqs. (1.26) and (1.30) are shown in Fig. 1.6a. This figure illustrates that the shape of a stress–strain loop computed for the closed stress cycle drastically depends on the temperature difference, $\delta T \equiv T - T_{MS}$. The theoretical loops reproduce the observed ones in several experimental consequences of the δT increasing. These consequences are (1) the elevation of plateau at the stress–strain loop; (2) the narrowing of the plateau that is

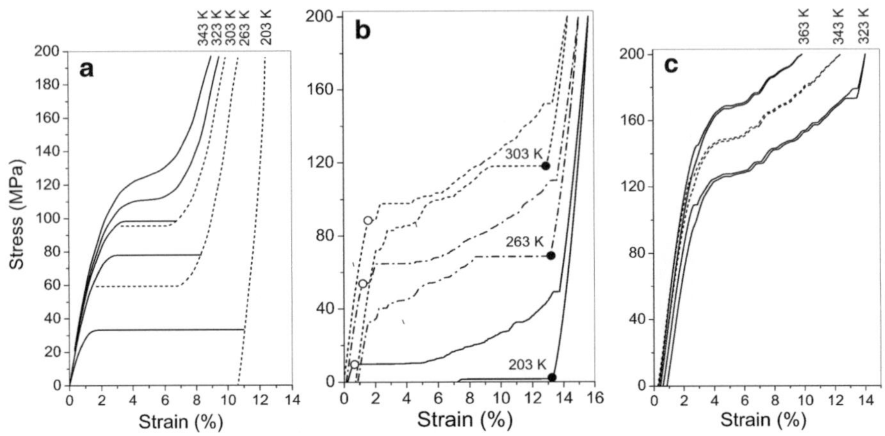

Fig. 1.6 Theoretical stress–strain dependencies, (**a**), and experimental stress–strain loops demonstrating the subcritical, (**b**), and postcritical, (**c**), deformational behavior of $Ni_{50}Fe_{19}Co_4Ga_{27}$ (at %) alloy. The *open* and *closed circles* correspond to the starts of forward and reverse MTs, respectively. See Ref. [23]

the diminishing of the jump of strain, $\Delta\varepsilon_{zz}$, at fixed stress value; and (3) the lowering of hysteresis of the stress-induced MT.

In addition to the commonly known features (1)–(3) of stress–strain loops, Fig. 1.6a shows an absence of the hysteresis and strain jump in the case when the stress–strain loop is computed for the temperature $T > T^*$. It should be emphasized that, namely, the jump of strain and hysteresis of stress indicate that the martensitic transformation is a first-order phase transition. The disappearance of these characteristics indicates either the change of the character of phase transition from the first to the second order or disappearance of phase transition. The defining feature of the second-order cubic–tetragonal phase transition is the appearance/disappearance of the diagonal strain tensor components at the phase transition point, because the order parameter is a linear combination of these components. However, even the extremely small stress σ_{zz} (corresponding to the beginning of the stress–strain cycle) induces the nonzero strains ε_{zz}, ε_{yy}, and ε_{xx}, and therefore, nonstrain component appears/disappears during the cycle. Therefore, the second-order phase transition does not happen. As so, the absence of plateaus at the stress–strain curves taken at elevated temperatures shows that the stress-induced MT is possible only below the critical temperature T^*; above this temperature only a continuous nonlinear deformation of crystal lattice is possible. It means that the critical point (T^*, σ^*) at the phase diagram in Fig. 1.5 is the endpoint of the phase transition line. Incidentally, this conclusion was drawn a long time ago from the consistent general Landau theory of the ferroelastic phase transitions [35]). Figure 1.5 illustrates that the vertical line corresponding to the isothermal stress cycle performed above the temperature T^* does not cross any line of the phase diagram which means that the stress-induced phase transition does not happen.

A continuous anhysteretic deformation of SMA in the temperature range above the critical temperature is referred to as the postcritical deformational behavior. To the best of our knowledge, the first experimental demonstration of the presence of critical point at the stress–temperature diagram of SMA has been presented very recently in Ref. [23]. As it was shown, the hysteresis width of the stress-induced MT observed experimentally at the temperatures shown in Fig. 1.6a noticeably depends on the temperature in the low-temperature range (see Fig. 1.6b). It becomes small for the deformation cycle performed at 323 K and remains very small at 343 and 363 K (see Fig 1.6c). The "residual" hysteresis depends on the duration of stress–strain cycle and is caused by the slow spatial redistribution of defects in the course of deformation process [23, 36].

It is worth to compare the observed in Ref. [23] temperature dependence of hysteresis of stress-induced MT with the determined in Ref. [25] dependence of the thermal hysteresis width on the middle eigenvalue of martensitic transformation matrix, denoted as λ_2. The λ_2 value characterizes the compatibility of crystal lattices of martensitic and austenitic phases. In the case of ideal compatibility, $\lambda_2 = 1$. The λ_2 value was tuned in Ref. [25] by the variation of the chemical composition of Ti–Ni–Au SMAs, and the sharp drop of the hysteresis width was observed only for the narrow interval of λ_2 values, close to $\lambda_2 = 1$. In the case of $Ni_{50}Fe_{19}Co_4Ga_{27}$ (at %) alloy, the hysteresis of the stress-induced MT reduces in a wide temperature range until its full disappearance indicating the absence of stress-induced MT in the high-temperature range.

1.3.5 Concluding Remarks

Two remarks are appropriate for this section.

First, it should be noted that several factors promoting the large anhysteretic deformation of the shape memory alloys are known at present; these factors are:

(1) A compatibility of crystal lattices of spatial domains of austenite and martensite, which enables the existence of undistorted interfaces in the mixed austenite–martensite state [24–26](2) The quasi-second-order character of martensitic phase transformation [20, 22, 31](3) A pronounced softening of the shear elastic modulus in the wide temperature range around the MT temperature, which enables an observation of the large quasi-linear deformation before the occurrence of stress-induced MT and large nonlinear deformation in the absence of stress-induced MT (which is postcritical deformation) [22, 23, 36](4) A lowering of the concentration and mobility of the crystal defects [36]

Second, it should be emphasized that the reduction of thermal hysteresis of MT favors the improvement of functional stability of SMAs under the multiple cyclic loading [26]. This kind of functional stability of the alloys is obviously related to the concentration, type, and mobility of crystal defects in the real alloy. Therefore, the influence of crystal defects on the functional properties of SMAs deserves special consideration.

1.4 Stability of Deformational Behavior of SMA

1.4.1 Problem Statement

Exploiting the SMA-based devices presumes, first, a long-term keeping of SMA in the martensitic state and, second, the repeated thermomechanical cycling of an alloy, which causes the reversible martensite reorientation or reversible martensitic transformation in the course of each cycle.

The long-term keeping of an alloy in the martensitic state is referred to as the *martensite aging*. An excellent description of the main martensite aging effects is presented in Ref. [37]. Basically, these effects follow from the stabilization of martensitic phase and change of the martensite microstructure with a time. It is commonly recognized now that both the martensite stabilization and structural changes are closely related to the spatial redistribution of defects in the crystal, such as point defects, dislocations, precipitates, etc. The defect reconfiguration approaches the real crystal to the energy-optimal state.

As opposed to aging, a repeated thermomechanical cycling withdraws the martensitic alloy from the energy-optimal state leading to martensite destabilization. The martensite stabilization results in the increase of MT temperature (see Ref. [37] and references therein), while the destabilization manifests itself as the MT temperature decrease (see, e.g., [26, 38, 39] and review article [33]).

The increase/decrease of MT temperature is one of the most important consequences of the martensite stabilization/destabilization, because the deformational properties of SMA drastically depend on the difference, δT, between the temperature of deformed alloy and MT temperature. On one hand, this difference can be changed most easily by changing the alloy temperature. The resultant changes in the deformational behavior of SMAs were discussed in Sect. 1.3 (see Fig. 1.6 and explanations to it). On the other hand, the δT value varies due to the shift of MT temperature during the martensite stabilization/destabilization processes. Due to this, the analogy between the change of the alloy temperature and martensite stabilization/destabilization arises: both factors noticeably change the stress–strain loops obtained in the course of cyclic deformation of an alloy. The changes in the deformation behavior of SMAs reflect the martensite stabilization/destabilization processes, therefore.

This section is intended to elucidate the fundamental aspects of interrelation between the martensite stabilization/destabilization processes and deformation behavior of SMAs.

1.4.2 Internal Stress and Pressure as the Factors of Martensite Stabilization

The most evident explanation of physical effects accompanying the martensite aging is based on the symmetry-conforming short-range-order (SC-SRO) principle [32, 40], which predetermines a tendency to the spatial redistribution of the point defects in an alloy after the martensitic transformation. According to this principle, the energy of the crystal with defects is minimal when the probabilities of finding point defects in the certain crystallographic positions have the same symmetry as the crystal lattice has. In other words, the probabilities are the same for the physically equivalent defect sites and different for the nonequivalent ones. The forward martensitic transformation lowers the crystal symmetry. According to SC-SRO principle, the high-symmetry distribution of point defects should adopt the lowered symmetry after MT. Thus, a reconfiguration of the point defects during martensite aging is driven by the symmetry mismatch between the host crystal lattice and the spatial distribution of point defects.

The defects distort the crystal lattice and induce the elastic strain and mechanical stress; the latter may be referred to as "internal stress," to emphasize that it can arise even in the absence of external forces applied to the single- and polycrystalline specimens. The internal stress produced by the point defects varies with time due to the defect generation, annihilation, and/or diffusion. These processes can take place during the martensite aging in the free specimens or under the influence of external factors (mechanical load, magnetic field, radiation, etc.). The characteristic times of defect generation, annihilation, and diffusion noticeably exceed the time of the elastic response of the crystal lattice to the external force application, and so, the deformational and magnetic properties of the crystal slowly evolve even under the steady external conditions (see, e. g., Refs. [41, 42] and references therein). The physical effects caused by an evolution of the defect subsystem are inherent to different solids but especially pronounced in the shape memory alloys, because of the thermodynamic instability of their crystal lattice, which transforms martensitically.

The internal stress can be subdivided into internal pressure $P^{(i)}(t)$ and internal shear stress $\sigma_{ik}^{(sh)}(t)$ as

$$\sigma_{ik}^{(i)}(t) = P^{(i)}(t)\delta_{ik} + \sigma_{ik}^{(sh)}(t), \tag{1.36}$$

where t is time and δ_{ik} is Kronecker symbol. The internal pressure is interrelated with the isotropic expansion/contraction of a crystal lattice by defect generation/annihilation, while the internal shear stress describes the shear deformation of the crystal lattice by defects. The introduction of internal stress enables a thermodynamic description of the martensite aging process and the elaboration of Landau theory of MTs in the real crystals, that is, the crystals with defects [43].

It should be emphasized that only the time-dependent shear stress is governed by the SC-SRO principle. The internal pressure is isotropic and, therefore, independent on the symmetry of crystal lattice. It was shown very recently that the influence of slowly evolving internal pressure on the MT temperature and deformation behavior of SMA may be comparable or even more pronounced than the influence of internal shear stress [43, 44]. This result is not in line with the traditional views on the thermoelastic MTs. According to these views, MTs are characterized mainly by a shear deformation of the crystal lattice, while the comparatively small volume change during MT is considered as the secondary effect, which can be disregarded when the basic transformational properties of the shape memory alloys are analyzed. However, a careful consideration of the classic experimental works makes these views questionable. It was observed, for example, that the hydrostatic pressure of 1 GPa applied to different SMAs results in the small volume changes, $\Delta V/V \sim -0.1\ \%$, but a noticeable shift of the MT temperatures, $\Delta T_M \sim 4 - 20$ K, can be observed (see [45–47] and references therein). Other examples illustrating an influence of the volume change on the properties of SMAs are given in Ref. [33].

The significant influence of a small volume change on the transformation behavior of SMA can be explained as follows. The volume change, $|\Delta V/V| \sim 10^{-3} - 10^{-2}$, results in the energy density change, $|\Delta F_1| \sim |\Delta V/V| B \sim 0.1 - 1$ GPa ($B \sim 100$ GPa is bulk elastic modulus). The spontaneous shear of a crystal lattice during MT results in the energy density change, $|\Delta F_2| \sim |\varepsilon_M| C' \sim 0.1 - 1$ GPa. In the last estimation, the MT strain, $|\varepsilon_M| \sim 10^{-2} - 10^{-1}$, and shear modulus, $C' \sim 10$ GPa, are acceptable if the temperature of SMA is close to MT temperature. The energy changes ΔF_1 and ΔF_2 are of the same order of magnitude, and, therefore, both shear strain and volume change must be taken into account when the transformational and elastic properties of shape memory alloys are considered.

A strong influence of the hydrostatic pressure P on the cubic–rhombohedral MT in Au–Cd alloys was reported long ago: the linear dependence $T_M(P)$ with a slope of $\mathrm{d}T_M/\mathrm{d}P \approx 20$ K/GPa was observed [45]. Moreover, MT leads to the comparatively small contraction of a crystal lattice in the [111] crystallographic direction, which is characterized by the shear strain, $\varepsilon_M \approx -8 \times 10^{-3}$. This suggests that a stabilization of the martensitic phase in the course of Au–Cd aging is caused mainly by the internal pressure. A stabilization of the martensitic phase of Au–Cd alloy during the martensite aging was observed in Ref. [48]. The austenite finish temperature, $T_{AF}(t)$, that is, the temperature of finish of the reverse martensite–austenite transformation, was determined as a function of the time of martensite aging. The experimental values of a time-dependent shift of the austenite finish temperature, $\Delta_{AF}(t) = T_{AF}(t) - T_{AF}(0)$, are shown in Fig. 1.7 by the open circles.

A modified version of the Landau theory, which is referred to as the symmetry-conforming Landau theory, was advanced in Ref. [43] for the theoretical description of the martensite stabilization. The changes of the characteristic temperatures of the reverse MT were considered. The austenite finish temperature was computed for Au–Cd alloy assuming that the martensite stabilization is caused by the time-dependent internal pressure:

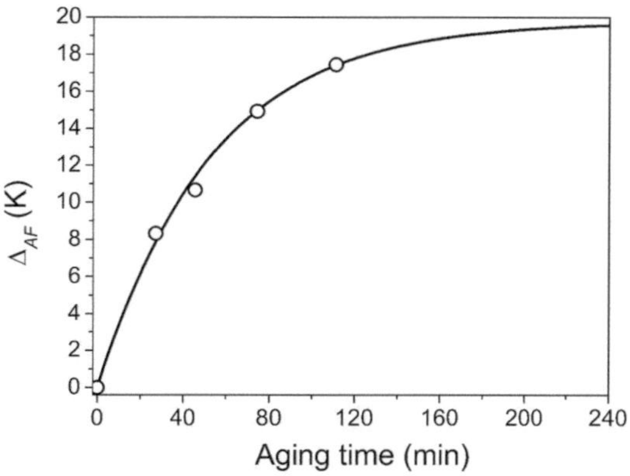

Fig. 1.7 The time variation of austenite finish temperature measured in Ref. [48] (*circles*) and theoretical curve obtained in Ref. [43] (*line*)

$$P^{(i)}(t) = P^{(i)}(\infty)[1 - \exp(-t/\tau_P)]. \tag{1.37}$$

This dependence is typical for the relaxation processes. A good agreement between the theory and experiment was achieved for the relaxation time and limiting value of the internal pressure equal to $\tau_P = 50$ min and $P(\infty) = 3.4$ GPa, respectively. The computations showed, however, that the limiting pressure, $P(\infty)$, strongly depends on the MT strain, transformation volume change, and shear elastic modulus, C_{44}. These values depend, in turn, on the composition, degree of chemical ordering, and alloy treatment. It may be concluded, nevertheless, that the internal pressure gives noticeable contribution to the martensite stabilization effect. This conclusion was directly confirmed in Ref. [44] by the measurements of elastic modulus of a Cu–Al–Ni martensite.

The elastic modules are among the most important characteristics of the functional materials. It is physically clear that the martensite stabilization/destabilization must be accompanied by the increase/decrease of the elastic modules and first of all, the shear modulus, which undergoes a most pronounced softening in the course of MT, that is, C' for the cubic–tetragonal and C_{44} for the cubic–rhombohedral MT. The dynamic mechanical analysis (DMA) is often used for a determination of the so-called storage elastic modulus, which characterizes the elastic response of the solid to the small external force application. In the case of pronounced softening of a shear modulus (i.e., if $C' << B$), the storage modulus measured in the axial tension/compression mode is close in value to Young's modulus.

A slow variation of the storage elastic modulus, *E*(*t*), of Cu–Al–Ni martensite under the permanent external conditions was observed in the course of the continuous DMA measurement [44]. A time dependence of Young's modulus was computed in this work using the symmetry-conforming Landau theory of martensite stabilization. The computations were carried out taking into account, first, the axial internal stress only, second, the internal pressure only, and, third, both the axial internal stress and the internal pressure. A relaxational character of the time dependencies of both the internal pressure (Eq. (1.37)) and axial stress,

$$\sigma_A^{(i)}(t) = \sigma_A^{(i)}(\infty)[1 - \exp(-t/\tau_\sigma)], \tag{1.38}$$

was assumed. It should be noticed that according to the symmetry-conforming Landau theory, the internal pressure renormalizes the coefficients of the second-order and third-order terms in the Landau expansion for the free energy, Eq. (1.19), while the axial stress is involved in the linear term in the expression for Gibbs potential, Eq. (1.22). The influence of pressure and axial stress on Young's modulus of martensite was studied in Ref. [44] by the mathematical analysis of the minimum conditions for a Gibbs potential (Eq. (1.22)). The numerical calculations were performed using the values of MT strain, $\varepsilon_M = -0.085$, and transformation volume effect, $\Delta V/V = -3 \times 10^{-3}$, estimated from the experimental values of lattice parameters reported in Ref. [49] for Cu–Al–Ni alloy. Moreover, the inequality $\tau_P < \tau_\sigma$ was assumed because the elastic forces promoting a minimization of the volume were accepted to be larger than the forces assisting the achievement of the optimal shape of the unit cells of the crystal lattice.

The main experimental and theoretical results obtained for the Cu–Al–Ni martensite using the DMA technique and symmetry-conforming Landau theory are shown in Fig. 1.8. Figure 1.8a shows the ratio, $E(t)/E(0)$, which was obtained at the different time moments, t, of the continuous DMA measurement. The theoretical curves in Fig. 1.8a can be explained in the following way. If the internal pressure is disregarded, the theoretical time dependence of Young's modulus is caused by the axial internal stress; this dependence can be fitted to the experimental one only in the long-time limit, since the inequality $\tau_P < \tau_\sigma$ is accepted, the best fit is achieved for $\tau_\sigma = 180$ min. If the axial stress is disregarded, the theoretical time dependence of Young's modulus is caused by the initial pressure, and only an initial part of the theoretical curve can be fitted to the experimental points, the best fit is achieved for $\tau_P = 18$ min.

It should be emphasized, first, that the experimental values shown in Fig. 1.8a were measured at the alloy temperature of 363 K and the determined relaxation

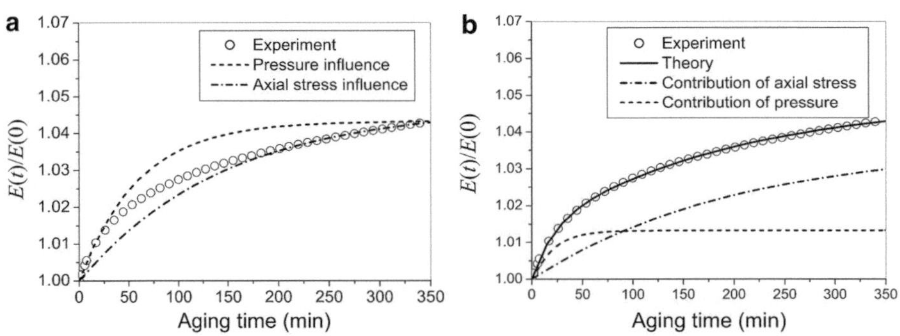

Fig. 1.8 Time dependencies of the storage elastic modulus (experiment) and Young's modulus (theory) of Ni–Al–Ni martensite. Theoretical curves are obtained taking into account: (**a**) only axial internal stress or only internal pressure and (**b**) both axial internal stress and internal pressure

times correspond to this temperature. The temperature change resulted in the noticeable changes of the relaxation times [44]. It can be concluded, second, that the experimental time dependence of Young's modulus cannot be described taking into account only one of the values, $\sigma^{(i)}(t)$ or $P^{(i)}(t)$, and therefore, the values $\sigma^{(i)}(\infty)$ and $P^{(i)}(\infty)$ used for the computation of theoretical dependencies presented in Fig. 1.8a do not correspond to the real values of the axial stress and pressure induced by defects. To determine real values, both axial internal stress and internal pressure were taken into account in the course of computations, and the excellent fit of the whole theoretical curve to the experimental points was achieved for the values $\sigma(\infty) = -13.5$ MPa and $P(\infty) = 0.4$ GPa (Fig. 1.8b). Besides that, the cooperative experimental and theoretical study of the time dependence of Young's modulus enabled a separation of the aging effects caused by the axial internal stress and internal pressure. As it is seen from Fig. 1.8b, the contributions of these values to the time-dependent part of Young's modulus are of the same order of magnitude.

The time dependence of a storage modulus shows a tendency to the saturation, and therefore, this tendency is inherent to the time-dependent parts of the axial strain and volume change. The saturation values of time-dependent strain and volume change can be estimated from the relationships $\delta\varepsilon_M(\infty) \sim \sigma(\infty)/E(\infty)$ and $\delta[\Delta V(\infty)/V(\infty)] \sim P(\infty)/B$, where the Young's modulus is about 12 GPa [44], and the value of bulk modulus is $B \sim 130$ GPa, resulting from the elastic modules measured in Ref. [50]. These relationships give $\delta\varepsilon_M(\infty) \sim 10^{-3} << \varepsilon_M$ and $\delta[\Delta V(\infty)/V(\infty)] \sim 3 \times 10^{-3} \sim \Delta V/V$. The estimations show that the defect subsystem is an important contributor to the volume change accompanying martensitic transformation of a real crystal. As so, the volume changes during MT may be noticeably different even for the alloys that have practically the same compositions.

To summarize, the internal pressure appearing in the course of Cu–Al–Ni martensite aging was noticeably smaller than the stabilizing internal pressure estimated above for the Au–Cd martensite. It is worth noting that the evaluation of an axial internal stress, $\sigma(\infty) \approx -30$ MPa, strictly follows from the experimental data reported for $Au_{50}Cd_{50}$ alloy [51]. Moreover, a rational change of the lattice parameter, $\Delta c/c \approx 6 \times 10^{-4}$, caused by the $Au_{49.5}Cd_{50.5}$ martensite aging was observed [52]. Large values of the internal stress and pressure imply a strong influence of aging on the elastic modules of Au–Cd alloys. This prediction should be verified experimentally.

1.4.3 Martensite Destabilization by Thermomechanical Cycling

As it was mentioned above, the concentration, type, and mobility of crystal defects influence the stability of the deformational behavior of the Ni–Mn–Ga ferromagnetic SMAs under the multiple thermomechanical cycling [21]. Recent studies discovered the very pronounced changes in the characteristics of temperature- and stress-induced MTs in Ni–Mn–Ga and Ti–Ni alloys subjected to the cyclic temperature or/and stress variation [26, 53]. Moreover, a strong influence of thermomechanical cycling on the storage elastic modulus of the Ni–Mn–Ga martensite with $c/a > 1$ was observed by DMA technique [53]. To study this influence, the temperature dependence of a storage elastic modulus of the $Ni_{57.5}Mn_{22.5}Ga_{20}$ alloy sample (specimen 1) was measured. The dependence obtained on cooling is depicted by curve 1 in Fig. 1.9. The characteristic temperatures of the forward MT, $T_{MS} = 579$ K and $T_{MF} = 543$ K, were determined using this curve. The temperatures of the reverse MT, $T_{AS} = 588$ K and $T_{AF} = 623$ K, were determined in the same way from the $E(T)$ dependence obtained on a heating of the specimen 1. The 15 stress–strain cycles were performed then at different temperatures, and an evolution of the forward and reverse stress-induced MT in the specimen 1 was monitored. Note that after each cycle, the specimen was heated to 673 K and held at this temperature for 10 min. The experimental temperature dependence of a storage elastic modulus, $E(T)$, measured on cooling of the specimen 1 after the cycling is depicted by the curve 2 in Fig. 1.9. Comparing the curves 1 and 2, one can see that the cycling reduced the storage modulus by a factor of about 0.5. The sharp decrease of MT temperatures ($\Delta T_{MS} \sim 50$ K, $\Delta T_{MF} \sim 20$ K) together with a drastic reduction of the storage modulus clearly indicate a destabilization of the martensitic phase by the thermomechanical cycling. The computations, carried out using the symmetry-conforming Landau theory, showed that the observed changes in the temperature dependence of the storage modulus can be caused by the high internal pressure, such as $P^{(in)} \approx 6$ GPa. This value of isotropic dilatation provided a *quantitative* agreement between the experimental and theoretical values of the elastic modulus in the whole temperature range shown in Fig. 1.9 [39]. The high internal pressure value may be explained, in principle, by the enhanced transformation volume effect in high-temperature Ni–Mn–Ga alloys and large distortion of a crystal lattice ($c/a \approx 1.2$) in the course of MT in these alloys, which lead to a non-compatibility of the different elements of martensitic structure and a high defect concentration.

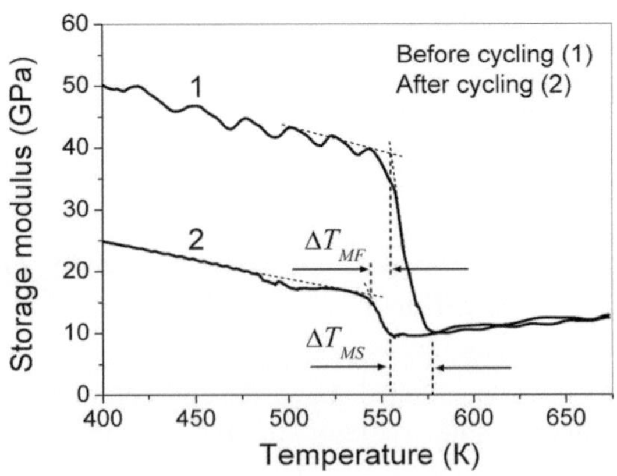

Fig. 1.9 Temperature dependencies of the storage elastic modulus measured on cooling of Ni–Mn–Ga alloy before and after 15 thermomechanical cycles

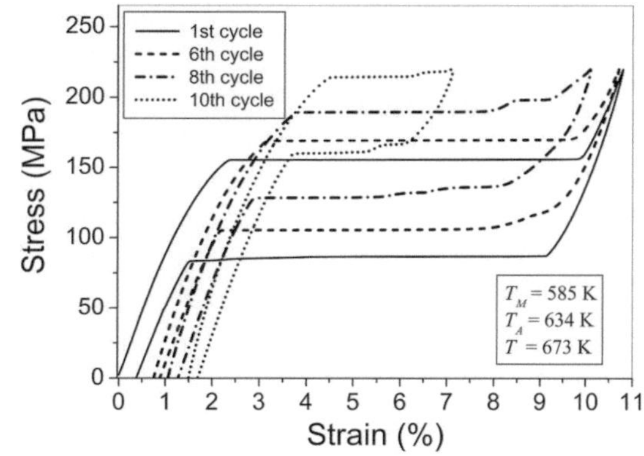

Fig. 1.10 The stress–strain loops obtained in the course of cyclic stress-induced MTs of the Ni–Mn–Ga alloy. The cycle numbers, forward MT temperature (T_M), reverse MT temperature (T_A), and alloy temperature (T) are shown in the legends

Another $Ni_{57.5}Mn_{22.5}Ga_{20}$ sample (specimen 2) was prepared for the observation of stress-induced MT, and its characteristic temperatures, $T_M = (T_{MS} + T_{MF})/2$ and $T_A = (T_{AS} + T_{AF})/2$, were determined by a standard differential calorimeter [53]. These temperatures and experimental stress–strain loops obtained for the specimen 2 are presented in Fig. 1.10. It is seen that an evolution of the stress–strain curves in the course of mechanical cycling looks like that caused by the elevation of alloy temperature (compare Figs. 1.6 and 1.9): the cycling leads to the pronounced elevation and shortening of the plateaus at the stress–strain loops accompanied by the narrowing of hysteresis.

A resemblance between the series of stress–strain loops obtained during cycling and the increasing sequence of temperatures is explained in Ref. [39] as follows: the mechanical cycling generates and rearranges the crystal defects; the generation and rearrangement of defects lead to the crystal lattice expansion conjugated with the internal pressure, which destabilizes the martensitic phase and lowers the MT temperature, therefore. The MT temperature lowering results in the increase of the interval δT between the MT temperature and the alloy temperature. The heating of an alloy also leads to the increase of this interval, whereby changes the deformational properties of an alloy similarly to the mechanical cycling. The symmetry-conforming Landau theory showed that the changes in the deformational behavior of the specimen 2 observed in the course of a mechanical cycling can be caused by the internal pressure $P^{(in)} \approx 2$ GPa, which is lower by factor 1/3 than the internal pressure induced in the specimen 1 after the 15 thermomechanical cycles. A careful consideration of the stress–strain curves obtained for the specimen 2 shows, however, that the mechanical cycles with the numbers from 1 to 6 changed the deformational behavior of an alloy not so strongly as the next four cycles. It may be assumed, therefore, that the 15 mechanical cycles can induce the internal pressure comparable in value with that induced in the specimen 1 by the thermomechanical cycling.

It is interesting to compare the series of stress–strain loops depicted in Figs. 1.6 and 1.9 with the experimental stress–strain curves obtained in Ref. [54] for the oxygen-doped Ti–Nb alloy. These loops are shown in Fig. 1.11 to illustrate that the

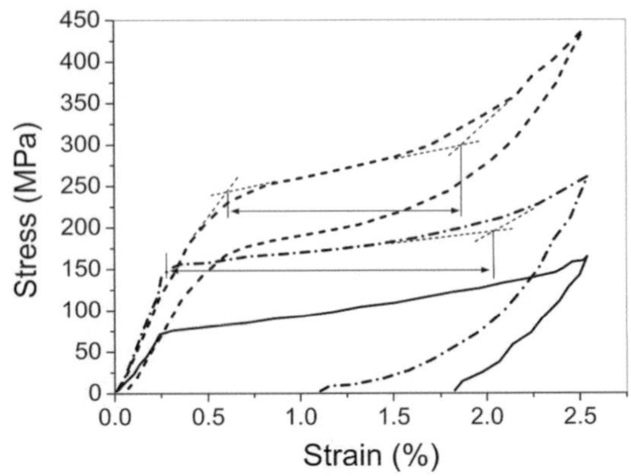

Fig. 1.11 Stress–strain curves obtained in Ref. [54] for the Ti–$Nb_{36-x}O_x$ (at %) alloys with $x = 0$ (*solid*), 0.3 (*dash-doted*), and 0.5 (*dashed*) at the temperature of 233 K

oxygen atoms located in the interstitial crystallographic sites induce the following changes in the stress–strain curves: (a) a smoothing of the sharp flexures, (b) the increase of the slope of plateau, (c) the elevation of plateau, (d) the narrowing of plateau as shown by the two-side arrows in Fig. 1.11, and (e) the lowering of hysteresis.

The features (a) and (b) of the deformational behavior of Ti–Nb–O alloy were described in Ref. [54] using the random-field Ising model. It was shown that the interstitial oxygen provokes a formation of the randomly oriented nanometer-sized martensite variants. The randomness of martensitic structure formed by these variants leads to the smoothing of stress–strain dependencies. In the same time, the stress–strain curves computed using the Ising model do not exhibit the features (c) and (d), while the feature (e) is hardly found. In contrast to this, the symmetry-conforming Landau theory adequately describes all the features of stress–strain loops mentioned above [36, 39]. According to this theory, both the elevation and shortening of a plateau at the stress–strain loops measured for the oxygen-doped Ti–Nb may be caused by the negative internal pressure, which arises when the oxygen atoms penetrate into the host crystal lattice expanding it. It should be mentioned, in addition, that the symmetry-conforming Landau theory was successfully modified for the description of both a process of the hydrogen absorption in vanadium and the order–disorder phase transition in the vanadium–hydrogen system [55].

Thus, both shear strain and volume evolution in the course of the martensite aging and thermomechanical cycling have a considerable influence on the functional properties of SMAs. It suggests that in the case of ferromagnetic SMAs, these properties can also be sensitive to the volume magnetostriction.

1.5 Influence of Volume Magnetostriction on the Elastic and Transformational Properties of Ferromagnetic SMAs

1.5.1 Influence of Volume Magnetostriction on the Elastic Modules of Ferromagnetic SMAs

Phase transition from the paramagnetic to ferromagnetic phase is accompanied by a spontaneous volume magnetostriction, that is, specific volume change of SMA. The relative value of magnetostrictive volume change is $v_{me} \equiv (V_{FM} - V_{PM})/V_{PM}$, where V_{FM} and V_{PM} are the volumes of an alloy in the ferromagnetic and paramagnetic phases, respectively. As it was mentioned in Sect. 1.2, the volume magnetostriction can be described in terms of the magnetoelastic pressure, P_{me}, which is proportional both to the dimensionless magnetoelastic constant, δ_{ex}, and squared magnetization, $M^2(T)$ (see Eq. (1.11)). It means that the magnetostrictive volume change can be expressed as:

$$v_{me} = P_{me}/B = -\delta_{ex}M^2(T)/3B, \quad (1.39)$$

where B is the bulk modulus of SMA. As it was observed long ago, the high hydrostatic pressure noticeably shifts the MT temperature. It means that the transformational properties of an alloy depend on its volume. Therefore, the characteristics of MTs in ferromagnetic SMAs should depend on the volume magnetostriction, which, in turn, depends on the temperature through the temperature dependence of magnetization value. More generally, it may be stated that the transformational properties of SMA should depend on its magnetic state.

To clarify an influence of the volume magnetostriction on the transformational properties of SMA, the transformation of cubic lattice into z-variant of the tetragonal lattice may be considered, for the sake of definiteness. The expression for the free energy, Eq. (1.19), corresponding to this MT, must be supplemented with the terms relating a volume change to the MT strains. The supplemented expression is

$$\begin{aligned} F = &\frac{1}{2}c_1u_1^2 + \frac{1}{2}c_2(T)u^2 + \frac{1}{2}a_2u_1u^2 + \frac{1}{3}a_4u^3 \\ &+ \frac{1}{4}b_4u^4 + \frac{1}{2}b_7u_1u^3 - \delta_{ex}M^2(T)u_1, \end{aligned} \quad (1.40)$$

where $c_1 = 9B$, a_2 and b_7 are phenomenological constants denoted according to Refs. [35, 36], and u_1 variable is related to the spur of the strain tensor by Eq. (1.2) in such a way that the relative volume change, v, of deformed alloy is related to this value as $v = 3u_1$.

In the case when the volume change is caused by the volume magnetostriction, a substitution of the value $u_1 = 3P_{me}/B$ into Eq. (1.40) shows that the magnetostriction results in a renormalization of the elastic constants, where the constants corresponding to the paramagnetic (*pm*) and ferromagnetic (*fm*) phases are interrelated as

$$
\begin{aligned}
c_2^{(fm)}(T) &= c_2^{(pm)}(T) - \frac{3a_2}{c_1} P_{me}(T), \\
a_4^{(fm)}(T) &= a_4^{(pm)} - \frac{9\Omega b_7 P_{me}(T)}{2c_1}, \\
b_4^{(fm)} &= b_4^{(pm)} \equiv b_4.
\end{aligned}
\tag{1.41}
$$

This renormalization does not change the expression for the free energy quantitatively. It means that the observable elastic properties of a ferromagnetic solid are formed under the influence of the ferromagnetic ordering. In the case of "linear" elastic media, the coefficient $c_2 = C'/3$ is large, and therefore, the third-order and forth-order terms are immaterial in most cases, as well as the magnetoelastic renormalization of c_2. In contrast to this, the abnormally low values of C' are inherent to ferromagnetic SMAs, and "magnetic contribution" to c_2 may be noticeable. The renormalization of elastic constants results in the change of theoretical value of the MT strain. According to Eq. (1.40), the spontaneous strain characterizing MT, which goes in the ferromagnetic state of SMA, is related to the renormalized elastic constants as $\varepsilon_{MT} = u(T_{MF})/3 = -a_4^{(fm)}/3b_4$. However, the estimation of the role of ferromagnetic order in the elastic properties of SMA and characteristics of MT is a difficult task, because the ferromagnetic ordering does not lead to qualitative changes in the expression for the free energy. It means that the quantitative numerical calculations are needed. The calculations of this kind meet with difficulties because of an uncertainty in the values of coefficients a_2 and b_7. An example of successful overcoming of these difficulties is presented below.

1.5.2 Volume Magnetostriction Contributing to the Heat Evolution During MT

The forward MT is exothermic in most cases and, therefore, is characterized by the negative value of the latent heat $\Delta Q = T(S_M - S_A)$, where the S_A and S_M are the entropy values of the austenitic and martensitic phases, respectively. It is convenient to consider the positive entropy change $\Delta S \equiv S_A - S_M$ and positive heat $q \leq T\Delta S$ evolved in a course of the forward MT. The evolved heat is less than the absolute value of latent heat if the free energy of alloy does not reach the minimum value after MT, that is, if the nonequilibrium martensitic state arises. The difference $T\Delta S - q$ is often referred to as the "stored elastic energy."

Experiment shows that both the MT temperature and evolved heat strongly depend on a deviation of the alloy composition from a stoichiometry. The open circles in Fig. 1.12 show the experimental values of evolved heat obtained by different authors for the nonstoichiometric Ni–Mn–Ga alloys with different

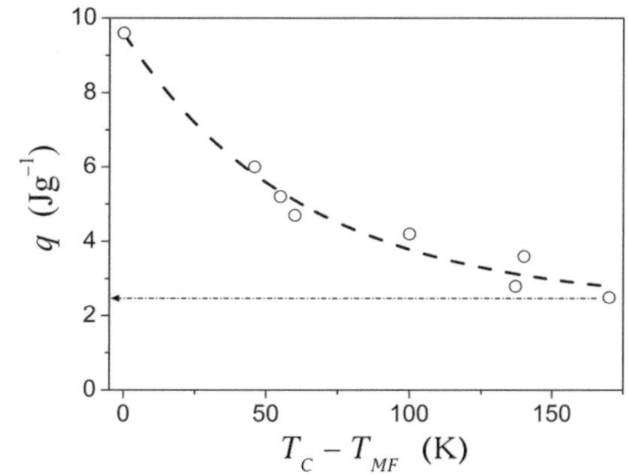

Fig. 1.12 Experimental values of evolved heat measured for different Ni–Mn–Ga alloys (*circles*). The *arrow* points to the minimum evolved heat value measured for these alloys. The *dashed line* is a guide for the eye

martensite finish temperatures, T_{MF}. The dashed line is a guide for the eye. The experimental data show that the closer is MT temperature interval to the Curie point, the larger is the evolved heat value. For the sake of obviousness, the arrow in Fig. 1.12 points to the minimum evolved heat value, which is less than maximum one by factor 3.8. In view that the entropy change is related to the free energies of two phases as

$$\Delta S = -\frac{\partial}{\partial T}(F_A - F_M), \tag{1.42}$$

the experimental data suggest that the ferromagnetic order is responsible for the sharp increase of heat evolution on approach of martensite finish temperature to Curie point: indeed, the free energy of ferromagnetic SMA depends on the squared magnetization, $M^2(T)$; the closer is the alloy temperature to Curie point, the larger are both the derivative $\mathrm{d}M^2(T)/\mathrm{d}T$ and magnetic contribution to the entropy change. This tentative conclusion has been supported by the quantitative theoretical analysis in Ref. [56].

Free energy of the ferromagnetic SMA was presented in Ref. [56] as a sum of the elastic, magnetic, and magnetoelastic terms,

$$F = F_e(u_\alpha) + F_m(m_i) + F_{me}(u_\alpha, m_i), \tag{1.43}$$

where u_α are the linear combinations of strains and m_i are the magnetic vector components (see Eqs. (1.2)–(1.4)).

The magnetic term

$$F_m = \frac{1}{2}J(T)M^2(T) + F_A(m_i) \tag{1.44}$$

involves the magnetic anisotropy energy, F_A, and spin exchange energy, characterized by the parameter $J(T)$. The sum of elastic and magnetoelastic terms is described in Eq. (1.40). The magnetic term does not depend on strains and describes the magnetic energy of undeformed alloy in a cubic phase. Therefore, this term does not contribute to the entropy change caused by the appearance of spontaneous strains in the course of MT. By this reason, Eqs. (1.42) and (1.43) give

$$\Delta S = -\frac{\partial}{\partial T}(\Delta F_e + \Delta F_{me}), \tag{1.45}$$

where ΔF_e and ΔF_{me} are the elastic and magnetoelastic parts of the energy change caused by MT. Equation (1.45) can be presented in the form

$$\Delta S = \Delta S_e + \Delta S_{mag}, \tag{1.46}$$

where

$$\begin{aligned} \Delta S_e &= -\partial \Delta F_e / \partial T, \\ \Delta S_{mag} &= -\partial \Delta F_{me} / \partial T. \end{aligned} \tag{1.47}$$

The terms ΔS_e and ΔS_{mag} are *conventionally* referred to as the elastic and magnetic part of the entropy change, respectively. In essence, ΔS_e corresponds to the martensitic transformations occurring in the paramagnetic state of alloy. In the case if MT goes in the ferromagnetic state, this value should be determined as the difference between the total entropy change ΔS and its magnetic part ΔS_{mag} [57]. The ΔS_{mag} is that part of total entropy change, which depends on the magnetization of ferromagnetic alloy, $M(T)$.

Let us consider the case of MT, which goes in the ferromagnetic state of an alloy in the absence of external forces. In the austenitic phase, the equilibrium strain values are related to the spontaneous magnetostriction, while in the martensitic phase both magnetostrictive and MT strain exist. As far as the MT strain exceeds the magnetostriction by the order of magnitude, the values $u = u_1 = 0$ can be accepted for austenitic phase and $u = 3\varepsilon_M, u_1 = v_{MT}/3$ for martensitic phase, where $v_{MT} = (V_M - V_A)/V_A$ is the relative volume change caused by MT.

Taking into account that only the second and last terms in Eq. (1.40) explicitly depend on the temperature, one can calculate the partial derivatives involved in Eq. (1.46) and obtain the relationships:

$$\Delta S_e = \frac{1}{2} u_0^2(T_{MF}) \left[\frac{\partial c_2(T)}{\partial T}\right]_{T=T_{MF}}, \tag{1.48}$$

$$\Delta S_{mag} = -\frac{2}{3} \delta_{ex} v_{MT} M(T_{MF}) \left[\frac{\partial M(T)}{\partial T}\right]_{T=T_{MF}}, \tag{1.49}$$

where $u_0(T_{MF}) = 2[c(T_{MF})/a(T_{MF}) - 1]$. It was argued in Ref. [56] that the volume magnetostriction gives the main contribution to the magnetic entropy change, while the axial magnetostriction can be disregarded.

The magnetic entropy change proved to be proportional to the dimensionless parameter:

$$\kappa \equiv (2/3)\delta_{ex}v_{MT}. \tag{1.50}$$

As so, the evaluation of the magnetic entropy change is possible if the magnetoelastic constant, δ_{ex}, and volume change are measured for the same alloy. However, these measurements meet serious difficulties [47, 58]. In this connection, another way of the evaluation was proposed [56]. It was taken into account that the sum of spin-exchange energy and magnetoelastic energy is expressed as

$$F_m + F_{me} = \frac{1}{2}J^*(T)M^2(T), \tag{1.51}$$

where

$$J^*(T) = J(T) - \delta_{ex}u_1 = \begin{cases} J(T) - \kappa, & \text{if } T_C \le T_{MF}, \\ J(T), & \text{if } T_C \ge T_{MS}. \end{cases} \tag{1.52}$$

as it follows from Eqs. (1.40) and (1.44). The exchange parameter, $J*(T)$, is one of the values predetermining the temperature dependence of magnetization in the saturating magnetic field. Due to this, the parameter κ proved to be determinable from the equation for magnetization [56]. Moreover, the equation for $M(T)$ enabled a determination of the derivative $\partial M/\partial T$ and magnetic entropy change, Eq. (1.49). The only requirement for the equation is a good fit of its solution with the experimental dependence $M(T)$. (For more details see Refs. [33, 56]).

Figure 1.13 shows the theoretical value of evolved heat as the function of the difference between the Curie temperature and martensite finish temperature. The function was computed using the experimental dependence $M(T)$ measured for $Ni_{52.6}Mn_{23.5}Ga_{23.9}$ alloy (alloy 1) and theoretical $M(T)$ dependence fitted to experimental one by the proper choice of κ value. The presented data in Fig. 1.13 confirm the tentative conclusion about the main contribution of magnetic subsystem to the heat evolution during MTs in the ferromagnetic SMAs with close values of martensite finish and Curie temperatures (see above). It should be noticed, however, that the evolved heat values are not small for the Ni–Mn–Ga alloys with $T_{MS} > T_C$: the typical values measured for the number of alloys are about 10 Jg^{-1} [53, 59]. This fact can be explained using Eq. (1.48), which shows that the elastic part of the evolved heat is proportional to the factor $u_0^2 = 4(1 - c/a)^2$. This factor is of the order of 0.15 for Ni–Mn–Ga alloys with $T_{MS} > T_C$ and about of 0.01 for those with $T_{MS} < T_C$. Therefore, the elastic part of the evolved heat of MTs in the alloys with $T_{MS} > T_C$ is much larger than that of MTs in the alloys with $T_{MS} < T_C$.

A dominating magnetic contribution to the transformation entropy change was also proved for MT in the metamagnetic SMAs [60].

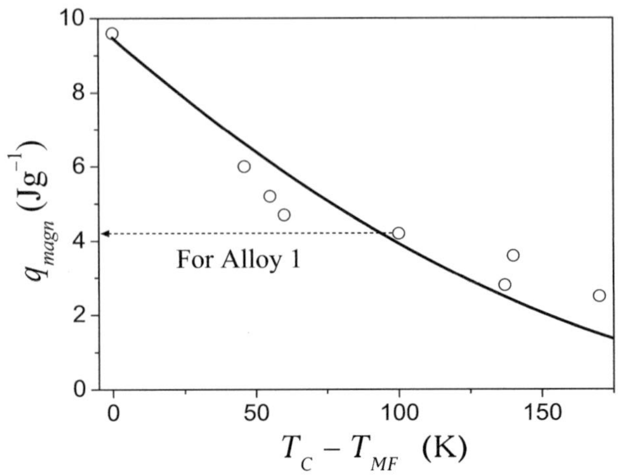

Fig. 1.13 Theoretical (*line*) and experimental (*open circles*) values of the evolved heat as the function of difference between the Curie temperature and martensite finish temperature

1.6 Summary

It can be summarized that the factors such as (a) magnetostriction, (b) lattice misfit in the mixed austenitic–martensitic state, and (c) lattice distortion caused by crystal defects exert strong influence on the deformational and thermodynamic properties of shape memory alloys:

(a) *Axial magnetostriction* provides physical mechanism of twinning–detwinning process, which, in turn, leads to the giant magnetic field-induced deformation of a ferromagnetic SMA. The magnetic field influence on the twin structure of an alloy is characterized by magnetostress. A consistent consideration of magnetoelastic coupling, which causes the magnetostriction, enables the determination of magnetostress value as a function of the magnetic field. This function predetermines the value and direction of the external magnetic field needed for the observation of giant magnetically induced deformation. *Volume magnetostriction* has a strong influence on the martensitic transformation strain, transformation heat, and elastic constants of ferromagnetic shape memory alloy.

(b) The systematic experimental and theoretical studies show that the minimization of *lattice misfit* in the mixed austenitic–martensitic state enables the observation of practically anhysteretic martensitic transformations in the course of thermal cycling of an alloy. Moreover, the minimization of lattice misfit improves the stability of transformational behavior of an alloy during the multiple thermal cycling. Recently obtained experimental and theoretical data discovered the possibility of large anhysteretic deformation of SMA in the course of stress–strain cycles performed in a *postcritical* regime.

(c) The lattice distortion caused by the *crystal defects* noticeably changes the martensitic transformation temperatures. The latter ones essentially shift in the course of martensite aging. The adjustment of the symmetry of a spatial distribution of the crystal defects to the symmetry of a crystal lattice in martensitic phase was traditionally considered as the main mechanism of martensite aging effects. The recent studies showed, however, that another mechanism of martensite aging effect of martensite aging exists. This mechanism is related to the slow variation of the specific volume of an alloy after the martensitic transformation. The *contributions of volume change* to the martensite aging effects occurred comparable in value with the contribution of symmetry change of the defect system.

Acknowledgments This work was supported by Spanish Ministry of Science and Innovation (MAT2014-56116-C4-3-4-R) and by the University of the Basque Country, UPV/EHU (Grupos Consolidados GIC12/10).

References

1. Sozinov, A., Lanska, N., Soroka, A., Zou, W.: Appl. Phys. Lett. **102**, 021902 (2013)
2. Ullakko, K., Huang, J.K., Kantner, C., Kokorin, V.V., O'Handley, R.C.: Appl. Phys. Lett. **69**, 1966 (1996)
3. Likhachev, A.A., Ullakko, K.: Eur. Phys. J. B **14**, 263 (2000)
4. Likhachev, A.A., Ullakko, K.: Phys. Lett. A **275**, 142 (2000)
5. Chernenko, V.A., L'vov, V.A., Cesari, E., McCormick, P.: Mater. Transact. JIM. **41**, 928 (2000)
6. Chernenko, V.A., L'vov, V.A., Müllner, P., Kostorz, G., Takagi, T.: Phys. Rev. B **69**, 134410 (2004)
7. Heczko, O., Sozinov, A., Ullakko, K.: IEEE Trans. Mag. **36**, 3266 (2000)
8. Murray, S.J., Marioni, M., Allen, S.M., O'Handley, R.C.: Appl. Phys. Lett. **77**, 886 (2000)
9. Chernenko, V.A., L'vov, V.A.: Mater. Sci. Forum **583**, 1 (2008)
10. Murray, S.J., O'Handley, R.C., Allen, S.M.: J. Appl. Phys. **89**, 1295 (2001)
11. Heczko, O., Straka, L.: J. Appl. Phys. **94**, 7139 (2003)
12. L'vov, V.A., Glavatska, N., Aaltio, I., Söderberg, O., Glavatskiy, I., Hannula, S.-P.: Eur. Phys. J. **158**, 199 (2008)
13. Sasso, C.P., L'vov, V.A., Chernenko, V.A., Barandiaran, J.M., Pasquale, M., Kono, Y.: Phys. Rev. B **81**, 224428 (2010)
14. Likhachev, A.A., Sozinov, A., Ullakko, K.: Mech. Mater. **38**, 551 (2006)
15. O'Handley, R.C., Paul, D.I., Allen, S.M., Richard, M., Feuchtwanger, J., Peterson, B., Techapiesancharoenkij, R., Barandiaran, M., Lazpita, P.: Mat. Sci. Eng. A **438–440**, 445 (2006)
16. Straka, L., Novak, V., Landa, M., Heczko, O.: Mater. Sci. Eng. A **374**, 263 (2004)
17. L'vov, V., Chernenko, V.: Mater. Sci. Forum **684**, 31 (2011)
18. Aaltio, I., Söderberg, O., Ge, Y., Hannula, S.-P.: Scr. Mater. **62**, 9 (2010)
19. Chernenko, V., L'vov, V., Cesary, E., Pons, J., Portier, R., Zagorodnyuk, S.: Mat. Trans. JIM **43**, 856 (2002)
20. Fukuda, T., Yamamoto, M., Yamaguchi, T., Kakeshita, T.: Acta Mater. **62**, 182 (2014)
21. Aaltio, I., Soroka, A., Ge, Y., Söderberg, O., Hannula, S.-P.: Smart Mater. Struct. **19**, 075014 (2010)
22. Fukuda, T., Kakeshita, T.: Scr. Mater. **69**, 89 (2013)
23. Kosogor, A., L'vov, V.A., Chernenko, V.A., Villa, E., Barandiaran, J.M., Fukuda, T., Terai, T., Kakeshita, T.: Acta. Mater. **66**, 79 (2014)

24. James, R.D., Zhang, Z.: A way to search for multiferroic materials with 'unlikely' combinations of physical properties, vol. 79. In: Mañosa, L., Planes, A., Saxena, A.B. (eds.) The interplay of magnetism and structure in functional materials. Springer series in materials science. Springer, Berlin (2005)
25. Zhang, Z., James, R.D., Müller, S.: Acta Mater. **57**, 4332 (2009)
26. Zarnetta, R., Takahashi, R., Young, M.L., et al.: Adv. Funct. Mater. **20**, 1917 (2010)
27. Buenconsejo, P.J.S., Zarnetta, R., König, D., Savan, A., Thienhaus, S., Ludwig, A.: Adv. Funct. Mater. **20**, 1917 (2010)
28. Lei, C.H., Li, L.J., Shu, Y.C., Li, J.Y.: Appl. Phys. Lett. **96**, 141910 (2010)
29. Liberman, D.S., Read, T.A., Wechsler, M.S.: J. Appl. Phys. **28**, 532 (1957)
30. Landau, L.D., Lifshitz, E.M.: Course of theoretical physics; Vol 5: statistical physics. Pergamon, New York (1980)
31. Danilevich, A.G., L'vov, V.A.: J. Magn. Magn. Mater. **333**, 108 (2013)
32. Otsuka, K., Ren, X.: Mat. Sci. Eng. A **273–275**, 89 (1999)
33. Chernenko, V.A., L'vov, V.A., Cesari, E., Kosogor, A., Barandiaran, J.M.: Metals **3**, 237 (2013)
34. Kustov, S., Corró, M.L., Pons, J., Cesari, E.: Appl. Phys. Lett. **94**, 191901 (2009)
35. Gomonaj, E.V., L'vov, V.A.: Phase Transit. **47**, 9 (1994)
36. Kosogor, A., Matsishin, N.J., L'vov, V.A.: Phase Transit. **86**, 796 (2013)
37. Otsuka, K., Ren, X.: Mater. Sci. Eng. A. **312**, 207 (2001)
38. Miyazaki, S., Igo, Y., Otsuka, K.: Acta Metall. **34**, 2045 (1986)
39. L'vov, V.A., Kosogor, A., Barandiaran, J.M., Chernenko, V.A.: Acta Mater. **60**, 1587 (2012)
40. Otsuka, K., Ren, X.: Scr. Mater. **50**, 207 (2004)
41. Nix, W.D., Gibeling, J.C., Hughes, D.A.: Metall. Transact. A **16**, 2215 (1985)
42. Hardy, V., Maignan, A., Hébert, S., Yaicle, C., Martin, C., Hervieu, M., Lees, M.R., Rowlands, G., Paul, D., Mc, K., Raveau, B.: Phys. Rev. B. **68**, 220402 R (2003)
43. Kosogor, A., L'vov, V.A., Soderberg, O., Hannula, S.-P.: Acta Mater. **59**, 3593 (2011)
44. Kosogor, A., Xue, D., Zhou, Y., Ding, X., Otsuka, K., L'vov, V.A., Sun, J., Ren, X.: J. Phys. Condens. Matter **25**, 335402 (2013)
45. Gefen, Y., Halwany, A., Rosen, M.: Philos. Mag. **28**, 1 (1973)
46. Kakeshita, T., Yoshimura, Y., Shimizu, K., Endo, S., Akahama, Y., Fujita, F.E.: Mater. Trans. JIM. **29**, 781 (1988)
47. Chernenko, V.A., L'vov, V.A.: Philos. Mag. **A73**, 999 (1996)
48. Murakami, Y., Nakajima, Y., Otsuka, K., Ohba, T., Matsuo, R., Ohshima, K.: Mater. Sci. Eng. A **237**, 87 (1997)
49. Otsuka, K., Shimizu, K.: Mater. Trans. JIM. **15**, 103 (1974)
50. Sedlák, P., Seiner, H., Landa, M., Novák, V., Šittner, P., Mañosa, L.: Acta Mater. **53**, 3643 (2005)
51. Otsuka, K., Ren, X., Murakami, Y., Kawano, T., Ishii, T., Ohba, T.: Mater. Sci. Eng. A **273–275**, 558 (1999)
52. Ishibashi, H., Kogachi, M., Ohba, T., Ren, X., Otsuka, K.: Mater. Sci. Eng. A **329–331**, 568 (2002)
53. Chernenko, V.A., Villa, E., Besseghini, S., Barandiaran, J.M.: Phys. Procedia **10**, 94 (2010)
54. Nii, Y., Arima, T., Kim, H.Y., Miyazaki, S.: Phys. Rev. B **82**, 214104 (2010)
55. Kosogor, A., Hjörvarsson, B., L'vov, V.A., Wolff, M.: Phys. Rev. B **89**, 014207 (2014)
56. L'vov, V.A., Cesari, E., Recarte, V., Pérez-Landazábal, J.I.: Acta Mater. **61**, 1764 (2013)
57. Recarte, V., Pérez-Landazáball, J.I., Gómez-Polo, C., Sánchez-Alarcos, V., Cesari, E., Pons, J.: J. Phys. Condens. Matter **22**, 416001 (2010)
58. Buchelnikov, V.D., Khovailo, V.V., Takagi, T.: J. Magn. Magn. Mater. **300**, 459 (2006)
59. Chernenko, V.A., Cesari, E., Kokorin, V.V., Vitenko, I.N.: Scr. Met. Mat. **33**, 1239 (1995)
60. Comtesse, D., Gruner, M.E., Ogura, M., Sokolovskiy, V.V., Buchelnikov, V.D., Grunebohm, A., Arroyave, R., Singh, N., Gottschall, T., Gutfleisch, O., Chernenko, V.A., Albertini, F., Fähler, S., Entel, P.: Phys. Rev. B **89**, 184403 (2014)

Chapter 2
Magnetic, Magnetocaloric, Magnetotransport, and Magneto-optical Properties of Ni–Mn–In-Based Heusler Alloys: Bulk, Ribbons, and Microwires

I. Dubenko, N. Ali, S. Stadler, Arcady Zhukov, Valentina Zhukova, B. Hernando, V. Prida, V. Prudnikov, E. Gan'shina, and A. Granovsky

Abbreviations

MT	Martensitic transition
MST	Magnetostructural transition
FM	Ferromagnetic
AFM	Antiferromagnetic
AP	Austenitic phase
MP	Martensitic phase
AHE	Anomalous Hall effect
MO	Magneto-optical

The original version of the book was revised because Arcady Zhukov's name was misspelled. An erratum explaining this can be found at DOI 10.1007/978-3-319-26106-5_11

I. Dubenko (✉) • N. Ali
Department of Physics, Southern Illinois University Carbondale, Carbondale, IL 62901, USA
e-mail: igor_doubenko@yahoo.com

S. Stadler
Department of Physics and Astronomy, Louisiana State University, Baton Rouge, LA 70803, USA

A. Zhukov
UPV/EHU, Basque Country University, San Sebastian, Spain

V. Zhukova
Faculty of Chemistry, Basque Country University, San Sebastian 20080, Spain

B. Hernando • V. Prida
Department de Física, Universidad de Oviedo, Calvo Sotelo s/n, 33007 Oviedo, Spain

V. Prudnikov • E. Gan'shina • A. Granovsky
Faculty of Physics, M.V. Lomonosov Moscow State University, Moscow 119991, Russia
e-mail: granov@magn.ru

A. Zhukov (ed.), *Novel Functional Magnetic Materials*, Springer Series in Materials Science 231, DOI 10.1007/978-3-319-26106-5_2

MR Magnetoresistance
TKE Transverse Kerr effect
SOI Spin–orbit interaction

2.1 Introduction

Ternary stoichiometric intermetallics with the general formula XYZ (called half Heusler) or X_2YZ (full Heusler), in which X and Y are typically transition metals and Z a main group element, were named collectively as Heusler compounds. Nowadays this term is used for nonstoichiometric ternary compounds as well as for quaternary, i.e., this doped with extra metal or metalloid elements. Due to this definition, it is an endless class of compounds [1], and, depending on composition, Heusler alloys exhibit a wide diversity of magnetic, electrical, optical, and mechanical properties important for fundamental research and promising for applications (for a brief review, see [1] and references therein). In this chapter we will discuss a relatively novel subclass of Heusler alloys, namely, those based on Ni–Mn–In compounds with nearly 15 at.% concentration of In, which exhibit magnetostructural transitions (MST) and, related to the MST, multifunctional properties, such as giant magnetocaloric effects (MCE), large magnetoresistance (MR), anomalous Hall effects (AHE), strong mechanical deformations, magnetic shape memory effects, exchange bias, kinetic arrest, etc.

An MST is a structural martensitic transition that results in a simultaneous change of the structure and magnetic state of the material. A martensitic transition (MT) is well-known in materials science as a first-order phase transition from high-temperature austenitic (AP) to low-temperature martensitic phase (MP). The thermal hysteretic behavior of a MT and its characteristic temperatures are shown in Fig. 2.1. In the Ni–Mn–In-based alloys, the magnetic states of the high-temperature

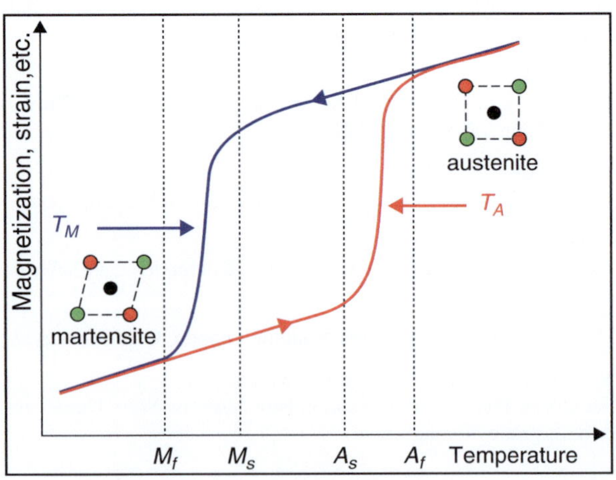

Fig. 2.1 A schematic of a martensitic transition (MT) and corresponding thermal hysteresis. T_A is the temperature of the inverse MT and T_M the temperature of the direct MT. M_s and M_f are the temperatures of the "start" and "finish" of the direct MT. A_s and A_f are the temperatures of the "start" and "finish" of the inverse MT

austenitic (AP) and low-temperature martensitic (MP) phases can be quite different and inhomogeneous. Moreover, in some cases the MST can be induced by an external magnetic field. In spite of the significant progress made in recent years in understanding the multifunctional properties related to MSTs in Ni–Mn–In based Heusler alloys, the detailed mechanisms responsible for the MST are far from being well understood. Due to the delicate balance between electronic, ionic, vibration, and magnetic energies in the vicinity of the MST, the properties of these alloys are extremely sensitive to any changes in intrinsic parameters, such as chemical composition, type of crystal structure, and type and volume fraction of the extra doping elements, as well as on extrinsic parameters, such as fabrication techniques and conditions, annealing temperature, applied magnetic field, pressure, rate of heating and cooling, sequence of measurements, and cycling. From one hand it presents an opportunity to search for desirable properties at ambient temperatures and at accessible magnetic fields but, on the other hand, it makes such research extremely challenging.

2.2 Bulk Ni–Mn–In-Based Alloys

2.2.1 Sample Fabrication

Bulk polycrystalline samples were prepared by arc melting the constituent elements of purity better than 99.99 % under a constant flow of "ultrahigh" purity argon using a water-cooled massive bronze crucible and tungsten electrode. It should be noted here that the higher purity of argon and starting metals, the better samples that can be produced. The total mass of the metallic components must be much less than the mass of crucible to provide the required heat dissipation. In our particular case, the mass of the components and crucible were less than 5 g and 1.5 kg, respectively. This method is commonly used to produce metallic alloys and compounds, to prevent contamination from the crucible and electrode material (the temperature of the crucible and, therefore, the bottom sample surface must be less than 50 °C during melting. The electrode must be kept below the melting point of tungsten, 3422 °C). The samples were melted four times to ensure homogenization. The loss of the elements during the melting was controlled by the measuring of the total sample mass using analytical balance with an accuracy of 5×10^{-5}. The samples with losses less than 0.2 % of the total mass were hold for the studies. The melted samples had been wrapped in tantalum foil and annealed in high vacuum (~10^{-5} Torr) for 1–2 days at 850 °C.

2.2.2 Examples of the Crystal Structure

The martensitic transition (MT) is described as a temperature-induced first-order structural phase transition from a high-temperature austenitic phase (cubic $L2_1$ or B_2 crystal structure) to a low-temperature martensitic phase, characterized by a crystal cell of lower symmetry (tetragonal, orthorhombic, or monoclinic) [2]. Magnetic states of austenitic (AP) and martensitic (MP) phases in magnetic off-stoichiometric Heusler alloys can be quite different and inhomogeneous. The stoichiometric Ni_2MnX with X = In Heusler alloy has been reported to be collinear ferromagnetic below $T_C = 317$ K, with a cubic $L2_1$ structure [3]. The coupling between Mn magnetic moments is altered from that of stoichiometric systems when Mn atoms occupy the X sites in off-stoichiometric Ni–Mn–X Heusler alloys. A possible antiferromagnetic (AFM) coupling of Mn magnetic moments in X sites relative to properly located Mn moments (i.e., antisite disorder) has been reported for $Ni_2Mn(Mn_xSn_{1-x})$ and for Ni–Mn–Ga compounds in Ref. [4–6]. This AFM coupling results in a decrease in magnetic moment in the MP compared to the AP. The reduction in magnetization depends on composition and can reach 60 % in high magnetic fields [7]. The substitution of In for Mn atoms also results in structural instability, and the first-order MT has been observed for In concentration of ~15 at.% [2, 7, 8] at temperatures $T < 350$ K. Examples of XRD patterns for cubic $L2_1$ AP and different types of MP crystal cells are shown in Fig. 2.2 [7]. The alloys of interest are those that show an MST near room temperature. In such compounds, it is typical to observe a mixture of high-temperature cubic AP and low-temperature MP at $T = 300$ K [7–9]. This behavior results from the first-order structural transition and, therefore, from the structural temperature hysteresis.

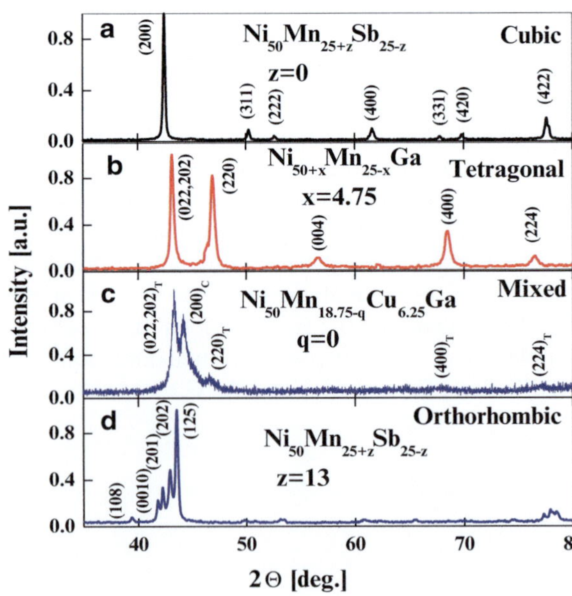

Fig. 2.2 Typical XRD patterns for Ni–Mn–X (X = Ga, In, Sb, and Sn), at $T = 300$ K

2.2.3 Temperature-Induced Phase Transitions and the Ground State of the Austenitic and Martensitic Phases

The magnetic phase transitions of the off-stoichiometric Heusler alloys can be characterized by the three transition temperatures: T_{CM}, T_M, and T_C (see Fig. 2.3, where the $M(T)$ curves for some Heusler alloys are shown as an example [10–13]). Ferromagnetic and paramagnetic types of $M(H)$ curves were observed for $T < T_{CM}$, and $T_A/T_M < T < T_C$, and above T_C, respectively. The sharp change in magnetization at T_A/T_M is associated with the MT from a magnetic state characterized by low magnetic moment (AFM or paramagnetic phase) to a ferromagnetic AP phase. The difference in the ZFC and FC $M(T)$ curves in the low-temperature region ($T < T_{CM}$) is typical for many Ni–Mn–X-based compounds and is related to the magnetic heterogeneity that results in a shift of the field-cooled (FC) $M(H)$

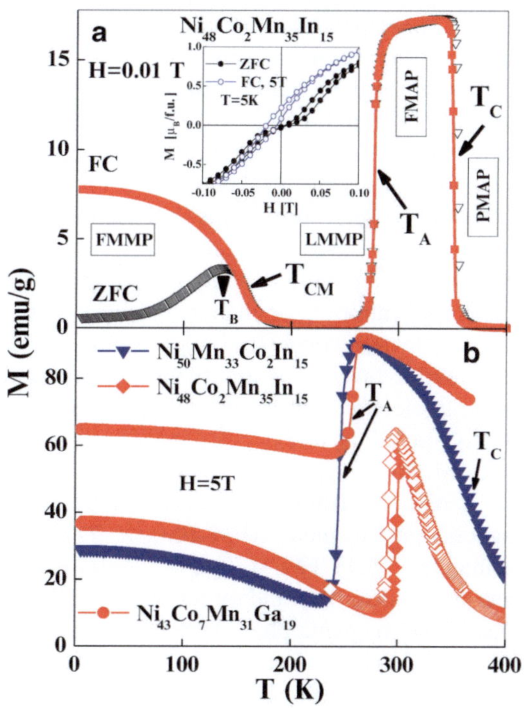

Fig. 2.3 (**a**) FC and zero field-cooled (ZFC) $M(T)$ curves obtained at $H = 0.01T$ for $Ni_{48}Co_2Mn_{35}In_{15}$. *Inset*: ZFC and FC ($H = 5T$) hysteresis loops at 5 K. (**b**) $M(T)$ curves in $H = 5T$ for the some representative In- and Ga-based Heusler alloys. $M(T)$ data collected during heating and cooling cycles are shown by *open* and *closed symbols*, respectively. The labels *FMMP*, *LMMP*, *FMAP*, and *PMAP* are ferromagnetic MP, low magnetization MP, ferromagnetic AP, and paramagnetic AP, respectively, and T_B, T_{CM}, T_C, and T_A are blocking temperature, ferromagnetic ordering temperature of MP and AP, and temperature of inverse MT/MST, respectively

loop from $H = 0$ (exchange bias phenomena) as shown in the inset of Fig. 2.3a [14–17]. Thus, the ground state (at $T = 5$ K) of the Ni–Mn–X Heusler alloys undergoing the MST depends on the thermomagnetic history and can be explained in terms of FM–AFM heterogeneity.

Such heterogeneity originates from an AFM coupling of the FM austenitic (clusters below T_M) and martensitic phases, or from AFM correlations in the ferromagnetic MP. In the latter case, the AFM-ordered Mn in the X positions provide the "pinning" of FM clusters of the MP phase (below T_B) [2, 14, 15]. It is worth noting here that the ZFC $M(H)$ curves demonstrate nearly magnetically compensated (AFM-like) behavior (see the insert of Fig. 2.3 for ZFC $M(H)$). The magnetization process of the MP in the interval $T < T_{CM}$ and at a magnetic field H greater than the saturation field of $H_{sat} \sim 0.1$ T was found to be of the FM type. The ground state of the AP was found to be FM, with a magnetic moment, (100–140) emu/g (4–5μ_B/Mn). The applied magnetic field is a driving force for temperature stabilization of the FM state of the AP compared to that of the low magnetization state of MP and results in the shift of MST temperature to lower temperature (see examples in Fig. 2.3b). The temperature shift can vary from 5 to 40 K for a 5 T the applied field. The phase compositions of some Ni–Mn–X Heusler alloys and the magnetization at 5 K (with $H = 5$ T), along with phase transition temperatures and other parameters, are collected in Table 2.1 (see also in Refs. [7, 8, 10, 18]). The average magnetic moment per Mn atom is generally smaller in the MP compared to that of AP. The magnetic moment of MP was found to vary from (1–3) μ_B/Mn at 5 K for $H = 5$ T. As can be seen from Table 2.1, the transition temperatures of Ni–Mn–X Heusler alloys can be tuned through a relatively large range of temperatures (143–400 K), depending on the composition and doping scheme. The variation in component concentration, or chemical composition, certainly provides the change in the valence electron concentration per atom, e/a, Mn–Mn interatomic distances, and Ni3d–Mn3d hybridization and therefore affects all alloy characteristics through a change in electronic band structure [2, 7, 8, 16].

The change in relative positions of T_{CM}, T_M, and T_C can result in different types of magnetic phase transitions at T_M [2, 11, 19–21, 23–25]. With decreasing temperature, these are transitions from paramagnetic–paramagnetic (PM–PM, if $T_C < T_M$), ferromagnetic–paramagnetic (FM–PM, if $T_{CM} \sim T_M \sim T_C$), and low magnetization–ferromagnetic (LM–FM, if $T_{CM} < T_M < T_C$) states that correspond to disorder–disorder, order–disorder, and order–order types of magnetic phase transitions. Examples of the magnetic phase transitions are shown in Fig. 2.3 (see for LM–FM transitions) and Fig. 2.4 (see for FM–PM and LM–PM transitions and Refs. [22, 23] for details). It is worth emphasizing here that the transitions under consideration are temperature-induced first-order transitions (see the hysteresis of $M(T)$ curves in Figs. 2.3 and 2.4) originating from structural MTs. The magnetization isotherms were found to show, in the vicinity of the MT temperature, behavior associated with a field-induced martensitic metamagnetic transition at $H = H_M$ (see Fig. 2.5, $Ni_{50}Mn_{36.5}In_{13.5}$ and $Ni_{48}Co_2Mn_{35}In_{15}$ for LM–PM and LM–FM transitions, respectively). The jump in magnetization at T_M was found to depend on the

Table 2.1 Annealing parameters and some crystal and magnetic characteristics of In- and Ga-based Heusler alloys

$Ni_2Mn_{1+x}In_{1-x}$, (*TA*, *t*): (850 °C, 24 h)							
x	Phases 300 K	M [μ_B/Mn] 5 K, 5 T	$T_C/T_M/T_{CM}$ [K], ZFC H = 0.01 T	H_{ex} [*T*] 5 K FC *H* = 5 T	ΔS_{5T} at T_C (J/kgK)	ΔS_{5T}at T_M (J/kgK)	MR_{5T} at T_M [%]
0.352	Cub.	4.4	326/–/–	–	−6.2	–	
0.36	Mixed	2.8	324/143/>143	–	−6.8	5.3	−80
0.38	Mixed	2.9	321/148/>148		−6.9	4.5	−15
0.392	Mixed	1.4	323/218/>218	−0.003	−7	13	−47
0.398	Mixed	1.1	321/284/204	−0.012	−6.6	23	−80
0.4	Orth.	1.1	323/319/187	−0.013	−6	37.5	−55
0.42	Mixed		–/344/148				
0.43	Mixed		–/360/125				
0.44	Mixed	1.0	–/380/115/	–	–	–	–
y/*z*	$Ni_{2-z}Co_zMn_{1.4-y}Co_yIn_{0.6}$ (TA, t): (850 °C, 24 h.)						
0.04/0	Mixed	1.14	332/296/187	–	−6	18	
0.08/0	Mixed	1.38	338/276/186	–			
0.12/0	Mixed	1.30	333/281/195	–	−5.5	27	
0/0.04	Mixed	1.0	354/277/157	−0.012	−5.5	23	−77
Z	$Ni_2Mn_{1.4}In_{0.5}Z_{0.1}$, Z = Al, Ge (TA, t): (850 °C, 24 h)						
Al	Ortho.	1.1	187/296/332	−0.013	−6	10	−30
Ge	Ortho.	1.3	186/276/338	−0.013	−6.3	50	−40
y	$Ni_2Mn_{1.4}In_{0.6-y}Si_y$(TA, t): (850 °C, 48 h)						
0.04	Cub.	1.24	307/281/208	−0.014	−6.6	82	−44
0.08	Cub.	1.32	293/272/209	−0.013	−6	112	−47
0.12	Cub.	1.42	276/218/250	−0.014	−5	124	−42
0.16	Cub.	1.40	268/236/>236	−0.017	−4	27	−26
0.20	Cub.	1.2	272/235/220	−0.018	−4	15	−26
v	$Ni_{1.68}Co_{0.32}Mn_{1.28-v}Fe_vGa_{0.72}$ (TA, t): (850 °C, 72 h.)						
0	tet.	1.6	>400/400/300	–	–	–	
0.04	tet.	1.82	>400/352/312	–	–	–	
0.6	tet.	2.00	>400/304/>304	–	–	–	
0.8/2	tet.	2.24	>400/227/>227	–	–	31	−20

type of the transition and vary from ~6 [23] to 100 emu/g [26]. Since field-induced magneto-responsive properties such as magnetoresistance (MR) and MCE result from magnetization changes, the remarkable MR and MCE have been reported for the "order–order" type of MST (Table 2.1).

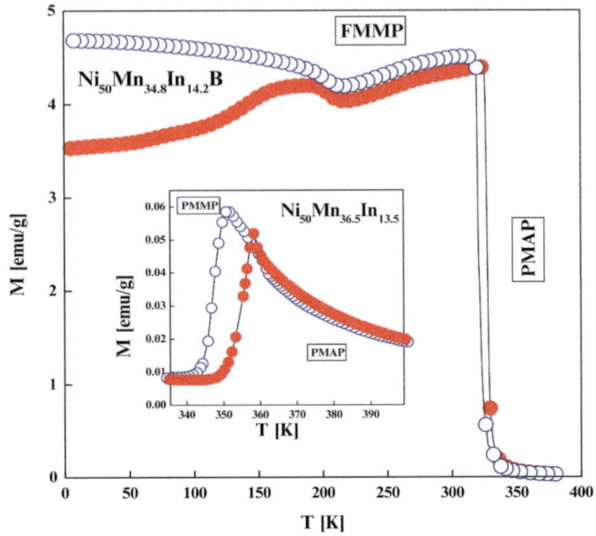

Fig. 2.4 $M(T)$ curves characteristic for FMMP–PMAP and PMMP–PMAP (*inset*) transitions obtained for $H = 0.001\,T$ for $Ni_{50}Mn_{34.8}In_{14.2}B$ and $N_{50}Mn_{36.5}In_{13.5}$. *Open* and *closed symbols* indicate the FCC and ZFC, respectively

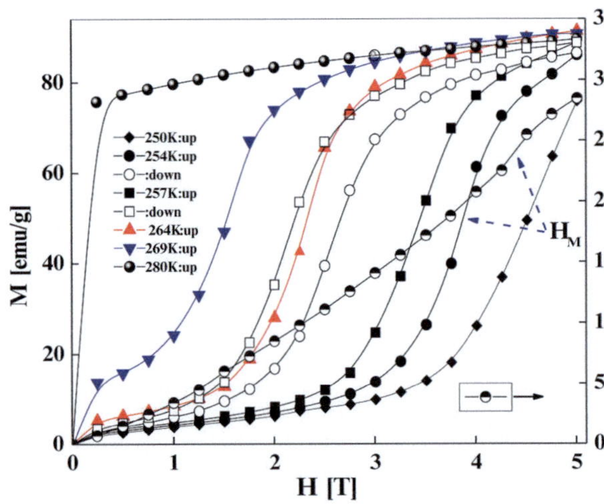

Fig. 2.5 M(H) isotherms in the vicinity of martensitic transition temperature shown for $Ni_{48}Co_2Mn_{35}In_{15}$ (*left panel*) and for $N_{50}Mn_{36.5}In_{13.5}$ (*right panel*). The *arrows* indicate the critical field of metamagnetic transitions (H_M): for LM–FM and PM–PM transitions, *left* and *right panel*, respectively

2.3 Magnetocaloric Effect

The magnetocaloric effect (MCE) is defined as a reversible, magnetic-field-induced temperature change in a magnetic material. The discovery of the effect is credited to E. Warburg for his observation of the phenomenon in pure iron in 1881. One of the original applications of the MCE was formulated by William F. Giauque in 1927 (the actual experiment was carried out in 1933) in the adiabatic demagnetization of paramagnetic salts to achieve temperatures below one Kelvin, resulting in the Nobel Prize in Chemistry in 1949. For the next few decades, the MCE was primarily employed in low-temperature physics until the mid-1970s, when Brown developed a near-room temperature magnetic refrigerator that exploited the phenomenon in metallic gadolinium. The engineering aspects of gadolinium-based refrigerators, as well as those employing other working materials, have been aggressively developed ever since.

In the past two decades, the interest in magnetocaloric materials has significantly expanded. The number of papers published yearly on the topic has increased by more than fivefold since 1990, reflecting the discovery of new materials and advances in device engineering. We refer readers to the recent reviews on this field [7, 8, 27, 28] and original papers in which it was shown that Ni–Mn–In-based alloys are very promising MCE materials [29]. Indeed, the Ni–Mn–In-based Heusler alloys are ecologically friendly materials and do not contain expensive rare-earth metals. The compounds also provide an opportunity to utilize both normal and inverse MCE in a Carnot cycle. Moreover, the MCE of quaternary alloys may be enhanced, tuned, and adjusted to required technical parameters. Recently it has been shown that it is possible to provide reversible MCE in spite of thermal hysteresis [30]. These features make these alloys quite competitive with the best giant MCE materials.

There are several indirect methods to determine the isothermal change of magnetic entropy $\Delta S_M(T,H)$ [31]. In most reports this has been done by means of the Maxwell equation [31–33]

$$\Delta S_{\mathrm{M}}(T,H) = \int_0^H \left[\frac{\partial M(T,H))}{\partial T}\right]_H \mathrm{d}H \tag{2.1}$$

and magnetization isotherm data. It is clearly seen from Eq. (2.1) that the sign and value of MCE are directly related to the change in magnetization.

Magnetic entropy change data $\Delta S_M(T)$ for some Ni–Mn–In-based Heusler alloys studied with the increase of temperature are shown in Table 2.1 and in Fig. 2.6 (see also [7–12, 18, 23]). These systems span a wide range of temperatures and most exhibit both "normal" (at T_C) and inverse (at T_M) entropy changes. The MCE is rather large in the vicinity of both transitions and comparable to those for Gd-based systems. The ΔS_M was found to be ~−7 J/kg·K, varied from 20 to 40 J/kg·K, and ~−(5–6) J/kg·K for magnetostructural (FMMP–PAAP), (LM–FM), and magnetic transitions, respectively, for $\Delta H = 5$ T. The maximum refrigeration capacity (RC)

$$RC = \int_{T_{\mathrm{cold}}}^{T_{\mathrm{hot}}} [\Delta S_M(T)]_{\Delta H} \mathrm{d}T \tag{2.2}$$

of 300 J/kg and −280 J/kg for $\Delta H = 5$ T in the vicinity of T_M and T_C, respectively, has been reported for $Ni_{48}Co_2Mn_{35}In_{15}$ [10]). The RC values are also comparable to those reported for well-known MCE materials as $Gd_5Si_2Ge_2$ (240 J/kg) and Gd (400 J/kg) in Refs. [34–36]. Since the MT for $Ni_{50}Mn_{37.5}In_{13.5}$ has been reported to

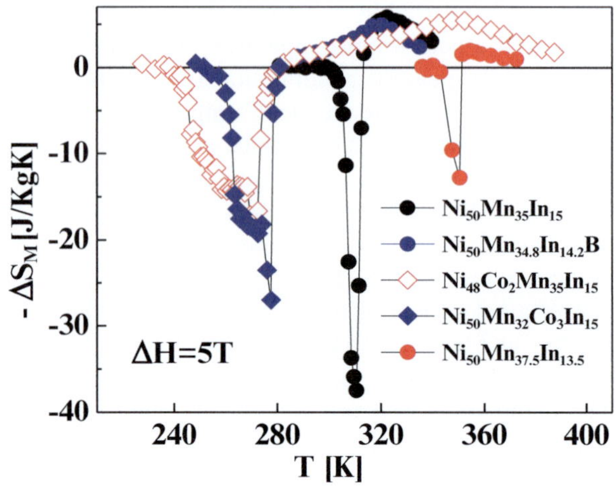

Fig. 2.6 The typical for Ni–Mn–X system temperature variations of entropy estimated for $\Delta H = 5\,T$ using thermomagnetic curves and Maxwell relations. The heat capacity measurements, C (H,T), had been used to calculate ΔS for $Ni_{50}Mn_{35}In_{14.2}B$ [22]

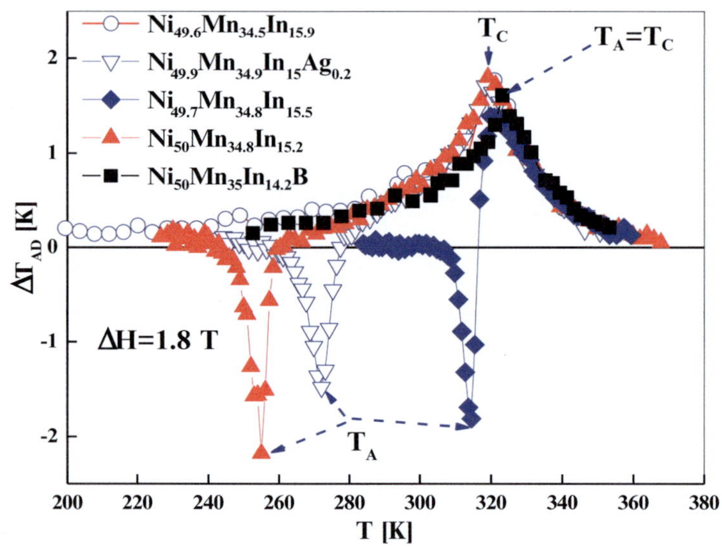

Fig. 2.7 The examples of temperature dependences of adiabatic temperature changes of Ni–Mn–In-based Heusler alloys obtained at $\Delta H = 1.8$ T and on heating cycle

be a paramagnetic–paramagnetic transition, the observed change in entropy (see Fig. 2.7) can be attributed to a "pure" field-induced structural transition from a tetragonal martensitic to a cubic austenitic phase. Thus, the structural contribution to total entropy changes of Ni–Mn–X Heusler alloys can be estimated as −13 J/kg.

The understanding of the magnetocaloric properties of a material requires knowledge of both the adiabatic temperature change (ΔT_{ad}) and the magnetic entropy change (ΔS_M). Although these quantities can be obtained from magnetization isotherms and heat capacity measurements, such indirect measurements do not always yield accurate results and should be verified. In the case of the Ni–Mn–In-based Heusler alloys, a direct method had been used to measure ΔT_{ad} in the vicinity of the Curie and MT temperatures [8, 18, 22, 25, 37, 38]. The measurements were carried out using an adiabatic magnetocalorimeter in a temperature interval of 250–350 K and for applied magnetic field changes upto $\Delta H = 1.8$ T. The largest observed changes were $\Delta T_{ad} = -2$ K and 2 K near the martensitic (first-order) and ferromagnetic (second-order) transitions for $\Delta H = 1.8$ T, respectively. Notably, $|\Delta T_{ad}| \approx 1$ K [37, 38] for relatively small changes in field ($\Delta H = 1$ T) for both transitions. The impact of these results lies in the demonstration of significant adiabatic temperature changes occurring in these materials at relatively small fields. That ΔT_{ad} was measured directly removes the ambiguity present in indirect methods. It is also worth mentioning that the comparison of the value of ΔT_{ad} measured directly and that calculated from $C_p(T, H)$ data has been reported in [22]. The results are in surprisingly good agreement. Such a close agreement is likely to be the case for most polycrystalline materials, such as the samples under investigation, which do not show a truly discontinuous first-order phase transition. In such cases, it is therefore justified to employ the Maxwell relations to estimate MCE parameters. This is perhaps especially true in our case, since the materials in question exhibit a pronounced crystal-phase temperature hysteresis.

It is worth mentioning that, in the case of bulk Ni–Mn–X Heusler alloys, there is still an opportunity for MCE enhancement since the theoretical limit is given by $\Delta S_M = R\ln(2J + 1)$, where R is the universal gas constant and J is the average total angular moment per unit cell. In this case, the theoretical limit is about 90 J/kg·K, i.e., about two times larger than that observed in the experiments.

2.4 Ni–Mn–In-Based Ribbons

By using melt-spun techniques and optimized thermal treatments, it is possible to significantly improve magnetic properties of Heusler alloy ribbons. In this section, we analyze the effect of the annealing temperature on the MT, crystal structure and microstructural features, and magnetocaloric properties in $Ni_{50.0}Mn_{35.5}In_{14.5}$ melt-spun ribbons. This composition has been selected taking in account that its In content is close to 15 at.%, giving rise to an MT near room temperature along with a sharp change in the magnetization and related multifunctional properties [18].

A polycrystalline master alloy with nominal composition $Ni_{50.0}Mn_{35.5}In_{14.5}$ was fabricated under pure Ar atmosphere from its pure constituent elements (>99.98 %) using a standard arc-melting technique. Ingots were melted several times to ensure a good homogeneity. Moreover, to compensate for Mn losses during the melting, an excess of a few wt.% Mn was added to each ingot. The master alloys were then induction melted in quartz tubes in a melt-spinning system and ejected in an argon environment onto the polished surface of a copper wheel rotating at a linear speed of 48 m/s. Ribbon flakes of about 1.5–2.0 mm width were obtained. Some ribbons were wrapped in Ta foil and sealed in quartz tubes in an argon atmosphere and annealed for 10 min at $T = 1048$, 1073, 1098, and 1123 K, followed by quenching in ice water. The crystal structures were analyzed by X-ray diffraction (XRD) at room temperature (RT) with a D8 Discover (Bruker), in the range between $20° \leq 2\theta \leq 100°$ with Cu-$K\alpha$ radiation ($\lambda = 1.5418$ Å). The ribbons' composition was determined by energy-dispersive X-ray spectroscopy (EDX). The sample morphology and microstructure was observed with scanning electron microscopy (SEM, JEOL 6100). Magnetic measurements ($M(H)$ and $M(T)$) were carried using ZFC, field-cooling (FC), and field-heating (FH) protocols, in the temperature range from 50 K up to 400 K and in applied magnetic field up to 30 kOe using a vibrating sample magnetometer (VSM-VersaLab, QD). Isothermal $M(H)$ hysteresis loops were recorded in a Physical Property Measurement System (PPMS-QD) in the temperature range of 4–350 K.

In order to study the effects of annealing on the crystalline structure in comparison with the as-quenched sample, we have collected the XRD patterns for all samples at RT (See Fig. 2.8). In the as-quenched sample, a pure AP was observed with a face-centered cubic (fcc) structure, consistent with previous results stating that the MT was showing at 257 K [39].

Moreover, the superlattice reflection peaks reveal the presence of second-neighbor order, such as (111) and (311), indicating that an ordered $L2_1$ AP is developed with a lattice parameter $a_c = 0.5987(3)$ nm. After annealing ribbons at 1048, 1073, and 1098 K, an increase in the intensity of the superlattice reflections, i.e., (111), (311), and (331), was detected with increasing annealing temperature. This indicates a higher ordering in the annealed samples relative to the as-quenched one. Also, an increase in the lattice parameter was found, lager for the ribbon annealed at 1098 K ($a_c = 0.5995(2)$ nm). It should be noted that high degree of order has been obtained in the ribbons subjected to only 10 min of annealing. This is considerably shorter than in bulk samples, where normally more than 2 h of annealing is required for obtaining a high degree of ordering [2]. However, after annealing at 1123 K for 10 min, the structure at RT has changed to the *10 M* orthorhombic structure, indicating a lower degree of ordering than the other samples, corresponding to the MP, which is textured in the (125) direction. The calculated lattice parameters was $a = 0.431(5)$ nm, $b = 0.58$ (6) nm, and $c = 2.115$ (1) nm.

The RT microstructure of all samples is shown in Fig. 2.9. Columnar grains, which grow perpendicular from the ribbon surface in contact with the wheel during the quenching process, can be appreciated in all figures. On this surface the annealing effect reinforces the grain structure. For the as-spun ribbon, a grain size at about 2.2 μm was observed in the columnar structure. However, for annealed ribbons, recrystallization and higher order are developed, and then the columnar structure shows a grain size increase when the annealing T rises, reaching values around 8.7 μm.

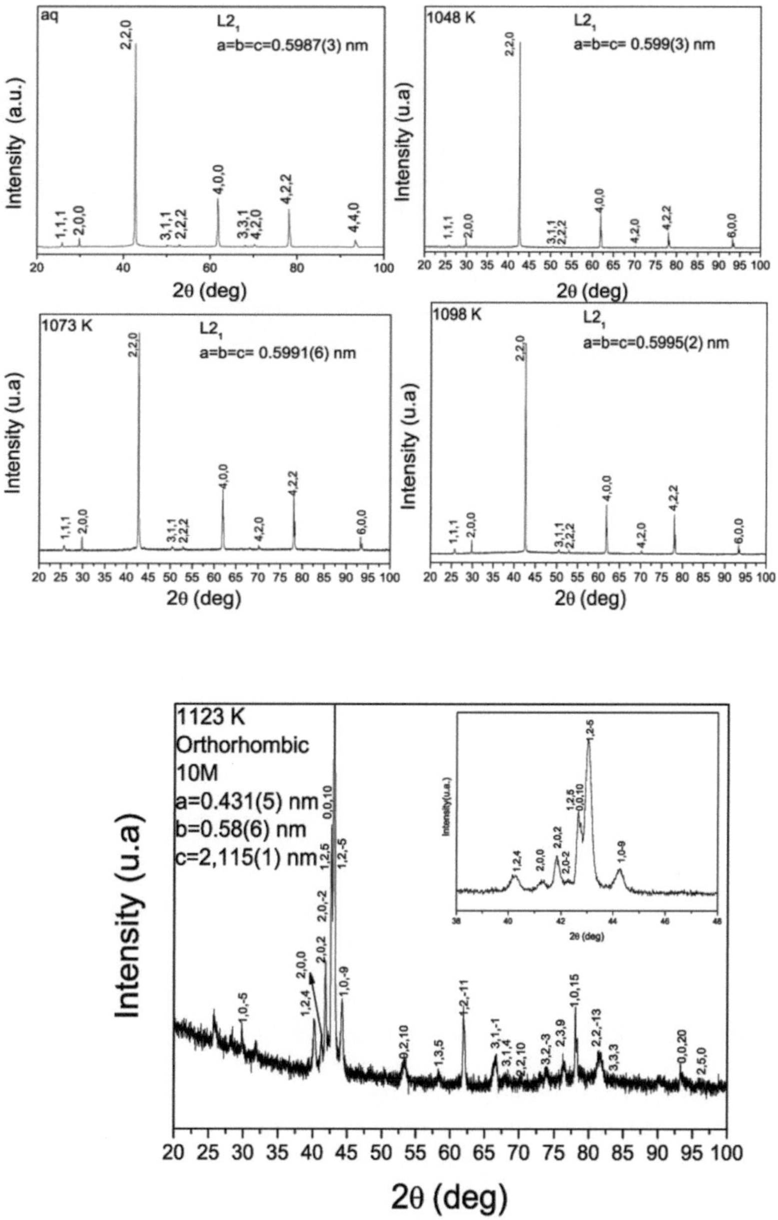

Fig. 2.8 XRD diffraction patterns of the $Ni_{50.0}Mn_{35.5}In_{14.5}$ as-quenched ribbon and annealed ribbons during 10 min for: 1048, 1073, 1098, and 1123 K. The *inset* zooms in on the diffraction pattern for the sample annealed at 1123 K

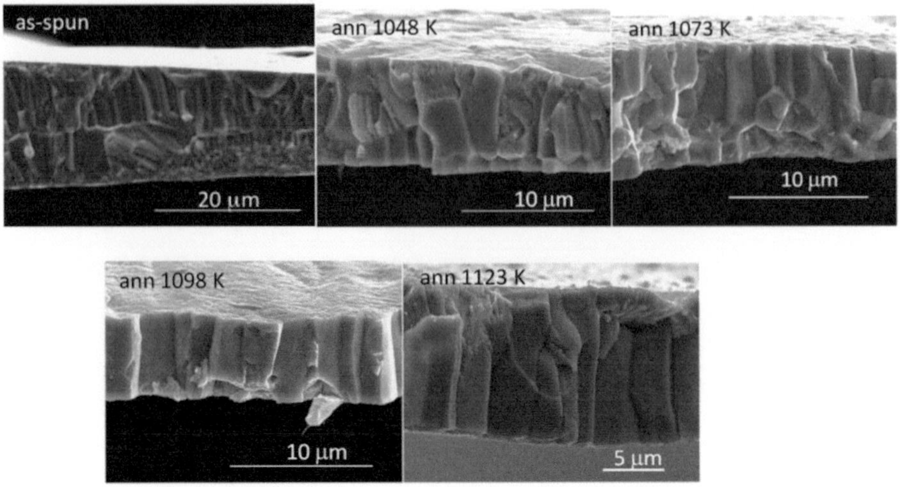

Fig. 2.9 Micrographs of $Ni_{50.0}Mn_{35.5}In_{14.5}$ as-spun ribbon and annealed ribbons during 10 min at $T = 1048$, 1073, 1098, and 1123 K

After an exhaustive study by EDX microanalysis on different regions at the ribbons' surface and transversal section, averaged compositions of $Ni_{50.0}Mn_{35.5}In_{14.5}$ for either the as-quenched and/or annealed samples were determined. The estimated error in determining the concentration of each element is of ± 0.1 %. The microanalysis confirms the chemical homogeneity of the alloy without influence of the grain shape on composition.

Figure 2.10 shows thermomagnetic curves, *M*(*T*), measured using ZFC, FC, and FH protocols for all $Ni_{50.0}Mn_{35.5}In_{14.5}$ ribbons under an applied magnetic field of 50 Oe. All curves show a change of magnetization typical of the first-order MT observed in In-based Heusler alloys and a second-order magnetic phase transition. The MT occurred near 250 K, whereas high-temperature second-order magnetic transition in the AP was observed around 280 K.

Before starting the ZFC measurements, the samples were heated up to 350 K, in AP paramagnetic phase at zero applied magnet field. In the case of the annealed samples, the MP transformation was sharper than that observed in the as-spun sample, which can be characterized by a large change in the magnetization. Around $T = 200$ K, a splitting of the ZFC and FC curves can be seen, reflecting the coexistence of AFM and FM interactions. Above this temperature, the magnetic moment decreases to values close to zero, and, in such a case, we can have either a PM or AFM state [37, 40]. The phase transition temperatures are collected in Table 2.2.

The Curie temperatures (T_C) of the MP and AP increase for annealed ribbons, compared to the as-spun sample, but significant differences in annealed samples have not been observed. However, fluctuations appear in the MT temperatures in annealed samples (M_s). These fluctuations can be ascribed to both internal stress and structural relaxation with atomic ordering increase after annealing process.

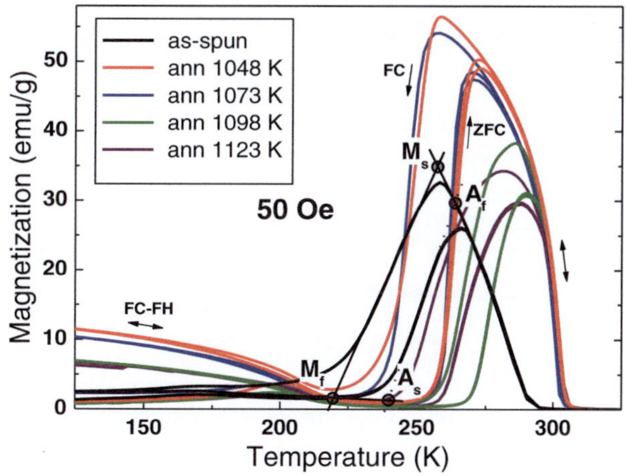

Fig. 2.10 ZFC, FC, and FH temperature dependence of magnetization of as-spun and annealed $Ni_{50}Mn_{35.5}In_{14.5}$ ribbons obtained at a field of 50 Oe. *Arrows* indicate cooling and heating scans

Table 2.2 Characteristic structural and magnetic phase transition temperatures for as-spun and annealed ribbons at $T = 1048$, 1073, 1098, and 1123 K

Ribbon	A_s (K)	A_f (K)	M_s (K)	M_f (K)	T_C^M (K)	T_C^A (K)
As-spun	239	266	257	221	185	285
Ann1048K	257	270	256	237	206	302
Ann1073K	257	265	253	238	207	299
Ann1098K	270	288	282	259	207	302
Ann1123K	258	282	276	244	210	302

The valence electron concentration per atom for $Ni_{50}Mn_{35.5}In_{14.5}$ is $e/a = 7.9$

Figure 2.11 displays the ZFC, FC, and FH temperature dependence of the magnetization for the as-spun $Ni_{50.0}Mn_{35.5}In_{14.5}$ sample and the alloy annealed at 1073 K for different applied fields. Before starting the ZFC measurements, the samples were heated to 350 K in AP paramagnetic phase at zero applied magnetic field. Afterwards, they were cooling to 50 K where the magnetic field was applied. Qualitatively similar thermomagnetic behaviors have been obtained for all annealed samples. These samples have some common features: (1) all samples undergo two main phase transitions near RT associated with the MT, from the low-T MP to the high-T AP and with the magnetic transformation from the low-T ferromagnetic AP to the high-temperature paramagnetic AP; (2) positive values of the change in the magnetization $\Delta M = M_{AP} - M_{MP}$ were obtained for all samples, where M_{AP} and M_{MP} are the magnetization values of the AP and MP, respectively. These ΔM values vary between 35 and 55 emu/g for the studied samples and take place over a temperature interval of 25–50 K, where there is AP and MP coexistence. Furthermore, this phase coexistence and metastability are also evidenced in the presence of $H = 50$ kOe at low T when the MT is kinetically arrested [41].

To characterize the MCE in the as-spun and annealed $Ni_{50}Mn_{35.5}In_{14.5}$ ribbons, the temperature (Fig. 2.5) and field (Fig. 2.6) dependences of $\Delta S_M(T, H)$ have been

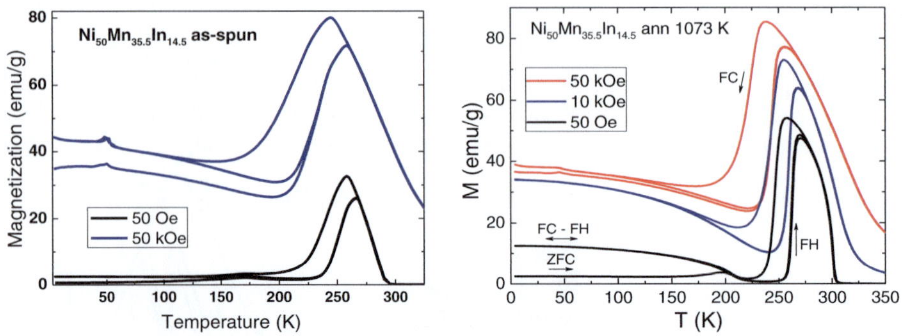

Fig. 2.11 ZFC, FC, and FH temperature dependence of magnetization of as-spun and annealed at $T = 1073$ K $Ni_{50}Mn_{35.5}In_{14.5}$ ribbons obtained at different fields

calculated using Eq. (2.1) in the whole experimental T and magnetic field (H) ranges of the ZFC mode.

Taking into account that the characteristic sharp discontinuity of the magnetization in the MT can be totally rounding due to impurities and/or inhomogeneity [42], and hence the discontinuity is eliminated being then continuous in the temperature range where AP and MP coexist, the Maxwell relationship can be used as a first approximation. However, it is important to know that strictly speaking, this relation is not valid for metastable hysteretic behavior near the MT. The fact that the MT at M_s cannot be totally field induced in an isothermal process using the ZFC mode leads us to use Eq. (2.1) as a first attempt to know the magnetic entropy change in this MST.

Figure 2.12 shows the ΔS_M (T, H) curve thermal treatments, indicating the existence of inverse and direct (normal) MCE $\Delta S_M^{peak} = 13.0$ JK^{-1} kg^{-1} and 5.0 JK^{-1}kg^{-1}, respectively, for $\Delta H = 3$ T. The ΔS_M values have been displayed only near the MT and magnetic transitions. In general, the ΔS_M^{peak} values of the as-spun sample, together with their peak temperatures at both transitions, reach higher values and evolve toward higher temperatures, after annealing. The reason the increase in the ΔS_M^{peak} values for the Ann1123 sample is not as large as the other annealed samples could be due to the experimental transformation in the Ni–In superlattice around this annealing temperature, indicating that the orthorhombic structure could remain frozen during the quenching process [39, 43]. The maximum temperature of M_s for the MT occurs in the Ann1098 sample and can be related to the fact that the order parameter takes a maximum value in this annealed sample with a cubic $L2_1$ structure. As we will address below, the magnetocaloric responses, ΔS_M (T), of the Ann1048 and Ann1073 samples are similar.

The field dependence of $|\Delta S_M^{peak}|$ associated with the MT and magnetic transition (Fig. 2.13) indicates that, unlike for the MT, the field dependence of ΔS_M^{peak} proposed as a power law [44] cannot be used in the Ann1048 and Ann1073 samples. Unexpectedly, the field dependence of ΔS_M^{peak} can be expressed by this

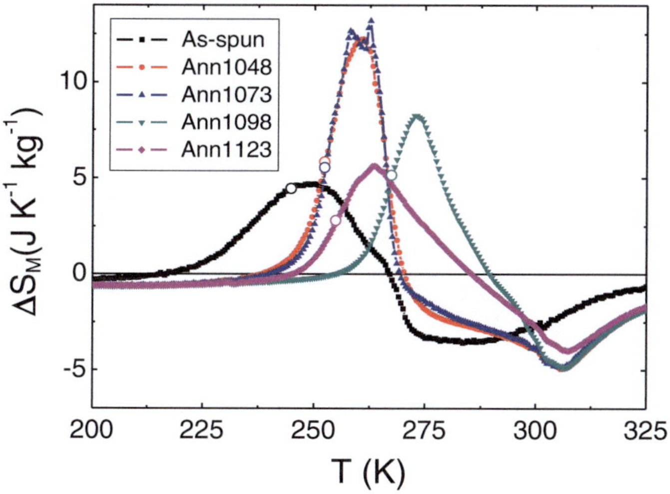

Fig. 2.12 Temperature dependence of ΔS_M (T) from Eq. (2.1), measured at 30 kOe around MT and magnetic phase transition for as-spun and annealed ribbons

Fig. 2.13 Field dependence of the absolute value of maximum entropy change obtained from Eq. (2.2) around the magneto-structural and magnetic phase transitions for the as-spun and annealed samples. *Lines* are guides for the eyes

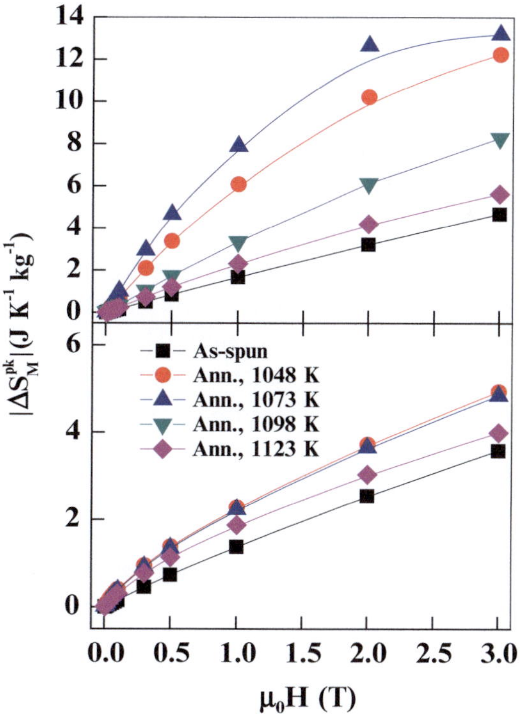

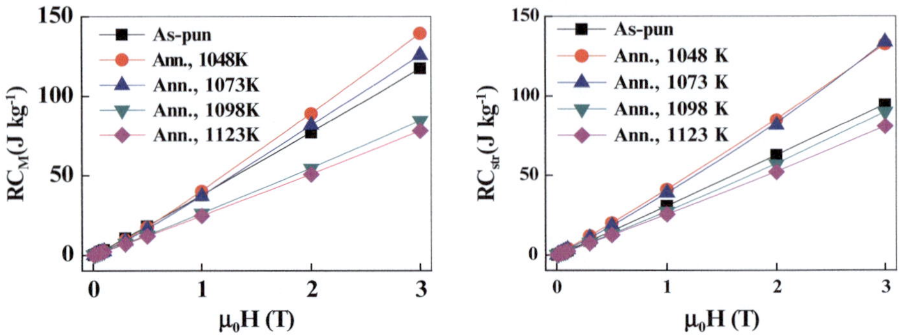

Fig. 2.14 Field dependence of the refrigerant capacity RC_{str} and RC_M for the as-spun and annealed samples associated to the MT and magnetic transitions, respectively. *Dotted lines* are guides for the eyes

power law in the as-spun, Ann1098, and Ann1123 sample, in the whole experimental magnetic field range.

An estimation of the thermal treatment dependence of refrigerant capacity (*RC*) has been obtained using Eq. (2.2) for all $Ni_{50.0}Mn_{35.5}In_{14.5}$ samples.

Following our recent work (see in Ref. [45]), where the thermal and magnetic hysteresis was reported, we have obtained the field dependence of *RC* at both structural and magnetic transition, RC_{str} and RC_M, for all as-spun and annealed samples, and a near linear behavior was obtained (Fig. 2.14). These results show that the annealing process produces an enhancement in *RC* of the as-spun sample for the Ann1048 sample.

It is important to note that the strategy to enhance the refrigerant capacity by annealing is clearly unsuccessful for high thermal treatment temperatures (Ann1098 and Ann1123 ribbons).

In conclusion, it has been shown that large enhancements of the magnetic entropy change and refrigerant capacity of as-spun $Ni_{50}Mn_{35.5}In_{14.5}$ ribbons can be obtained by short thermal treatments (10 min), tuning the critical temperatures to near room temperature. There is an optimal annealing temperature (1073 K) that maximizes the ΔS_M (T) values. These values evolve from $\Delta S_M^{peak} = 4.7$ JK^{-1}kg^{-1} up to 13.0 JK^{-1}kg^{-1} and from $\Delta S_M^{peak} = 3.5$ JK^{-1}kg^{-1} up to 5.0 JK^{-1}kg^{-1} for the structural and magnetic transitions, respectively, for $\Delta H = 3\,T$. Their corresponding values of *RC* have been optimized for the annealed sample at 1048 K, with a net improvement of 25 and 20 %, respectively. The values of ΔS_M (T) corresponding to the Ann1048 and the Ann1073 annealed samples have been calculated using both the Maxwell relation and the Clausius–Clapeyron equation, which are in reasonable agreement within their respective structural and magnetic transformation intervals.

2.5 Ni–Mn–In-Based Microwires

One method used to synthesize thin metallic wires coated by glass is the so-called Taylor–Ulitovsky technique. This technique allows for the fast fabrication of a few km long glass-coated metallic microwires with typical metallic nucleus diameters, d, ranging from 1 to 30 μm, and the thickness of the insulating glass coating between 0.5 and 20 μm. This method involves rapid quenching from the melt, and therefore microwires with amorphous, nanocrystalline, microcrystalline, or granular structures can be prepared [46].

One of the peculiarities of the Taylor–Ulitovsky technique for glass-coated microwires is that it involves the simultaneous solidification of a composite microwire consisting of a ferromagnetic nucleus surrounded by glass coating. The drastically mismatched thermal expansion coefficients of the glass and metallic alloys introduce considerable internal stresses inside the ferromagnetic nucleus during the fast solidification [47–50]. Strengths of these internal stresses depend on the ρ-ratio defined as $\rho = d/D$, where d is the metallic nucleus diameter and D is the total microwire diameter. The estimated values of the internal stresses in these glass-coated microwires arising from the difference in thermal expansion coefficients are on the order of 100–1000 MPa, depending on the ρ-ratio [48–50], increasing with the glass thickness. Moreover, in most of the volume of the metallic nucleus, tensile internal stresses dominate.

It is worth mentioning that residual stresses arising during simultaneous solidification have been estimated from simulations of the process of simultaneous solidification of a metallic nucleus inside of a glass tube [47–50]. Direct experimental determination of such residual stresses is rather complex. An example of experimental evidence of such stresses is the dependence of hysteresis loops and magnetic properties (coercivity, remanent magnetization) on the $\rho = d/D$ ratio [51, 52] as well as the applied stress dependence of the hysteresis loops [53] and effect of chemical etching of the glass on hysteresis loops [54, 55].

Direct confirmation of tensile internal stresses of observed changes of the hysteresis loops is the dependence of the switching field, H_s, on applied stresses, σ_a. We have observed monotonic growth of H_s with σ_a [53]. The same tendency has been observed when we measured hysteresis loops of the same composition of metallic nucleus but with different ρ-ratios: a considerable increase in the coercivity was observed with decreasing ρ-ratio [51].

Additionally, after chemical etching of the glass coating, we observed regular changes in the hysteresis loop from inclined to almost rectangular in Co-rich microwires (Fig. 2.15). This dependence of the hysteresis loops on the thickness of the glass coating confirms its influence on ferromagnetic nucleus, induced internal stresses by the external, nonmagnetic glass coating.

Typically, Heusler alloys are prepared by arc-melting methods, followed by high-temperature (up to 1073 K) and long-duration (up to 3 months) annealing. From the point of view of emerging technological applications, miniaturization of MSM materials for the preparation of wires, ribbons, films, and multilayers and

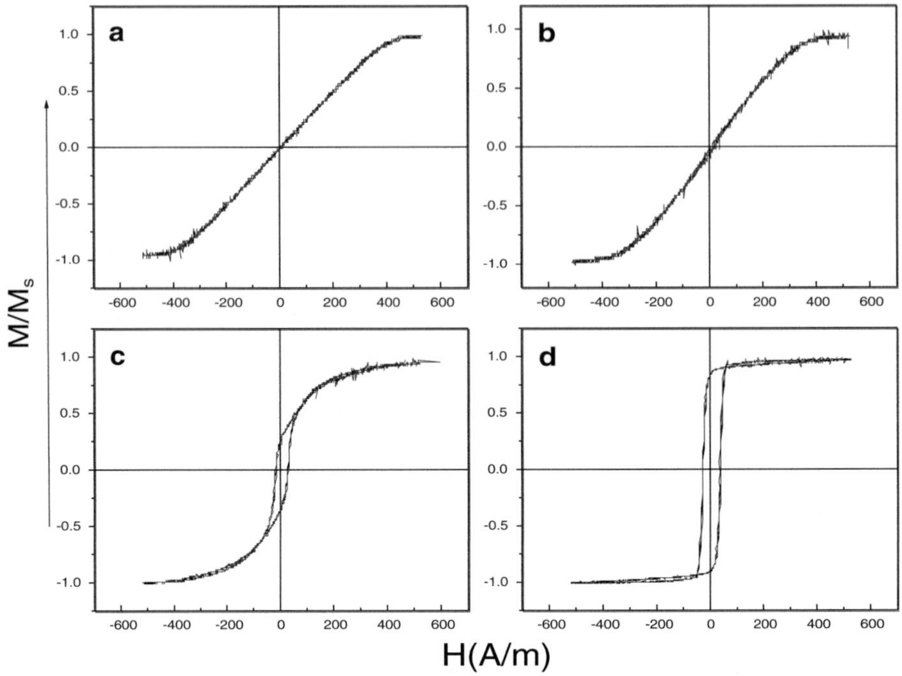

Fig. 2.15 Hysteresis loops of a $Co_{70.5}Mn_{4.5}Si_{10}B_{15}$ microwire (**a**) in the as-cast state and chemically treated for (**b**) 5 min, (**c**) 10 min, and (**d**) 50 min. Reprinted from Garcia Prieto, M. J., Pina, E., Zhukov, A.P., Larin, V., Marin, P., Vázquez, M., Hernando, A.: Glass coated co-rich amorphous microwires with improved permeability. Sensor. Actuat. A. **81**,(1–3), 227–231 (2000)

development of methods allowing the fast production of large quantities of materials are quite important [56]. In recent years, the rapid quenching method has been applied for the fast production of large amount of alloy without requiring long thermal treatment [57–59].

The Heusler-type microwires might be promising for magnetic cooling near room temperature owing to the high surface area-to-volume ratio [60, 61].

With this in mind, we have prepared Ni–Mn–In microwires using the Taylor–Ulitovsky method which allows the fabrication of glass-coated micron-dimensional composite wires and studied their magnetic properties [60].

Since the magnetization reversal in the martensitic phase is related with variant rearrangement and causes substantial change of volume, internal stresses apparently can affect magnetic properties of the Heusler-alloy-based microwires.

Temperature dependences of the magnetic moment (M) of an as-prepared $Ni_{50}Mn_{35}In_{15}$ microwire with metallic nucleus diameter $d \approx 23$ μm and total diameter $D \approx 55$ μm measured at different magnetic fields are presented in Fig. 2.16a. As-prepared glass-coated microwires are paramagnetic at RT. The Curie temperature, T_c, evaluated from the temperature dependence of the magnetic moment, $M(T)$, was estimated as $T_C \approx 240$ K.

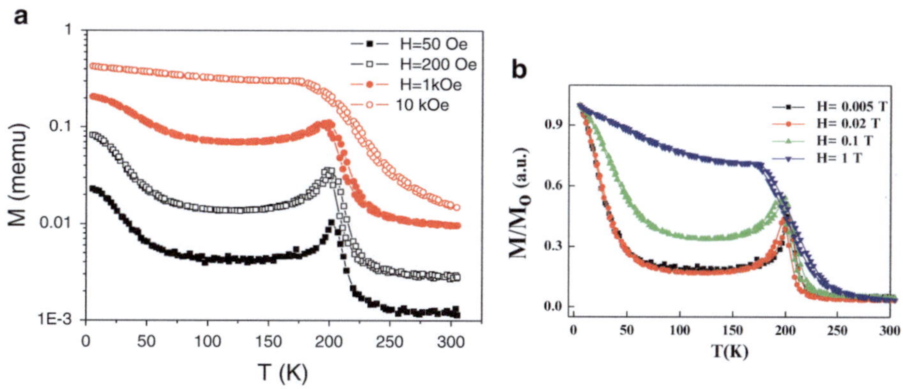

Fig. 2.16 (**a**) Temperature dependence of the magnetic moment (M) and (**b**) the normalized magnetic moment (M/M_0) for an as-prepared $Ni_{50}Mn_{35}In_{15}$ glass-coated microwire measured at different applied fields

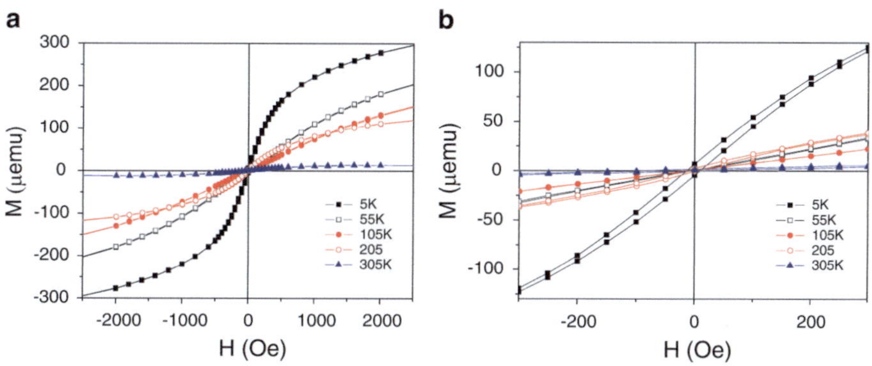

Fig. 2.17 The hysteresis loops at different temperatures measured in an as-prepared $Ni_{50}Mn_{35}In_{15}$ glass-coated microwire (**a**) at elevated and (**b**) at low magnetic fields

The value of the magnetic moment depends considerably on the magnetic field (H) applied during the measurements. Non-monotonic $M(T)$ dependence observed for studied $Ni_{50}Mn_{35}In_{15}$ microwires even at high magnetic field, considerable $M(H)$ dependence, and some hysteresis of $M(T)$ in Figs. 2.16 might be attributed to the coexistence of two crystalline phases and magnetic transition. On the other hand, the magnetization increase in the vicinity of Curie temperature can also be related to the Hopkinson effect (see hysteresis loops in Fig. 2.16b). The hysteresis loops measured at different temperatures exhibit soft magnetic character (Fig. 2.17a, b), but we did not observe magnetic softening at $T = 205$ K (the temperature corresponding to the maximum of magnetic moment on $M(T)$).

Similarly, the $M(T)$ and $M/M_0(T)$ of as-prepared $Ni_{42.5}Mn_{37.5}In_{12.5}Co_{7.7}$ with metallic nucleus diameter $d \approx 13$ μm and total diameter $D \approx 47$ μm (sample 2)

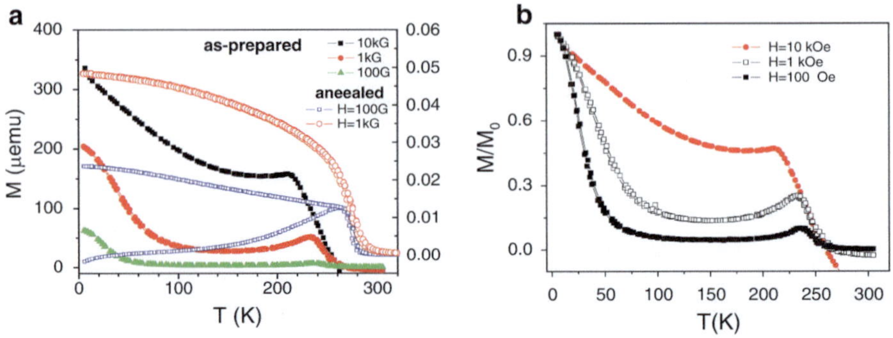

Fig. 2.18 (**a**) $M(T)$ dependence of as-prepared and annealed $Ni_{45}Mn_{36.5}In_{13.5}Co_5$ (823 K for 30 min) glass-coated microwire measured at different fields and (**b**) $M/M_0(T)$ in an as-prepared $Ni_{45}Mn_{36.5}In_{13.5}Co_5$ glass-coated microwire

measured at different magnetic fields are presented in Fig. 2.4. In this sample the estimated values of the Curie temperature was $T_C \approx 250$ K. As in the case of the $Ni_{50}Mn_{35}In_{15}$ glass-coated microwire, we note the considerable dependence of the magnetic moment (M and M/M_0) on magnetic field and the increase of M and M/M_0 in vicinity of the Curie temperature. In contrast with the $Ni_{50}Mn_{35}In_{15}$ sample, the maximum $M(T)$ and $M/M_0(T)$ dependences remain even at high magnetic field ($H = 5$ kOe).

The as-prepared $Ni_{45}Mn_{36.5}In_{13.5}Co_5$ microwires show similar features (see Fig. 2.18), i.e., a strong dependence of magnetic moment on magnetic field and a Curie temperature of about 260 K (slightly higher than in $Ni_{50}Mn_{35}In_{15}$). $Ni_{45}Mn_{36.5}In_{13.5}Co_5$ microwires annealed at 823 K for 30 min show higher Curie temperature (about 280 K). As can be seen in Fig. 2.18, $M(T)$ dependences of as-prepared and annealed samples 1 show considerable differences. Similarly to the as-prepared $Ni_{50}Mn_{35}In_{15}$ and $Ni_{45}Mn_{36.5}In_{13.5}Co_5$ $M(T)$ of the annealed $Ni_{45}Mn_{36.5}In_{13.5}Co_5$ microwire show considerable increase in the M in vicinity of T_C. We have observed considerable differences in the $M(T)$ dependences of the annealed $Ni_{45}Mn_{36.5}In_{13.5}Co_5$ sample when heating and cooling the sample under the field of 100 Oe. This difference disappears at higher magnetic field (1 kOe).

We also observed a considerable increase of the magnetic susceptibility (both real (χ') and imaginary (χ'') parts) in both as-prepared and annealed $Ni_{45}Mn_{36.5}In_{13.5}Co_5$ glass-coated microwires ($d = 20$ μm, $D = 50$ μm) near the Curie temperature (Fig. 2.19). The observed maximum near the Curie temperature must be related to the Hopkinson effect. The temperature corresponding to the magnetic susceptibility maximum is higher for the annealed sample. This difference must be attributed to the higher Curie temperature of annealed $Ni_{45}Mn_{36.5}In_{13.5}Co_5$ sample.

The observed qualitative $M(T)$ dependences are similar to those observed in similar Heusler-type alloys fabricated by conventional arc melting [29, 38]. For Ni–Co–Mn–In alloys produced by conventional arc melting, the $M(T)$ dependences are

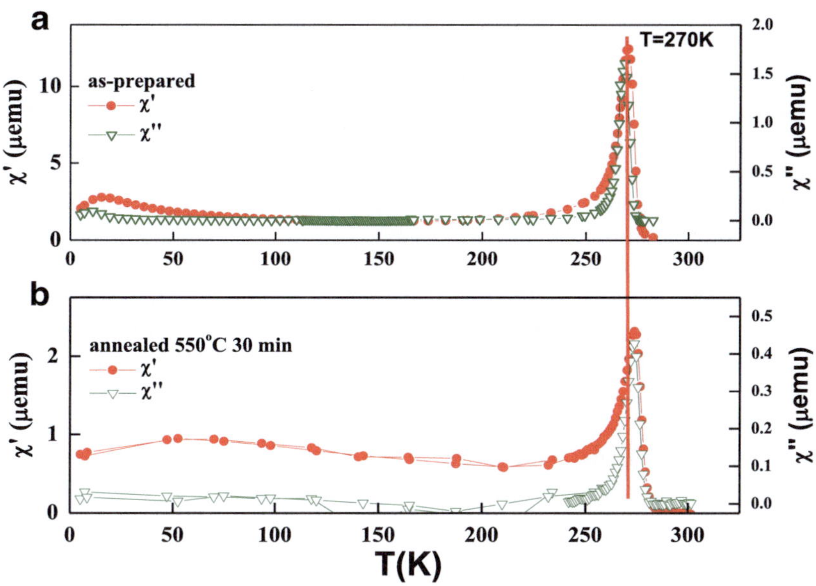

Fig. 2.19 Temperature dependence of the magnetic susceptibility measured in an (**a**) as-prepared and (**b**) $Ni_{45}Mn_{36.5}In_{13.5}Co_5$ annealed (823 K for 30 min) glass-coated microwire (sample 1)

explained considering three phase transition temperatures: T_{CM} (Curie temperature of MP), T_M (temperature of MT), and T_C, the Curie temperature of the austenitic (high-temperature) phase (AP). The AP is generally in a ferromagnetic state below its Curie temperature T_C and $T_C > T_M$ [38]. In the case of the studied Ni–Co–Mn–In microwires, the T_C values for all samples are lower ($T_C \approx 270$ K in as-prepared microwires estimated from Fig. 2.5). As observed in Fig. 2.19, after annealing, there was considerable increase in T_C (by about 20 K). The magnetization increase observed in Ni–Co–Mn–In alloys produced by conventional arc melting at about $T_M \approx 270$ K was associated with a MT from the magnetic state characterized by a low magnetic moment (antiferromagnetic or paramagnetic state) to a ferromagnetic AP state. The change in the ZFC magnetization in the low-temperature region ($T < T_{CM}$) previously observed in Ni–Mn–In-based compounds has been attributed to the magnetic heterogeneity in this temperature range [38]. In the present case, the observed T_M values (between 200 and 220 K) are considerably lower. Additionally all studied samples (especially the annealed $Ni_{45}Mn_{36.5}In_{13.5}Co_5$ sample) exhibit a dependence of the magnetic moment on magnetic field, even near Curie temperature (Figs. 2.16 and 2.18).

The origins of the field-dependent magnetization near or above the Curie temperature have previously been discussed in terms of induced magnetization, sample inhomogeneity, and local spin fluctuations [62, 63]. We assume that one of the reasons of such significant magnetic field dependence of magnetization might be related to the aforementioned magnetic heterogeneity.

We can assume that the main peculiarity of the Taylor–Ulitovsky technique employed for the Ni–Mn–In and Ni–Mn–In–Co microwire preparation is the simultaneous rapid solidification of metallic nucleus surrounded by the glass coating [47–50]. This process induces strong internal stresses distributed in a complex way within the microwires. Moreover it is essentially important that the internal stress distribution along the microwire's radius is not homogeneous: near the axis of the metallic nucleus, the tensile stresses are strongest. Closer to the interlayer with the glass coating, the compressive stresses are dominant. Recently we showed that the phase composition and magnetic properties of the Ni–Mn–Ga microwires annealed with and without glass coating are rather different [64].

Consequently internal stress relaxation induced by annealing strongly affects both the magnetic properties and structure of the microwires.

For further clarification of the observed differences in the magnetic properties of Ni–Mn–In and Ni–Mn–In–Co microwires as compared with Ni–Co–Mn–In alloys produced by conventional arc-melting method, the studies of the crystal structure depending on temperature are needed. These studies are in progress.

2.6 Magnetotransport Properties

The Ni–Mn–In-based alloys belong to the family of high-resistivity metals with specific resistivity $\rho = 80 - 350\ \mu\Omega \cdot \text{cm}$. In fact, this is rare because, as a rule, the resistivity of crystalline or amorphous ferromagnetic metals does not exceed $150\ \mu\Omega \cdot \text{cm}$. Therefore it makes possible to use these alloys as a probe to study magnetotransport in high-resistivity ferromagnetic metals. On the other hand, magnetotransport properties can provide important information about phase transitions, electronic structure, and scattering mechanisms.

2.6.1 Resistivity

Examples of the temperature dependence of the resistivity ρ(T) for some quaternary alloys are shown in Figs. 2.20 and 2.21. Resistivity is very high in the MP and much smaller in the AP and exhibits hysteretic behavior in the same range as the magnetization. The width of thermal hysteresis seen in the ρ(T) dependencies is consistent with those revealed by magnetic measurements. Temperatures at which the resistivity shows a well-defined change in the slope upon heating and cooling are in qualitative agreement with the A_f and M_s temperatures determined from low-field magnetization data. The temperature dependence of the resistivity in the ferromagnetic AP is quite common for ferromagnetic metals, specifically that it increases with temperature and its slope slightly diminishes at the Curie

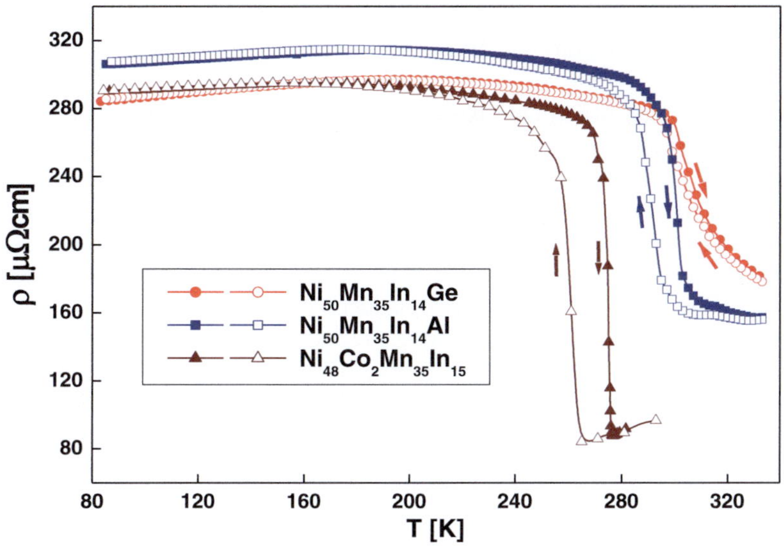

Fig. 2.20 Examples of $\rho(T)$ during heating and cooling

Fig. 2.21 Temperature dependence of resistivity and magnetoresistance for $Ni_{50}Mn_{35}In_{12}Si_3$

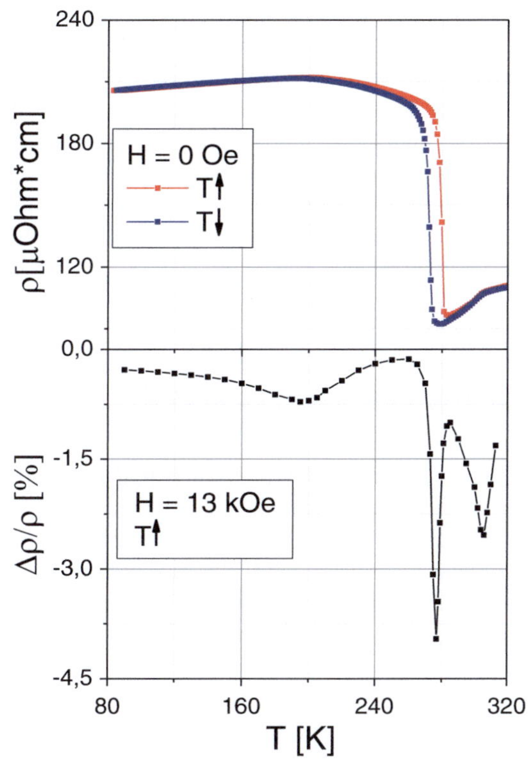

temperature T_C^{au}. By contrast, in MP resistivity slightly increases with temperature up to the Curie temperature of MP T_C^{m} and then decreases in the temperature range $T_C^{m} < T < A_s$.

According to the Mooij correlation (for a review, see Ref. [65]), $\partial\rho/\partial T < 0$ for high-resistivity metals for which $\rho > \rho^*$, where ρ^* is usually equal to $\approx 150\ \mu\Omega \cdot \text{cm}$ but can vary from 100 to 300 $\mu\Omega \cdot \text{cm}$. This correlation has recently been successfully explained by the weak localization in high-resistivity metals [66]. Since the weak localization is suppressed by external or internal magnetic fields, the Mooij correlation is valid only for para- or diamagnetic metals. Our results are in agreement with this explanation because $\partial\rho/\partial T > 0$ in ferromagnetic state despite the high resistivity.

The most intriguing question is why the resistivity is so high in the MP, but then decreases in the AP. The most popular explanation is that the density of electronic states at the Fermi energy $N(E_F)$ in the AP is high, making the electronic contribution to the total energy too large compared to that of the MP when temperature approaches MT, and therefore $N(E_F)$ decreases significantly at MT. Since resistivity $\rho \sim [N(E_f)]^{-2}$, it might explain the high resistivity in the MP. This explanation coincides with widely discussed point of view on the MT that the transitions are induced by strong changes in electronic structure [67, 68] but does not agree with experimental Hall effect data or magneto-optical spectra (see below). The high resistivity can be connected with crystalline grain structure of polycrystalline samples due to tunneling between grains. During heating, huge deformations arise in the MP in the vicinity of the MT which may diminish the width and height of the tunnel barriers and therefore strongly affect the resistivity. This point of view is partly supported by the much smaller resistivity in single crystals. It might also be possible that scattering potential in the cubic AP is much smaller than in the MP with lower symmetry. Finally, the magnetic contribution to the resistivity in Ni–Mn–In alloys can be extremely large in MP due to AFM correlations. So one can see that there is no definite answer to this question and it needs further investigations.

2.6.2 Magnetoresistance

Figure 2.21 shows that the MR is negative and large at the first- and second-order phase transitions, specifically in the vicinity of the MST, and near the Curie temperatures of the MP and AP. The behavior of the MR near the Curie temperature is typical for ferromagnets because the magnetic field suppresses spin disorder and more clearly indicates the Curie temperature positions than the change of slope of $\rho(T)$. The mechanism of large negative MR in the vicinity of MT is analogous to that for colossal MR in manganites. In both cases the external magnetic field changes the relative volume fractions of high-resistivity (MP) and low-resistivity (AP) phases. Apparently, the larger the difference in resistivity of these phases, the lager the MR. Therefore, in some cases the MR in Heusler alloys reaches values up to 60–80 % but in high magnetic fields (see Table 2.1). For example, the observation of MR up to −80 % in value at $H = 5T$ has been reported in [10, 12] for $Ni_{48}Co_2Mn_{35}In_{15}$.

Remarkable features of the MST are crystal/magnetic phase coexistence and metastability that can result in field-induced irreversibility, i.e., "kinetic arrest" of the AP [41], and also in asymmetric behavior of MR [69]. The asymmetry between the forward and reverse metamagnetic transitions resulted in a large switching-like, low-field MR (16 % for a field change of $B = 0 \rightarrow 0.25\,T$ at $T = 304$ K) in the bulk Heusler alloys $Ni_{50}Mn_{35}In_{15-x}B_x$ ($1 < x < 2$) [69].

The MCE and MR are even function of the magnetization and reach maximum values in the phase transition region; one could thus expect the existence of a correlation between MCE and MR, at least within a limited interval of fields and temperatures. On the one hand, such a correlation would provide a possibility to study MCE in nano- and micro-objects, in which determination of the MCE directly by measuring the adiabatic variation of temperature under magnetization or indirectly, using data on magnetization, is cumbersome enough or impossible altogether because of the small volume of the samples. On the other hand, such correlation also permits to study the relation between the MR and the degree of spin disorder in the immediate vicinity of phase transitions.

Two possible forms of correlation between the MR and MCE have recently been proposed [70, 71]:

$$F[\rho(T,H) - \rho(T,H=0)] = [S_M(T,H) - S_M(T,H=0)], \tag{2.3}$$

$$\Delta S_M(T,\Delta H) = -\alpha \int_0^H \left[\frac{\partial \ln \rho(T,H))}{\partial T}\right]_H dH, \tag{2.4}$$

where $\rho(T, H)$ is the resistivity of a sample in a field H, $\Delta S_M(T, H)$ is the magnetic part of the entropy, and F and α are empirical coefficients assumed to be independent of temperature T and field H. Both types of correlations have been studied using measurements of the resistivity, MR, and magnetization of the $Ni_{50}Mn_{35}In_{12}Si_3$ and $Ni_{50}Mn_{35}In_{11}Si_4$ Heusler alloys. It has been shown that, although large values of MCE and MR are observed at the same temperatures, the correlation of type (2.3) is not universal. The empirical parameter F depends strongly on temperature in the regions of the first-order and second-order phase transitions. Relation (2.4) is satisfied at the qualitative level for both the first-order and second-order phase transitions but with different values of parameter α. It is even possible to estimate the value of this parameter from magnetization data [72]. Thus, measurements of MR offer a possibility of revealing the temperature and field intervals in which a significant magnetocaloric effect persists, without employing any direct or indirect methods for evaluation of this effect.

2.6.3 *Hall Effect*

It is generally accepted that the anomalous Hall effect (AHE) in ferromagnets is one of the most important magnetotransport phenomena, and interest in this phenomenon is constantly growing [73–75]. This is related to both extensive investigations of the spin Hall effect, which is important for spintronics and has nature and mechanisms in common with AHE, and the attempts to revise the AHE theory [73] using new terms and theoretical schemes. Nevertheless, despite a more than 130-year history of studying AHE, most questions regarding the dominating AHE mechanisms are still debatable, especially in the case of high-resistivity and heterogeneous alloys ([76] and references therein). In this section we will discuss the obtained experimental data on Ni–Mn–In-based alloys [38, 77–79] and show that the modern theories do not describe the revealed dependences.

Hall resistivity in ferromagnets ρ_{H} is usually represented as a sum of two terms,

$$\rho_{\mathrm{H}} = R_0 B_z + 4\pi R_{\mathrm{s}} M_z \tag{2.5}$$

where the first term describes ordinary or normal Hall effect (NHE) induced by the Lorentz force, and the second term characterizes the AHE related to spin–orbit interaction (SOI). In Eq. (2.5), M_z is the magnetization component along the z axis, and B_z is the magnetic induction component

$$B_z = H_z + 4\pi M_z(1 - N), \tag{2.6}$$

where $0 \leq N \leq 1$ is the demagnetizing factor of the sample and R_0 and R_{s} are called NHE and AHE coefficients, respectively. According to Eq. (2.1), we have

$$\begin{aligned} R_s &= \frac{\sigma_{xy}(M_z)}{4\pi M_z\left(\sigma_{xx}^2 + \sigma_{xy}^2\right)} \approx \frac{\sigma_{xy}(M_z)}{4\pi M_z}\rho^2, \\ R_o &= \frac{\sigma_{xy}(B_z)}{B_z\left(\sigma_{xx}^2 + \sigma_{xy}^2\right)} \approx \frac{\sigma_{xy}(B_z)}{B_z}\rho^2, \end{aligned} \tag{2.7}$$

where σ_{xy} is the off-diagonal conductivity (which is usually much smaller than diagonal conductivity σ_{xx}) and $\rho = 1/\sigma_{xx}$ is the resistivity. Note that the NHE is proportional to the magnetic induction rather than the magnetic field; therefore, the NHE can depend on the magnetization if $N < 1$. Hirsch [80] showed that the second term in Eq. (2.5), which is linear in magnetization, can have an electrodynamic nature in itinerant ferromagnets rather than being related to SOI. For the Hirsch mechanism, the AHE coefficient is lower than or comparable with the NHE coefficient, which is in conflict with experimental data; therefore, this mechanism is usually not taken into account. We will consider the AHE as an effect that is induced only by SOI. At present, the following three competitive AHE mechanisms are considered: the Karplus–Luttinger (KL) mechanism, the skew scattering mechanism, and the side-jump mechanism [73–75]. The KL mechanism was proposed in the first work dealing with the AHE theory in 1954 [81], i.e., 74 years after the discovery of this effect. Karplus and Luttinger showed that a SOI-linear correction to the velocity appears in an ideal periodic lattice in the presence of periodic intrinsic SOI (i.e., interaction of an electron spin with its orbital motion). As a result of this correction, $\sigma_{xy}(M)$ is independent of the impurity concentration and the type and magnitude of scattering potential V. Therefore, according to Eq. (2.7), for any ferromagnets and at any temperatures, we have

$$\left(\sigma_{xy}\right)^{\mathrm{KL}} \infty\, \sigma_{xx}{}^{0} \quad \text{and} \quad (R_{\mathrm{s}})^{\mathrm{KL}} = A\rho^2 \tag{2.8}$$

This mechanism was then called intrinsic AHE in order to emphasize that scattering is insignificant in this mechanism (as in the Hirsch mechanism). The Berry phase mechanism, which was first positioned as a new AHE mechanism, is identical to the KL mechanism: only the final equations are formulated in terms of the Berry phase and the Berry curvature [73].

Since the authors of [81] did not take into account scattering, Smit [82] soundly criticized them: he thought that the KL mechanism contribution should be fully compensated by other terms in the solutions to a kinetic equation when scattering is taken into account. Smit proposed the following skew scattering mechanism: the electron scattering probability to the left or right (along axis y) from the direction of electron motion (along axis x) becomes dependent on the electron spin direction (along axis z) in the presence of SOI, which can be intrinsic, non-intrinsic (interaction of an electron spin with the orbital motion of another electron), periodic, or nonperiodic [83]. In the lower orders in impurity concentration, this mechanism leads to the dependence

$$(R_{\mathrm{s}})^{\mathrm{sc}} = a\rho_0 + b\rho_0{}^2 \tag{2.9}$$

where ρ_0 is the residual resistivity ($\rho_0 \sim c$) and the second term is lower than the first term and has the opposite sign, for impurity scattering, and $T = 0$ if the impurity concentration is low $c \ll 1$ and the impurity scattering is weak, i.e., $V/E_{\mathrm{F}} \ll 1$. For the case of strong scattering ($V/E_{\mathrm{F}} \leq 1$) and the impurity concentration is not very low, both terms in Eq. (2.9) can be of the same order of magnitude and have the same sign [82]. Equation (2.9) does not work in concentrated alloys [82], and $(R_{\mathrm{s}})^{\mathrm{sc}}$ changes its sign at moderate concentrations under weak scattering [83].

When analyzing the scattering of spin-polarized carriers by impurities, Luttinger [84], who performed the most complete and consistent calculation of residual AHE upon impurity scattering in the case of weak scattering and periodic SOI, and Smit [82] found another mechanism. Berger [85] calculated the scattering of a wave packet by impurities and interpreted this mechanism as a side jump, i.e., a jump-like shift in an electron trajectory upon impurity scattering. As in the KL mechanism, this mechanism is formally independent of both the scattering potential and the type of scattering (impurity, phonon, magnon, etc.) and should result in the relation

$$(R_s)^{sj} = B\rho^2 \tag{2.10}$$

Therefore, the KL and side-jump mechanisms were not distinguished in early works. Luttinger showed that, according to Smit [82], coefficients A and B in Eqs. (2.8) and (2.10) have the same order of magnitude and opposite signs. However, complete cancellation does not take place even in the simple model considered in [84].

For low-resistivity metals and alloys, the dominant mechanism is skew scattering. With increasing resistivities, all three mechanisms become important, and it is widely believed and demonstrated in many papers [73–75] that $R_s \sim \rho^2$ if the resistivity is larger than 1–10 μΩ · cm but less than 100–200 μΩ · cm, usually related to intrinsic and side-jump mechanisms. In the case of very high-resistivity metals, so-called "bad" metals, with resistivity as high as $\rho > 100$–200 μΩ · cm, the relationship

$$(R_s) = D\rho^{0.4-0.2} \quad \text{or} \quad (\sigma_{xy}) \infty {\sigma_{xx}}^{1.6-1.8} \tag{2.11}$$

was experimentally observed but not explained [73]. Nevertheless, it should be emphasized that for all modern theories, the AHE coefficient increases with resistivity.

There are several attempts to study the Hall effect in the alloys undergoing MTs [38, 77–79, 86]. The authors of [86] measured Hall resistivity ρ_H in $Ni_{50}Mn_{17}Fe_8Ga_{25}$ alloy ribbons and obtained $\rho_H \infty \rho^{2.1}$ in the AP, which was related to the side-jump mechanism. In the MP, they found $\rho_H \infty \rho^{4.2}$ and assumed (without any support from the AHE theory) that this dependence is associated with the side-jump mechanism during scattering by clusters rather than single impurities. It should be noted that they did not take into account the NHE, did not distinguish the AHE and NHE coefficients, and did not study the Hall effect in the immediate vicinity of the MT. The authors of [77] observed a giant Hall effect in a ternary $Ni_{50}Mn_{50-x}In_x$ alloy with $x = 15.2$ in the immediate vicinity of the MT. However, they did not determine the NHE and AHE coefficients, since the magnetic field was insufficient for saturation and the samples failed mechanically by cracking during temperature and field cycling. Finally, the authors of [38, 77] investigated the Hall effect in a high-resistance $Ni_{48}Co_2Mn_{35}In_{15}$ alloy, qualitatively estimated the NHE and AHE coefficients from the low- and high-field slopes of the Hall curves, respectively, far from the MT, and found that the AHE coefficient cannot be described by a relationship similar to Eq. (2.11).

Figure 2.22 shows typical Hall resistivity curves for temperatures corresponding to the ferromagnetic MP ($T = 110$ K), ferromagnetic AP ($T = 293$ K), and in close vicinity to the MT ($T = 263$ K) for the Ni–Co–Mn–In sample. The curves at 110 and 293 K are quite standard for the Hall effect in ferromagnets because the AHE saturates at high magnetic field. However, it is not the case in the vicinity of the MT. The change of slope of the Hall curve from negative to positive at 10 kOe clearly indicates the occurrence of a the field-induced phase transition at 263 K accompanied by the appearance of the AP with larger magnetic moment than that of

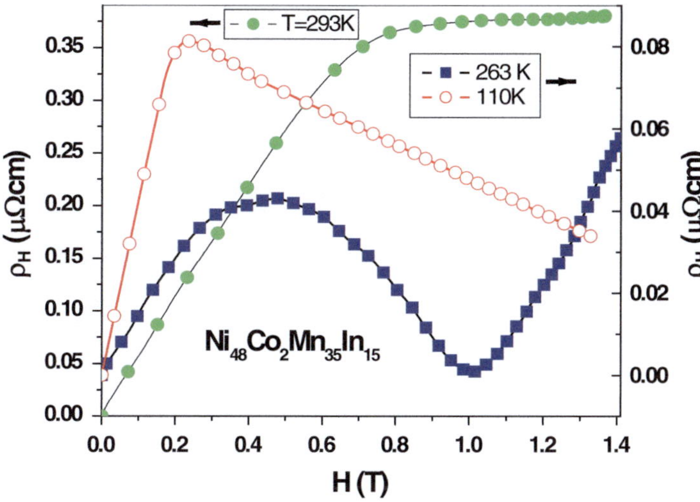

Fig. 2.22 Hall resistivity of $Ni_{1.92}Co_{0.8}Mn_{1.4}In_{0.6}$ as a function of external magnetic field (H) at given temperatures (T)

the MP. Therefore the field dependence of Hall resistivity can be used to determine the MST parameters.

As it is follows from Eqs. (2.5)–(2.7) to determine the AHE and NHE coefficients, one needs to measure resistivity, Hall resistivity, MR, and magnetization for the same sample and in the same geometry and fields. It is possible that these coefficients depend on magnetization as

$$\begin{aligned} R_s &= R_s(0)\left[1+\alpha\frac{M_z{}^2}{M_s{}^2}\right]\left[1+\beta\frac{M_z{}^2}{M_s{}^2}\right]^2, \\ R_0 &= R_0(0)\left[1+\gamma\frac{M_z{}^2}{M_s{}^2}\right]\left[1+\beta\frac{M_z{}^2}{M_s{}^2}\right]^2, \end{aligned} \tag{2.12}$$

where parameters α and γ are connected with a possible nonlinearity in the Hall coefficients, and the MR is

$$\frac{\Delta\rho}{\rho}(H_z) = \frac{\rho(H_z)-\rho(0)}{\rho(0)} = \beta\frac{M_z{}^2}{M_s{}^2}$$

The obtained results are shown for two compositions in Figs. 2.23 and 2.24.

The ordinary Hall coefficient $R_s(M_z = 0)$ is negative, slightly decreases in value with increasing temperature, and does not change significantly at MT. There is no abrupt change in $R_s(M_z = 0)$ at MT as for the resistivity. It is interesting that the temperature dependence of $R_s(M_z = 0)$ is much stronger in the low-temperature ferromagnetic MP than in the vicinity of the MT. Such temperature dependence can be explained in the frame of a two-band model. Perhaps, it is due to a redistribution

Fig. 2.23 Temperature dependences of (**a**) the magnetization (M), (**b**) the electrical resistivity, and (**c**) the coefficients of the normal and anomalous Hall effects in an $Ni_{50}Mn_{35}In_{14}Si$ alloy

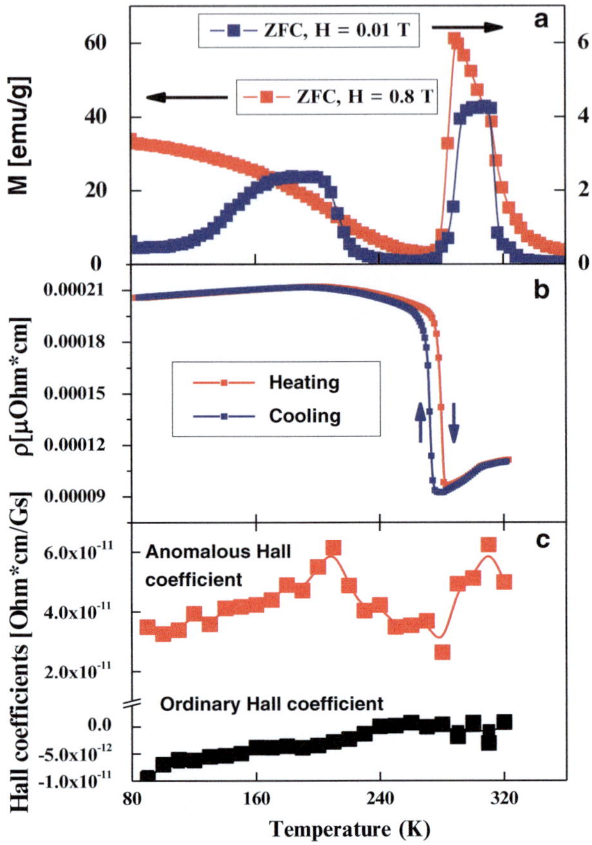

of spin-up and spin-down electrons near the Fermi surface in the MP with increasing temperature. It is worth noting that in the vicinity of MT, the parameter γ reaches a large value of 0.2 (Fig. 2.24) meaning that the ordinary Hall conductivity (see Eq. (2.12)) becomes a nonlinear function of magnetization. This nonlinearity of $\sigma_{xy}(B_z)$ and, as a result of the nonlinearity of R_s, can be a consequence of two-band current conduction or due to a possible shift in energy levels.

It is clearly seen from Figs. 2.23 and 2.24 that the AHE coefficients do not follow any known AHE mechanism. Indeed, in all theories the AHE coefficient should increase with resistivity. However, in the studied samples, it increases while resistivity decreases twice nearby MT or increases much stronger in the MP, where resistivity is almost constant. The reason for such behavior is not clear and needs to be studied in detail. Perhaps, it is due to a redistribution of spin-up and spin-down electrons, which is more important for the AHE than for the NHE. We cannot exclude that antiferromagnetic correlations in Heusler alloys play an important role in the AHE, but they were not taken into account in the AHE theory. We can conclude that the obtained data on the NHE and AHE do not provide evidence for abrupt changes in electronic structure at the MT.

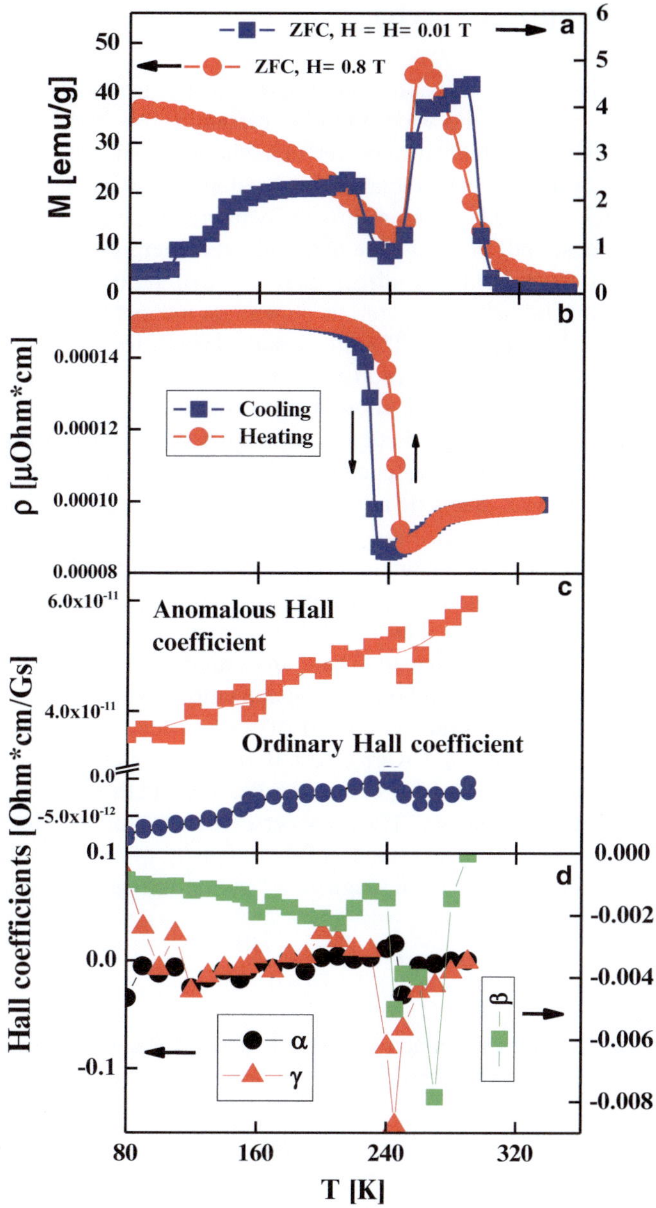

Fig. 2.24 Temperature dependences of the (**a**) magnetization (M), (**b**) electrical resistivity (ρ), (**c**) coefficients of the normal and anomalous Hall effects, and (**d**) coefficients α, β, and γ (see Eq. (2.12)) in an $Ni_{50}Mn_{35}In_{12}Si_3$ alloy

2.7 Magneto-optical Spectra

Absorption, reflection, transmission, and emission of electromagnetic radiation by matter strongly depend on the wavelength of radiation: this is the foundation for optical spectroscopy, one of the most powerful methods in materials science. Optical spectroscopic analysis has been crucial in the development of the most fundamental theories in physics, since it is the most direct way to probe energy levels of electrons in atoms, molecules, and, in solid-state materials, the energies of interband transitions. In spite of these being no direct interaction of light with magnetization, interband transitions in magnetic materials depend on magnetization because of the influence of spin–orbit interaction. Therefore, optical transitions change under application of an external magnetic field or in magnetization. This is grounds for magneto-optical (MO) magnetometry, which allows the measurements of magnetic properties as a function of an applied magnetic field, and magneto-optical spectroscopy, which allows the study of spectral dependence of the MO response to obtain information about electronic structure, magnetic microstructure, magnetic phases, and so on. It is worth noting that, due to high sensitivity and good spatial resolution, MO methods are very efficient in studying the magnetic properties of thin films, interfaces, surface layers, and magnetic inhomogeneous materials. In addition, the MO response can provide a time resolution better than 10^{-12} s and can be used to determine all three components of the magnetic moment. There are numerous MO effects in reflection and transmission modes, even and odd in magnetization, existing in a wide spectral range from THz to X-ray band [87]. Each of these effects can be used for MO spectroscopic analysis. The MO Kerr effect is linear in magnetization, relatively strong at visible wavelengths, and operates in reflection and therefore is very convenient in investigation of metals. The transverse (or equatorial) MO Kerr effect (Fig. 2.25) consists of a relative change of the intensity of the *p*-wave of linearly polarized light upon reflection from magnetized and non-magnetized samples:

$$TKE = \frac{I(H) - I(0)}{I(0)},$$

where $I(H)$ and $I(0)$ are the reflected light intensities measured with and without applied magnetic field. The spectral and field dependences of the transverse Kerr

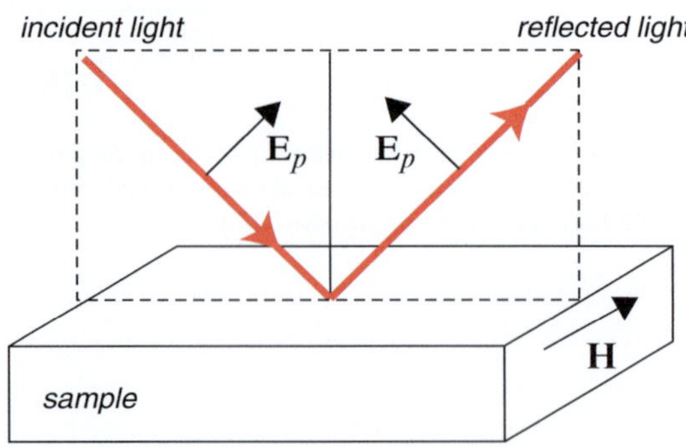

Fig. 2.25 Schematic of a transverse Kerr effect (TKE)

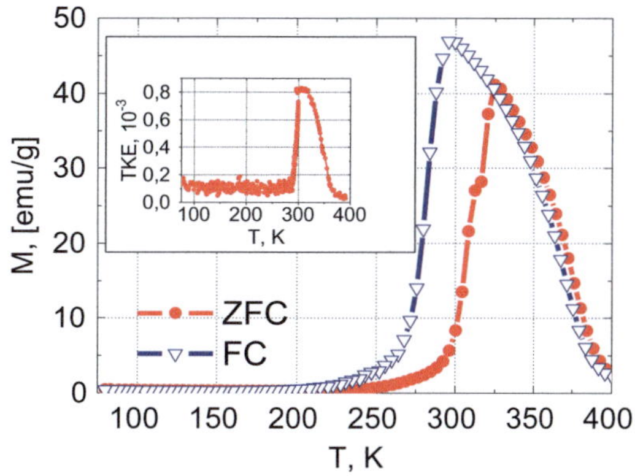

Fig. 2.26 Temperature dependence of the magnetization (M) in comparison with that for the TKE signal in a $Ni_{45}Mn_{36.7}In_{13.3}Co_5$ single crystal during heating. The magnetic field value was $H = 2.5$ kOe

effect parameters in Heusler alloys were measured using an automated MO spectrometer in the range of incident photon energies E from 0.5 to 4.0 eV. The measurements were performed for several angles of incidence of the p-polarized light beam. The amplitude of the applied alternating magnetic field reached up to 3.5 kOe.

The response signal was detected using a dynamic technique, which allowed measuring the relative changes of the reflected light intensity as small as 10^{-5} with an error not exceeding 5 %.

Figure 2.26 shows the temperature dependence of the magnetization measured by VSM in comparison with that for the TKE signal in the single crystal $Ni_{45}Mn_{36.7}In_{13.3}Co_5$ measured during heating. The characteristic temperature for the martensitic transformation measured by both methods was identical in this case, and the TKE signal reproduces all features of the MT transition. Since the MO response is formed in metal by surface layers with depth about 30–50 nm, in this case the quality of the sample was very high and the properties of the surface and bulk were very similar. This is not usually the case. For example, for the samples $Fe_{48}Mn_{24}Ga_{28}$ [88], the TKE signal clearly shows that the MT at the surface occurs at temperatures very different from the bulk temperatures. Moreover, we could not observe the TKE signal from the surface of polycrystalline samples after mechanical polishing of the surface. This means that, in general, the magnetic state of the surface of Heusler alloys can be quite different from that of the bulk.

Figure 2.27 shows the TKE spectra for the same monocrystal in austenitic (AP) and martensitic (MP) phases. The signal in MP is much smaller because of weak magnetization but its shape is approximately the same as for the AP. This result is more clearly seen from the data for $Ni_{43.7}Mn_{43.6}In_{12.7}$ ribbons (Fig. 2.28) with larger magnetization and the TKE signal in MP. Close to the Curie temperature of the AP, the TKE signal is small ($T = 300$ K), and then, with decrease in temperature, it increases with magnetization in AP ($T = 250$ K) and decreases in MP ($T = 60$ K) (see Figs. 2.28 and 2.29). But in all cases, the spectrum profile, namely, the spectral positions of positive and negative peaks, is almost identical. It means that interband transitions do not change at the MT, at least in the energy

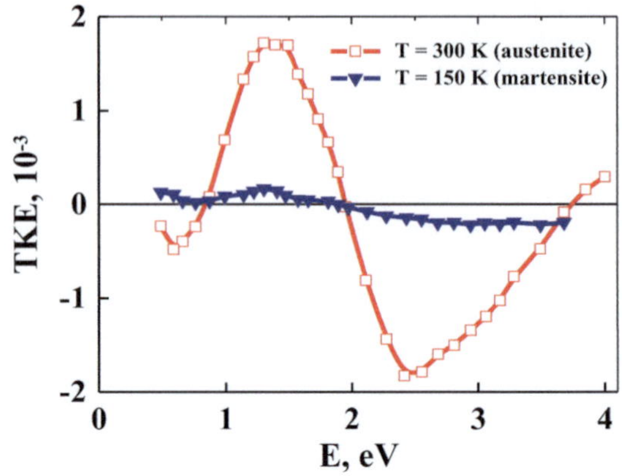

Fig. 2.27 TKE spectra for the single-crystal $Ni_{45}Mn_{36.7}In_{13.3}Co_5$ in austenitic ($T = 300$ K) and martensitic ($T = 150$ K) phases. Magnetic field $H = 2.7$ kOe

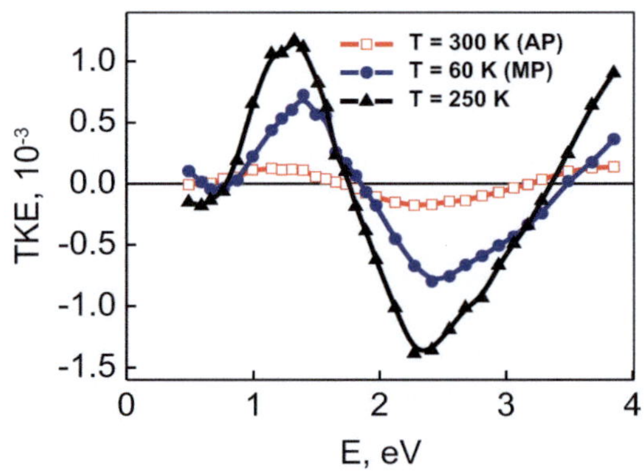

Fig. 2.28 TKE spectra for the $Ni_{43.7}Mn_{43.6}In_{12.7}$ ribbons in austenitic ($T = 300$ K) and martensitic ($T = 60$ K) phases. Magnetic field $H = 2.8$ kOe

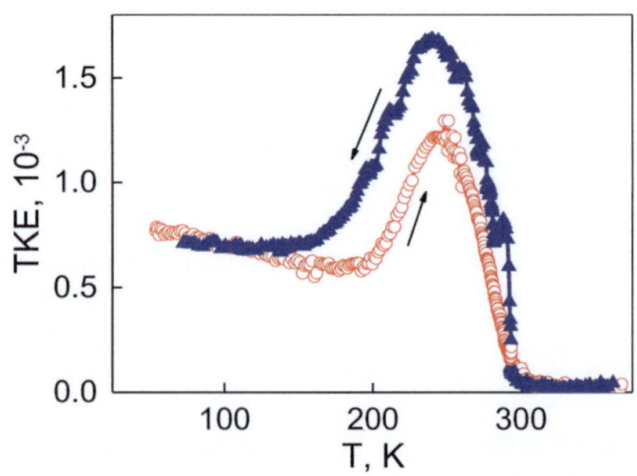

Fig. 2.29 Temperature dependence of TKE signal for the $Ni_{43.7}Mn_{43.6}In_{12.7}$ ribbons. Magnetic field $H = 2.8$ kOe

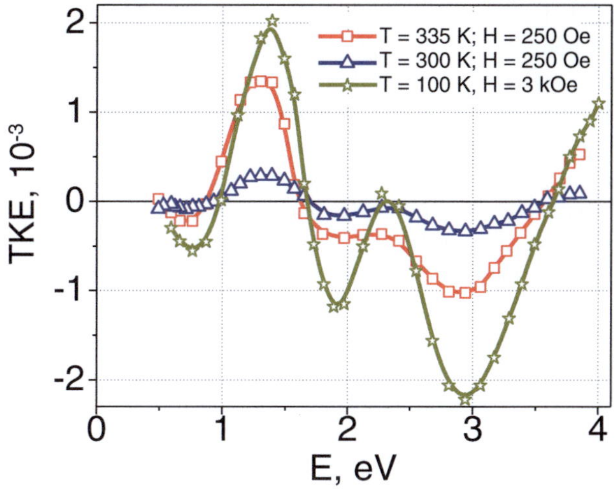

Fig. 2.30 TKE spectra for the single-crystal Ni_2MnGa in AP ($T = 335$ K) and MP ($T = 300$ K) in a magnetic field $H = 250$ Oe. The spectrum for $T = 100$ K (H = 3 kOe) is also shown

range 0.5–4.0 eV. So MO investigations at visible do not support the widely accepted model of strong changes in electronic structure at the MT. It is necessarily to extend the MO spectral range of investigation to infrared and X-ray range to obtain more details about electronic spectra transformation at the MT.

It is interesting to compare MO spectra for different compositions. Fig. 2.30 shows the TKE spectra for Ni_2MnGa. First it should be mentioned that they are in a good agreement with data for the same compositions reported in [89]. It means that, indeed, MO spectra can characterize the composition. The spectra for Ni_2MnGa and Ni_2MnIn are very similar [3, 90] but differ from that for alloys with different Ni–Mn ratio (see Figs. 2.27 and 2.28). We can conclude that the TKE spectra are determined mostly by the Ni and Mn partial densities of states, but not by In or Ga.

This is not surprising because the magnetic moments are associated with Ni and Mn. Unfortunately, presently the quantitative explanation of MO spectra in Heusler alloys is impossible [91], because calculations of electronic structure for three-component alloys with strong electronic correlations and spin–orbit interaction are an extremely difficult problem to solve, but will hopefully be solved in the future.

2.8 Conclusions

The Ni–Mn–In Heusler alloy is the origin of a cornucopia of rich physics with implications for applications. The exotic behaviors observed in materials based on this compound are derived foremost from magnetostructural transitions connected with the, sometimes confluent, magnetic and martensitic transformations. Since such abrupt, concurrent changes in magnetic and structural properties often affect many physical behaviors simultaneously, many of the derived alloys exhibit multifunctional behaviors, i.e., magnetocaloric and giant magnetoresistance.

In this review, we have demonstrated the flexibility of the Ni–Mn–In-based alloys in their properties, compositional modifications, and forms (ribbons, wires, etc.). The throng of observed properties includes normal and inverse magnetocaloric effects, giant magnetoresistance, exchange bias, and anomalous Hall effects, among others, many of which occur near room temperature. These properties can be optimized or tuned in temperature through variations in stoichiometry, chemical substitution, and fabrication parameters and post-processing and depend on external parameters such as magnetic field, stress, and pressure. We have presented a sampling of results of bulk, ribbon, and microwire Ni–Mn–In-based systems, illustrating the potential of these materials for application and as an opportunity to study the physics from which these properties are derived.

There are numerous issues in this field which need to be studied in detail. There is no definite answer on an origin of martensitic transition since experimental data are controversial. Further work should be done to improve the quality of microwires and ribbons. There is no up to now high-quality Ni–Mn–In single crystals and thin films which exhibit martensitic transition. The current theory fails to explain even qualitatively the behavior of anomalous Hall effect. Further work should be done also to suppress a thermal hysteresis, to obtain robust and reversible properties at cycling of magnetic field, and to study frequency and size-effect dependences of related to martensitic transition phenomena.

In the end, we hope that we have provided motivation for others to explore this rich vista of research.

Acknowledgment This work was supported by the Office of Basic Energy Sciences, Material Science Division of the US Department of Energy (DOE, Grant No. DE-FG02-06ER46291 (SIU) and DE-FG02-13ER46946 (LSU)), by the Russian Foundation for Basic Research (MSU), by the Spanish MINECO, and by the projects MAT2013-48054-C2-2-R and MAT2013-4731-C2-1-P.

References

1. Graf, T., Parkin, S., Felser, C.: Heusler compounds – a material class with exceptional properties. IEEE Trans. Magn. **47**, 367–372 (2011)
2. Krenke, T., Acet, M., Wassermann, E.F., Moya, X., et al.: Ferromagnetism in the austenitic and martensitic states of Ni−Mn−In alloys. Phys. Rev. B **73**, 174413–10 (2006)
3. Kudryavtsev, Y.V., Lee, Y.P., Rhee, J.Y.: Dependence of the optical and magneto-optical properties and electronic structures on the atomic order in Ni_2MnIn Heusler alloys. Phys. Rev. B **69**, 195104–195109 (2004)
4. Brown, P.J., Gandy, A.P., Ishida, K., et al.: The magnetic and structural properties of the magnetic shape memory compound $Ni_2Mn_{1.44}Sn_{0.56}$. J. Phys. Condens. Matter **18**, 2249–2259 (2006)
5. Richard, M.L., Feuchtwanger, J., Allen, S.M., et al.: Chemical order in off-stoichiometric Ni–Mn–Ga ferromagnetic shape-memory alloys studied with neutron diffraction. Philos. Mag. **87**, 3437–3447 (2007)

6. Stager, C.V., Campbell, C.C.M.: Antiferromagnetic order in the Heusler alloy $Ni_2Mn(Mn_xSn_{1-x}$. Can. J. Phys. **56**, 674–677 (1978)
7. Dubenko, I., Khan, M., Pathak, A.K., et al.: Magnetocaloric effects in Ni-Mn-X based Heusler alloys with X=Ga, Sb, and In. J. Magn. Magn. Mater. **321**, 754–757 (2009)
8. Dubenko, I., Samanta, T., Pathak, A.K., et al.: Magnetocaloric effect and multifunctional properties of Mn-Based Heusler alloys. J. Magn. Magn. Mater. **324**, 3530–3534 (2012)
9. Quetz, A., Muchharla, B., Samanta, T., et al.: Phase diagram and magnetocaloric effects in $Ni_{50}Mn_{35}(In_{1-x}Cr_x)_{15}$ and $(Mn_{1-x}Cr_x)NiGe_{1.05}$ alloys. J. Appl. Phys. **115**, 17A922–17A923 (2014)
10. Pathak, A.K., Dubenko, I., Xiong, Y., et al.: Effect of partial substitution of Ni by Co on the magnetic and magnetocaloric properties of $Ni_{50}Mn_{35}In_{15}$ Heusler alloys. J. Appl. Phys. **109**, 07A916–07A913 (2011)
11. Pathak, A.K., Khan, M., Dubenko, I., et al.: Large magnetic entropy change in $Ni_{50}Mn_{50\text{-}x}In_x$ Heusler alloys. Appl. Phys. Lett. **90**, 262504–262503 (2007)
12. Pathak, A.K., Dubenko, I., Xiong, Y., et al.: Effect of partial substitution of Ni by Co on the magnetic and magnetocaloric properties of $Ni_{50}Mn_{35}In_{15}$ Heusler alloys. IEEE. Trans. Mag. **46**, 1444–1446 (2010)
13. Pathak, A.K., Dubenko, I., Karaca, H.E., et al.: Large inverse magnetic entropy changes and magnetoresistance in the vicinity of a field-induced martensitic transformation in $Ni_{50-x}Co_xMn_{32-y}Fe_yGa_{18}$. Appl. Phys. Lett. **97**, 062505–062503 (2010)
14. Pathak, A.K., Khan, M., Gautam, B.R., et al.: Exchange bias in bulk Ni–Mn–In-based Heusler alloys. J. Magn. Magn. Mater. **321**, 963–965 (2009)
15. Khan, M., Dubenko, I., Stadler, S., Ali, N.: Exchange bias in bulk Mn rich Ni–Mn–Sn Heusler alloys. J. Appl. Phys. **102**, 113914–113913 (2007)
16. Prudnikov, V.N., Kazakov, A.P., Titov, I.S., et al.: Quasi- magnetism and exchange anisotropy in Ni-Mn-Co-In Heusler alloys. Phys. Solid State **53**, 3490–3493 (2011)
17. Pathak, A.K., Dubenko, I., Stadler, S., Ali, N.: Exchange bias in bulk $Ni_{50}Mn_{35}In_{(15\ -x)}Si_x$ Heusler alloys. IEEE Trans. Magn. **45**, 3855–3857 (2009)
18. Dubenko, I., Quetz, A., Pandey, S., et al.: Multifunctional properties related to magnetostructural transitions in ternary and quaternary Heusler alloys. J. Magn. Magn. Mater. **383**, 183–189 (2015)
19. Priolkar, K.R., Lobo, D.N., Bhobe, P.A., et al.: Role of Ni-Mn hybridization in the magnetism of the martensitic state of Ni-Mn-In shape memory alloys. Eur. Phys. Lett. **94**, 38006–p6 (2011)
20. Khan, M., Dubenko, I.S., Stadler, S., et al.: Enhancement of ferromagnetism by Cr doping in Ni-Mn-Cr-Sb Heusler alloys. Appl. Phys. Lett. **102**, 112402–112404 (2013)
21. Khan, M., Jung, J., Stoyko, S.S., et al.: The role of Ni-Mn hybridization on the martensitic phase transitions in Mn-rich Heusler alloys. Appl. Phys. Lett. **100**, 172403–172404 (2012)
22. Dubenko, I., Samanta, T., Quetz, A., et al.: The comparison of direct and indirect methods for determining the magnetocaloric parameters in the Heusler alloy $Ni_{50}Mn_{34.8}In_{14.2}B$. Appl. Phys. Lett. **100**, 192402–192404 (2012)
23. Pathak, A.K., Dubenko, I., Pueblo, C., et al.: Magnetoresistance and magnetocaloric effect at a structural phase transition from a paramagnetic martensitic state to a paramagnetic austenitic state in $Ni_{50}Mn_{36.5}In_{13.5}$ Heusler alloys. Appl. Phys. Lett. **96**, 172503 (2010)
24. Stadler, S., Khan, M., Mitchell, J., et al.: Magnetocaloric properties of $Ni_2Mn_{1-x}Cu_xGa$. Appl. Phys. Lett. **88**, 192511–192513 (2006)
25. Dubenko, I., Samanta, T., Quetz, A., et al.: The adiabatic temperature changes in the vicinity of the first-order paramagnetic-ferromagnetic transition in the Ni-Mn-In-B Heusler alloy. IEEE Trans. Magn. **48**, 3738–3741 (2012)
26. Kainuma, R., Imano, Y., Ito, W., et al.: Magnetic-field-induced shape recovery by reverse phase transformation. Nature **439**, 957–960 (2006)

27. Zimm, C., Jastrab, A., Sternberg, A., et al.: Description and performance of a near-room temperature refrigerator. In: Kittel, P. (ed.) Adv Cryog Eng 43 (Parts A and B), pp. 1759–1766. Plenum, New York (1988)
28. Franco, V., Blásquez, J.S., Ingale, B., Conde, A.: The magnetocaloric effect and magnetic refrigeration near room temperature: materials and models. Annu. Rev. Mater. Res. **42**, 305–342 (2012)
29. Liu, J., Gottschall, T., Skokov, K.P., et al.: Giant magnetocaloric effect driven by structural transitions. Nat. Mater. **11**, 620–626 (2012)
30. Titov, I., Acet, M., Farle, M., et al.: Hysteresis effects in the inverse magnetocaloric effect in martensitic Ni-Mn-In and Ni-Mn-Sn. J. Appl. Phys. **112**, 073914–073915 (2012)
31. Pecharsky, V.K., Gschneidner Jr., K.A., Pecharsky, A.O., Tishin, A.M.: Thermodynamics of the magnetocaloric effect. Phys. Rev. B **64**, 144406–144413 (2001)
32. Gschneidner Jr., K.A., Pecharsky, V.K., Tsokol, A.O.: Recent developments in magnetocaloric materials. Rep. Prog. Phys. **68**, 1479–1539 (2005)
33. Pathak, A.K., Dubenko, I., Pueblo, C., et al.: Magnetism and magnetocaloric effects in $Ni_{50}Mn_{35-x}Co_xIn_{15}$ Heusler alloys. J. Appl. Phys. **107**, 09A907–09A903 (2010)
34. Yucel, A., Lerman, Y., Aksoy, S.: Changes in the phase structure and magnetic characteristics of $Gd_5Si_2Ge_2$ when alloyed with Mn. J. Alloys Compd. **420**, 182–185 (2006)
35. Shull, R.D., Provenzano, V., Shapiro, A.J., et al.: The effects of small metal additions (Co, Cu, Ga, Mn, Al, Bi, Sn) on themagnetocaloric properties of the $Gd_5Ge_2Si_2$ alloy. J. Appl. Phys. **99**, 08K908–08K9083 (2006)
36. Phan, M.H., Yu, S.-C.: Review of the magnetocaloric effect in manganite materials. J. Magn. Magn. Mater. **308**, 325–340 (2007)
37. Kazakov, A.P., Prudnikov, V.N., Granovsky, A.B.: Direct measurements of field-induced adiabatic temperature changes near compound phase transitions in Ni-Mn-In based Heusler alloys. Appl. Phys. Lett. **98**, 131911–131913 (2011)
38. Kazakov, A., Prudnikov, V., Granovsky, A., et al.: Phase transitions, magnetotransport and magnetocaloric effects in a new family of quaternary Ni–Mn–In–Z Heusler alloys. J. Nanosci. Nanotechnol. **12**, 7426–7431 (2012)
39. Sánchez, T., Turtelli, S., Grössinger, R., et al.: Exchange bias behavior in $Ni_{50.0}Mn_{35.5}$ $In_{14.5}$ ribbons annealed at different temperatures. J. Magn. Magn. Mater. **324**, 3535–3537 (2012)
40. Khovaylo, V.V., Kanomata, T., Tanaka, T., et al.: Magnetic properties of $Ni_{50}Mn_{34.8}In_{15.2}$ probed by Mössbauer spectroscopy. Phys. Rev. B **80**, 144409–144407 (2009)
41. Sharma, V.K., Chattopadhyay, M.K., Roy, S.B.: Kinetic arrest of the first order austenite to martensite phase transition in Ni50Mn34In16: dc magnetization studies. Phys Rev B **76**, 140401–140404 (2007)
42. Imry, Y., Wortis, M.: Influence of quenched impurities on first-order phase transitions. Phys. Rev. B **19**, 3580–3585 (1979)
43. Perring, L., Kuntz, J.J., Bussy, F., Gachon, J.C.: Heat capacity on the equiatomic compounds in Ni-X (X = Al, In, Si, Ge and Bi) and M-Sb (with M = Ni, Co and Fe) systems. Intermetallics **7**, 1235–1239 (1999)
44. Shen, T.D., Schwarz, R.B., Coulter, J.Y., Thompson, J.D.: Magnetocaloric effect in bulk amorphous $Pd_{40}Ni_{22.5}Fe_{17.5}P_{20}$ alloy. J. Appl. Phys. **91**, 5240–5245 (2002)
45. Caballero-Flores, R., Sánchez, T., Rosa, W.O., et al.: On tuning the magnetocaloric effect in Ni-Mn-In Heusler alloy ribbons. J. Alloys Compd. **545**, 216–221 (2012)
46. Vazquez, M., Chiriac, H., Zhukov, A.: On the state-of-the-art in magnetic microwires and expected trends for scientific and technological studies. Phys. Status Solidi A **208**(3), 493–501 (2011)
47. Antonov, A.S., Borisov, V.T., Borisov, O.V., et al.: Residual quenching stresses in glass-coated amorphous ferromagnetic microwires. J. Phys. D. Appl. Phys. **33**, 1161–1168 (2000)
48. Chiriac, H., Ovari, T.A., Zhukov, A.: Magnetoelastic anisotropy of amorphous microwires. J. Magn. Magn. Mater. **254–255**, 469–471 (2003)

49. Chiriac, H., Ovari, T.A., Pop, G.: Internal stress distribution in glass-covered amorphous magnetic wires. Phys. Rev. B **42**, 10105–10113 (1995)
50. Velázquez, J., Vazquez, M., Zhukov, A.: Magnetoelastic anisotropy distribution in glass-coated microwires. J. Mater. Res. **11**(10), 2499–2505 (1996)
51. Zhukov, A., Blanco, J.M., Ipatov, M., et al.: Manipulation of domain wall dynamics in amorphous microwires through the magnetoelastic anisotropy. Nanoscale Res. Lett. **7**, 223–228 (2012). doi:1556-276X-7-223/1556-276X-7-223
52. Zhukov, A.P., Vázquez, M., Velázquez, J., et al.: The remagnetization process of thin and ultrathin Fe-rich amorphous wires. J. Magn. Magn. Mater. **151**, 132–138 (1995)
53. Zhukov, A., Ipatov, M., Blanco, J.M., et al.: Fast magnetization switching in amorphous microwires. Acta Phys. Pol. A **126**, 7–11 (2014)
54. Garcia Prieto, M., Pina, E., Zhukov, A.P., et al.: Glass coated Co-rich amorphous microwires with improved permeability. Sensor. Actuat. A **81**(1-3), 227–231 (2000)
55. Zhukov, A., Gonzalez, J., Blanco, J.M., et al.: Induced magnetic anisotropy in Co-Mn-Si-B amorphous microwires. J. Appl. Phys. **87**, 1402–1408 (2000)
56. Dunand, D.C., Müllner, P.: Size effects on magnetic actuation in Ni-Mn-Ga shape-memory alloys. Adv. Mater. **23**, 216–232 (2011)
57. Varga, R., Ryba, T., Vargova, Z., et al.: Magnetic and structural properties of Ni–Mn–Ga Heusler-type microwires. Scr. Mater. **65**, 703–706 (2011)
58. Garcıa, C., Prida, V.M., Vega, V., et al.: Magnetic characterization of $Cu_{56}Ga_{28}Mn_{16}$ microwires. Phys. Status Solidi A **206**(4), 644–647 (2009)
59. Craciunescu, C.M., Ercuta, A., Mitelea, I., et al.: Rapidly solidified ferromagnetic shape memory alloys. Eur. Phys. J. Spec. Top. **158**, 161–165 (2008)
60. Zhukova, V., Ipatov, M., Granovsky, A., Zhukov, A.: Magnetic properties of Ni-Mn-In-Co Heusler-type glass-coated microwires. J. Appl. Phys. **115**, 17A939 (2014)
61. Kuz'min, M.D.: Factors limiting the operation frequency of magnetic refrigerators. Appl. Phys. Lett. **90**, 251916 (2007)
62. Belov, K.P.: Magnetic Transitions. Consultants Bureau, New York (1959)
63. Nishihara, H., Furutani, Y., Wada, T., et al.: Magnetization process near the Curie temperature of a ferromagnetic Heusler alloy Co2VGa. J. Supercond. Nov. Magn. **24**, 679–681 (2011)
64. Aronin, A.S., Abrosimova, G.E., Kiselev, A.P., et al.: The effect of mechanical stress on $Ni_{63.8}Mn_{11.1}Ga_{25.1}$ microwire crystalline structure and properties. Intermetallics **43**, 60–64 (2013)
65. Gunnarsson, O., Calandra, M., Han, E.: Saturation of electrical resistivity. Rev. Mod. Phys. **75**, 1085–1098 (2003)
66. Gantmakher, V.F.: Mooij rule and weak localization. JETP Lett. **94**, 626–628 (2011)
67. Vasiliev, A.N., Heczko, O., Volkova, O.S., et al.: On the electronic origin of the inverse magnetocaloric effect in Ni–Co–Mn–In Heusler alloys. J. Phys. D. Appl. Phys. **43**, 055004–055011 (2010)
68. Fuji, S., Ishida, S., Asano, S.: Electronic structure and lattice transformation in Ni_2MnGa and Co_2NbSn. J. Phys. Soc. Japan **58**, 3657–3665 (1989)
69. Samanta, T., Saleheen, A.U., Lepkowski, D.L., et al.: Asymmetric switchinglike behavior in the magnetoresistance at low fields in bulk metamagnetic Heusler alloys. Phys. Rev. B **90**, 064412–064416 (2014)
70. Sakamoto, N., Kyomen, T., Tsubouchi, S., Itoh, M.: Proportional relation between magnetoresistance and entropy suppression due to magnetic field in metallic ferromagnets. Phys. Rev. B **69**, 092401–092404 (2004)
71. Xiong, C.M., Sun, J.R., Chen, Y.F., et al.: Relation between magnetic entropy and resistivity in $La_{0.67}Ca_{0.33}MnO_3$. IEEE Trans. Magn. **41**, 122–125 (2005)
72. Rodionov, I.D., Mettus, D.E., Kazakov, A.P., et al.: Correlation between magnetoresistance and magnetic entropy at first-order and second-order phase transitions in Ni–Mn–In–Si Heusler alloys. Phys. Solid State **55**, 1861–1865 (2013)

73. Nagaosa, N., Sinova, J., Onoda, S., et al.: Anomalous Hall effect. Rev. Mod. Phys. **82**, 1539–1592 (2010)
74. Jungwirth, T., Sinova, J., Mašek, J., et al.: Theory of ferromagnetic (III, Mn)V semiconductors. Rev. Mod. Phys. **78**, 809–867 (2006)
75. Sinova, J., Jungwirth, T., Cerne, J.: Magneto-transport and magneto-optical properties of ferromagnetic (III, Mn)V semiconductors: a review. Int. J. Mod. Phys. B. **18**, 1083–1118 (2004)
76. Mikhailovsky, Y.O., Mettus, D.E., Kazakov, A.P., et al.: Anomalous Hall effect in $(Co_{41}Fe_{39}B_{20})_x(Al\text{-}O)_{100-x}$ nanocomposites. JETP Lett. **97**, 473–477 (2013)
77. Dubenko, I., Quetz, A., Pandey, S., et al.: Multifunctional properties related to magnetostructural transitions in ternary and quaternary Heusler alloys. J. Magn. Magn. Mater. 383, 183–189 (2015)
78. Prudnikov, V.N., Kazakov, A.P., Titov, I.S., et al.: Hall effect in a martensitic transformation in Ni-Co-Mn-In Heusler alloys. JETP Lett. **92**(10), 666–670 (2010)
79. Granovskii, A.B., Prudnikov, V.N., Kazakov, A.P., et al.: Determination of the normal and anomalous Hall effect coefficients in ferromagnetic $Ni_{50}Mn_{35}In_{15-x}Si_x$ Heusler alloys at the martensitic transformation. J. Exp. Theor. Phys. **115**, 805–814 (2012)
80. Hirsch, J.C.E.: Overlooked contribution to the Hall effect in ferromagnetic metals. Phys. Rev. B **60**, 14787–14792 (1999)
81. Karplus, R., Lutinger, J.M.: Hall effect in ferromagnetics. Phys. Rev. **95**, 1154–1160 (1954)
82. Smit, J.: Theory of the Hall effect in ferromagnetic substances I. Physica (Amsterdam) **21**, 877–887 (1955)
83. Vedyaev, A.V., Granovskii, A.B., Kotelnikova, O.A.: Transport Phenomena in Disordered Ferromagnetic Alloys, p. 158. Moscow State University, Moscow (1992). In Russian
84. Luttinger, J.M.: Theory of the Hall effect in ferromagnetic substances. Phys. Rev. **112**, 739–751 (1958)
85. Berger, L.: Side-jump mechanism for the Hall effect of ferromagnets. Phys. Rev. B **2**, 4559–4566 (1970)
86. Zhu, S., Or, W., Wu, G.: Anomalous Hall effect in quarternary Heusler-type effect in Ni50Mn17Fe8Ga25 melt-sdpun ribbons. Appl. Phys. Lett. **95**, 032503–032504 (2009)
87. Antonov, V., Harmon, B., Yaresko, A.: Electronic Structure and Magneto-Optical Properties of Solids. Springer, New York (2004). 528 pages
88. Gan'shina, E.A., Novikov, A.I., Zhykov, G.S., et al.: Magneto-optical spectroscopy of the martensitic transition in $Fe_{48}Mn_{24}Ga_{28}$ Heusler alloys. Phys. Solid State **55**, 1866–1870 (2013)
89. Novikov, A., Gan'shina, E., Granovsky, A., et al.: Magneto-optical spectroscopy of Heusler alloys: bulk samples, thin films and microwires. Solid State Phenom. **190**, 335–338 (2012)
90. Lee, S.J., Lee, Y.P., Hyun, Y.H., Kudryavtsev, Y.V.: Magnetic, magneto-optical, and transport properties of ferromagnetic shape-memory Ni_2MnGa alloy. J. Appl. Phys. **93**, 6975–6981 (2003)
91. Novikov, A.I., Gan'shina, E.A., Gonzalez-Legarreta, L., et al.: Magnetic and magneto-optical research of $Ni_{43.7}Mn_{43.6}In_{12.7}$ alloy ribbons. Solid State Phenom. **233–234**, 200–203 (2015)

Chapter 3
Heusler Alloy Ribbons: Structure, Martensitic Transformation, Magnetic Transitions, and Exchange Bias Effect

L. González-Legarreta, R. Caballero-Flores, W.O. Rosa, Mihail Ipatov, L. Escoda, J.J. Suñol, V.M. Prida, J. González, and B. Hernando

3.1 Introduction: Heusler Alloys – Martensitic Transformation

Half-metals, first postulated by de Groot in the early 1970s, can be considered as hybrids between metals and semiconductors, since the majority spin band structure shows a metallic behavior, while the minority spin band structure exhibits a gap at the Fermi level such as in a semiconductor. Due to the gap for one spin direction, the density of states at the Fermi level has, theoretically, 100 % spin polarization. Such fully spin-polarized ferromagnetic materials are of considerable interest in the newly developed area of spin electronics (or spintronics).

Among the most cited candidates for 100 % spin polarization are the semi-Heusler alloys [1–9], full Heusler alloys [10, 11], zinc blende structure materials [12–14], colossal magnetoresistance materials ($La_{1-x}Sr_xMnO_3$ [15, 16], Sr_2FeMoO_6 [17]), and semimetallic magnetic oxides (CrO_2 [18–21] and Fe_3O_4 [22, 23]). A number of other materials have also been suggested as half-metallic ferromagnets [24, 25].

An important part of the scientific effort has been focused on the study and fabrication of half-metallic Heusler alloys. The first family of Heusler alloys studied was of the form X_2YZ, crystallizing in the $L2_1$ structure, which consists

L. González-Legarreta (✉) • R. Caballero-Flores • V.M. Prida • B. Hernando (✉)
Department of Physics, University of Oviedo, Av. Calvo Sotelo s/n, 33007 Oviedo, Spain
e-mail: lorena.glegarreta@gmail.com; grande@uniovi.es

W.O. Rosa
Centro Brasileiro de Pesquisas Físicas, CEP 22290-180 Urca, Rio de Janeiro, Brazil

M. Ipatov • J. González
Department of Materials Physics, University of the Basque Country, Paseo Manuel de Lardizábal 3, 20018 San Sebastián, Spain

L. Escoda • J.J. Suñol
Girona University, Campus Montilivi, ed. PII, Lluís Santaló s/n, 17003 Girona, Spain

A. Zhukov (ed.), *Novel Functional Magnetic Materials*, Springer Series in Materials Science 231, DOI 10.1007/978-3-319-26106-5_3

of four fcc sublattices, where X is a high-valence transition or noble metal atom, Y a low-valence transition metal atom, and Z an sp element [26, 27]. X is usually a transition metal 3d (Fe, Co, Ni, Cu, Zn), 4d (Ru, Rh, Pd, Ag, Cd), or 5d (Ir, Pt, Au). The position of Y is usually occupied by 3d (Ti, V, Cr, Mn), 4d (Y, Zr, Nb), or 5d (Hf, Ta) or by lanthanides (Gd, Tb, Dy, Ho, Er, Tm, Yb, Lu) or actinides (U). The Z is a group B element: IIIB (Al, Ga, In, Tl), IVB (Si, Ge, Sn, Pb), or VB (As, Sb, Bi). These Heusler compounds have attracted a lot of interest due to the possibility to study in the same family of alloys a series of interesting diverse magnetic phenomena like itinerant and localized magnetism, antiferromagnetism, helimagnetism, Pauli paramagnetism or heavy-fermionic behavior [26, 27], and giant magnetocaloric effect (MCE) [28–30]. The Heusler alloys of the second class are of the form X–Y–Z, crystallizing in the *C1b* structure and consisting of three fcc sublattices; they are often called half- or semi-Heusler alloys in the literature, while the $L2_1$ compounds are referred to as full-Heusler alloys. Otherwise, off-stoichiometry alloys were produced, and sometimes a fourth element is also introduced to check their influence on functional properties as magnetic superelasticity, large inverse magnetocaloric effect, and large magnetoresistance change.

Currently, the search for an affordable magnetic material which exhibits a large MCE that works in the temperature range of 100 up to 300 K using a magnetic field variation of $\Delta H = 10$ kOe has started. The most extensively studied Heusler alloys that have those properties are based on the Ni–Mn–Ga system; nevertheless to overcome some of the problems related with practical applications (such as the high cost of gallium and the usually low martensitic transformation temperature), the search for Ga-free alloys has been recently attempted. To reduce such costs and to improve the martensitic transition temperature, the substitution of Ga is proposed by introducing In, Sb, or Sn [31–34] instead. These ferromagnetic shape memory alloys (FSMA) exhibit ferromagnetism and shape memory effect simultaneously.

The ferromagnetic shape memory effect can be controlled mainly by temperature, stress, and magnetic field. Recently, Ni–Mn–In Heusler alloys have drawn much attention due to their potential as ferromagnetic shape memory alloys, which undergo a thermoelastic martensitic transformation (MT) from parent austenitic phase to a martensitic one on cooling [35]. By lowering the temperature, a cubic high-temperature parent austenite phase transforms into a tetragonal, orthorhombic, or monoclinic martensite. Furthermore, orthorhombic and monoclinic martensites can be also structurally twinned or modulated. Sometimes modulation is found non-commensurate. Furthermore, structural transformation is accompanied by a drastic jump in the magnetization. These alloys exhibit notable sensitivity of MT to the applied magnetic field, and they seem to be among the most suitable for the room temperature (RT) applications, for example, in micro- and nanomechanics devices and in alternative energy technologies, due to the giant magnetocaloric effect (MCE) observed in this alloys [36, 37]. Moreover, some authors have reported that such FSMAs also present giant magnetoresistance (GMR) due to the first-order phase transition, which can undergo a GMR variation of around 80 % [38]. The transformation temperatures of shape memory alloys strongly depend on the composition, and their values spread in a very wide range [39].

In this chapter, we discuss the Heusler alloy ribbon production and properties in the Ni–Mn–Z (Z = Ga, In, Sn, Sb) ternary and quaternary (minor addition of other elements as Co or Fe) samples. We also present the influence of different annealing treatments on the crystal structure, martensitic transition, magnetic behavior, and exchange bias effect in off-stoichiometric Ni–Mn–Z (Z = Sn, In) Heusler alloy ribbons. Along with the scientific interest of the results here reported, these materials could be exploited in refrigeration by using positive and negative magnetic entropy changes. These materials are candidates for specific future applications due to their shape and texture, phenomena that favors different behavior between bulk and ribbon alloys [40].

3.2 Heusler Alloy Ribbon Production

Heusler alloy ribbons are usually produced in a three-step procedure: (a) bulk production by arc melting, (b) controlled annealing, and (c) ribbon production by melt spinning (ribbon annealing is an option).

3.2.1 Arc Melting and Annealing

The first step is the preparation of polycrystalline ingots of the systems Ni–Mn–Z (Z = In, Sn, Ga) from high-purity elements by arc melting under protective argon (Ar) atmosphere in water-cooled Cu crucibles. Samples were subsequently remelted in order to ensure their homogeneity. This procedure is also applied to obtain only bulk samples. Thus, considerations in this subsection are the same for bulk and ribbon shape Heusler alloys, and we also introduce references of works where ribbons were finally not produced.

Schlagel et al. suggest that Mn should be purified by sublimation and metals to be cleaned with acid [41]. Additional manganese is normally added ($\approx$5 wt%) to compensate minor evaporation losses. Posteriorly, the ingots are encapsulated under argon (or vacuum) in quartz ampoules and annealed into specific conditions of temperature and time. Nevertheless, up to now, the selection of the different conditions found in the bibliography is unclear. For instance, we can find annealing temperatures from 1073 K [42] to 1273 K [43] during 2 h or at 1173 K during 24 h [44, 45] to 1273 K during 72 h [46]. Ishikawa also suggests that, before annealing, the samples with gallium and aluminum have to be wrapped in molybdenum foil to prevent reaction with the quartz ampoules [46]. The annealing process can be followed either by quenching into an ice water or by slow cooling with the furnace, depending on the required features of crystal structure.

One of the key parameters that control the martensitic transformation temperature in Ni–Mn–Z systems is the alloy composition [39, 42]. Thus, the change of the degree of long-range atomic order also affects both the transformation and the magnetic properties due to the modification of the electronic configuration and crystalline structure. The atomic order can be modified by thermal treatments such as annealing or quenching, and their influences have been studied in the Ni–Mn–Z systems by different authors [41, 45, 47, 48].

3.2.2 Melt Spinning

Studies on Ni–Mn–Z alloys have so far mostly focused on ingot polycrystalline materials produced by arc melting with subsequent heat treatment at high temperature for compositional homogenization. However, this production process causes different problems, such as decomposition [49] or precipitation of unexpected phases, which negatively affect the functional properties of the alloy [50].

Melt Spinning has been proven to be an effective method to obtain synthesized ribbons with homogeneous chemical composition avoiding prolonged annealing stages [51]. Due to the high cooling rate (10^4–10^6 Ks^{-1}), it is possible to produce nearly a single-phase and highly textured microstructure. Furthermore, rapid solidification from the liquid phase allows for obtaining the atoms in a metastable state. This permits to change the atomic order, and therefore this fact is of great interest to investigate.

In the literature, the analysis of ribbon samples shows important differences in the behavior of martensitic transformation if compared with bulk ingots of the same composition, probably due to their microstructure and processing conditions [52–56]. For instance, the martensitic transformation of the ribbons occurs about 150 K below the as-cast ingot alloy for $Ni_{50}Mn_{41}Sn_9$ alloy, although this difference is lower between ribbons and annealed ingots [57]. It is also found that crystalline structure of the martensite depends on processing conditions, for example, Santos et al. [52] observed a seven-layered orthorhombic martensite, *14 M*, in $Ni_{50}Mn_{37}Sn_{13}$ with melt-spinning technique, whereas Muthu et al. [58] found a four-layered orthorhombic martensite, *4O*, in the conventional arc-melted ingot $Ni_{50}Mn_{37}Sn_{13}$. Hernando et al. reported a modulated *14 M* orthorhombic martensite for $Ni_{50}Mn_{37}Sn_{13}$ ribbons and a modulated *10 M* orthorhombic martensite for $Ni_{50}Mn_{37}Sn_{13}$ ingots [55].

Additionally, it has also been shown that post-annealing for a short time (e.g., 10–20 min) can improve some physical properties of ribbons [50].

As-cast ingots can be melted by induction in quartz crucibles with a circular or rectangle nozzle and ejected applying an argon overpressure on the polished surface of a cooper wheel rotating at a specific speed [50, 55].

In the bibliography we can also find other techniques of production of Mi–Mn–Z alloys, for instance, mechanical ball milling, chemical deposition, or electrodeposition. Mechanical ball-milling process permits to obtain nanoparticles, but for Heusler nanoparticles dedicated for magnetic shape memory alloys, the ball-milling approach might be not favorable since substantial amounts of defects and stresses are introduced that might substantially mediate the nucleation and growth of martensitic phase. Chemical deposition has been used to prepare the so-called "pseudo" binary Heusler nanoparticles and nanowires. Up to now, this approach has not been attempted for ternary intermetallic nanowires [59]. Finally, triple layers of nickel, manganese, and gallium can be electrodeposited from chemical solutions onto tungsten and molybdenum refractory metal substrates. These layered films are subsequently annealed in order to form a Ni–Mn–Ga Heusler alloy by diffusion [60].

3.3 Characterization

3.3.1 Crystalline Structure (XRD) and Microstructure (SEM-EDX) of Austenitic and Martensitic Phases

Rapid solidification on melt-spinning process favors the suppression of the superstructural order of the cubic $L2_1$ structure. Thus, less-ordered cubic structures as *A2* or *B2* can be formed. The same problem is well-known for the stoichiometric Ni_2MnGa bulk alloy; the ordered $L2_1$ structure can be reached either from the less-ordered body-centered cubic phase, *A2*, following $A2 \rightarrow L2_1$ process or through a partially ordered phase, *B2*, by $A2 \rightarrow B2 \rightarrow L2_1$ process [61]. In Ni_2MnGa alloy, the martensite transformation is around 202 K being the magnetic transition around 376 K. Finding new Ni–Mn–Ga compositions is being required in order to obtain the magnetic transition at 300 K. In Ni–Mn–(In, Sn)-based ribbon samples, usually cubic structure is *B2* or $L2_1$. If *B2* structure is found in the as-quenched samples, controlled annealing induces the transformation from *B2* to $L2_1$ [62] or the increase of the intensity of the superlattice reflections of $L2_1$ structure [63]. The main difference between both crystallographic structures is that in *B2* it is not possible to distinguish elements in Y and Z sites. The confirmation of the $L2_1$ phase presence is the detection of superlattice reflections such as 1 1 1, 3 3 1, 3 1 1, or 5 1 1. Other features associated to rapid quenching are (a) the increase of atomic disorder and (b) high internal stress [40]. It has been reported that the degree of atomic order induces the formation of *B2* structure in as-quenched samples, whereas bulk sample develops $L2_1$ structure [64]. Martensite structure can be tetragonal, orthorhombic, or monoclinic. Furthermore, martensite can be four layered (*4O*), five layered (*10 M*), or seven layered (*14 M*) modulated [65, 66]. Also, it is possible to produce alloys with the coexistence of two modulated structures [67]. Likewise, crystalline structure sometimes has some preferential orientation favored by the existence of columnar grains between both sides of ribbons (side in contact with the wheel and opposite free side) [68]. Different authors have found a correlation between the average grain size of a crystalline shape memory alloys and the martensitic transformation [69].

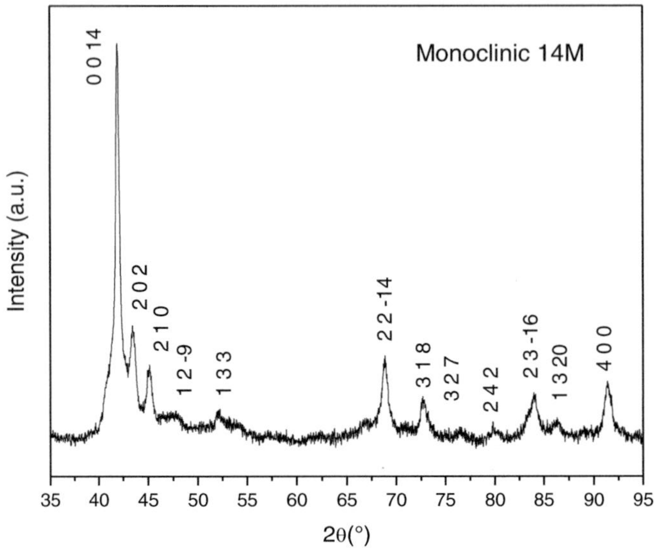

Fig. 3.1 X-ray diffraction pattern at room temperature for $Mn_{49.1}Ni_{43.1}In_{7.8}$ alloy

XRD patterns are usually collected at room temperature. Figures 3.1 and 3.2 show the diffraction patterns of two alloys, with martensite and austenite structures, respectively. The XRD structure lattice parameters are given in Table 3.1. Alloy with 7.8 In atomic content develops a modulated monoclinic *14 M* structure. It is a textured sample as shown by 0 0 1 4 reflection intensity. Likewise, alloy with 10.5 In atomic content has an austenite $L2_1$ structure, as confirmed by the 3 1 1 peak (as shown in Fig. 3.2). One of the problems to obtain a well-defined Heusler structure is due to the non-stoichiometry of the alloy with a low content of Z elements favoring weak minor peaks of superstructure.

XRD at different temperatures can be also performed to show the austenite and martensite crystallographic structures. Figure 3.3 shows the structure of $Ni_{45.5}Mn_{43}In_{11.5}$ alloy at different temperatures.

Heusler alloys produced by melt spinning show a typical columnar structure in the fracture cross section. Figure 3.4 shows scanning electron microscopy (SEM) micrographs of the fracture section of alloys $Ni_{50}Mn_{35.5}In_{14.5}$ (left) and $Ni_{45.5}Mn_{43}In_{11.5}$ (right). All ribbon flakes show a similar morphology which consists of fully crystalline and granular columnar-type microstructure. This is a sign of the fast crystallization and growth kinetics of the samples. It suggests that the heat removal during rapid solidification process induces the directional growth of the crystalline phase formed. Energy-dispersive X-ray spectroscopy (EDX) is usually coupled with SEM, and it is applied to obtain the exact composition of every sample to calculate average electron density parameter, *e*/*a*.

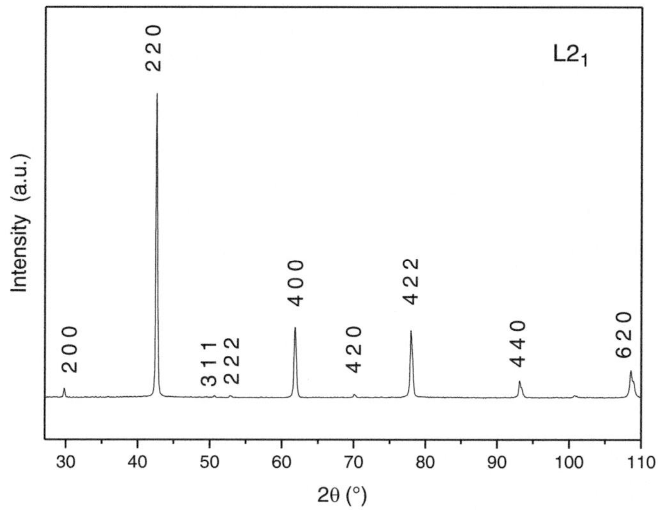

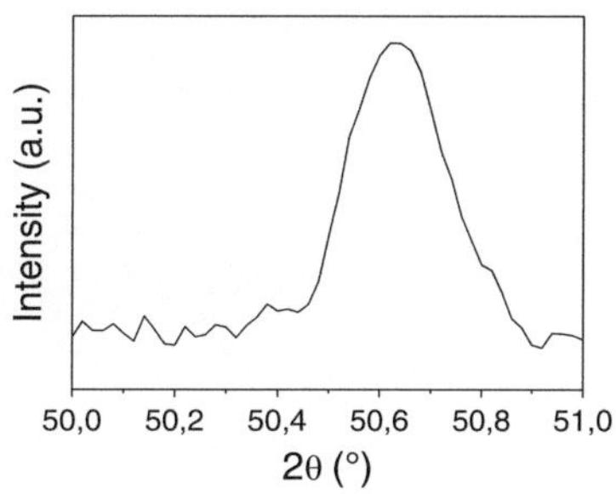

Fig. 3.2 X-ray diffraction pattern at room temperature for $Mn_{49.3}Ni_{40.2}In_{10.5}$ alloy: *up* (full pattern), *down* (50–51 2θ zone)

Table 3.1 Structure and lattice parameters at room temperature of as-quenched samples

Alloy	a(Å)	b(Å)	c(Å)	Structure
$Mn_{49.1}Ni_{43.1}In_{7.8}$	4.286 ± 0.001	5,78 ± 0.04	30.116 ± 0.001	Monoclinic *14 M*
$Mn_{49.3}Ni_{40.2}In_{10.5}$	6.0013 ± 0.0001	–	–	Cubic *L2$_1$*

The structure of the $Mn_{49.1}Ni_{43.1}In_{7.8}$ sample is monoclinic with $\beta = 93.61° \pm 0.03°$

3.3.2 Calorimetric Characterization of the Martensitic Transformation (DSC)

To determine the thermal analysis conditions, it is important to have the knowledge of crystal structure at room temperature. If the detected phase is cubic, the austenitic–martensitic transition must be found below room temperature. If the

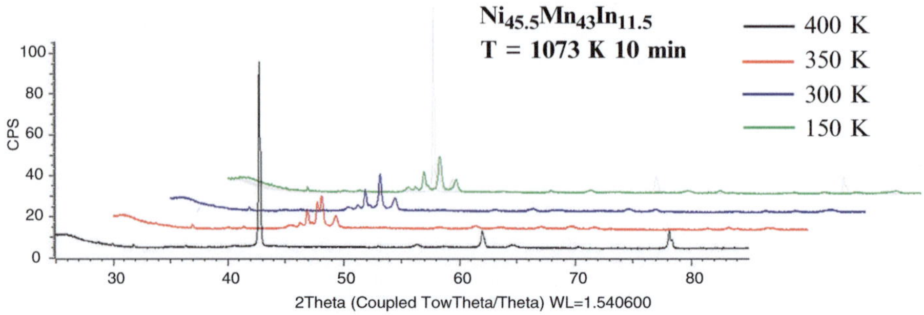

Fig. 3.3 X-ray diffraction patterns of $Ni_{45.5}Mn_{43}In_{11.5}$ alloy (after annealing at 1073 K during 10 min) at different temperatures

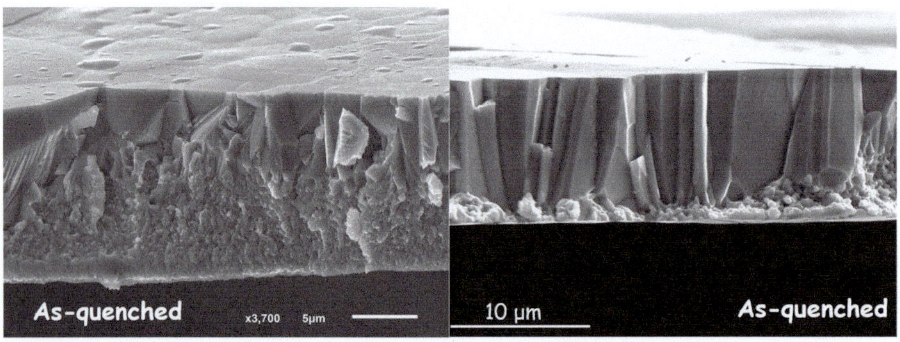

Fig. 3.4 Micrographs of alloys $Ni_{50}Mn_{35.5}In_{14.5}$ (*left*) and $Ni_{45.5}Mn_{43}In_{11.5}$ (*right*)

detected phase is orthorhombic, tetragonal, or monoclinic, the austenite–martensite transformation must be found by heating the alloy.

From XRD diffraction patterns, it is clear that DSC scan of alloy $In_{7.8}/In_{10.5}$ should be performed by heating/heating from RT in order to detect the martensite–austenite transformation (see Figs. 3.5 and 3.6). Cyclic experiments are done due to the hysteresis of the transformation. The characteristic transformation temperatures at which martensite start and finish, M_S and M_F, and austenite start and finish, A_S and A_F, are collected in Table 3.2. The hysteresis is due to the increase of the elastic and the surface energy during the martensite formation. Thus, the nucleation of the martensite implies supercooling. For it, it is determined the width of the hysteresis, ΔT, as the difference of temperatures corresponding to each peak position. In this study, this parameter range decreases as increasing In content (Table 3.2). The transition region can be also characterized by the martensite transformation temperature, T_0, as the temperature at which Gibbs energies of martensite and parent phases are equal.

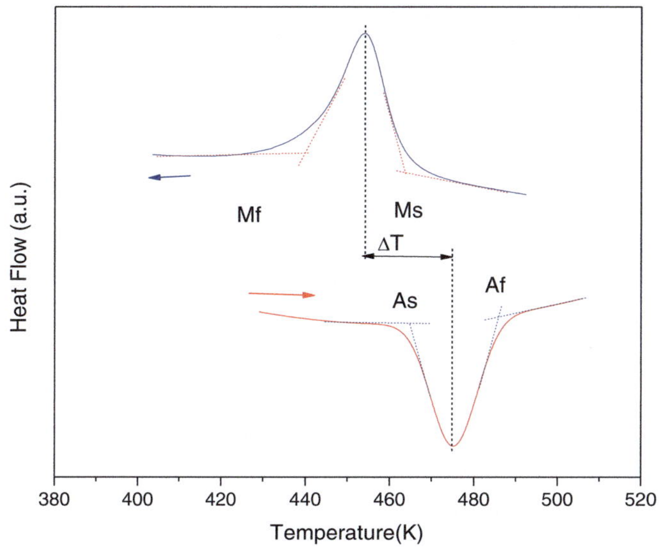

Fig. 3.5 DSC cyclic scan for the as-quenched alloy $Mn_{49.1}Ni_{43.1}In_{7.8}$ at a heating/cooling rate of 10 K/min. *Arrows* indicate cooling (*up*, austenite to martensite) and heating (*down*, martensite to austenite)

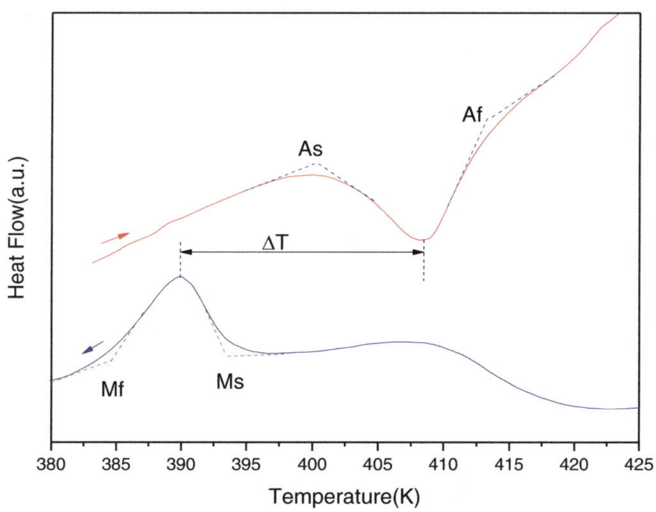

Fig. 3.6 DSC cyclic scan for the as-quenched alloy $Mn_{49.3}Ni_{40.2}In_{10.5}$ at a heating/cooling rate of 10 K/min. *Arrows* indicate cooling (*up*, austenite to martensite) and heating (*down*, martensite to austenite)

Table 3.2 The characteristic transformation temperatures of as-quenched samples: martensite start and finish, austenite start and finish, and thermal hysteresis

Alloy	M_S(K)	M_F(K)	A_S(K)	A_F(K)	ΔT(K)
$Mn_{49.1}Ni_{43.1}In_{7.8}$	462	437	464	485	22
$Mn_{49.3}Ni_{40.2}In_{10.5}$	288	223	230	301	18

Table 3.3 ΔH and ΔS of the Mn–Ni–In as-quenched ribbons as a function of composition

	ΔH(J/mol)		ΔS(J/mol/K)	
Alloy	Heating (endo)	Cooling (exo)	Heating	Cooling
$Mn_{49.1}Ni_{43.1}In_{7.8}$	668	635	1.48	1.42
$Mn_{49.3}Ni_{40.2}In_{10.5}$	52.8	64.1	0.12	0.16

$$T_0 = \frac{M_S + A_F}{2} \tag{3.1}$$

T_0 values are 463 and 259 K for $In_{7.8}$ and $In_{10.5}$ alloys, respectively.

The entropy (ΔS) and enthalpy (ΔH) changes of the structural transformations are calculated from calorimetric data using the equations:

$$\Delta H = \int_{T_i}^{T_f} \left[\frac{dQ}{dt} \left(\frac{dT}{dt} \right)^{-1} \right] dT \tag{3.2}$$

$$\Delta S = \int_{T_i}^{T_f} \left[\frac{1}{T} \frac{dQ}{dt} \left(\frac{dT}{dt} \right)^{-1} \right] dT \tag{3.3}$$

where T_i and T_f are initial and final temperatures of the transition. The entropy and enthalpy change values are included in Table 3.3.

The entropy (ΔS) and enthalpy (ΔH) changes of the structural transformations are calculated from calorimetric data. The values decrease as increasing (decreasing) the In content (ratio, *e*/*a*). The same trend has been reported [12]. It has been argued that the electron contribution to ΔS is negligible for these Heusler alloys and the character of the *e*/*a* dependence is related to a magnetic contribution that relies on the difference between the magnetic exchange below and above M_S.

One of the most suitable ferromagnetic shape memory alloy phase diagrams is the graphical representation of the martensite start temperature as a function of the *Z* element content or as a function of the average valence electron density, *e*/*a*.

A new approach to analyze the thermal response is to apply a magnetic field during differential scanning calorimetry scans [70].

For potential technological applications, for instance, magnetic refrigeration near room temperature, the controlled tuning of the transformation temperatures is an important factor. There are several ways to modify these temperatures: (a) by modifying elemental composition between X–Y–Z atoms [39], (b) by fourth element addition (interstitial or substitutional) [71, 72], (c) by modifying processing conditions [73], or (d) by controlled annealing [74].

3.3.3 Thermomagnetic Curves and Hysteresis Loops

The influence of composition on martensitic and magnetic transformation temperatures in Ni_2MnZ (Z = Ga, In, Sn) Heusler alloys not only depends on the valence electron concentration per atom (ratio, *e*/*a*), but also the Z element and the chemical disorder also play an important role [75]. In these intermetallic compounds, the magnetic moment is localized at the Mn sites, and the chemical disorder is responsible for the magnetic interaction (ferromagnetic vs antiferromagnetic) competition depending on lattice sites occupied by Mn atoms that interact via indirect Ruderman–Kittel–Kasuya–Yosida (RKKY) interaction. This competition leads to the characteristic drop of the magnetization at the martensitic transformation temperature, as can be observed by the magnetization vs temperature plots obtained for two applied magnetic field, *H*, values, in $Mn_{50}Ni_{40}In_{10}$ and $Ni_{50}Mn_{37}Sn_{13}$ Heusler alloy ribbons shown in Figs. 3.7a, 3.7b, and 3.8, respectively.

As-quenched alloys were single phase with ferromagnetic bcc $L2_1$ austenite as high-temperature parent phase as it has been previously mentioned. At low temperatures austenite transforms into a structurally modulated martensite with lattice symmetry 7-layer modulated orthorhombic for $Ni_{50}Mn_{37}Sn_{13}$ and 14-layer modulated monoclinic for $Mn_{50}Ni_{40}In_{10}$ [55]. Magnetization measurements were performed by means of a physical properties measurement system (PPMS-9 T) using the vibrating sample magnetometer module. *H* has been applied along two characteristic directions of the ribbon: the rolling direction (ribbon axis) for both ribbon samples and the perpendicular to the ribbon plane for $Ni_{50}Mn_{36}In_{14}$ ribbon. Zero-field-cooling (ZFC), field-cooling (FC), and field-heating (FH) magnetization

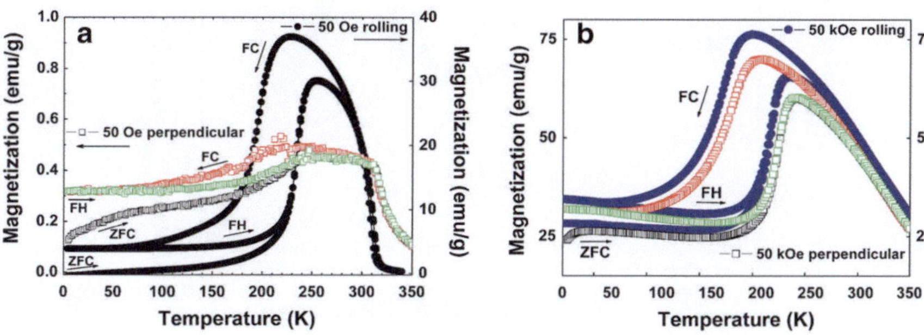

Fig. 3.7 As-quenched $Ni_{50}Mn_{40}In_{10}$ ribbon: ZFC, FC, and FH *M*(*T*) curves measured at a magnetic field (**a**) 50 Oe and (**b**) 50 kOe applied along the rolling direction (ribbon axis) and perpendicular to the ribbon plane

as a function of temperature curves $M(T)$ were recorded at 50 Oe and 50 kOe, with a temperature heating or cooling rate of 1.5 K/min. Curie temperature, T_C, was inferred from the minimum in the dM/dT vs T curve.

Figure 3.7a, b shows thermomagnetic $M(T)$ curves at 50 Oe and 50 kOe. The large abrupt change in magnetization defines two well distinct ferromagnetic regions [56]. Transition temperatures were $M_S = 213$ K, $M_F = 173$ K, $A_S = 222$ K, and $A_F = 243$ K. MT occurs in a broad temperature interval with a large thermal hysteresis that was not significantly dependent on the applied field ($\Delta T = 38$ K). Curie point of austenitic phase is 311 K. The heating and cooling thermal dependences of saturation magnetization measured at a high field value of 50 kOe show two well distinct ferromagnetic regions and illustrate the reversible and abrupt change in M_S as well as the field dependence of the martensitic transformation. A significant decrease in the characteristic temperatures of the structural transformation is observed, with estimated values: $\Delta M_S = -20$ K, $\Delta M_F = -40$ K, $\Delta A_S = -22$ K, and $\Delta A_F = -22$ K, indicating that the applied magnetic field stabilizes the austenitic phase in this ribbon alloy. When $H = 50$ Oe was applied perpendicular to the ribbon plane, the influence of the abovementioned crystallographic texture along with the demagnetizing factor can be observed in $M(T)$ curves (see Fig. 3.7a) and less significantly in Fig. 3.7b when $M(T)$ were recorded at 50 kOe. In fact, direct and inverse magnetocaloric properties of this $Mn_{50}Ni_{40}In_{10}$ Heusler alloy ribbons show a rather weak dependence on its crystallographic texture arising from the relatively small difference in magnetic anisotropy along both ribbon axis and perpendicular to the ribbon plane directions [37].

$M(T)$ curves recorded at 50 Oe and 50 kOe for $Ni_{50}Mn_{37}Sn_{13}$ alloy ribbon are plotted in Fig. 3.8. FC and FH curves show a similar behavior. Both of them reflect the occurrence of the phase transformation. At 50 Oe the magnetization is nearly invariable from the lowest temperature up to around 170 K. As the temperature

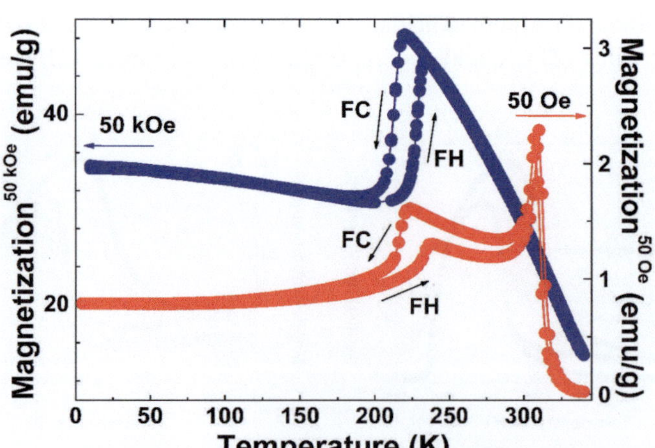

Fig. 3.8 FC and FH M(T) curves measured at 50 Oe and 50 kOe for the as-quenched $Ni_{50}Mn_{37}Sn_{13}$ ribbon

approaches to the transition region, both in FH and FC regimen, an abrupt change in the slope is observed; the transformation shows a thermal hysteresis of $\Delta T = 16$ K. The start and finish temperatures of the structural transformation were $M_S = 221$ K, $M_F = 209$ K, $A_S = 222$ K, and $A_F = 236$ K. They are well below to those reported for the bulk $Ni_{50}Mn_{37}Sn_{13}$ alloy [18, 29]. Above the MT, the magnetization shows a Hopkinson maximum followed by a sharp drop. Curie temperature of austenite is 313 K. The lower saturation magnetization of martensite leads to the characteristic steplike behavior of the $M(T)$ curves measured at 50 kOe. Transition temperatures were shifted to lower values by effect of the field, and the $M(T)^{50\ \mathrm{kOe}}$ curve emphasizes that $T_C{}^M$ is well above A_S. The latter explains the moderate jump in magnetization exhibited by the sample upon the martensitic transformation.

Magnetization isotherms $M(H)$ curves increasing and decreasing the magnetic field were measured following a well-defined thermal protocol: prior to measuring each curve, the sample was heated in zero field up to 350 K (being austenite in paramagnetic state), cooled down to 150 K (100 K in case of $Mn_{50}Ni_{40}In_{10}$ sample), and then heated up again to the measuring temperature. This protocol ensures that before heating the sample up to each measuring selected temperature, martensite is the existing phase.

Hysteresis loops at 270 and 150 K for $Mn_{50}Ni_{40}In_{10}$ ribbon are shown in Fig. 3.9. It can be appreciated the magnetically soft character of austenitic phase, related to its highly symmetric cubic structure and low magnetocrystalline anisotropy [55]. Coercive field for austenite was negligible within the uncertainty of the measurement (±2 Oe). In contrast, martensite shows a coercivity of 93 Oe, as can be seen in the inset in Fig. 3.9.

Hysteresis loops measured at 270 and 150 K for $Ni_{50}Mn_{37}Sn_{13}$ underline the ferromagnetic ordering of both phases; see Fig. 3.10a, where the inset zooms into

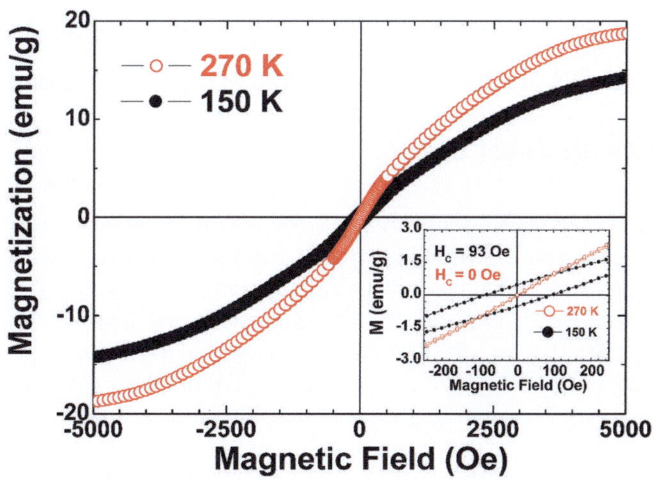

Fig. 3.9 Hysteresis loops measured at 270 and 150 K for as-quenched $Mn_{50}Ni_{40}In_{10}$ ribbon. *Inset*: zoom into the low-field region

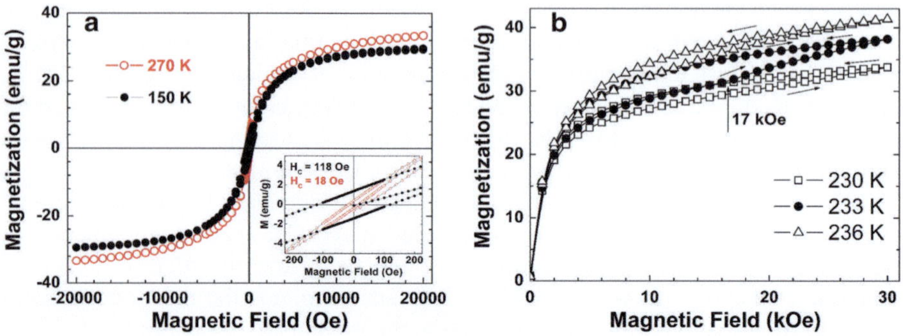

Fig. 3.10 As-quenched $Ni_{50}Mn_{37}Sn_{13}$ ribbon: (**a**) hysteresis loops measured at 270 and 150 K. Inset: zoom into the low-field region; (**b**) *M*(*H*) at selected temperatures in the interval where the magnetic field-induced reverse martensitic transformation occurs

the low-field region. Coercive fields of 18 and 118 Oe were measured. As expected, austenite phase is magnetically softer than martensite (at 270 K, *M*(*H*) reaches the saturation at a lower field value, a behavior that reflects the lower magnetic anisotropy of austenite). Magnetization isotherms measured increasing and decreasing the magnetic field (up to 30 kOe) at 230 K ($T > A_S = 222$ K), 233 K, and 236 K ($T = A_F$) are displayed in Fig. 3.10b. A noticeable hysteresis due to the magnetic field-induced reverse MT is found. In addition, at 233 K a well-defined discontinuity in the field-up *M*(*H*) curve is observed at a critical field of about 17 kOe. The magnetization value achieved at 30 kOe is below the magnetization of austenite, indicating that a higher field is required to complete the field-induced reverse transformation.

3.4 Martensitic Transformation and Magnetic Transitions

3.4.1 Magnetic Field Effect

Heusler alloys based on Ni–Mn–Z (Z = In, Sn, Ga, or in general a group IIIA–VA element) may undergo the following phase transitions: a magnetic second-order phase transition (SOPT) that takes place at temperature T_{CM} in the low-temperature martensitic phase (MP); followed by a magnetostructural first-order phase transition (FOPT), known as the reverse martensitic transformation, from the low-temperature MP to the high-temperature austenitic phase (AP) at the temperature T_{str}, involving a change in both structural (exhibiting a noticeable change in the lattice parameters) and magnetic (commonly a significant change in the magnetization *M*) properties of the material; and finally, a second magnetic SOPT in the high-temperature AP that occurs at the temperature T_C^A. The difference between the critical temperature values at which these phase transitions occur (T_C^A, T_{str} and T_C^M)

is an important feature exhibited by these compounds. It is important to note that T_C^M can be either lower or higher than T_C^A depending on the composition and the Z element [76].

Although the aforementioned change in magnetization is attributed to the martensitic transformation (MP↔AP) at temperature T_{str}, it takes place, as it is usually referred in the literature [77], over the field-dependent temperature range given by the characteristic temperatures of the martensitic transformation: (←) martensite start M_S and martensite finish M_F and (→) austenite start A_S and austenite finish A_F, defined as the extreme values of $M(T)$. This temperature range becomes field independent for a high enough magnetic field value characteristic of the material.

The interdependence of thermal and magnetic properties of magnetic materials, known as the magnetocaloric effect (MCE), is commonly studied by the application or removal of an external magnetic field (ΔH) and measured as an adiabatic change in temperature (ΔT_{ad}) or as an isothermal change in entropy (ΔS) [78]. Since the MCE exhibits maximum values of ΔT_{ad} or ΔS at the temperatures at which phase transitions take place, it is well justified that Ni–Mn–Z Heusler alloys have been recently presented as an important class of magnetocaloric materials for enhancing both the direct ($\Delta T_{ad} > 0$ or $\Delta S < 0$) and inverse ($\Delta T_{ad} < 0$ or $\Delta S > 0$) MCE [76, 79–82].

The magnetocaloric response of these materials can allow us to study the effect of the magnetic field on the martensitic transformation and magnetic transitions. When the martensitic FOPT takes place, the Clausius–Clapeyron equation tells us how the temperature T changes as a function of magnetic field H and vice versa, along the coexistence curve of the MP and AP phases (with AP being the high temperature phase) [83]:

$$\frac{\mathrm{d}T}{\mu_0 \mathrm{d}H} = -\frac{\Delta M}{\Delta S} = -\frac{\Delta M}{L} T \tag{3.4}$$

where μ_0 is magnetic permeability of vacuum and ΔM the change in magnetization in the martensitic transformation accompanied by a latent heat L.

The effect of the magnetic field on the martensitic transformation can be studied taking into account that L must always be positive (heat is absorbed) in transitions from MP to AP, and hence positive change in $\Delta M = M_{AP} - M_{MP} > 0$ (or $\Delta M < 0$) indicates that the phase transition temperature T_{str} shifts toward lower (higher) temperatures as the magnetic field increases.

For the magnetic SOPTs, in which $\Delta M = L = 0$, the indetermination in Eq. (3.4) can be solved by Ehrenfest's relations and the relation between ΔM and ΔS given by the Maxwell relation [84–87],

$$\Delta \mathrm{S}(T, H) = \mu_0 \int_0^{\mathrm{H}} \partial M_{T,H} \mathrm{d}H \tag{3.5}$$

For the studied martensitic FOPT (MP ↔ AP), the chemical potential $\mu_g(T, H)$, and hence the Gibbs free energy G, must be equal for both AP and MP phases. Since no restriction is placed on the derivatives $M = -\partial G_{\mathrm{H,T}}$ and $S = -\partial G_{\mathrm{T,H}}$, the discontinuity at the transition temperature $T_{\mathrm{str}}(H)$ in the derivatives $\partial G_{\mathrm{H,\ T}}$ and $\partial G_{\mathrm{T,\ H}}$ means that both M and S have different values in the two phases and the increments (discontinuities) $\Delta M = M_{\mathrm{AP}} - M_{\mathrm{MP}}$ and $\Delta S = S_{\mathrm{AP}} - S_{\mathrm{MP}}$ can be either positive or negative.

According to thermodynamics [84], the stability of the equilibrium state places certain conditions on the sign of the thermal, mechanical, and chemical response functions. The condition for thermal stability tells us that $C_{\mathrm{H}} > 0$ and $\partial S_{\mathrm{T,H}} > 0$, with C_{H} the heat capacity at constant field. If a small excess of heat is added to a volume element of the Heusler alloy ribbon when a FOPT takes place, then the temperature of the volume element must increase relative to its surroundings so that some of the heat will flow out again. This requires that the heat capacity C_{H} be positive. If the heat capacity was negative, the temperature would decrease and even more heat would flow in, thus leading to instability [84].

Although we cannot use entirely thermodynamics to describe the state of the system during the FOPT because the system is not being in equilibrium, we always can use it to relate the initial and final states as it occurs so often in physics (for instance, the Joule–Kelvin effect). However, as it has been mentioned in the literature [88, 89], the discontinuities in M and S can be totally rounded due to inhomogeneities, being $\partial M_{\mathrm{T,H}}$ and $\partial S_{\mathrm{T,H}}$ continuous in the temperature range where the FOPT takes place. Therefore, the discontinuities in M and in S transform into peaks and can be treated as continuous functions, resulting in the fact that in general Eq. (3.5) can be used both for FOPTs and SOPTs, thus validating the use of the Maxwell relation, which is the most commonly used technique in the study of the MCE and the effects of the magnetic field on the phase transitions.

3.4.2 Magnetic Phase Analysis in the Martensite and Austenite Phases

Within the Ehrenfest scheme [90], the overheating (undercooling) of the MP (AP) arises naturally in the FOPTs, but neither their temperature and field limits nor metastable nature is specified. The existence of these overheating and undercooling gives rise to the thermal and magnetic hysteresis of the thermodynamic variables S and M in the martensitic transformation.

In this transformation, a phase coexistence of a high-temperature metastable AP and a low-temperature equilibrium MP is also another consequence of the nature of this FOPT. This AP–MP phase coexistence that appears below the magnetostructural transition at T_{str} can be extended to the reverse martensitic transformation above T_{str}, as it is confirmed by numerous experiments [91]; thus, there exists phase coexistence of the MP having a secondary AP with spontaneous magnetization $M_{0\mathrm{A}}$ ($T < T_{\mathrm{str}}$) and in the same sense that of AP having a secondary MP with spontaneous magnetization $M_{0\mathrm{M}}$ ($T > T_{\mathrm{str}}$), in the temperature and field limits that should be specified. This MP–AP phase coexistence or metastability and the values of the spontaneous magnetization M_{0A} and M_{0M} must be taken into account in order to study the thermomagnetic properties of these Ni–Mn–Z compounds. It is important to note that M_{0M} can be either lower or higher than M_{0A}.

A study of the magnetic phases in the martensite and austenite can be done, for instance, in the intermetallic Ni–Mn–Sn compounds, due to its complex behavior [92], where the magnetic moments are located at the Mn sites with a ferromagnetic (FM) exchange interaction in both the MP and AP phases [93, 94]. On the other hand, as part of Sn content is replaced by Mn in $Ni_2Mn_{1+x}Sn_{1-x}$, the excess of Mn atoms occupies Sn sites reducing the Mn–Mn stoichiometric ($x = 0$) distance that, together with the change of the lattice parameters in the martensitic transformation, makes the antiferromagnetic (AFM) exchange arise in the MP, located in the environment of the Mn ions at both the Mn and Sn sites.

In this scenario, displaying the MP and AP (with higher crystalline symmetry and less effective anisotropy) phase coexistence with AFM and FM orderings, respectively, the Ni–Mn–Sn sample can be treated, in the simplest case, via two magnetic phases with opposite-aligned magnetic moments. Therefore, in addition to the magnetic anisotropies present in a single phase, a significant initial magnetic permeability μ variation can be introduced due to the induced anisotropy by the secondary phase with different magnetic order, i.e., secondary AP with FM ordering and spontaneous magnetization M_{0A} (higher than M_{0M}) in the MP ($T < T_{str}$) and vice versa, MP with AFM ordering and lower spontaneous magnetization value in the AP ($T > T_{str}$).

According to the so-called Hopkinson effect [95], the initial permeability μ of many ferromagnetic materials can increase with increasing temperature due to the influence of the effective magnetic anisotropy induced by the secondary phase and exhibiting a sharp peak below the critical temperature T_C followed by a drop to a small value; see Fig. 3.8. In others words, when the temperature approaches to T_C, the decrease rate of the induced effective anisotropy energy density (e_A) (which prevents the alignment of the magnetization with the field) is much faster than that magnetic energy density (e_H) (with the opposite result), and, therefore, the following magnetic orderings can take place: (1) ($\uparrow\downarrow$) opposite-aligned magnetic moments in both phases with net opposite component M ($\downarrow$) with H ($\uparrow$) (when $e_A > e_H$); (2) ($\uparrow\downarrow$) opposite-aligned magnetic moments with null magnetization M at the compensate temperature T_{cp} (when the opposite-aligned driving forces are compensated $e_A = e_H$); and (3) ($\uparrow\uparrow$) parallel alignment of magnetic moments with local maximum of the $M(T)$ curves or an abrupt change in its temperature derivative at the so-called spin reorientation temperature T_R (when $e_A = 0$).

For the sake of best understanding, if the magnetic field H is kept constant and the temperature increases, the opposite magnetic moments of the two phases (when $e_A > e_H$) start to rotate from their initial directions ($\uparrow\downarrow$) toward the direction of the applied magnetic field. As a result, at T_{cp} both magnetic moments can compensate each other ($\uparrow\downarrow$) [89, 96, 97], and at T_R they become parallel ($\uparrow\uparrow$) to the applied magnetic field when $e_A = 0$ (local maximum of the $M(T)$ curves for $T < T_{str}$, or an abrupt change in its temperature derivative for $T > T_{str}$, that specify the temperature metastability limits) [98].

Analogously, in terms of the magnetic energy, if the temperature T is kept fixed and the magnetic field is increased, the opposite magnetic moments of the two phases start to rotate from the easy magnetization axis toward the direction of the applied magnetic field. At the so-called compensate field H_{cp} (where $e_A = e_H$), these two sublattice magnetic moments cancel each other, and at a given critical magnetic field, called the spin reorientation magnetic field H_R, they become parallel aligned with the applied magnetic field direction when $e_A = 0$.

According to these arguments (similarly to diamagnetic phenomenology and, therefore, named in the literature as quasi-diamagnetic (QD)-like behavior [97]), a negative value (↑↓) of the initial permeability μ can be obtained due to a strong induced effective anisotropy ($e_A > e_H$) in the temperature range below the compensate point $T_{cp}(H)$ with $H < H_{cp}$. At higher magnetic field values, the magnetic moments of the two sublattices become parallel to the applied magnetic field, and, once it happens, the AFM order in the MP changes to FM one, the zero-field-cooling (ZFC) and field-heating (FH) protocol curves become coincident, the induced effective anisotropy energy density is null $e_A = 0$, and the Hopkinson effect and the metastability range disappear.

On increasing the temperature (→), Ni–Mn–Sn can present a two-step martensite transformation since the applied field (or e_H) at T_{str} is not strong enough to drive the phase transition (first step), and, therefore, an extra contribution of the thermal energy e_T is required (second step at higher temperature). As the magnetic field increases, e_H becomes high enough to drive a one-step martensite transformation at T_{str}.

3.4.3 Annealing Influence

To study the annealing effect on the martensitic transformation and magnetic transitions of the $Ni_{45.5}Mn_{43.0}In_{11.5}$ and $Ni_{50.0}Mn_{36.5}Sn_{13.5}$ as-quenched and annealed ribbons, we have performed thermomagnetic measurements, $M(T)$, with ZFC, FC, and FH routines in the temperature range of 50–400 K applying in the ribbon plane different magnetic fields up to 30 kOe by vibrating sample magnetometry (VSM, Versalab, QD). The heating and cooling rate was 5 K/min.

Figure 3.11 shows a comparison among the thermomagnetic measurements at two magnetic fields for as-quenched and annealed $Ni_{45.5}Mn_{43.0}In_{11.5}$ ribbons at 923, 973, 1023, and 1073 K during 10 min. As it can be observed, the martensitic transformation and reverse transformation temperatures of the Heusler alloy ribbon are significantly shifted to higher temperatures as a consequence of the annealing treatment temperature of the ribbon.

The characteristic temperatures of the martensitic structural transformation of the as-quenched and annealed $Ni_{45.5}Mn_{43.0}In_{11.5}$ ribbons are displayed in Table 3.4.

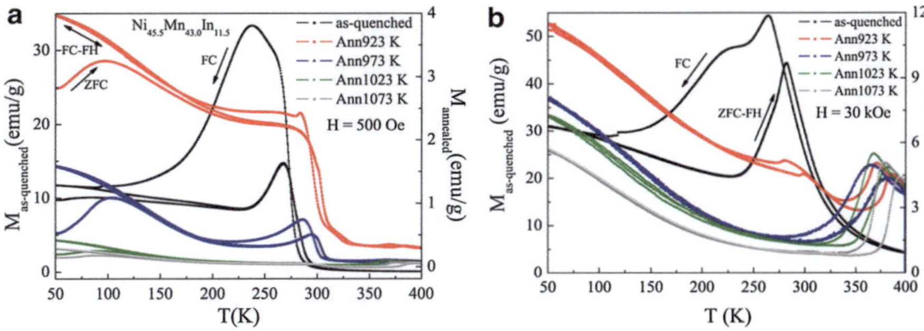

Fig. 3.11 Temperature dependence of magnetization of as-quenched and annealed $Ni_{45.5}Mn_{43.0}In_{11.5}$ ribbons at 923, 973, 1023, and 1073 K obtained at 500 Oe (**a**) and 30 kOe (**b**). The *arrows* are indicative of cooling and heating runs in ZFC, FC, and FH protocols

Table 3.4 Characteristic temperatures of the structural phase transition for $Ni_{45.5}Mn_{43.0}In_{11.5}$ as-quenched and ribbons annealed at 923, 973, 1023, and 1073 K determined by *M*(*T*) measurements at 500 Oe

$Ni_{45.5}Mn_{43.0}In_{11.5}$	M_S (K)	M_F (K)	A_S (K)	A_F (K)	ΔT (K)
As-quenched	239	166	237	268	29
Ann923 K	370	348	367	384	14
Ann973 K	364	311	330	380	16
Ann1023 K	367	336	351	383	16
Ann1073 K	376	359	376	391	15

In Table 3.4 it can be seen that the MT appears below room temperature (300 K) for the as-quenched ribbon, while for annealed ribbons, MT takes place above RT. Hence, at RT the austenite is the main phase observed in the as-quenched sample. However, for annealed ribbons, martensite is the main phase observed in the respective magnetic measurements.

A drastic decreasing of the magnetization comparing with the as-quenched ribbon can also been spotted in Fig. 3.11. This behavior could be ascribed to the annealing influence characterized by an increasing in the grain size and crystal defects, along with the relaxation of internal stresses induced during the ribbon quenching process. The sharp increase of *M* at the Curie temperature, T_C^A, of austenite and the sharp drop of *M* after the martensitic transformation in the as-quenched ribbon are absent in treated samples. For all ribbons, the *M*(*T*) behavior can be explained by considering the variation of the magnetic exchange parameters across the martensitic transformation from the paramagnetic cubic high-temperature austenite to the low-temperature ferromagnetic *10 M* monoclinic martensite [99]. In the martensitic phase, the Mn-excess atoms occupying the In sublattice sites interact antiferromagnetically with the Mn atoms on the Mn sublattice sites. Then, an enhancement of the AFM interactions occurs at low temperatures, being also responsible for the drop in the magnetization during the martensitic transformation [77, 100].

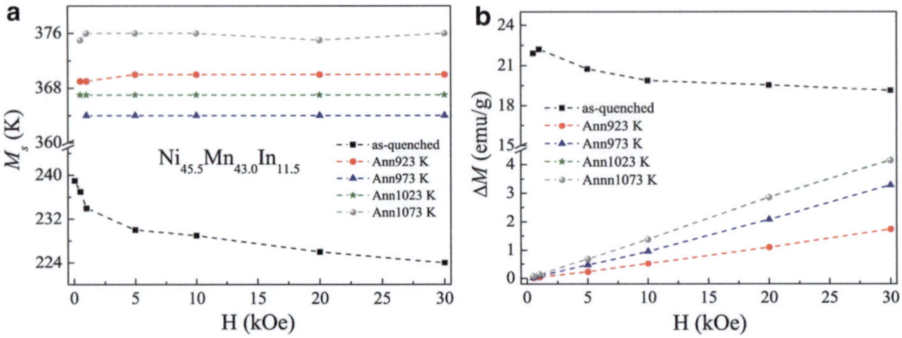

Fig. 3.12 M_S temperature as a function of external cooling field (**a**) and field dependence of $\Delta M(H) = M_{AP}(H) - M_{MP}(H)$ (**b**) for as-quenched and annealed ribbons. *Symbols* denote experimental data and *lines* are guides for the eye

It is worthwhile to state that for as-quenched and annealed $Ni_{45.5}Mn_{43.0}In_{11.5}$ ribbons, the splitting between ZFC and FC curves is shifted to lower temperatures with increasing the magnetic field. A similar tendency has been observed, for instance, in bulk polycrystalline $Ni_{49.5}Mn_{43.5}In_{16.0}$ alloy being attributed to a few ferromagnetic components embedded in an antiferromagnetic matrix [100].

Now, we analyze the field dependence of the martensite start temperature, M_S, and the magnetization difference between austenite phase at martensite start temperature and martensitic phase at martensite finish temperature for the as-quenched and annealed ribbons. The results show that as the applied magnetic field increases to 30 kOe, M_S keeps unchanged in annealed ribbons, while in the as-quenched one, it is shifted to lower temperatures; see Fig. 3.12a. On the other hand, it is detected that the magnetization difference between austenite and martensite rises as the applied magnetic field increases in the annealing ribbons, while ΔM remains almost unchanged in the as-quenched sample; see Fig. 3.12b.

Subsequently, we study the annealing influence on Heusler $Ni_{50.0}Mn_{36.5}Sn_{13.5}$ ribbon. For this, a piece of the sample was kept as reference (as-quenched ribbon), and some pieces of the same ribbon were annealed for 10 min at 923 and 1073 K. Annealing was performed in vacuum quartz tubes, and tantalum foil was used for wrapping each sample before introducing it in the quartz container for avoiding Si contamination. The quartz tubes, containing the ribbons to be treated, were introduced in a furnace after reaching the appropriate annealing temperature. Ribbon flakes were afterward quenched in ice water. A comparison among the thermomagnetic measurements at 400 Oe and 30 kOe for the three samples is shown in Fig. 3.13a, b.

After annealing, a shift of the martensitic transformation toward higher temperatures, which signifies the stabilization of the martensitic phase at the expense of the parent one, is observed. The corresponding MT characteristic temperatures for the annealed ribbons along with the as-quenched one are collected in Table 3.5.

Furthermore, in comparison with both as-quenched and Ann923 K samples that show the same magnetic transition temperature T_C^A at the austenitic phase, the

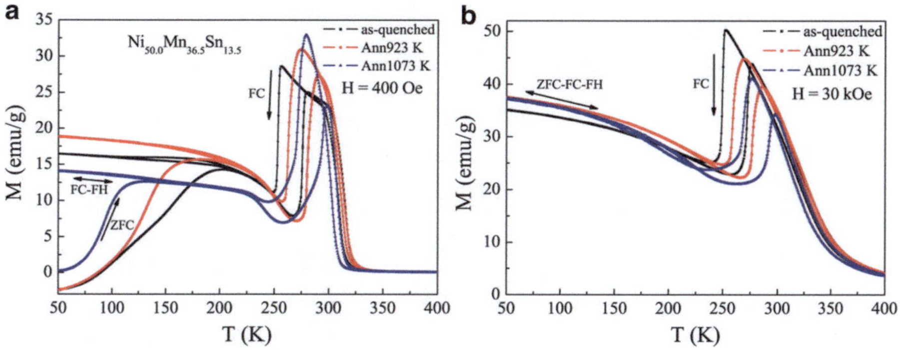

Fig. 3.13 Magnetization temperature dependence *M(T)* obtained at a field of 400 Oe (**a**) and 30 kOe (**b**) for as-quenched and annealed $Ni_{50.0}Mn_{36.5}Sn_{13.5}$ alloy ribbons. *Arrows* are indicative for cooling and heating runs in ZFC, FC, and FH protocols

Table 3.5 Martensitic transformation temperatures, thermal hysteresis $\Delta T = A_F - M_S$, and Curie point for austenite for as-quenched and annealed $Ni_{50.0}Mn_{36.5}Sn_{13}.5$ ribbons at 923 and 1073 K determined from *M(T)* measurements at 50 Oe

$Ni_{50.0}Mn_{36.5}Sn_{13.5}$	M_S (K)	M_F (K)	A_S (K)	A_F (K)	ΔT (K)	$T_C{}^A$ (K)
As-quenched	258	250	270	275	17	312
Ann923 K	271	259	277	290	19	312
Ann1073 K	278	266	284	299	21	306

corresponding $T_C{}^A$ to the Ann1073 K sample decreases. This can be explained by the shift in the compositions, since the Ann1073 K ribbon has a higher Mn content than the other ones. For Ni–Mn–Sn Heusler alloys, it is well-known that $T_C{}^A$ is mainly determined by the ferromagnetic Mn–Mn interaction strength, while the Mn-excess content leads to an antiferromagnetic coupling with the Mn located at the Ni/Sn sites, which in turn reduces the strength of the FM interactions. As a result, the $T_C{}^A$ decreases with increasing Mn content [58]. Moreover, in Fig. 3.13a, we observe that the annealed ribbons display a splitting between the ZFC and FC curves, which suggests an FM–AFM coupling at low temperatures [93]. The quasi-diamagnetic (QD)-like behavior [97], mentioned in Sect. 3.4.2, is also clearly observed in Fig. 3.13a for as-quenched and Ann923 K samples, as negative *M* values in ZFC curve at low temperature.

Moreover, studying the field dependence of the martensite start temperature for the as-quenched and annealed ribbons (see Fig. 3.14), we note that M_S presents a monotonous decrease as increasing the magnetic field for the three ribbons. The shift of M_S is estimated as $dM_S/dH \approx -2.3\ KT^{-1}$ for as-quenched ribbon and $-2.7\ KT^{-1}$ for annealed ones.

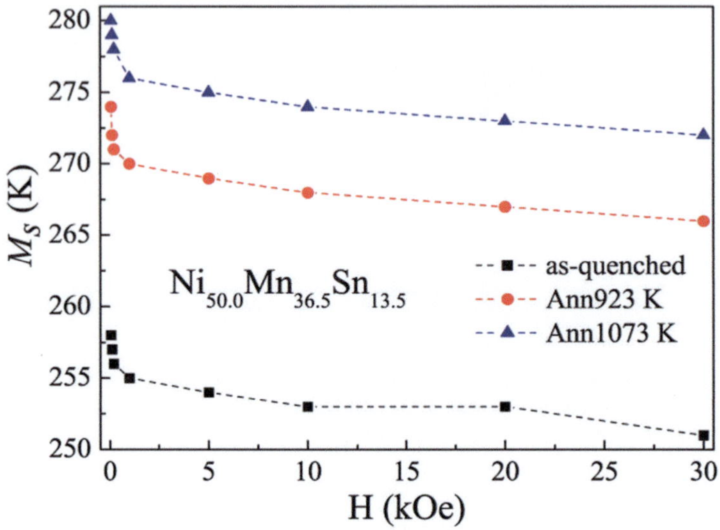

Fig. 3.14 M_S temperature as a function of external cooling field for as-quenched and annealed $Ni_{50.0}Mn_{36.5}Sn_{13.5}$ ribbons. *Symbols* denote experimental data and *lines* are guides for the eyes

3.5 Exchange Bias Effect: Annealing Influence

The observation of exchange bias in Ni–Mn-based martensitic Heusler alloys [101–105] provided strong evidence for the presence of AFM interactions in these systems, since such interactions over large spatial extensions coexisting with the FM matrix are required for the exchange bias effect to be observed [106]. In the off-stoichiometric Heusler alloys, the appearance of the exchange bias effect on the hysteresis loops at low temperatures is well reported, and this effect is assigned to the FM/AFM exchange interactions. The EB effect can be interpreted in terms of magnetic phase separation into distinct FM and AFM regions in the martensite phase, as FM clusters embedded in an AFM matrix [104]. This interaction between FM and AFM regions is modified after annealing by the decrease of the interatomic distances, since the cell parameters of the as-quenched ribbon crystalline structure at temperatures below M_S are larger than the respective for the annealed samples. The general process to evaluate the EB is field cooling the system from a high temperature (over T_N and T_C) for the reconfiguration of the FM spins at the interface between different magnetic phases [107].

To explore the low-temperature magnetism of the $Ni_{45.5}Mn_{43.0}In_{11.5}$ and $Ni_{50.0}Mn_{36.5}Sn_{13.5}$ as-quenched and annealed ribbons, we have performed hysteresis loops, $M(H)$, in a temperature range from 5 to 100 K after field cooling the samples with an applied field of +10 kOe from 375 K. A maximum value of ±30 kOe has been applied.

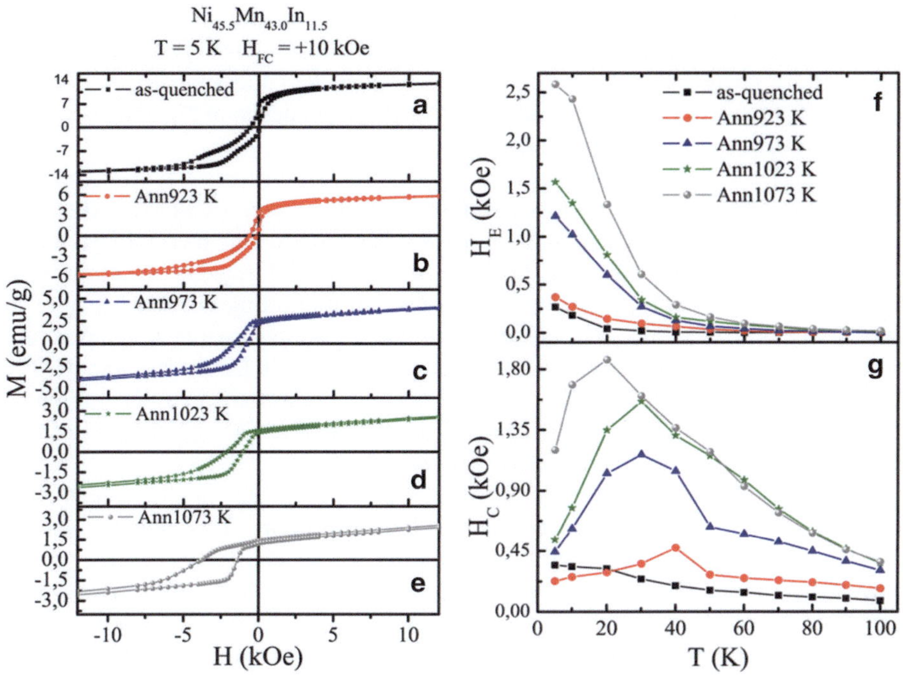

Fig. 3.15 Isothermal magnetization hysteresis loops for as-quenched and for annealed $Ni_{45.5}Mn_{43.0}In_{11.5}$ alloy ribbons performed at 5 K after FC (H_{FC} = +10 kOe) from 375 K are represented in (**a**)–(**e**). In (**f**) and (**g**), the evolution of the exchange bias (H_E) and coercivity (H_C) field with respect to the temperature for as-quenched and annealed ribbons are plotted. *Lines* are guides to the eyes

First, the exchange bias effect on $Ni_{45.5}Mn_{43.0}In_{11.5}$ Heusler ribbons is studied; the hysteresis loops at 5 K for as-quenched and for the annealing samples at 923, 973, 1023, and 1073 K during 10 min are represented in Fig. 3.15a–e. Though the data has been collected up to ±30 kOe, for better clarity it is shown only in the range of −15 to +15 kOe.

In Fig. 3.15a–e, a shift toward negative field axis which clearly shows the existence of EB effect in as-quenched and annealed $Ni_{45.5}Mn_{43.0}In_{11.5}$ ribbons is displayed. Measurement of shift corresponds to the exchange bias field, H_E, in all $Ni_{45.5}Mn_{43.0}In_{11.5}$ ribbons. The value of H_E and coercivity, H_C, are defined as $H_E = -(H_1 + H_2)/2$ and $H_C = -(H_1 - H_2)/2$, respectively, where H_1 and H_2 are left and right fields at which the magnetization equals to zero. Up to now, the highest value of H_E reported for any Heusler alloy system is 2230 Oe at 10 K in Ni–Mn–Sn by Wang et al. [108]. However, in quaternary Heusler $Ni_{55}Mn_{19}Al_{24}Si_2$ alloy, Singh et al. [109] have reported a value of $H_E \approx 2520$ Oe at 2 K. In the present work, we reveal our results of $H_E \approx 2584$ Oe at 5 K for $Ni_{45.5}Mn_{43.0}In_{11.5}$ Heusler ribbons annealed at 1073 K. We have observed that H_E is enhanced after annealing off-stoichiometric Heusler $Ni_{45.5}Mn_{43.0}In_{11.5}$ alloy ribbons.

Now, the dependence of both H_E and H_C with the temperature, for that we have calculated the corresponding values from the hysteresis loops performed in the temperature range from 5 to 100 K for the as-quenched and annealed ribbons, is evaluated. Figure 3.15f, g shows the temperature dependence of H_E and H_C for all $Ni_{45.5}Mn_{43.0}In_{11.5}$ ribbon samples.

In Fig. 3.15f, it is observed that for all samples, H_E decreases with increasing temperature and gradually disappears around a temperature, usually referred to as the blocking temperature, T_B. With increasing temperature, the thermal fluctuation reduces the exchange coupling between the AFM–FM regions resulting in a decrease in the H_E value. As abovementioned, an interesting observed feature in this figure is that at 5 K, H_E enhances, as the annealing treatment temperature increases, from 270 Oe for the as-quenched ribbon to 2584 Oe for the Ann1073 K sample. In addition, the EB effect in the annealed ribbons exits in a wider temperature range up to 80 K, probably due to the modification of FM and AFM exchange interaction parameters after annealing processes.

Studying the coercivity in Fig. 3.15g, it can be seen that for the as-quenched ribbon, H_C decreases with the increment of the temperature. This behavior should be noted if the martensitic phase were dominated by the AFM order, and the decreasing of both H_E and H_C would be due to thermal instability of the nanoscopic FM clusters but not to the AFM order parameter [110]. This situation results similar to FM/AFM bilayers with Curie and Néel temperatures such that $T_C < T_N$ [111]. In addition, if the thermal stability of the FM component is low relative to the AFM one, lower FM volume and/or magnetocrystalline anisotropy, the magnetization reversal in the FM would be easier and would not imply a significant rearrangement of the AFM spin structure, and thus H_C would decrease around T_N [112]. This behavior is consistent with a martensitic phase where nanoscopic FM clusters near the thermal stability would be embedded in a long-range order AFM matrix [110]. On the other hand, for the annealed ribbons, the coercivity (see Fig. 3.15g) increases in the beginning and then starts decreasing as H_E becomes near zero. The peak that appears for the H_C as a function of the temperature is situated at T_B, and as the annealing temperature increases, it is shifted to lower temperatures. The fact that H_C exhibits an increasing with T has been previously observed in several off-stoichiometric Heusler $Ni_{50}Mn_{25+y}Z_{25-y}$ (Z = Sn, Sb, In) alloys [102, 104, 113]. This coercivity increase is common in exchange bias systems and occurs at FM/AFM interfaces existent between regions with Curie and Néel temperatures such as $T_N < T_C$ and T_B next to T_N [107]. Thus, as T_N is approached from lower temperature, the decreasing anisotropy in the AFM region originates an increasingly significant modification of the AFM interfacial spin structure induced by the magnetization reversal in the FM region, leading to a significant energy loss in the AFM region and, as a consequence an enhancement on H_C when the annealing temperature increases [107]. Furthermore, it can be observed that the coercive field increments with increasing the annealing temperature, i.e., at 20 K the H_C varies from 320 Oe for the as-quenched sample to 1870 Oe for the Ann1073 ribbon.

Subsequently, we explore the low-temperature magnetic behavior of the $Ni_{50.0}Mn_{36.5}Sn_{13.5}$ for as-quenched and samples annealed at 923 and 1073 K during

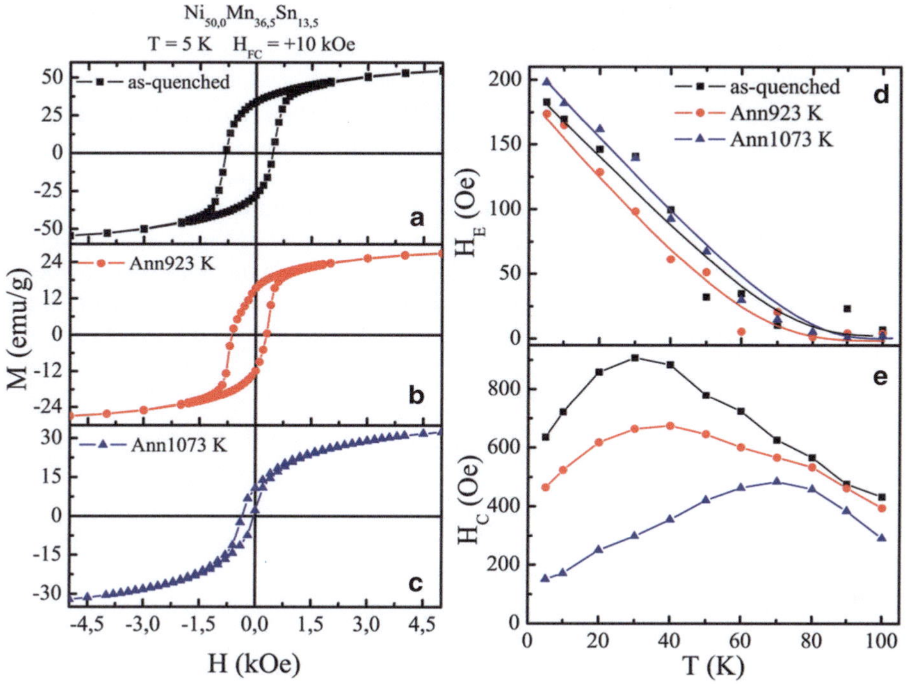

Fig. 3.16 Isothermal magnetization hysteresis loops for as-quenched and for annealed $Ni_{50.0}Mn_{36.5}Sn_{13.5}$ alloy ribbons performed at 5 K after FC (H_{FC} = +10 kOe) from 375 K are represented in (**a**)–(**c**). In (**d**) and (**e**), the evolution of the exchange bias (H_E) and coercivity (H_C) field with respect to the temperature for as-quenched and annealed ribbons are plotted. *Lines* are guides to the eyes

10 min. We have performed FC hysteresis loops' measurements at low temperatures within the range from 5 to 100 K. In Fig. 3.16a–c, hysteresis loops at 5 K measured from −5 to +5 kOe for as-quenched and annealed ribbons are represented.

In Fig. 3.16a–c, it is clearly shown that magnetization hysteresis loops significantly shift to the negative field values, which evidences that the exchange bias effect indeed exists in the as-quenched and annealed $Ni_{50.0}Mn_{36.5}Sn_{13.5}$ Heusler alloy ribbons. The temperature dependence of H_E and the coercivity, H_C, evaluated from hysteresis loops at various temperatures for each ribbon, is shown in Fig. 3.16d–e. In fact, H_E decreases with increasing the temperature and almost vanishes at 80 K for the three samples, which is referred like blocking temperature T_B. Above this T_B the AFM regions cannot pin the FM regions, and magnetic moments in FM regions start to rotate with the applied field [107]; it means the AFM–FM coupling is weakened. At 5 K H_E is around 183, 174, and 198 Oe for the as-quenched $Ni_{50.0}Mn_{36.5}Sn_{13.5}$ ribbon, the Ann923 K and the Ann1073 K ones, respectively. These values are in good agreement with other Ni–Mn–Sn systems of similar composition [101, 114] and reveal that H_E remains unchanged by the annealing treatment, unlike of $Ni_{45.5}Mn_{43.0}In_{11.5}$ ribbons.

Concerning the behavior of H_C, it initially increases with temperature and then decreases for the three $Ni_{50.0}Mn_{36.5}Sn_{13.5}$ Heusler ribbons. This occurs because as the anisotropy of AFM regions decreases with increasing temperature, then the rotation of FM spin can drag more and more AFM spin; thus, H_C becomes maximum, and the EB effect is very small as temperature approaches to T_B [115]. Furthermore, unlike of $Ni_{45.5}Mn_{43.0}In_{11.5}$ ribbons, in Fig. 3.16e, it can be observed that the coercive field decreases when increasing the annealing temperature, i.e., at 20 K H_C varies from 858 Oe for the as-quenched sample to 250 Oe for the Ann1073 ribbon.

3.6 Conclusions

Results here described prove that rapid solidification by melt spinning is a suitable route to produce directly from the melt single-phase ribbons of Heusler alloys in the Ni–Mn–In and Ni–Mn–Sn systems. These systems-based shape memory alloys are already recognized as smart materials with useful applications based on their functional properties. A complete understanding and tuning of the magnetic field-induced martensitic transformation in these Ni–Mn–In and Ni–Mn–Sn ribbons may even turn them out to be also useful magnetocaloric and magnetoresistive materials and not only magnetic shape memory materials.

Acknowledgments Financial support under Spanish MINECO research projects MAT2013-47231-C2-1-P, MAT2013-47231-C2-2-P, and MAT2013-48054-C2-2-R is acknowledged. Scientific support from the University of Oviedo SCT is also recognized.

References

1. de Groot, R.A., Mueller, F.M., van Engen, P.G., Buschow, K.H.J.: New class of materials: half-metallic ferromagnets. Phys. Rev. Lett. **50**, 2024–2027 (1983)
2. de Groot, R.A., Mueller, F.M., van Engen, P.G., Buschow, K.H.J.: Half-metallic ferromagnets and their magneto-optical properties. J. Appl. Phys. **55**, 2151 (1984)
3. Kübler, J.: First principle theory of metallic magnetism. Physica B **127**, 257–263 (1984)
4. Fang, C.M., de Wijs, G.A., de Groot, R.A.: Spin-polarization in half-metals. J. Appl. Phys. **91**, 8340 (2002)
5. Hanssen, K.E.H.M., Mijnarends, P.E.: Positron-annihilation study of the half-metallic ferromagnet NiMnSb: theory. Phys. Rev. B **34**, 5009 (1986)
6. Galanakis, I., Ostanin, S., Alouani, M., Dreysse, H., Wills, J.M.: Ab initio ground state and L2,3 x-ray magnetic circular dichroism of Mn-based Heusler alloys. Phys. Rev. B **61**, 4093 (2000)

7. Kang, J.-S., Park, J.-G., Olson, C.G., Youn, S.J., Min, B.I.: Valence band and Sb 4d core level photoemission of the XMnSb-type Heusler compounds (X = Pt, Pd, Ni). J. Phys. Condens. Matter **7**, 3789 (1995)
8. Kang, J.-S., Hong, S.H., Jung, S.W., Lee, Y.P., Park, J.-G., Olson, C.G., Youn, S.J., Min, B.I.: Electronic structures of the half-metallic Heusler alloys: NiMnSb and PtMnSb. Solid State Commun. **88**, 635–657 (1993)
9. Galanakis, I.: Surface properties of the half-and full-Heusler alloys. J. Phys. Condens. Matter **14**, 6329 (2002)
10. Galanakis, I., Dederichs, P.H., Papanikolaou, N.: Slater-Pauling behavior and origin of the half-metallicity of the full-Heusler alloys. Phys. Rev. B **66**, 174429 (2002)
11. Galanakis, I., Dederichs, P.H., Papanikolaou, N.: Origin and properties of the gap in the half-ferromagnetic Heusler alloys. Phys. Rev. B **66**, 134428 (2002)
12. Liu, B.G.: Robust half-metallic ferromagnetism in zinc-blende CrSb. Phys. Rev. B. **67**, 172411 (2003)
13. Xie, W.H., Xu, Y.Q., Liu, B.G., Pettifor, D.G.: Half-metallic ferromagnetism and structural stability of zincblende phases of the transition-metal chalcogenides. Phys. Rev. Lett. **91**, 037204 (2003)
14. Galanakis, I.: Surface half-metallicity of CrAs in the zinc-blende structure. Phys. Rev. B **66**, 012406 (2002)
15. Pickett, W.E., Singh, D.J.: Electronic structure and half-metallic transport in the La1 − xCaxMnO3 system. Phys. Rev. B **53**, 1146 (1996)
16. Singh, D.J., Pickett, W.E.: Pseudogaps: Jahn-Teller distortions, and magnetic order in manganite perovskites. Phys. Rev. B **57**, 88 (1998)
17. Kobayashi, K.-I., Kimura, T., Sawada, H., Terakura, K., Tokura, Y.: Room-temperature magnetoresistance in an oxide material with an ordered double-perovskite structure. Nature **395**, 677–680 (1998)
18. Schwarz, K.: CrO2 predicted as a half-metallic ferromagnet. J. Phys. F. Met. Phys. **16**, L211 (1986)
19. van Lueken, H., de Groot, R.A.: Electronic structure of the chromium dioxide (001) surface. Phys. Rev. B **51**, 7176 (1995)
20. Korotin, M.A., Anisimov, V.I., Khomskii, D.I., Sawatzky, G.A.: CrO2: a self-doped double exchange ferromagnet. Phys. Rev. Lett. **80**, 4305 (1998)
21. Lewis, S.P., Allen, P.B., Sasaki, T.: Band structure and transport properties of CrO2. Phys. Rev. B **55**, 10253 (1997)
22. de Groot, R.A., Buschow, K.H.J.: Recent developments in half-metallic magnetism. J. Magn. Magn. Mater. **54–57**, 1377 (1986)
23. Penicaud, M., Silberchoit, B., Sommers, C.B., Kübler, J.: Calculated electronic band structure and magnetic moments of ferrites. J. Magn. Magn. Mater. **103**, 212–220 (1992)
24. Irkhin, V.Y., Katsnelson, M.I.: Half-metallic ferromagnets. Phys. Usp. **37**, 659 (1994)
25. Mazin, I.I.: Robust half metallicity in FexCo1 − xS2. Appl. Phys. Lett. **77**, 3000 (2000)
26. Webster, P.J., Ziebeck, K.R.A.: Alloys and compounds of d-elements with main group elements. Part 2. In: Wijn, H.R.J. (ed.) Landolt-Börnstein, New Series, Group III (Vol 19) Pt.c, pp. 75–184. Springer, Berlin (1988)
27. Ziebeck, K.R.A., Neumann, K.U.: Magnetic properties of metals. In: Wijn, H.R.J. (ed.) Landolt-Börnstein, New Series, Group III (vol 32/c), pp. 64–414. Springer, Berlin (2001)
28. Stadler, S., Khan, M., Mitchell, J., Ali, N., Gomes, A.M., Dubenko, I., Takeuchi, A.Y., Guimarães, A.P.: Magnetocaloric properties of Ni2Mn1 − xCuxGa. Appl. Phys. Lett. **88**, 192511 (2006)
29. Krenke, T., Duman, E.M., Wassermann, E.F., Moya, X., Mañosa, L., Planes, A.: Inverse magnetocaloric effect in ferromagnetic Ni-Mn-Sn alloys. Nat. Mater. **4**, 450–454 (2005)
30. Han, Z.D., Wang, D.H., Zhang, C.L., Xuan, H.C., Gu, B.X., Du, Y.W.: Low-field inverse magnetocaloric effect in Ni50 − xMn39 + xSn11 Heusler alloys. Appl. Phys. Lett. **90**, 042507 (2007)

31. Sánchez Llamazares, Y.W., Hernando, B., Prida, V.M., García, C., González, J., Varga, R., Ross, C.A.: Magnetic field influence on the structural transformation in ferromagnetic shape memory alloy Mn50Ni40In10 melt spun ribbons. J. Appl. Phys. **105**, 07A945 (2009)
32. Wang, B.M., Wang, L., Liu, Y., Zhao, B.C., Zhao, Y., Yang, Y., Zhang, H.: Strong thermal-history-dependent magnetoresistance behavior in Ni49.5Mn34.5In16. J. Appl. Phys. **106**, 063909 (2009)
33. Ma, S.C., Xuan, H.C., Zhang, C.L., Wang, L.Y., Cao, Q.Q., Wang, D.H., Du, Y.W.: Investigation of the intermediate phase and magnetocaloric properties in high-pressure annealing Ni–Mn–Co–Sn alloy. Appl. Phys. Lett. **97**, 052506 (2010)
34. Umetsu, R.Y., Ito, K., Ito, W., Koyama, K., Kanomata, T., Ishida, K., Kainuma, R.: Kinetic arrest behavior in martensitic transformation of NiCoMnSn metamagnetic shape memory alloy. J. Alloys Compd. **509**, 1389–1393 (2011)
35. Han, Z.D., Wang, D.H., Zhang, C.L., Xuan, H.C., Zhang, J.R., Gu, B.X., Du, Y.W.: The martensitic transformation and the magnetocaloric effect in Ni50 − xMn38 + xIn12 alloys. Solid State Commun. **146**, 124–127 (2008)
36. Ito, W., Imano, Y., Kainuma, R., Sutou, Y., Oikawa, K., Ishida, K.: Martensitic and magnetic transformation behaviors in Heusler-type NiMnIn and NiCoMnIn metamagnetic shape memory alloys. Metall. Mater. Trans. A **38**, 759–766 (2007)
37. Hernando, B., Sánchez Llamazares, J.L., Prida, V.M., Baldomir, D., Serantes, D., Ilyn, M., González, J.: Magnetocaloric effect in preferentially textured Mn50Ni40In10 melt spun ribbons. Appl. Phys. Lett. **94**, 222502 (2009)
38. Yu, S.Y., Liu, Z.H., Liu, G.D., Chen, J.L., Cao, Z.X., Wu, G.H., Zhang, B., Zhang, X.X.: Large magnetoresistance in single-crystalline Ni50 Mn50 − xInx alloys (x = 14 − 16) upon martensitic transformation. Appl. Phys. Lett. **89**, 162503 (2006)
39. Coll, R., Escoda, L., Saurina, L., Sánchez-Llamazares, J.L., Hernando, B., Suñol, J.J.: Martensitic transformation in Mn-Ni-Sn Heusler alloys. J. Therm. Anal. Calorim. **99**, 905–909 (2010)
40. Khovaylo, V.V., Rodionova, V.V., Shevyrtalov, S.N., Novosad, V.: Magnetocaloric effect in "reduced" dimensions: thin films, ribbons, and microwires of Heusler alloys and related compounds. Phys. Status Solidi B **251**, 2104–2113 (2014)
41. Schlagel, D.L., McCallum, R.W., Lograsso, T.A.: Influence of solidification structure on the magnetic properties of Ni-Mn-Sn Heusler alloys. J. Alloys Compd. **463**, 38–46 (2008)
42. Moya, X., Mañosa, L., Planes, A., Krenke, T., Acet, M., Wassermann, E.F.: Martensitic transition and magnetic properties in Ni-Mn-X alloys. Mater. Sci. Eng. A **438–440**, 911–915 (2006)
43. Krenke, T., Duman, E., Acet, M., Moya, X., Mañosa, L., Planes, A.: Effect of Co and Fe on the inverse magnetocaloric properties of Ni-Mn-Sn. J. Appl. Phys. **102**, 033903 (2007)
44. Chen, L., Hu, F.X., Wang, J., Shen, J., Sun, J.R., Shen, B.G., Yin, J.H., Pan, L.Q., Huang, Q. Z.: Effect of post-annealing on martensitic transformation and magnetocaloric effect in Ni45Co5Mn36.7In13.3 alloys. J. Appl. Phys. **109**, 07A939 (2011)
45. Sánchez-Alarcos, V., Recarte, V., Pérez-Landazábal, J.L., Gómez-Polo, C., Rodríguez-Velamazán, J.A.: Role of magnetism on martensitic transformation in Ni-Mn based magnetic shape memory alloys. Acta Mater. **60**, 459–468 (2012)
46. Ishikawa, H., Umetsu, R.Y., Kobayashi, K., Fujita, A., Kainuma, R., Ishida, K.: Atomic ordering and magnetic properties in Ni2Mn(GaxAl1-x) Heusler alloys. Acta Mater. **56**, 4789–4797 (2008)
47. Ito, W., Nagasato, M., Umetsu, R.Y., Kainuma, R., Kanomata, T., Ishida, K.: Magnetic field-induced reverse transformation in B2-type NiCoMnAl shape memory alloys. Appl. Phys. Lett. **93**, 232503 (2008)
48. Recarte, V., Pérez-Landazábal, J.L., Sánchez-Alarcos, V.: Dependence of the relative stability between austenite and martensite phases on the atomic order in a Ni-Mn-In metamagnetic shape memory alloy. J. Alloys Compd. **536S**, S308–S311 (2012)

49. Zheng, H.X., Xia, M.X., Liu, J., Huang, Y.L., Li, J.G.: Martensitic transformation of $(Ni55.3Fe17.6Ga27.1)_{100-x}Co_x$ magnetic shape memory alloys. Acta Mater. **53**, 5125–5129 (2005)
50. Zheng, H.X., Wu, D., Xue, S., Frenzel, J., Eggeler, G., Zhai, Q.: Martensitic transformation in rapidly solidified Heusler Ni49mn39Sn12 alloys. Acta Mater. **59**, 5692–5699 (2011)
51. Zhao, X.G., Hsieh, C.C., Lai, J.G., Cheng, X.J., Chang, W.C., Cui, W.B., et al.: Scr. Mater. **63**, 250 (2005)
52. Santos, J.D., Sánchez, T., Álvarez, P., Sánchez, M.L., Sánchez, M.L., Sánchez Llamazares, J.L., Hernando, B.: Microstructure and magnetic properties of Ni50Mn37Sn13 Heusler alloy ribbons. Appl. Phys. Lett. **103**, 07B326 (2008)
53. Xuan, H.C., Deng, Y., Wang, D.H., Zhang, C.L., Han, Z.D., Du, Y.W.: Effect of the annealing on the martensitic transformation and magnetoresistance in Ni-Mn-Sn ribbons. J. Phys. D. Appl. Phys. **41**, 215002 (2008)
54. Hernando, B., Sánchez Llamazares, J.L., Santos, J.D., Escoda, L., Varga, R., Baldomir, D., Serantes, D.: Thermal and magnetic field-induced martensite-austenite transition in Ni50.3Mn35.3Sn14.1 ribbons. Appl. Phys. Lett. **92**, 042504 (2008)
55. Hernando, B., Sánchez Llamazares, J.L., Santos, J.D., Sánchez, M.L., Escoda, L., Suñol, J.J., Varga, R., García, C., González, J.: Grain oriented NiMnSn and NiMnIn Heusler alloy ribbons produced by melt spinning: martensitic transformation and magnetic properties. J. Magn. Magn. Mater. **321**, 763–768 (2009)
56. Sánchez Llamazares, J.L., Sánchez, T., Santos, J.D., Pérez, M.J., Sánchez, M.L., Hernando, B., Escoda, L.L., Suñol, J.J., Varga, R.: Martensitic phase transformation in rapidly solidified Mn50Ni40In10 alloy ribbons. Appl. Phys. Lett. **92**, 012513 (2008)
57. Wang, W., Yu, J., Zhai, Q., Luo, Z., Zheng, H.: Origin of retarded martensitic transformation in Heusler Ni-Mn-Sn melt-spun ribbons. Intermetallics **42**, 126–129 (2013)
58. Esakki, M.S., Rama Rao, N.V., Maniel Raja, M., Raj Kumar, D.M., Mohan Radheep, D., Arumugan, S.: Influence of Ni/Mn concentration on the structural, magnetic and magnetocaloric properties in Ni50-xMn37 + xSn13 Heusler alloys. J. Phys. D. Appl. Phys. **43**, 425002 (2010)
59. Wang, C., Meyer, J., Teichert, N., Auge, A., Rausch, E., Balke, B., Hütten, A., Fecher, G.H., Felser, C.: Heusler nanoparticles for spintronics and ferromagnetic shape memory alloys. J. Vacuum Sci. Tech. B. **32**, 020802 (2014)
60. Gaitzsch, U., Drache, J., McDonald, K., Müllner, P., Lindquist, P.: Obtaining of Ni-Mn-Ga magnetic shape memory alloy by annealing electrochemically deposited Ga/Mn/Ni layers. Thin Solid Films **522**, 171–174 (2012)
61. Babita, I., Gopalan, L., Rajasekhar, M., Ram, S.: Studies on ordering temperature and martensite stabilization in Ni55Mn20-xGa25 + x alloys. J. Alloys Compd. **475**, 276–280 (2009)
62. Prasad, R.V.S., Srinivas, M., Manivel Raja, M., Phanikumar, G.: Microstructure and magnetic properties of Ni2(Mn, Fe)Ga Heusler alloys rapidly solidified by melt spinning. Metall. Mater. Trans. A **45A**, 2161–2170 (2014)
63. Liu, J., Woodcock, T.G., Scheerbaum, N., Gutfleisch, O.: Influence of annealing on magnetic field-induced structural transformation and magnetocaloric effect in Ni-Mn-In-Co ribbons. Acta Mater. **57**, 4911–4920 (2009)
64. Kreissl, M., Kanomata, T., Matsumoto, M., Neumann, K.U., Ouladdiaf, B., Stephens, T., Ziebeck, K.R.A.: The influence of atomic order and residual strain on the magnetic and structural properties of Ni2MnGa. J. Magn. Magn. Mater. **272**, 2033–2034 (2004)
65. Yu, S.Y., Hu, S.J., Kang, S.S., Gu, A.J.: Martensitic transformation in Ni-rich Ni55Mn25In20 Heusler alloy: Experiment and first-principles calculations. J. Alloys Compd. **633**, 18–21 (2015)
66. Rama Rao, N.V., Gopalan, R., Manivel Raja, M., Arout Chelvane, J., Majumdar, B., Chandrasekaran, V.: Magneto-structural transformation studies in melt-spun Ni-Mn-Ga ribbons. Scr. Mater. **56**, 405–408 (2007)

67. Cai, W., Feng, Y., Sui, F.H., Gao, Z.Y., Dong, G.F.: Microstructure and martensitic transformation behavior of the Ni50Mn36In14 melt-spun ribbons. Scr. Mater. **58**, 830–833 (2008)
68. Albertini, F., Besseghini, S., Paoluzi, A., Pareti, L., Pasquale, M., Passaretti, F., Sasso, C.P., Stantero, A., Villa, E.: Structural, magnetic and anisotropic properties of Ni2MnGa melt-spun ribbons. J. Magn. Magn. Mater. **1421**, 242–245 (2002)
69. Quintana-Nedelcos, A., Sánchez-Llamazares, J.L., Ríos-Jara, D., Lara-Rodríguez, A.G., García-Fernández, T.: Effect of quenching rate on the average grain size and martensitic transformation temperature in rapidly solidified polycrystalline Ni50Mn37Sn13 alloy ribbons. Phys. Status Solidi A **210**, 2159–2165 (2013)
70. Moya, X., Mañosa, L., Planes, A., Krenke, T., Duman, E., Acet, M., Wassermann, E.F.: Calorimetric study of the inverse magnetic effect in ferromagnetic Ni-Mn-Sn. J. Magn. Magn. Mater. **316**, e572–e574 (2007)
71. Dubenko, I., Samanta, T., Kumar, P.A., Kazakov, A., Prudnikov, V., Stadler, S., Granovsky, A., Zhukov, A., Ali, N.: Magnetocaloric effect and multifunctional properties of Ni-Mn-Sn Heusler alloys. J. Magn. Magn. Mater. **324**, 3530–3534 (2012)
72. Hu, F.X., Wang, J., Chen, L., Zhao, J.L., Sun, J.R., Shen, B.G.: Effect of the introduction of H atoms on magnetic entropy change in metamagnetic Heusler alloys Ni-Mn-In. Appl. Phys. Lett. **95**, 112503 (2009)
73. Bachaga, T., Daly, R., Khitouni, M., Escoda, L., Saurina, J., Suñol, J.J.: Thermal and structural analysis of Mn49.3Ni43.7Sn7.0 Heusler alloy ribbons. Entropy **17**, 646–657 (2015)
74. González-Legarreta, L., Rosa, W.O., García, J., Ipatov, M., Nazmunnahar, M., Escoda, L., Suñol, J.J., Prida, V.M., Somer, R.L., González, J., Leoni, M., Hernando, B.: Annealing effect on the crystal structure and exchange bias in Heusler Ni45.5Mn43.6In11.5 alloy ribbons. J Alloys Compd. **582**, 588–593 (2014)
75. Comtesse, D., Gruner, M.E., Ogura, M., Sokolovskiy, V.V., Buchelnikov, V.D., Grünebohm, A., Arróyave, R., Singh, N., Gottschall, T., Gutfleisch, O., Chernenko, V.A., Albertini, F., Fähler, S., Entel, P.: First-principles calculation of the instability leading to giant inverse magnetocaloric effects. Phys. Rev. B **89**, 184403 (2014)
76. Planes, A., Mañosa, L., Acet, M.: Magnetocaloric effect and its relation to shape-memory properties in ferromagnetic Heusler alloys. J. Phys. Condens. Matter. **21**, 233201 (2009)
77. Krenke, T., Acet, M., Wassermann, E.F., Moya, X., Mañosa, L., Planes, A.: Ferromagnetism in the austenitic and martensitic states of Ni−Mn−In alloys. Phys. Rev. B **73**, 174413 (2006)
78. Yu, B.F., Gao, Q., Zhang, B., Meng, X.Z., Chen, Z.: Review on research of room temperature magnetic refrigeration. Int. J. Refrig. **26**, 622 (2003)
79. Yan, J.L., Li, Z.Z., Chen, X., Zhou, K.W., Shen, S.X., Zhou, H.B.: Martensitic transition and magnetocaloric properties in $Ni_{45}Mn_{44-}xFexSn_{11}$ alloys. J. Alloy Compd. **506**, 516 (2010)
80. Chatterjee, S., Giri, S., De, S.K., Majumdar, S.: Giant magneto-caloric effect near room temperature in Ni–Mn–Sn–Ga alloys. J. Alloy Compd. **503**, 273 (2010)
81. Yu, H.J., Fu, H., Zeng, Z.M., Sun, J.X., Wang, Z.G., Zhou, W.L., Zu, X.T.: Phase transformations and magnetocaloric effect in Ni–Fe–Ga ferromagnetic shape memory alloy. J. Alloy Compd. **477**, 732 (2009)
82. Sasıoglu, E., Sandratskii, L.M., Bruno, P.: First-principles calculation of the intersublattice exchange interactions and Curie temperatures of the full Heusler alloys Ni2MnX (X = Ga, In, Sn, Sb). Phys. Rev. B **70**, 024427 (2004)
83. Tan, C.L., Huang, Y.W., Tian, X.H., Jiang, J.X., Cai, W.: Origin of magnetic properties and martensitic transformation of Ni-Mn-In magnetic shape memory alloys. Appl. Phys. Lett. **100**, 132402 (2012)
84. Reichl, L.E.: A modern course in statistical physics, 2nd edn. John Wiley, New York (1998)
85. Mukherjee, T., Michalski, S., Skomski, R., Sellmyer, D.J., Binek, C.: Overcoming the spin-multiplicity limit of entropy by means of lattice degrees of freedom: A minimal model. Phys. Rev. B **83**, 214413 (2011)
86. Pecharsky, V.K., Gschneidner Jr., K.A.: Some common misconceptions concerning magnetic refrigerant materials. J. Appl. Phys. **90**, 4614 (2001)

87. Pecharsky, V.K., Gschneidner Jr., K.A., Pecharsky, A.O., Tishin, A.M.: Thermodynamics of the magnetocaloric effect. Phys. Rev. B **64**, 144406 (2001)
88. Imry, Y., Wortis, M.: Influence of quenched impurities on first-order phase transitions. Phys. Rev. B **19**, 3580 (1979)
89. Caballero-Flores, R., Sánchez, T., Rosa, W.O., García, J., González-Legarreta, L., Serantes, D., Prida, V.M., Escoda, L., Suñol, J.J., Hernando, B.: On tuning the magnetocaloric effect in Ni–Mn–In Heusler alloy ribbons with thermal treatment. J. Alloy. Compd. **545**, 216 (2012)
90. Yeomans, J.M.: Statistical mechanics of phase transitions. Claredon, Oxford (1992)
91. Roy, S.B.: First order magneto-structural phase transition and associated multi-functional properties in magnetic solids. J. Phys. Condens. Matter. **25**, 183201 (2013)
92. Caballero-Flores, R., González-Legarreta, L., Rosa, W.O., Sánchez, T., Prida, V.M., Escoda, L., Suñol, J.J., Batdalov, A.B., Aliev, A.M., Koledov, V.V., Shavrov, V.G., Hernando, B.: Magnetocaloric effect, magnetostructural and magnetic phase transformations in Ni50.3Mn36.5Sn13.2 Heusler alloy ribbons. J. Alloy Compd. **629**, 332 (2015)
93. Krenke, T., Acet, M., Wassermann, E.F., Moya, X., Mañosa, L., Planes, A.: Martensitic transitions and the nature of ferromagnetism in the austenitic and martensitic states of Ni – Mn – Sn alloys. Phys. Rev. B **72**, 014412 (2005)
94. Shamberger, P.J., Ohuchi, F.S.: Hysteresis of the martensitic phase transition in magnetocaloric-effect Ni-Mn-Sn alloys. Phys. Rev. B **79**, 144407 (2009)
95. Hopkinson, J.: Magnetic properties of alloys of nickel and iron. Proc. R. Soc. A **48**, 1 (1890)
96. Sólyom, J.: Fundamentals of the physics of solids, 1st edn. Springer, Berlin (2007)
97. Prudnikov, V.N., Kazakov, A.P., Titov, I.S., Kovarskii, Y.N., Perov, N.S., Granovsky, A.B., Dubenko, I., Pathak, A.K., Ali, N., Gonzalez, J.: Quasi-diamagnetism and exchange anisotropy in Ni-Mn-In-Co Heusler alloys. Phys. Solid State **53**, 490 (2011)
98. de Oliveira, N.A., von Ranke, P.J.: Theoretical aspects of the magnetocaloric effect. Phys. Rep. **489**, 89 (2010)
99. Buchelnikov, V.D., Entel, P., Taskaev, S.V., Sokolovskiy, V.V., Hucht, A., Ogura, M., Akai, H., Gruner, M.E., Nayak, S.K.: Monte Carlo study of the influence of antiferromagnetic exchange interactions on the phase transitions of ferromagnetic Ni-Mn-X alloys (X = In, Sn, Sb). Phys. Rev. B **78**, 184427 (2008)
100. Wang, B.M., Liu, Y., Wang, L., Huang, S.L., Zhao, Y., Yang, Y., Zhang, H.: Exchange bias and its training effect in the martensitic state of bulk polycrystalline Ni49.5Mn34.5In16. J. Appl. Phys. **104**, 043916 (2008)
101. Khan, M., Dubenko, I., Stadler, S., Ali, N.: Exchange bias in bulk Mn rich Ni-Mn-Sn Heusler alloys. J. Appl. Phys. **102**, 113914 (2007)
102. Khan, M., Dubenko, I., Stadler, S., Ali, N.: Exchange bias behavior in Ni-Mn-Sb Heusler alloys. Appl. Phys. Lett. **91**, 072510 (2007)
103. Jing, C., Chen, J., Li, Z., Qiao, Y., Kang, B., Cao, S., Zhang, J.: Exchange bias behavior and inverse magnetocaloric effect in $Ni_{50}Mn_{35}In_{15}$ Heusler alloy. J. Alloy Compd. **475**, 1–4 (2009)
104. Wang, B.M., Liu, Y., Ren, P., Xia, B., Ruan, K.B., Yi, J.B., Ding, J., Li, X.G.: Large exchange bias after zero-field cooling from an unmagnetized state. Phys. Rev. Lett. **106**, 077203 (2011)
105. Machavarapu, R., Jakob, G.: Exchange bias effect in the martensitic state of Ni-Co-Mn-Sn film. Appl. Phys. Lett. **102**, 232406 (2013)
106. Acet, M., Mañosa, L., Planes, A.: Magnetic-field-induced effects in martensitic Heusler-based magnetic shape memory alloys. In: Buschow, K.H.J. (ed.) Handbook of magnetic materials, 19th edn, p. 231. Elsevier, Amsterdam (2011)
107. Nogués, J., Schuller, I.K.: Exchange bias. J. Magn. Magn. Mater. **192**, 203 (1999)
108. Wang, B.M., Liu, Y., Xia, B., Ren, P., Wang, L.: Large exchange bias obtainable through zero-field cooling from an unmagnetized state in Ni-Mn-Sn alloys. J. Appl. Phys. **111**, 043912 (2012)

109. Singh, R., Ingale, B., Varga, L.K., Khovaylo, V.V., Chatterjee, R.: Large exchange-bias in $Ni_{55}Mn_{19}Al_{24}Si_2$ polycrystalline ribbons. Physica B **448**, 143–146 (2014)
110. Bhatti, K.P., El-Khatib, S., Srivastava, V., James, R.D., Leighton, C.: Small-angle neutron scattering study of magnetic ordering and inhomogeneity across the martensitic phase transformation in $Ni_{50-x}Co_xMn_{40}Sn_{10}$ alloys. Phys. Rev. B **85**, 134450 (2012)
111. Cai, J.W., Liu, K., Chien, C.L.: Exchange coupling in the paramagnetic state. Phys. Rev. B **60**, 72 (1999)
112. Leighton, C., Fitzsimmons, M.R., Hoffmann, A., Dura, J., Majkrzak, C.F., Lund, M.S., Schuller, I.K.: Thickness-dependent coercive mechanisms in exchange-biased bilayers. Phys. Rev. B **65**, 064403 (2002)
113. Li, Z., Chao, J., Chen, J., Yuan, S., Cao, S., Zhang, J.: Observation of exchange bias in the martensitic state of Ni50Mn36Sn14 Heusler alloy. App. Phys. Lett. **91**, 112505 (2007)
114. Sánchez-Llamazares, J.L., Flores-Zúñiga, H., Ríos-Jara, D., Sánchez-Valdes, C.F., García-Fernández, T., Ross, C.A., García, C.: Structural and magnetic characterization of the intermartensitic phase transition in NiMnSn Heusler alloy ribbons. J. Appl. Phys. **113**, 17948 (2013)
115. Ray, M.K., Bagani, K., Banerjee, S.: Effect of excess Ni on martensitic transition, exchange bias and inverse magnetocaloric effect in Ni2 + xMn1.4-xSn0.6 alloy. J. Alloy Compd. **600**, 55–59 (2014)

Chapter 4
Magnetocaloric Materials

Julia Lyubina

4.1 Introduction

Energy consumption in residential and commercial buildings accounts for about 40 % of total primary energy use [1]. Cooling in its various forms takes up at least 15 % of this figure [2], and, thus, improving energy efficiency with respect to refrigeration plays an important role in reducing the overall energy consumption.

Magnetic refrigeration based on the magnetocaloric effect (MCE) is a cooling technology that offers significant energy-saving potential [3, 4]. It is estimated that a solid-state magnetic cooling system can provide energy saving of 20–30 % as compared to the conventional gas compression technology. Solid-state magnetic cooling technology can be used for both cooling and heating; it is environmentally friendly, as it does not use hazardous chemicals—water can be used as heat transfer fluid (Sect. 4.3). Due to a low number of movable parts, a quiet operation can be ensured. The magnetic refrigeration also allows design advantages compared to the gas compression technology. It is suitable for low to medium power devices working in the range from milliwatt to a few hundred watts, i.e. in the range where gas compressors are less competitive in terms of price. Moreover, less sophisticated lightweight (plastic) casing can be used, since it does not need to sustain high pressures as the pressure of the heat transfer fluid in the magnetic refrigerator is around ambient pressure.

The magnetocaloric effect has been discovered in Ni by Weiss and Piccard [5], and the first experimental demonstration of the magnetic cooling method for attaining very low temperatures using paramagnetic salts was provided by Giauque and MacDougall [6]. For a detailed historic review of the discovery and the early application of the magnetocaloric effect, the reader is referred to a paper by

J. Lyubina (✉)
Experimental Solid State Group, Blackett Laboratory, Department of Physics,
Imperial College London, Exhibition Road, London SW7 2AZ, UK
e-mail: y.lyubina@imperial.ac.uk

A. Zhukov (ed.), *Novel Functional Magnetic Materials*, Springer Series
in Materials Science 231, DOI 10.1007/978-3-319-26106-5_4

Smith [7]. In the late 1970s, it has been recognised that a much larger MCE can be obtained in a ferromagnet in the vicinity of its Curie temperature and the MCE effect can be used for heat pumping near room temperature [8]. The major breakthrough occurred in 1997, when the Ames Laboratory/Astronautics Corporation of America showed that a near room temperature magnetic refrigerator which uses Gd spheres as active magnetic refrigerant is competitive to conventional gas compression cooling [3]. Although the first industrial proof-of-concept wine cooler based on a magnetocaloric heat pump has been presented by Haier, Astronautics Corporation of America and BASF at the International Consumer Electronics Show (CES, Jan. 6–9, 2015) in Las Vegas, USA [9], the magnetic cooling technology has not been yet commercialised at the time of writing.

An increase in activity in the field of the magnetocaloric effect and magnetocaloric materials has started in 1997 with a report on a giant MCE in $Gd_5Si_2Ge_2$ by Pecharsky and Gschneidner [10]. It was Holtzberg et al. [11], who discovered $Gd_5(Si_xGe_{1-x})_4$ ternary intermetallic compounds in the Gd–Si–Ge system with Curie temperatures in the range between 295 and 335 K. Holtzberg et al. [11] measured the magnetisation of $Gd_5Si_2Ge_2$, which showed conventional second-order transition from the para- to ferromagnetic state. It should be noted, however, that the change of the transition from second to first order in $Gd_5(Si_xGe_{1-x})_4$ occurs in a very narrow composition range [12] and a slight deviation of the composition from $x = 0.5$ may result in the abrupt change of the transition type. For some compositions, Holtzberg et al. [11] even observed a discontinuity in the magnetisation as a function of temperature accompanied by a very low hysteresis and speculated whether the observed behaviour is due to the first-order transition. However, when tried to establish the order of the transition from heat capacity measurements, they came to a conclusion that the transition is of second order, which was later shown not to be the case [12]. As we know today, the reason for the incorrect conclusion lies in the highly non-trivial heat capacity measurements at first-order transitions (see Sects. 4.2.3.1 and 4.2.3.2). This example shows the importance of the adequate characterisation of potential magnetic refrigerant materials. Therefore, a part of this chapter is dedicated to methods for correct characterisation of magnetocaloric materials (Sect. 4.2).

In fact, the giant magnetocaloric effect has been reported even earlier in Fe–Rh, and the name "giant magnetocaloric effect" has been coined by Nikitin et al. [13]. Pecharsky and Gschneidner appreciate this fact in their seminal publication from 1997 as well as point to a high thermal hysteresis in Fe–Rh of about 10 K as being a significant hindrance for magnetic cooling applications. The seminal publication on the importance of the microstructure design in reducing the hysteresis and controlling the magnetic and thermal properties of magnetocaloric materials is due to Lyubina et al. [14, 15]. The implications of the hysteresis and other material parameters for the magnetic cooling/heat pumping application are discussed in Sect. 4.4.

There exist several reviews on magnetocaloric materials already [16–19]. In efforts to design high-performance, solid-state magnetic refrigerators, it has been recognised that such refrigerators will require magnetic materials that must fulfil requirements, whose implementation partially contradicts each other. This chapter will focus on challenges of the design of materials for magnetic heat pumping. It will provide an overview of the technologically relevant parameters and give an outline on how the understanding the fundamental phenomena helps to design the materials and improve their performance. The principle of the magnetic heat pumping is provided in Sect. 4.3. Section 4.5 gives an overview of materials that are considered to be a candidate or are already used in magnetic refrigerator prototypes. The aim of this book chapter is to provide an update in the field of solid-state magnetic cooling, as well as to summarise the main characterisation methods necessary for the adequate comparison of various materials in terms of their applicability for magnetic heat pumping, which still is and will remain an important task for fundamental and applied physics.

4.2 Magnetocaloric Effect

4.2.1 Thermodynamics of the Magnetocaloric Effect

The magnetic cooling is based on the magnetocaloric effect (MCE), which is the emission or absorption of heat in a magnetic material in response to a changing magnetic field. The magnetocaloric effect in solids is essentially the result of the influence of the magnetic field on a spin system with the corresponding entropy variation.

The total entropy of the solid S consists of the entropy of the phonon (lattice) system S_{ph} and the entropy of the electron system S_{el}. The phonon contribution exceeds the electronic entropy S_{el} already at very low temperatures, i.e. at a few percent of the Debye temperature [20]. In metals, however, the individual particle excitation of conduction electrons can be enhanced by the magnetic contribution, and, thus, this term becomes non-negligible, especially in the vicinity of the magnetic transitions [21]. In solids, where the magnetism is due to localised electrons, e.g. in rare-earth-based materials, the entropy can be represented as the sum of the phonon S_{ph}, electronic S_{el} and magnetic S_{mag} contributions. In itinerant systems, it is common to speak of the entropy of the system consisting of the phonon S_{ph} and magnetic S_{mag} terms.

The entropy is a measure of the disorder of a system: the larger the disorder, the higher the entropy. The entropy of the magnetic system is largely configurational and is associated with the randomness of the spins. The external magnetic field H acts on the magnetic system alone and reduces its entropy S_{mag}. Under adiabatic conditions, i.e. when the solid is thermally isolated and the magnetic field change ΔH is slow compared to processes leading to the establishment of equilibrium, the total entropy of the solid remains unchanged. The reduction in S_{mag}, therefore, must be compensated within the solid from other internal sources. The compensation can be achieved through an increase of the lattice vibrations, which leads to an increase of the temperature of the solid. Conversely, upon adiabatic demagnetisation, i.e. upon the removal of the magnetic field, the energy of the spin system and consequently the magnetic part of the entropy S_{mag} increases. The corresponding decrease in the phonon contribution S_{ph} leads to the temperature reduction from the initial temperature T_{i} to the final temperature T_{f}.

The principle of the magnetic cooling is illustrated in Fig. 4.1. The total entropy of a solid $S(T, H, P)$, the solid being, e.g. a paramagnet or a ferromagnet above the Curie temperature, is schematically shown in an initial field H_{i} and a final field H_{f}, where T is the temperature, P is the pressure and $H_{\mathrm{i}} < H_{\mathrm{f}}$, H_{i} often represents the zero magnetic field. The application of a sufficiently high magnetic field H_{f} leads to the increase of the magnetic order (Fig. 4.1b) and to the corresponding magnetic entropy reduction. This is schematically illustrated by the curve $S(T, H_{\mathrm{f}}, P)$. Thus, the isothermal application of the magnetic field (process AB) leads to the emission of heat. The heat should be removed from the system. The adiabatic (isentropic) demagnetisation, during which the solid is thermally isolated from the surroundings and the magnetic field is switched off, leads to the randomisation of the magnetic moments. The heat is absorbed from the phonon system and the solid cools.

The total entropy $S(T, H, P)$ is the quantity that allows to fully characterise the magnetocaloric effect. Since the majority of the MCE studies are performed at a constant pressure and the pressure is not changed substantially during a magnetic refrigeration cycle, the isobaric condition is tacitly assumed and is often not mentioned when referring to a measure of the MCE. In the following, the isobaric process will be assumed and the parameter $P = \mathrm{const}$ will be omitted for simplicity. The entropy can be determined by using the second and third law of thermodynamics as

$$S(T, H) = \int_0^T \frac{C_p(T', H)}{T'} dT', \tag{4.1}$$

where $C_p(T, H)$ is the heat capacity. Thus, to compute the entropy, one requires the knowledge of the heat capacity, which is obtainable from macroscopic measurements of the heat evolved or absorbed by a solid in dependence on the temperature and the magnetic field. Accurate measurements of heat capacity and the calculation of $S(T, H)$ from Eq. (4.1) can be a non-trivial task and will be discussed in Sect. 4.2.3 in more detail.

For most applications, it is sufficient to know only the entropy difference ΔS, i.e. values of the entropy measured with respect to some chosen (standard) state. According to the schematic thermodynamics of the MCE in Fig. 4.1, the isothermal entropy change in a field change $\Delta H = H_{\mathrm{f}} - H_{\mathrm{i}}$ is a difference $\Delta S = S_{\mathrm{B}} - S_{\mathrm{A}}$ and is determined by

$$\Delta S(T, \Delta H) = S(T, H_{\mathrm{f}}) - S(T, H_{\mathrm{i}}). \tag{4.2}$$

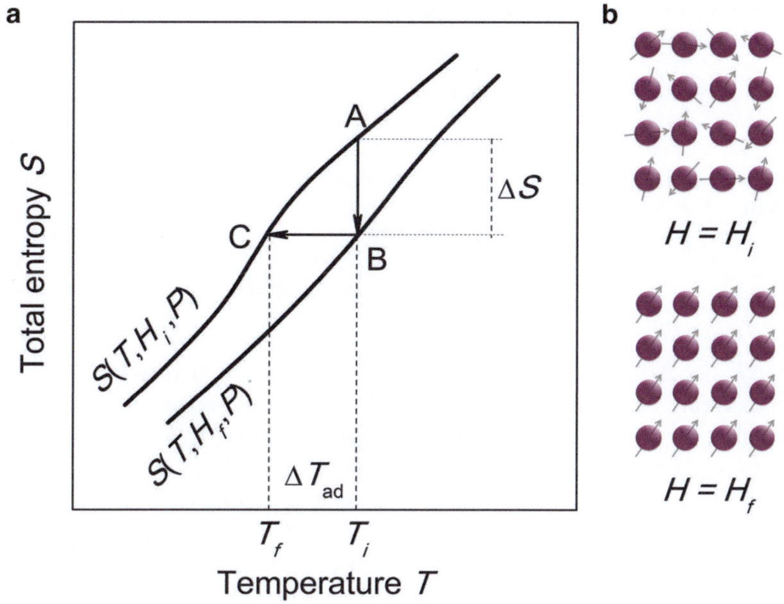

Fig. 4.1 Schematic representation of the magnetic cooling in terms of the magnetic field effect on the entropy (**a**) and on the spin system (**b**) of a solid. AB corresponds to the isothermal magnetisation; BC corresponds to the adiabatic demagnetisation

Under the condition of adiabaticity, $S = S_B = S_C$, the adiabatic temperature change upon the variation of the magnetic field from H_i to H_f, i.e. adiabatic magnetisation along the path C → B, is determined as

$$\Delta T_{ad}(T_{i,C}, \Delta H) = T_f(H_f) - T_i(H_i). \tag{4.3}$$

Along with the fields H_i and H_f, ΔT_{ad} is a function of temperature $T_{i,C}$, where the index C corresponds to the point C in Fig. 4.1 and stresses the fact that ΔT_{ad} depends on the particular initial temperature of the isentropic process. Note that since ΔS and ΔT_{ad} are non-linear functions of the field and scale with the magnetic field as H^n, where $n < 1$ [22–24], ΔS and ΔT_{ad} depend not simply on an arbitrary field difference ΔH but are functions of the specific initial H_i and final magnetic fields H_f, i.e. $\Delta S(T, \mu_0 \Delta H = (1-0)\mathrm{T}) \neq \Delta S(T, \mu_0 \Delta H = (2-1)\mathrm{T})$.

Results in Eqs. (4.2) and (4.3) can also be obtained analytically from the total differentials

$$dS = \left(\frac{\partial S}{\partial T}\right)_H dT + \left(\frac{\partial S}{\partial H}\right)_T dH \tag{4.4}$$

$$dT = \left(\frac{\partial T}{\partial S}\right)_H dS + \left(\frac{\partial T}{\partial H}\right)_S dH. \tag{4.5}$$

Integration of Eq. (4.4) at constant temperature with $\Delta H = H_\mathrm{f} - H_\mathrm{i}$ gives

$$\Delta S(T, \Delta H) = \int_{H_\mathrm{i}}^{H_\mathrm{f}} \left(\frac{\partial S}{\partial H}\right)_T dH, \tag{4.6}$$

where $\Delta S(T, \Delta H) = [S(T, H_\mathrm{f}) - S(T, H_\mathrm{i})]_T$. Integration of Eq. (4.5) at $S = \mathrm{const}$ results in

$$\Delta T_\mathrm{ad}(S, \Delta H) = \int_{H_\mathrm{i}}^{H_\mathrm{f}} \left(\frac{\partial T}{\partial H}\right)_S dH, \tag{4.7}$$

where $\Delta T_\mathrm{ad}(S, \Delta H) = [T(S, H_\mathrm{f}) - T(S, H_\mathrm{i})]_S$. In practice, it is more convenient to use temperature as the independent variable instead of entropy, i.e. $\Delta T_\mathrm{ad}(T, \Delta H)$ where temperature is the temperature of the system in the field H_i.

The two quantities, the isothermal entropy change ΔS and the adiabatic temperature change ΔT_ad, can be used in place of the total entropy $S(T, H)$ to *completely* characterise the magnetocaloric effect. In terms of the magnetic cooling applications, other material parameters can be of importance in addition to the knowledge of ΔS and ΔT_ad, and these will be discussed in Sect. 4.4.

4.2.2 Magnetocaloric Effect at Magnetic Phase Transitions

In solids, transitions between various types of magnetic order (magnetic phase transitions) can be of first- and second-order type. The type of the transition has implications on the experimental determination of the magnetocaloric effect, which is discussed in detail in Sect. 4.2.3.

Quantitative description of the phase transition can be made in terms of an order parameter [25], being magnetisation M in the case of the magnetic phase transitions. According to the Ehrenfest classification, at a second-order phase transition, the first derivatives of the thermodynamic potential with respect to temperature $\partial\Phi/\partial T$ and generalised force $\partial\Phi/\partial \mathrm{X}_i$ (X_i can be, e.g. magnetic field, pressure) are continuous, and the second derivatives $\partial^2\Phi/\partial T^2, \partial^2\Phi/\partial X_i^2, \partial^2\Phi/\partial T \partial X_i$ have discontinuities at the transition point. The differential of the free energy of a solid is

$$dF = -SdT - PdV - MdH, \tag{4.8}$$

where V is the volume. From Eq. (4.8) it follows that the entropy and magnetisation are

$$S = -\left(\frac{\partial F}{\partial T}\right)_{V,H} \tag{4.9}$$

$$M = -\left(\frac{\partial F}{\partial H}\right)_{V,T}, \tag{4.10}$$

respectively. Thus, second-order transition shows no discontinuity of the entropy and the magnetisation, and both are changing continuously with T from a finite value to zero, when approaching T_c from low temperatures (Fig. 4.2). Such type of transition is observed, e.g. in iron and gadolinium during the transition from the ferromagnetic to the paramagnetic state. Further, for second-order transitions, we expect a discontinuity in the derivatives of S and M. The specific heat, which is proportional to $(\partial S/\partial T)_p$ (Eq. 4.1), has a sharp break near T_c, a so-called λ-shape anomaly (Fig. 4.2). Such function does not present difficulties in the integration, and the entropy change can be readily determined by making use of Eqs. (4.1) and (4.2), if $C_p(T, H_i)$ and $C_p(T, H_f)$ curves are known:

$$\Delta S(T, \Delta H) = \int_0^T \frac{C_p(T', H_f)}{T'} dT' - \int_0^T \frac{C_p(T', H_i)}{T'} dT'. \tag{4.11}$$

There exist materials, where the transition from the paramagnetic to the ferromagnetic state can be of first order [26]. The so-called giant magnetocaloric effect in $Gd_5Si_2Ge_2$ is the consequence of the first-order magnetic phase transition coinciding with a structural transformation [10]. A spin reorientation transition in $Er_2Fe_{14}B$ and $Tm_2Fe_{14}B$ [27] and ferro- to antiferromagnetic transition in Fe–Rh [13, 28] are further examples of the first-order magnetic phase transitions.

At a first-order transition, the first derivatives $\partial\Phi/\partial T$ and $\partial\Phi/\partial X_i$ have a discontinuity. As a consequence, the magnetisation changes abruptly at the transition point (Fig. 4.2). The second-order phase transitions are also accompanied by a discontinuity in volume,

$$V = \left(\frac{\partial G}{\partial P}\right)_{T,H}, \tag{4.12}$$

where G is the Gibbs free energy. The volume jump ΔV at the first-order magnetic phase transition can be very significant with $\Delta V/V$ values exceeding 1 %, as, e.g. in the case of $La(Fe,Si)_{13}$-type alloys [29].

The first-order transition is accompanied by the emission or absorption of the latent heat L

$$L = T_c(S_2 - S_1), \tag{4.13}$$

where S_1 and S_2 are the entropies of the different magnetic states 1 and 2 corresponding to the free energies F_1 and F_2 (Fig. 4.2). Because of the discontinuity in the entropy as a function of temperature, there is an infinite spike in the heat capacity, a singularity, at the phase transition (Fig. 4.2). This singularity does not allow one to calculate the entropy and entropy change at the transition temperature

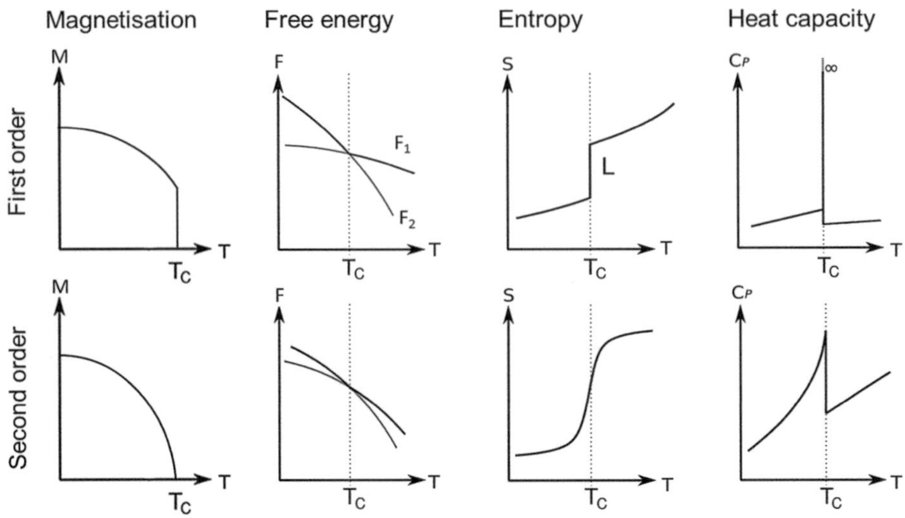

Fig. 4.2 Schematic temperature dependence of the magnetisation, free energy, entropy and heat capacity according to the classification of magnetic phase transitions

by integrating C_p in Eqs. (4.1) and (4.11), although both equations are valid for the first- and second-order transitions. To circumvent the problem, one needs to determine the latent heat L at the transition point (see Sect. 4.2.3.1).

4.2.3 Measurement of the Magnetocaloric Effect

4.2.3.1 Latent Heat Determination

An elegant way of the determination of the latent heat L was demonstrated by Pecharsky et al. [30] for an ultra-pure, homogenous polycrystalline dysprosium. In the temperature range below 300 K, dysprosium experiences two transitions. The high-temperature transition is of second order and is a transition from the helical paramagnetic to the antiferromagnetic state around 180 K upon cooling. At around 90 K, a first-order phase transition from the hexagonal antiferromagnetic phase to the orthorhombic ferromagnetic phase occurs. In accordance with the thermodynamics, at the first-order transition point, a singularity in the heat capacity is observed (Fig. 4.3). The entropy below and above the first-order transition can be determined using Eq. (4.1), whereas the value of the latent heat and the corresponding entropy jump at the transition cannot be determined from the $C_p(T, H)$ measurements, since the height of this jump is unknown. Therefore, at the transition point itself, another procedure should be adopted.

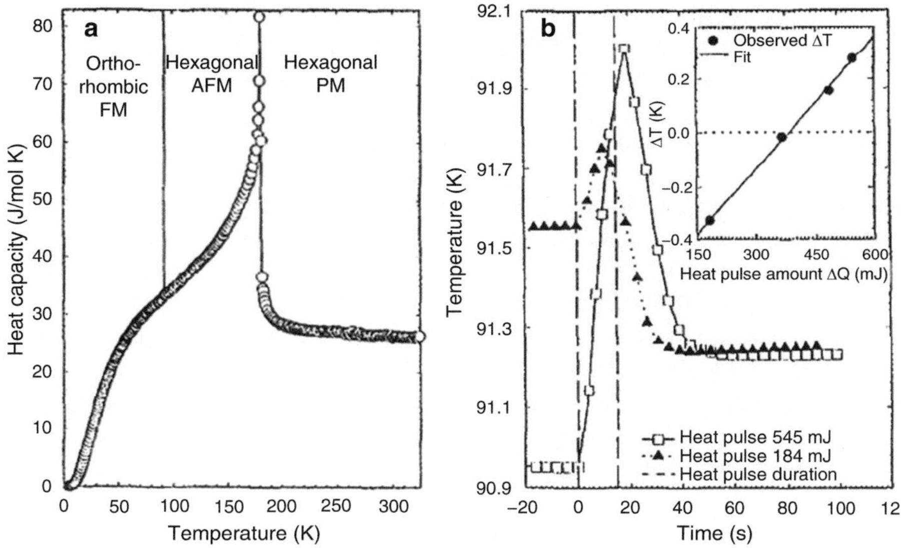

Fig. 4.3 Determination of the latent heat at the first-order magnetic phase transition after Pecharsky et al. [30]. (**a**) the zero magnetic field heat capacity of polycrystalline Dy. The height of heat capacity at 91 K is intentionally not shown, since it is infinite. (**b**) the time vs. temperature profiles near T_c of Dy for different heat pulse power. The inset shows the temperature rise ΔT on the amount of energy during the heat pulse. The solid line drawn represents a linear least squares fit of the data, which was used to determine the latent heat of transformation at $\Delta T = 0$

In the special case of the first-order transition in dysprosium, the following procedure for the determination of L can be used. It is known that the amount of heat ΔQ evolved or absorbed by a system is proportional to the temperature change ΔT:

$$\Delta Q = mC_p\Delta T, \tag{4.14}$$

where m is the mass. One of the methods for C_p measurement is an adiabatic heat pulse calorimetry, where a heat pulse is applied and a subsequent temperature rise and decay is recorded. Taking into account the latent heat, we may rewrite Eq. (4.14) as

$$\Delta Q_{\text{pulse}} = L + mC_p\Delta T. \tag{4.15}$$

If a heat pulse ΔQ_{pulse} provided to the system is larger than the heat required for the first-order transformation (latent heat), a positive ΔT is observed (Fig. 4.3b, heat pulse of 545 mJ). That means that after the transition is completed and the amount of heat equal to the latent heat has been absorbed, the not consumed heat results in the increase of the sample temperature ΔT. An application of a smaller heat pulse can be sufficient to trigger the phase transformation, but if the amount of heat is smaller than L, the additional heat required to finish the transformation is delivered by the sample causing it to cool. This is recorded as a negative ΔT (Fig. 4.3b, heat pulse of 184 mJ).

Two conclusions result from this behaviour [30]: (1) it is possible to adjust the amount of the heat pulse in the adiabatic heat pulse calorimeter to a value which is exactly equal to the latent heat of phase transition, so that the measured $\Delta T = 0$ and experimental evidence of the singularity of the heat capacity, i.e. the evidence for the first-order character of the phase transition, can be obtained, and (2) the latent heat L can be determined precisely by applying varying heat pulses, so that by interpolation in a $\Delta T(\Delta Q_{\text{pulse}})$ plot to $\Delta T = 0$, one can determine the point at which $\Delta Q_{\text{pulse}} = L$. That means that by integrating C_p in the temperature range from 0 to $T_c - \delta$ and from $T_c + \delta$ to T, i.e. prior to and after the transition at T_c and adding the latent heat at the transition temperature, we can construct the entropy curve in the whole temperature range:

$$S(T, \Delta H) = \int_0^{T_c - \delta} \frac{C_p(T', H)}{T'} dT' + \int_{T_c + \delta}^{T} \frac{C_p(T', H)}{T'} dT' + \frac{L(T_c, H)}{T_c}. \tag{4.16}$$

In contrast to the ultra-pure and homogeneous Dy described above, the majority of materials with a first-order transition do not show such a singularity in the heat capacity. Instead, the transition is spread over a temperature region, and the latent heat is evolved or absorbed not at a single point but is spread over a temperature region (Figs. 4.4 and 4.5a, b). As a consequence, in addition to the magnetic field, the latent heat also shows temperature dependence [32].

The reason for the spread of the transition can be ascribed to the non-uniform (also called continuous or sequential) first-order magnetic transition [14].

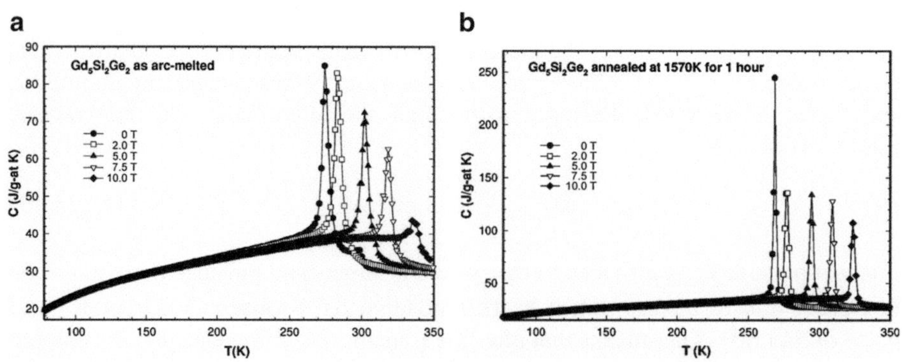

Fig. 4.4 Heat capacity of (**a**) as arc-melted and (**b**) homogenised $Gd_5Si_2Ge_2$ in various magnetic fields [31]

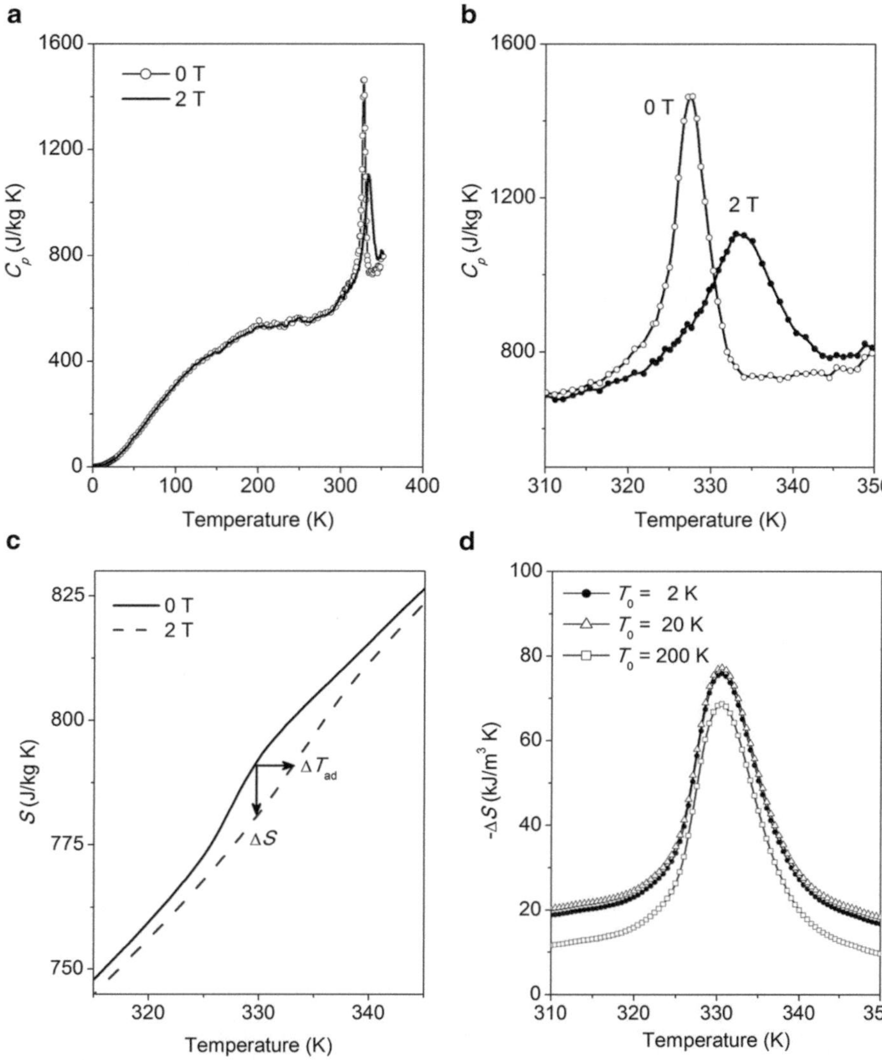

Fig. 4.5 Heat capacity of $LaFe_{11.6}Si_{1.4}H_{1.6}$ in zero magnetic field and in 2 T in the temperature range of 2–350 K (**a**) and 310–350 K (**b**) recorded on heating. Entropy S (**c**) and entropy change ΔS (**d**) calculated from the heat capacity data in (**a**) using Eqs. (4.17) and (4.20), respectively. $LaFe_{11.6}Si_{1.4}H_{1.6}$ undergoes a temperature-driven first-order transition from the para- to ferromagnetic state around 329 K and a field-induced itinerant-electron metamagnetic (IEM) transition above T_c

In polycrystalline materials consisting of grains or particles with a broad size distribution, the first-order transition occurs not at a single temperature but may be triggered and proceed in different crystallites at different temperatures. The transition is strongly influenced by impurities and inhomogeneity [31, 33], lattice strain, local stress acting on a particular crystallite [14], stray fields [34] and pressure [29]. For instance, the temperature-driven first-order transition from the para- to the ferromagnetic state in $LaFe_{11.6}Si_{1.4}$ around 193 K occurs first in smaller particles and interconnections between the particles (Fig. 4.6, images ❷ and ❸), that is, in the crystallites that are less constrained by the neighbouring grains/particles. As the temperature is lowered, larger particles transform to the ferromagnetic state (Fig. 4.6, images ❹ to ❻). A clear coexistence of the para- and ferromagnetic regions is observed in these large particles. Coexistence of the phases at the transition point is a fundamental property of the first-order transition. Thus, the smooth variation of the magnetisation in $LaFe_{11.6}Si_{1.4}$ (Fig. 4.6), typical for a second-order transition, is in fact due to a series of thermally induced first-order transitions. Such behaviour is observed other materials with a first-order transition as well [35, 36].

The heat capacity of materials exhibiting the *non-uniform* first-order magnetic phase transition is finite and has a fairly sharp peak near the transition point in a zero magnetic field. Exemplarily, the heat capacity of an as-cast $Gd_5Si_2Ge_2$ alloy undergoing a temperature-driven first-order magnetic phase transition around 277 K and the field-induced transformation above T_c in an external magnetic field H_m is shown in Fig. 4.4a [31]. These field- and temperature-induced magnetic transitions are coupled to a crystallographic phase transition, where the high-temperature monoclinic paramagnetic phase transforms to the low-temperature orthorhombic ferromagnetic phase. Application of the magnetic field has only a small effect on the heat capacity below the phase transition temperature. Near the transition point, the peak shifts appreciably to higher temperatures, and the peak width increases on the application of the magnetic field. The peak shift to a higher temperature can be explained by the stabilisation of the low-temperature ferromagnetic phase. The broadening of the C_p peak is due to the distribution of the critical fields of the transition H_m; the origin of the distribution can be, for example, impurities, inhomogeneity, lattice strain and local stress. If there would be no H_m distribution, the peak would stay sharp, and its height would be unchanged (the peak width approaches zero (singularity) in the case of a *perfect* first-order transition). This effect is to some extent observed in the homogenised $Gd_5Si_2Ge_2$ (Fig. 4.4b). In the homogenised $Gd_5Si_2Ge_2$, the zero-field heat capacity peak near T_c is significantly narrower as compared to the as-cast $Gd_5Si_2Ge_2$. Whereas the peak shift to a higher temperature with increasing magnetic field occurs in the homogenised $Gd_5Si_2Ge_2$ at the same rate as in the as-cast alloy, the magnetic field broadening of the C_p peak is less pronounced and the C_p peak height varies to a lesser extent in the applied magnetic fields of 2–10 T.

In materials with a magnetic phase transition of second order, the size of the discontinuity of the λ-type anomaly also depends appreciably on material purity and homogeneity [37, 38]. If a second-order transition from the ferro- to paramagnetic state is concerned, e.g. in Gd (Fig. 4.7), the application of the magnetic field favours the ferromagnetic state shifting the C_p peak to higher temperatures, albeit the shift rate with the magnetic field is significantly lower than at the first-order transition (cf. Fig. 4.4). At the same time, the magnetic field reduces the spontaneous cooperative alignment of the spins and, thus, results in the smearing out of the C_p peak over a wider temperature range and the decrease of its magnitude.

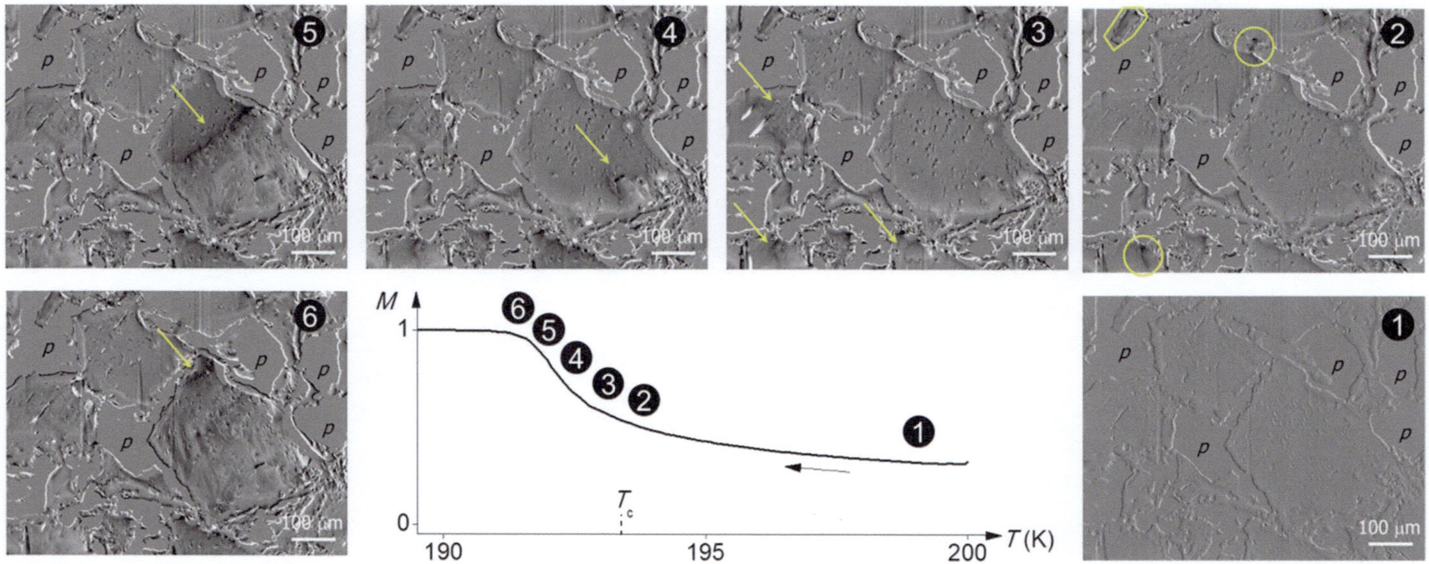

Fig. 4.6 Reduced magnetisation from macroscopic measurements and magnetic microstructure recorded by magneto-optic Kerr microscopy in $LaFe_{11.6}Si_{1.4}$ as a function of temperature. The sample architecture consists of interconnected $LaFe_{11.6}Si_{1.4}$ particles and pores (some larger pores are marked with p). Magneto-optic Kerr images of a polished section of pulverised and compacted $LaFe_{11.6}Si_{1.4}$ alloy were recorded during continuous cooling in the para- to ferromagnetic transition region: above (1–2) and below (3–6) T_c. A uniform grey contrast is seen on the particles in the paramagnetic state, while the ferromagnetic state is characterised by a multidomain pattern (marked by arrows and circles)

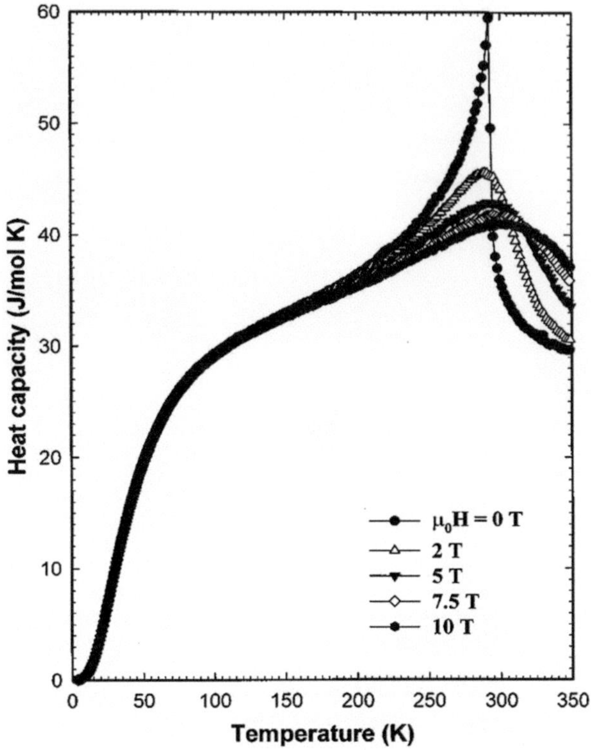

Fig. 4.7 Heat capacity of single-crystal Gd undergoing a second-order transition from the para- to ferromagnetic state near 295 K. The magnetic field is applied parallel to the [0001] direction [38]

4.2.3.2 Calculation of the Magnetocaloric Effect from Heat Capacity

The finite width of the C_p peak in the materials with the non-uniform magnetic phase transition allows one to calculate the entropy as a function of temperature using Eq. (4.1). We note that the lowest integration limit in Eq. (4.1) is 0 K. It is, however, clear that no measurement can be taken at absolute zero. Thus, Eq. (4.1) allows one to determine S to within a constant S_0:

$$S(T, H) = \int_{T_0}^{T} \frac{C_p(T', H)}{T'} dT' + S_0, \quad (4.17)$$

where T_0 is the lowest temperature of the measurement. This constant S_0 is not an arbitrary additive constant, since it is known that the entropy is a completely calculable quantity

$$S = k_B \ln \Omega, \quad (4.18)$$

where k_B is the Boltzmann constant and Ω is the number of possible configurations accessible to the system [39]. The value of S_0 defined as

$$S \xrightarrow{T \to 0^+} S_0, \tag{4.19}$$

where $T \to 0^+$ stands for practically achievable very low temperature and depends only on the type of atomic nuclei, but not on the energy or external parameters of the system [40], i.e. $S_0(H_i) = S_0(H_f)$. We can neglect S_0 in the calculation of the entropy change

$$\Delta S(T, \Delta H) = \int_{T_0}^{T} \frac{C_p(T', H_f)}{T'} dT' - \int_{T_0}^{T} \frac{C_p(T', H_i)}{T'} dT', \tag{4.20}$$

but one should be aware that the constant S_0 will introduce an error in the determination of S from Eq. (4.17) and correspondingly to $\Delta S(T, \Delta H)$ obtained from Eq. (4.20). This error is unknown and is small, as long as T_0 is sufficiently low. It is, thus, crucial to record heat capacity curves at as low temperature as possible. In practical laboratory measurements, the temperature T_0 is usually about 2 K.

The effect of selecting different T_0 is demonstrated in Fig. 4.5d, showing entropy change ΔS of a $LaFe_{11.6}Si_{1.4}H_{1.6}$ alloy calculated from the heat capacity data in Fig. 4.5a using Eq. (4.20). The lowest possible measurement temperature was 2 K. Increasing the first integration limit to $T_0 = 20$ K introduces a difference of 1 % in the maximum entropy change ΔS_{max}. Starting the integration at $T_0 = 200$ K, which is well below the temperature of the transition from the ferro- to paramagnetic state, leads to a ΔS_{max} difference of 10 % compared to $T_0 = 2$ K. Thus, the error increases significantly when T_0 is selected far away from $T \to 0^+$.

In the calculation of the entropy, it is important to take into account the fact that S is a large value (~10^2–10^3 J/kg K or ~10^3–10^4 J/m^3 K) and the magnetocaloric effect, i.e. the entropy change ΔS and the adiabatic temperature change ΔT_{ad}, is a small value. Therefore, the accuracy of the MCE calculations using heat capacity data is crucially dependent on the accuracy of the heat capacity measurements and data processing.

The accuracy of the MCE determination was analysed by Pecharsky and Gschneidner [41, 42]. The authors point to the intrinsic uncertainty of the heat capacity measurement in the heat pulse calorimeter in the vicinity of the first-order transition arising due the existence of the latent heat (see Sect. 4.2.3.1) and requiring judicious adjustment of the heat pulse ΔQ_{pulse}.

Yet another experimental procedure allowing to separate a latent heat contribution is a temperature-modulated heat capacity technique developed by Morrison et al. [43]. The technique allows to measure C_p of small samples (~microgram) and determine the entropy change ΔS by integrating the heat capacity data and adding in the latent heat, where there is one.

C_p measurements with reduced data collection time can be performed using a Peltier cell differential scanning calorimeter (DSC). In the design reported by Basso et al. [44, 45], the Peltier cells are used both as heat flow sensor and heat pump. The magnetic field is provided by an electromagnet; the sample mass is ~0.1 g. The total heat flow to the sample is a sum of the C_p component and the kinetic or latent heat component. This method also allows a *direct* entropy change measurement ΔS.

In order to fully characterise the MCE (ΔS and ΔT_{ad}) of a certain material, one, thus, needs precise C_p measurements over a large temperature range and in a number of magnetic fields, which can be a long and sophisticated procedure. On the other hand, the knowledge of the magnetisation as a function of field and temperature $M(T, H)$ equally allows the determination of ΔS and, due to the relative simplicity of the magnetisation measurements, is often preferred to the C_p method.

4.2.3.3 Calculation of the Magnetocaloric Effect from Magnetisation

Entropy Change from the Maxwell Relation

The calculation of the entropy change ΔS from magnetisation $M(T, H)$ is performed by using Maxwell relations. Maxwell relations provide the connection between the parameters on the right side of Eq. (4.8) and are obtained on the basis of the independence of the second derivatives on the order of differentiation, $\partial^2 F/\partial T \partial H = \partial^2 F/\partial H \partial T$; here we consider only H and T as external parameters [39]. The Maxwell relation connecting S, M, H and T can then be derived from Eq. (4.8) as

$$\left(\frac{\partial S}{\partial H}\right)_T = \left(\frac{\partial M}{\partial T}\right)_H. \tag{4.21}$$

The integration of Eq. (4.21) allows one to calculate the entropy change from the measured magnetisation $M(T, H)$ as

$$\Delta S(T, \Delta H) = \int_{H_i}^{H_f} \left(\frac{\partial M(T, H)}{\partial T}\right)_H dH \tag{4.22}$$

under the assumption that the following conditions are fulfilled:

(i) The sample is in the thermal equilibrium.
(ii) There exists the derivative of the magnetisation with respect to temperature $(\partial M/\partial T)_H$.

The first condition is a consequence of the fundamentals of thermodynamics: it stems from fact that the relations (4.21) and (4.22) are derived from the free energy and are, thus, only valid if the initial and final states are equilibrium states. The derivative of the magnetisation should exist, i.e. the magnetisation should not have a discontinuity, in order for $(\partial M/\partial T)_H$ to be integrable. Both requirements are fulfilled for materials with the second-order transition; the numerical integration of the magnetisation data is straightforward and allows one to obtain the entropy change ΔS as a function of temperature for various magnetic field changes.

In the case of materials with the first-order transition, several problems do arise when using the outlined calculation of relation (4.22). The first problem is the discontinuity in the magnetisation (Fig. 4.2). In the "textbook" example, the derivative of the magnetisation with respect to temperature does not exist. The magnetisation of real materials deviates from the behaviour expected from the thermodynamics: even in a high-purity single-crystal dysprosium, which shows a close to singularity sharp peak in the heat capacity [46], the magnetisation $M(T)$ in various magnetic fields is not discontinuous (Fig. 4.8a). As a consequence, the derivative of the magnetisation with respect to temperature exists even in very pure and homogeneous materials. That means the second condition for the calculation of the entropy change using Eq. (4.22) is fulfilled.

The numerical integration of Eq. (4.22) requires the knowledge of the isofield magnetisation $M(T)$ measured in different magnetic fields in order to construct the derivative $(\partial M/\partial T)_H$. However, high-accuracy isofield measurements are relatively slow and unreliable in zero and very low magnetic fields meaning that for isofield magnetisation measurements, the lowest integration limit in Eq. (4.22) is always $H_i > 0$. Therefore, isothermal magnetisation measurements as a function of the magnetic field are preferred. Similarly to the isofield magnetisation curves, the isothermal magnetisation in real materials does not show a discontinuity but is spread over some field region (Fig. 4.8b). The following procedure is adapted for the calculation of ΔS:

(a) $M(T)_H$ curves are constructed from $M(H)_T$ curves.
(b) $\partial M(T,H)/\partial T$ is calculated as a function of temperature T.
(c) $\partial M(T,H)/\partial T$ is constructed as a function of the field H.
(d) Eq. (4.22) is numerically integrated to obtain $\Delta S(T, \Delta H)$, where $\Delta H = H_f - H_i$.

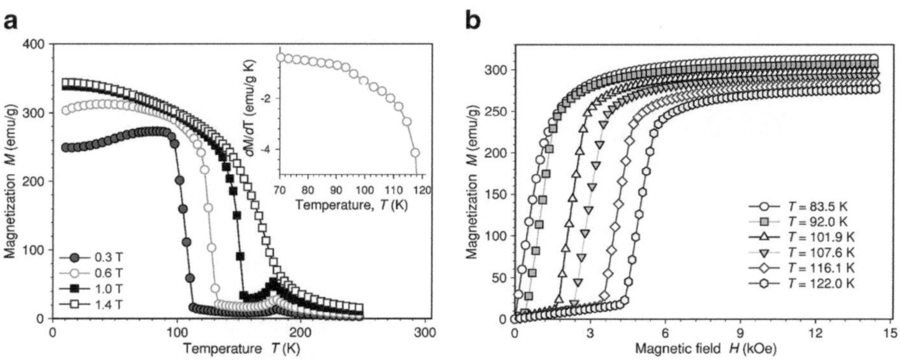

Fig. 4.8 Magnetisation M of single-crystal Dy; magnetic field applied parallel to the a-axis. (**a**) Isofield M(T) measured during heating. The inset shows the derivative of the magnetisation with respect to temperature between for $\mu_0 H = 0.6$ T. (**b**) Isothermal M(H) curves recorded after heating to 250 K, i.e. to the paramagnetic state, and subsequent cooling in a zero magnetic field to the measurement temperature (adapted from [46])

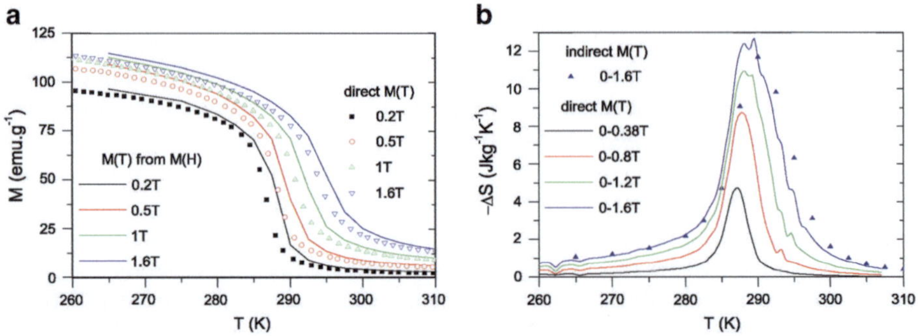

Fig. 4.9 Entropy change determined from isofield and isothermal magnetisation measurements in $LaFe_{11.384}Mn_{0.356}Si_{1.26}H_{1.52}$. (**a**) Magnetisation as a function of temperature obtained from M(T) curves measured in various constant magnetic fields and extracted from M(H) curves measured at various constant temperatures. (**b**) The entropy change ΔS calculated from the M(T) curves using Eq. (4.22) (adapted from [47])

The equivalency of magnetisation measurements recorded in dependence on the temperature and the magnetic field is illustrated in Fig. 4.9.

Note that if the magnetisation is measured in an open magnetic circuit, the magnetisation curves should always be corrected by the appropriate demagnetising factor N to obtain the internal field

$$H_{\text{int}} = H_{\text{appl}} - NM, \tag{4.23}$$

where H_{appl} is the applied magnetic field. The equation is strictly valid only for a uniformly magnetised sample having the form of an ellipsoid. In real samples, local magnetic fields are strongly inhomogeneous, and the correction using Eq. (4.23) gives only a rough estimate of demagnetising fields. It is, however, crucial to use a demagnetising factor to obtain approximate internal fields, even in non-ellipsoidal shapes where the demagnetizing field is not quite uniform. The demagnetising field can be neglected only in samples for which $N \approx 0$, e.g. in long needles with magnetisation direction parallel to the axis or in thin films with magnetisation direction parallel to the plane.

Whereas the first condition (i) is generally always valid for materials undergoing the second-order magnetic phase transition, in the vicinity of the first-order transition, the materials can be driven out of the equilibrium state during the isothermal magnetisation measurements. The violation of the equilibrium condition can lead to appreciable calculation errors, usually resulting in the overestimation of ΔS.

An example of the consequence of the violation of the equilibrium condition (i) is given for $Gd_5Si_2Ge_2$ in Fig. 4.10a, where ΔS calculated from magnetisation curves using Eq. (4.22) is compared to ΔS calculated from the heat capacity using Eq. (4.16). The heat capacity Cp was measured at different applied heat pulses ΔQ_{pulse} and for different sample masses. When the entropy discontinuity, i.e. the latent heat at the

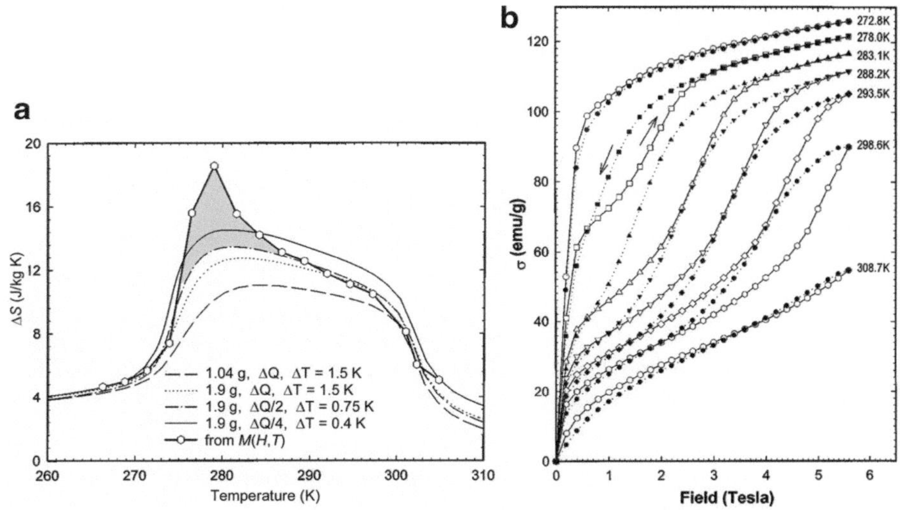

Fig. 4.10 (**a**) Entropy change ΔS in a field change $\mu_0 H = (0–5)$ T in $Gd_5Si_2Ge_2$ calculated from magnetisation M(T, H) data and heat capacity C_p measured at different applied heat pulses ΔQ_{pulse} leading to various temperature rise ΔT and for different sample masses [42]. (**b**) Isothermal magnetisation curves for $Gd_5Si_2Ge_2$ measured in the vicinity of the first-order transition on increasing and decreasing magnetic field [10]

first-order transition, is correctly determined by adjusting ΔQ_{pulse} (in this case corresponding to some value $\Delta Q/2$), ΔS calculated from $M(T, H)$ and Cp data should have the same value. However, in this case, there is a discrepancy between the ΔS values in the temperature range of about 275–285 K (shaded area in Fig. 4.10a).

In materials undergoing a series of first-order para- to ferromagnetic transitions, such as during the non-uniform magnetic phase transition, with each transition occurring at a particular Curie temperature T_{ci} or magnetic field H_{mi}, three equilibrium states can be distinguished: a paramagnetic (PM) state, where all the crystallites in the material are paramagnetic; a ferromagnetic state (FM), where all the crystallites are ferromagnetic; and a PM–FM state, where a mixture of the crystallites/phases being either in the PM or FM state is observed (Fig. 4.11).

To analyse the apparent thermal equilibrium states in $Gd_5Si_2Ge_2$, we follow the isothermal magnetisation curves recorded in increasing and decreasing magnetic field around $T_c \approx 277$ K (Fig. 4.10b). Above the Curie temperature and at zero field, the thermal equilibrium state of $Gd_5Si_2Ge_2$ is the PM state. On increasing magnetic field to 5 T, the material transforms to the FM state. In the temperature range below approximately 300 K and in the field of 5 T, the FM state is the equilibrium state. On the subsequent field removal, the sample transforms back to the PM state. At 278 K, i.e. still above T_c, the application and subsequent removal of the magnetic field does not bring the system to the equilibrium state, but some part of the sample has transformed to the FM state in the magnetic field. This is apparent from the largely disappeared metamagnetic-like behaviour, i.e. the S-shaped magnetisation

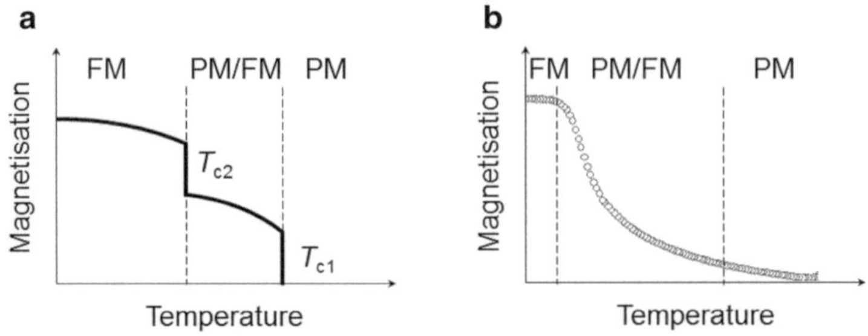

Fig. 4.11 Temperature variation of the magnetisation in ferromagnetic materials with three possible equilibrium state: paramagnetic (PM), ferromagnetic (FM) and their mixture. (**a**) Schematic representation of two ideal first-order transitions at temperatures T_{c1} and T_{c2}. (**b**) Magnetisation variation in real materials with a non-uniform magnetic phase transition

curve is less pronounced, recorded on reducing the field from 5 to 0 T. That means that a mixture of the PM and FM phases exists at a zero magnetic field above the Curie temperature. However, above the Curie temperature in a zero magnetic field, the PM state is the equilibrium state. Thus, during the experiment, the system was brought out of the equilibrium. This is a consequence of the finite hysteresis present at first-order phase transitions due to the presence of multiple local minima in the free energy separated by an energy barrier (hysteresis will be discussed in more detail in Sect. 4.4.3). Due to the thermal hysteresis, the part of the sample that has been transformed to the FM state does not return to the PM state once the magnetic field is reduced at the same temperature (here 278 K), but will need to be heated to a temperature above 278 K; the exact value of this temperature will depend on the hysteresis width.

It is common to use the isothermal magnetisation curves recorded during the field ramp-up for the calculation of ΔS. Obviously, there exists a sudden jump between the magnetisation at two temperatures (278 and 272.8 K) brought about by the non-equilibrium (zero field) initial state at which the magnetisation change during the field ramp-up at 272.8 K is recorded (Fig. 4.10b). Subsequent numerical integration of the data leads to the entropy change overestimation and the appearance of a "spike" in the ΔS curve (shaded area in Fig. 4.10a).

In the extreme cases, such as those observed in MnAs-based materials, the deviation from the thermal equilibrium results in what was called "colossal entropy change overestimation" [48] resulting in the appearance of extremely large, unphysical spike (Fig. 4.12, $M(H)_T$ "continuous" cooling). Thus, in materials, where the application and subsequent removal of the magnetic field induces a state other than equilibrium, a special procedure for the measurement of the isothermal magnetisation as a function of the magnetic field $M(H)_T$ is required. This procedure should allow to bring the sample to the thermal equilibrium state in order Eq. (4.22) to be applicable for the calculation of ΔS.

An often used measurement procedure is a so-called "continuous" cooling or heating in the vicinity of the transition temperature and consists of the following steps:

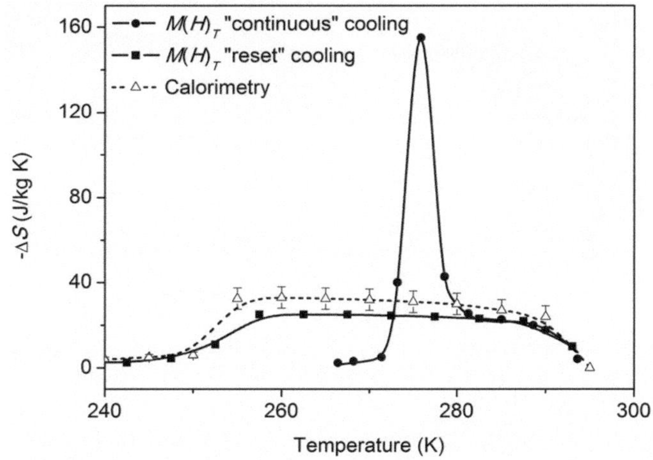

Fig. 4.12 Entropy change ΔS in $Mn_{0.985}Fe_{0.015}As$ in a field change $\mu_0\Delta H = (0–8)$ T obtained using Eq. (4.22) from magnetisation $M(H)_T$ recorded during "continuous" cooling resulting in the colossal entropy change overestimation and from $M(H)_T$ recorded during "reset" or also called "loop" cooling compared to the entropy change ΔS obtained from Cp measured in a calorimeter (data from [48])

(a) A magnetisation curve $M(H)_{T_i}$ is recorded at a temperature $T_i = T_1$, e.g. above the Curie temperature, $T_1 > T_c$. Alternatively, the initial measurement temperature T_1 can be selected below T_c.
(b) The temperature is reduced (increased) to a temperature $T_2 < T_1$ ($T_2 > T_1$) and the next magnetisation isotherm $M(H)_{T_2}$ is recorded.
(c) The temperature is reduced (increased) until the desired temperature range in the vicinity of the transition temperature is covered.

Such a procedure should only be used, if the application and subsequent removal of the magnetic field returns the system to the equilibrium state. Otherwise, it can result in erroneous data and appreciable ΔS overestimation (Figs. 4.10a and 4.12).

An alternative measurement procedure is what was called a "reset" or "loop" heating or cooling [48, 49]. The procedure is as follows:

(a) A magnetisation isotherm $M(H)_{T_i}$ is recorded at a temperature $T_i = T_1$, T_1 being either above or below T_c.
(b) After the isotherm has been recorded, the sample is not directly cooled (heated) to the next measurement temperature $T_2 < T_1$ ($T_2 > T_1$) but is heated above (cooled below) the transition temperature to ensure that the equilibrium state is reached. This equilibrium state can be, for example, PM above T_c or FM below T_c.

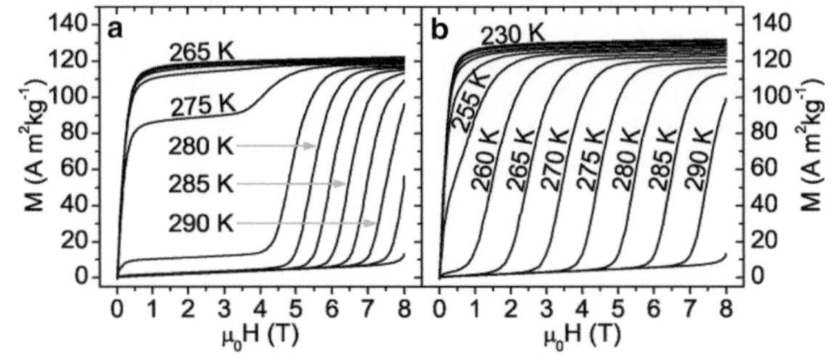

Fig. 4.13 Magnetisation isotherms of $Mn_{0.985}Fe_{0.015}As$ obtained by using (**a**) a "continuous" cooling procedure and (**b**) a "reset"/"loop" procedure [48]

(c) Only after the sample has been "reset" to the equilibrium state, is it cooled (heated) to the next measurement temperature T_2, at which an isotherm $M(H)_{T_2}$ is recorded.

(d) The steps (b) and (c) are repeated until the desired temperature range in the vicinity of the transition temperature is covered.

This procedure ensures that the sample is the equilibrium state prior to each measurement. An example of the influence of the selected measurement procedure on the field behaviour of the magnetisation isotherms is shown in Fig. 4.13 for $Mn_{0.985}Fe_{0.015}As$. The strikingly different shapes of the magnetisation curves in the vicinity of T_c are the result of different fractions of the para- and ferromagnetic phases at the initial point of the measurement ($T = T_i$, $H = 0$), being an equilibrium phase combination in the case of the "reset"/"loop" procedure (Fig. 4.13b) and out of equilibrium in the case of "continuous" cooling procedure (Fig. 4.13a).

The unphysical "colossal" ΔS peak obtained by using the magnetisation data from the "continuous" cooling procedure is not observed in ΔS calculated from the magnetisation data obtained by the "reset"/"loop" procedure (Fig. 4.12). Moreover, the ΔS peak resembles to within the experimental error the magnetic entropy change obtained from calorimetric measurements of the heat capacity with the corresponding latent heat determination (see also [50, 51]).

A point that should be stressed here is that the entropy change determined from the magnetisation measurements using Eq. (4.22) reflects the total entropy change and not only its magnetic part, as is often referred to in the literature. Although the application of the magnetic field has a noticeable effect only on the magnetic system of the solid, the result that is recorded in the magnetisation as well as in the heat capacity measurements contains the phonon (lattice) S_{ph} and, in the case of the localised moment magnets, the electronic part of the entropy S_{el} as well. The fact that to within the experimental error the entropy change values obtained from the magnetisation (as long as the applicability conditions of the Maxwell relations are not violated) and heat capacity data are equal serves an experimental evidence of this statement.

Entropy Change from the Clausius–Clapeyron Equation

The entropy change at first-order transitions can also be determined through the use of the Clausius–Clapeyron equation; here, we consider H and T as external parameters:

$$\frac{dH}{dT} = -\frac{\Delta S}{\Delta M}, \tag{4.24}$$

where $\Delta S = S_2 - S_1$ and $\Delta M = M_2 - M_1$, with M_1 and M_2 being the magnetisation of the different magnetic states 1 and 2 corresponding to the entropies S_1 and S_2. The Clausius–Clapeyron equation relates the slope of the phase-equilibrium line (Fig. 4.14a) to the entropy change and magnetisation change of the system undergoing a phase transition. This equation is derived from the free energy of a single-component system consisting of two phases in equilibrium and relies on the equality of the chemical potentials of both phases/states, $d\mu_1 = d\mu_2$, which is obviously fulfilled at the first-order transition [39]. The existence of the two phase-equilibrium lines is due to the hysteresis present at the first-order transition (Fig. 4.14), if the system is to transform from the phase 1 to the phase 2 in a finite amount of time [25]. Taking into account Eq. (4.13), we may also rewrite Eq. (4.24) as

$$\frac{dH}{dT} = -\frac{L}{T\Delta M}. \tag{4.25}$$

The phase-equilibrium lines can be readily constructed from the magnetisation measurements by measuring the field dependence of the transition temperature T_{ci} or the temperature dependence of the critical field H_{mi}. However, in real materials, the magnetisation at the first-order transition does not show a sharp jump—discontinuity—in the magnetic field (cf. to Fig 4.14b showing the discontinuity in M). This makes a high precision definition of the point, where the chemical potentials of both phases are equal, difficult. Thus, the determination of the magnetisation change ΔM at the transition is linked to a higher uncertainty.

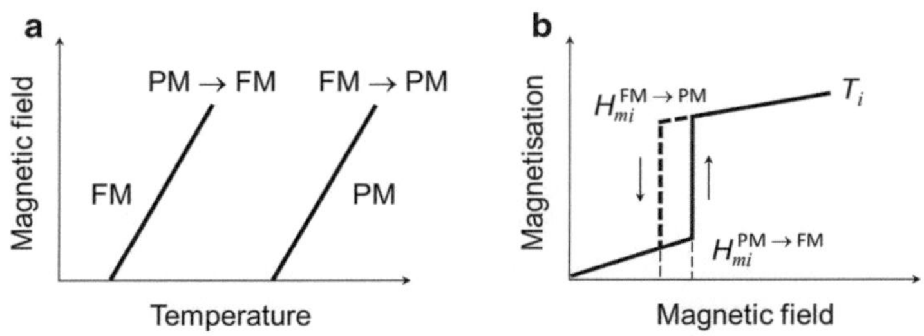

Fig. 4.14 A solid undergoing a first-order transition accompanied by a hysteresis. (**a**) Schematic phase diagram showing phase-equilibrium lines of different states, paramagnetic (PM) and ferromagnetic (FM), when crossing the line from the PM to the FM state and vice versa. (**b**) Schematic isothermal magnetisation at a temperature T_i showing critical fields of the first-order magnetic phase transition on increasing $H_{mi}^{PM \to FM}$ and decreasing $H_{mi}^{FM \to PM}$ magnetic field

From the form of the Clausius–Clapeyron Eqs. (4.24) and (4.25), it is clear that they provide the maximum entropy change ΔS_{max} that can be obtained in the system undergoing the first-order transition from the phase 1 to the phase 2 at a given temperature T_i, but does not provide the field dependence of ΔS. Thus, by using the Clausius–Clapeyron equation, one effectively obtains less information than from the Maxwell relation by measuring the same amount of data. Examples of the application of the Clausius–Clapeyron equation for the calculation of ΔS can be found, for example, in papers by Giguère et al. [52], Casanova et al. [53] and Fujita et al. [54].

Adiabatic Temperature Change from Combined Heat Capacity and Magnetisation Measurements

Combining Eqs. (4.1) and (4.21) and taking into account the total differential of ds in Eq. (4.4), the following expression for the infinitesimal adiabatic temperature rise is obtained:

$$dT = -\left(\frac{T}{C_p}\right)_H \left(\frac{\partial M}{\partial T}\right)_H dH. \tag{4.26}$$

The adiabatic temperature change is obtained by the integration of Eq. (4.26) as

$$\Delta T_{\mathrm{ad}}(T, \Delta H) = -\int_{H_\mathrm{i}}^{H_\mathrm{f}} \left(\frac{T}{C_p(T,H)}\right)_H \left(\frac{\partial M(T,H)}{\partial T}\right)_H dH, \tag{4.27}$$

where $\Delta H = H_\mathrm{f} - H_\mathrm{i}$. The numerical integration of Eq. (4.27) requires C_p(T, H) curves measured in a high number of fields in order to approach the resolution with which the magnetisation M(T, H) is usually measured. Since heat capacity measurements are slow, the time needed for data collection would be unacceptably long. On the other hand, if the C_p (T, H) data is available, there is no need to use Eq. (4.27) for the determination of ΔT_{ad}, as it can be readily obtained from Eq. (4.3) once the entropy curves for H_i and H_f have been constructed.

The simplification of Eq. (4.27) by considering C_p (T, H) to be independent of temperature and field and treating $T/C_p(T, H)$ in Eq. (4.27) as a constant that can be excluded from the integration leads to erroneous results, since C_p (T, H) is by no means constant, especially at low temperatures and in the vicinity of a phase transition ([41], also Figs. 4.4, 4.5 and 4.7).

Pecharsky and Gschneidner [41] suggested the following method allowing the combined use of the heat capacity measured at constant magnetic fields H_i and H_f and the magnetisation M(T, H) to correctly calculate the adiabatic temperature change:

(a) First, the entropy in the fields H_i and H_f, S(T, H_i) and S(T, H_f), is calculated from the heat capacity using Eq. (4.17).
(b) The magnetisation M(T, H) measurements are then used to calculate the entropy change $\Delta S(T, \Delta H)$ using Eq. (4.22).

(c) The missing $S(T, H_i)$ or $S(T, H_f)$ is calculated from Eq. (4.2).
(d) Determination of ΔT_{ad} is performed using Eq. (4.3) with the knowledge of $S(T, H_i)$ and $S(T, H_f)$.

4.2.3.4 Direct Adiabatic Temperature Change Measurement

Adiabatic temperature change ΔT_{ad} can also be determined directly in a device allowing the application of the magnetic field combined with the temperature measurement. For the determination of the direct adiabatic temperature change, mostly home-built equipment is used (reviewed recently by Smith et al. [55]). The measurement principle is straightforward: one needs to ensure adiabatic conditions during the measurement, provide the magnetic field that can be varied in time and record the sample temperature. The temperature can be recorded directly using a thermocouple or a semiconductor sensor (e.g. Cernox). The temperature change can also be detected by an infrared detector, which has an additional advantage of a spatial resolution [56], or by an acoustic detection of a temperature wave [57].

In the direct measurement of ΔT_{ad}, the application of the magnetic field must be performed fast enough so that the material under study does not exchange heat with the surrounding media. Note that by the definition, in the adiabatic process, external conditions must change slowly in comparison to the process of equilibration in a given material [25]. In a practical implementation, the adiabatic process is often referred to as "sufficiently fast". Implied is the condition, which ensures that the material is thermally isolated and the process occurs so fast that the material cannot exchange heat with the surroundings. The condition of slowness compared with the processes leading to the establishment of the equilibrium within the solid is tacitly assumed satisfied.

The practically used magnetic field rate dH/dt for the adiabatic temperature change determination depends on the type of the measurements. These can be divided into two categories: (1) a stationary sample and changing magnetic field or (2) a sample moving relatively to a source of the constant magnetic field. In the stationary sample method, a sample can be placed in vacuum and the adiabatic conditions are thus easier to realise. However, the magnetic field source should be chosen accordingly, in order not to compromise the adiabatic conditions. In this respect, a magnetic field change provided by a superconducting magnet (dH/dt ~0.01 T/s) and electromagnet (dH/dt ~0.1–1.0 T/s) may be too slow for ensuring adiabatic conditions, and instead moving permanent magnets, e.g. Halbach cylinders (dH/dt ~0.1–10 T/s), or pulsed magnetic fields (dH/dt ~10–100 T/s) are preferred. A fast dH/dt can be realised in a static magnetic field by a fast movement of the sample in and out of the high-field region by a piston [58].

An example of the direct adiabatic temperature ΔT_{ad} as a function of field and temperature is presented in Fig. 4.15 for a polycrystalline Gd. The experimental set-up consists of a sample rod with integrated heaters, a Hall sensor for magnetic field measurement and a sample platform insulated by heat shields (reports on direct ΔT_{ad} measurement of other materials using this set-up can be found, e.g., in

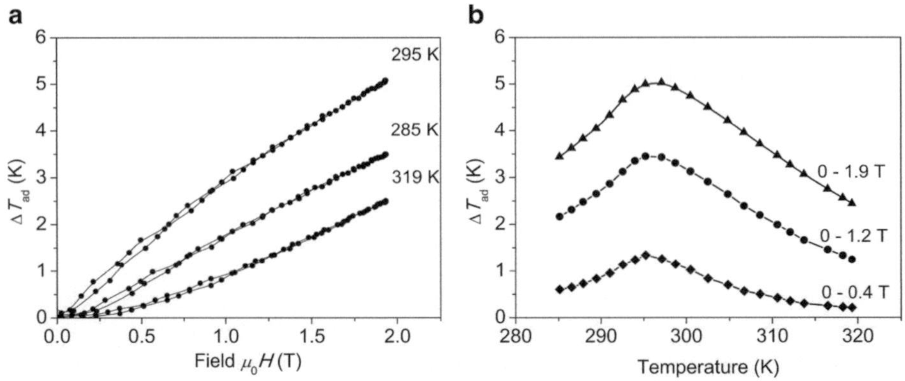

Fig. 4.15 Field (**a**) and temperature (**b**) dependence of the adiabatic temperature change ΔT_{ad} of polycrystalline Gd (purity 99.7 %) measured directly in the vicinity of T_c with a field change rate of 1.8 T/s

[14, 22, 47, 59, 60]). The sample rod is evacuated and inserted into a liquid nitrogen Dewar. ΔT_{ad} measurement is performed by a copper–constantan thermocouple sandwiched between two plates of the sample (size about $1 \times 5 \times 10$ mm^3) in the magnetic field produced by Halbach cylinders. The Halbach cylinders rotate at a speed of up to 2 T/s and produce a peak field of 1.93 T. The fast field change and good thermal insulation ensure almost adiabatic conditions (slight difference between the $\Delta T_{ad}(H)$ measured on increasing and decreasing the field in Fig. 4.15a is caused by an experimental uncertainty). The directly measured ΔT_{ad} is in good agreement with adiabatic temperature change calculated from Cp(T, H) measurements [47].

4.3 Magnetic Refrigeration

Heat pumping is a technologically important application of the magnetocaloric effect. Starting with the experimental demonstration of the adiabatic demagnetisation of the paramagnetic gadolinium sulphate octahydrate by Giauque and MacDougall [6] used for heat pumping within a few degrees of absolute zero, it was not until 1976, when it was realised that a much larger MCE can be obtained in a ferromagnet in the vicinity of its Curie temperature T_c and the effect can be used for heat pumping near room temperature, when a material with a T_c close to 295 K, such as Gd metal, is used [8]. In a heat pump, when a work W is done on it, the heat is transported from a reservoir at lower temperature T_{cold} to a reservoir at higher temperature T_{hot}. Two operation modes are possible: (1) a refrigerator, when the heat is removed from the colder reservoir T_{cold} and rejected to the surroundings T_{hot} (e.g. a domestic fridge, supermarket chillers and refrigerated display cabinet), and

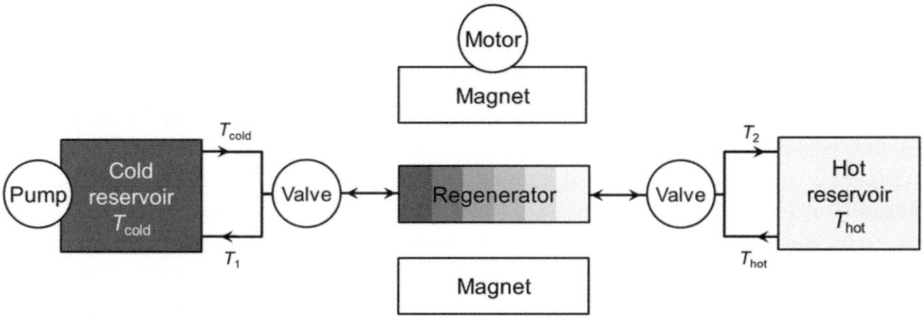

Fig. 4.16 Schematic diagram of the active magnetic regenerator (AMR)

(2) a heat pump, when the heat from the surroundings T_{cold} is transported to the interior T_{hot} (e.g. residential heating).

In the following, a brief introduction to the principle of the magnetic heat pumping near room temperature using the magnetocaloric effect is given. For more detailed information, the reader is referred to the papers by Brown [8], Barclay [61], Zimm et al. [4, 62], Gschneidner and Pecharsky [3], Rowe [63], Tura and Rowe [64], Engelbrecht et al. [65], Jacobs et al. [66] and references therein.

In the adiabatic demagnetisation, the operating temperature span of a refrigerator $\Delta T_{span} = T_{hot} - T_{cold}$ is limited by the adiabatic temperature change ΔT_{ad} of the magnetocaloric material, which is generally below 5 K in practical magnetic fields and near room temperature [8]. Therefore, in the first magnetic refrigerators that used the adiabatic demagnetisation, only a small temperature span $\Delta T_{span} \approx \Delta T_{ad}$ could be achieved. A magnetic regeneration technique demonstrated by Brown in 1976 was a major step towards the near room temperature refrigeration, as this technique allows obtaining temperature spans much larger than the maximum observed ΔT_{ad} [8]. In 1982, the active magnetic regenerator (AMR) concept was introduced, in which the regeneration is achieved by the active material itself, i.e. the magnetic material simultaneously acts as the magnetic refrigerant and the regenerator [61, 67].

A schematic diagram of the AMR is shown in Fig. 4.16. In its simplest configuration, the AMR consists of regenerator containing a magnetic solid (magnetic refrigerant), a source of the magnetic field and accessories (motor, pump, exchange fluid flow control valves and heat exchangers). The motor is required to provide a magnetic field change. The pump pushes the exchange fluid (e.g. water or inert gas) through the AMR to transport the heat generated or absorbed from magnetising and demagnetising the magnetic refrigerant to a hot and a cold reservoir (heat exchanger). For the fluid to be transportable, the magnetic refrigerant should have an arrangement permeable to the fluid and can be, for example, in the form of (spherical) particles, parallel plates or a block of the magnetocaloric material with channels inside.

A bed with the magnetic refrigerant is inserted into a magnetic field. The magnetic field can be produced by an electromagnet, superconducting magnet or a permanent magnet. The advantage of the permanent magnets is that no power and no cooling are required to generate the magnetic flux density. It should be noted, however, that electrical power, though low, is still required to produce the magnetic field change ΔH by either moving (rotating) the permanent magnet (Fig. 4.16) or a bed with the magnetic refrigerant. Magnet cost is often critical for applications, the magnetic cooling not being an exception. In fact, permanent magnet assembly represents the major cost of a magnetic cooling system. Therefore, optimised permanent magnet designs are crucial for the AMR and have been recently compared by Bjørk et al. [68]. With modern high-performance permanent magnets based on the $Nd_2Fe_{14}B$ phase, magnetic fields of about 2 T can be achieved in a Halbach-type array. However, cost considerations may push the use of simpler magnetic circuits with the maximum magnetic field of about 1 T. A geometry of the magnetic refrigerant should be selected so as to achieve as low demagnetising factor N as possible, thus maximising the usage of the available magnetic field; see Eq. (4.23).

The AMR cycle has four stages: (1) magnetisation, (2) hot blow, (3) demagnetisation and (4) cold blow [66]. During magnetisation (1), the fluid is kept stationary while a magnetic field is applied to the regenerator containing magnetocaloric material. This raises the regenerator temperature by the amount equal to ΔT_{ad}. In the step (2), the magnetic field is kept on and the fluid at a temperature T_{cold} is pushed from the cold to the hot side. The fluid picks up heat from the magnetocaloric material, and, thus, the temperature of the fluid rises as it passes through the regenerator. The fluid exits the regenerator at the temperature T_2 and can be circulated through a heat exchanger, where it gives up heat to the environment and returns to the temperature $T_{hot} < T_2$. In the demagnetisation step (3), the fluid flow is stopped and the magnetic field is turned off. This causes the regenerator temperature to drop. In the final step (4), the fluid at the temperature T_{hot} is pushed from the hot to the cold end. The fluid is cooled as it passes the regenerator with the magnetocaloric material reaching the temperature $T_1 \leq T_{cold}$. The colder fluid exiting the regenerator during the cold blow can be circulated through a cold-side heat exchanger, picking up heat from the refrigerator and allowing it to maintain its colder temperature. The fluid exits the cold-side heat exchanger at temperature T_{cold} and with this the AMR cycle is completed. The system performance depends crucially on the performance of the magnetocaloric material, heat transfer characteristics in the regenerator, the applied magnetic field strength and cycle parameters such as operating frequency and fluid flow rate [65].

The fluid moving back and forth through the magnetocaloric material bed builds up a temperature span ΔT_{span} between the cold and hot sides of the regenerator, which is larger than ΔT_{ad} of the material. A temperature gradient is created in the regenerator bed after a steady state has been reached. Since each regenerator segment sees only a small temperature span, a layered bed can be designed with the position of each layer chosen so that it contains a magnetocaloric material operating close to its Curie temperature. This is schematically illustrated by differently shaded areas of the regenerator in Fig. 4.16. The performance of an AMR containing the layered bed can be greatly optimised as compared to a bed containing a single material, and technologically relevant temperature spans of 20–30 K (and more) can be obtained. The majority of the reported AMR designs use gadolinium as magnetic refrigerant, a benchmark material against which performance of other materials is frequently compared [3, 69]. Magnetic refrigerators using Gd as refrigerant are capable of providing a cooling power in the range relevant for household fridge appliances, i.e. 100–1000 W [62, 64, 65]. The use of regenerator containing a layered bed with giant MCE materials, the so-called graded regenerator, can further improve the AMR performance in terms of the cooling power. For example, a rotary magnet magnetic refrigerator employing a layered bed containing $La(Fe,Si)_{13}H_y$ hydrides with different Curie temperatures has been shown to have superior performance compared to Gd: at 14 K span, the layered $La(Fe,Si)_{13}H_y$ regenerator provides 2.7 times higher cooling power of Gd (400 W vs. 150 W) [70]. Recently, a large-scale magnetic refrigerator, which uses a permanent magnet assembly with a peak field of 1.44 T and contains six layers of $La(Fe,Si)_{13}H_y$ with a total mass of 1.52 kg, was demonstrated to provide about 2 kW of cooling power over a temperature span of 12 K with electrical coefficient of performance >2 [66].

More than 40 prototype designs of magnetic refrigerators have been disclosed until 2010 [69]. A number of existing magnetic cooling (prototype) devices might well exceed this number: more than 140 patents and patent applications in the field of magnetic cooling devices are returned from the search in the patent collection of the World Intellectual Property Organization [71].

According to the principles of thermodynamics, it is possible to use a heat pump in a reverse way, i.e. as a heat engine [39]. The question is whether the MCE can be used for a direct generation of power. The early concept of a "thermomagnetic motor" has been proposed by Stefan [7, 72], where he indicated that in a ferromagnet (iron) the fall of the magnetisation to zero at its Curie temperature can be used to produce work. Thermomagnetic generators were patented by Tesla and Edison at the end of the nineteenth century [73, 74]. Thermomagnetic generators convert heat to electricity without a mechanical interface. A thermomagnetic generator in its simplest form consists of a shunt of magnetic (or magnetocaloric) material placed in the air gap of a magnet (Fig. 4.17, [75]). Means are provided to cycle the temperature of the shunt material. The magnetic material is further surrounded by a solenoid. As shunt temperature increases, magnetisation decreases and flux decreases. According to the Lenz's law, this will induce voltage in the surrounding coil. By continuously cycling shunt temperature, continuous electrical power production is achieved. By using the regeneration principle and materials with different Curie temperatures, the thermal efficiency and the operation temperature range of the generator can be increased [75, 76]. This thermomagnetic generation effect is, however, not related to the MCE. Apparently, the observation of the MCE at the Curie temperature and during the change of the magnetisation in general is the reason why the thermomagnetic generation is often mentioned in the literature in connection with the MCE [8, 77, 78]. On the other hand, Kirol and Mills [75] indicated that a large change in permeability μ_{mag} with temperature, $d\mu_{mag}/dT$, is favourable for improved efficiency of the thermomagnetic power conversion. Thus, materials with a first-order magnetic phase transition in the temperature range

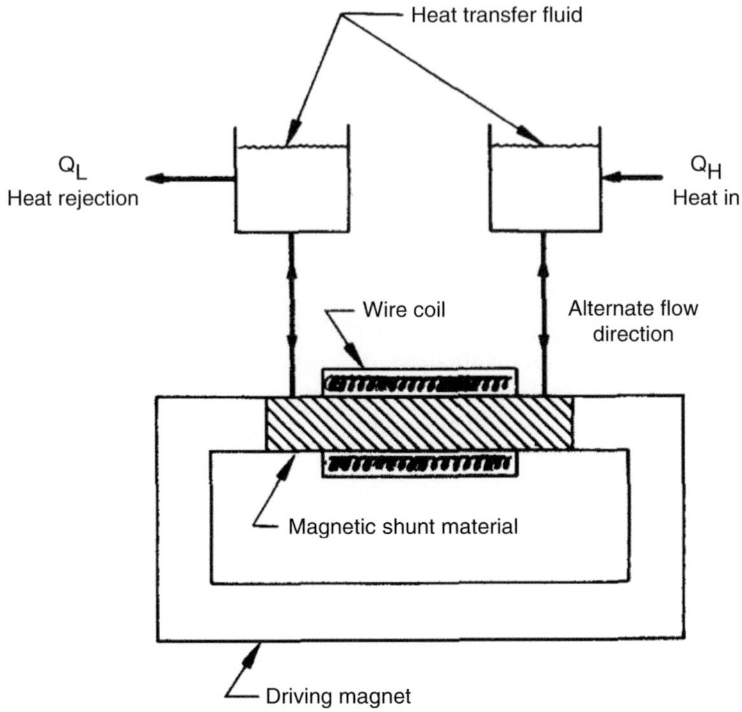

Fig. 4.17 Thermomagnetic generator (after [75])

between 300 and 450 K [77] can be suitable candidates for the purpose, especially for the conversion of low-grade waste heat, where the thermomagnetic power generation may even compete with thermoelectric generation [78].

4.4 Technologically Relevant Properties of Magnetocaloric Materials

When considering magnetic materials for the use as active magnetic refrigerants, a number of parameters should be taken into account. These parameters include both magnetic and non-magnetic properties. Moreover, the search for optimum magnetocaloric materials cannot prioritise specific material properties only; the assessment of the economic merits is equally important. Such factors as the cost of raw materials and their processing including shaping as well as toxicity of the material or its constituents are of high importance for the applications. From known materials, those based on 3d transition metals, Fe and Mn, are currently most technologically attractive materials both from the point of view of their performance and costs. Some of these materials use the most abundant light rare-earth

(RE) elements, La and Ce, in amounts typically below 20 at.%. The use of these light rare earths is currently less critical; La and Ce are usually by-products in the production of Nd used in Nd–Fe–B permanent magnets, which are a likely source of the magnetic field in the magnetic heat pumping. Heavy rare earths, e.g. Gd, are rarer and significantly more expensive than La and Ce.

4.4.1 Magnetocaloric Effect

To be useful for the magnetic refrigeration (heat pumping), a magnetic material should possess a large magnetocaloric effect. As is apparent from Eqs. (4.22) and (4.27), to maximise the magnetocaloric effect, one should maximise the magnetic field change ΔH or/and the derivative of the magnetisation with respect to temperature $\partial M(T,H)/\partial T$. Permanent magnet is the most likely magnetic field source; thus, the magnetic field change is usually limited to below 2 T. Moreover, the magnetocaloric effect cannot grow unlimitedly with $H \rightarrow \infty$ and will eventually saturate in high magnetic fields [23, 79]. Therefore, in order to increase the MCE, the magnetisation derivative should be increased and consequently a large MCE can be expected near magnetic phase transitions.

Whereas the magnetisation variation with temperature is smooth in the materials with a second-order transition, an abrupt change of the magnetisation is observed at a first-order transition (Fig. 4.2). This abrupt change of the magnetisation results in what was called a "giant" magnetocaloric effect [10, 13]. As a result, the entropy change ΔS is concentrated in a narrow temperature range, and the maximum entropy change ΔS_{max} can significantly exceed ΔS_{max} of materials with the second-order transition. This situation is illustrated in Fig. 4.18 for a giant MCE material, $Gd_5Si_2Ge_2$, having ΔS_{max} twice as large as the maximum entropy change observed in Gd. A first-order magnetic phase transition involving a structural change is common for all giant MCE materials (see Sect. 4.5).

At the second-order magnetic phase transitions, ΔS curves regarded as a function of temperature have the caret shape (Fig. 4.19a). By increasing the magnetic field, the height of the peak, i.e. the maximum entropy change ΔS_{max}, increases. The increase of ΔS_{max} can be described by the following dependence:

$$(-\Delta S)_{\max} \sim H^n, \tag{4.28}$$

where n = 2/3 was derived by Oesterreicher and Parker [24] from a mean field model. For materials undergoing the second-order transition, but not following the mean field model, Franco et al. [80] proposed the following expression for $n = 1 + (\beta - 1)/(\beta + \gamma)$, where β and γ are the critical exponents in the Arrott–Noakes equation of state; n takes different values below, above and at the Curie temperature.

With respect to materials with second-order transitions, it turns out that no simple proportionality relation of the form in Eq. (4.28) holds [81]. Rather, besides

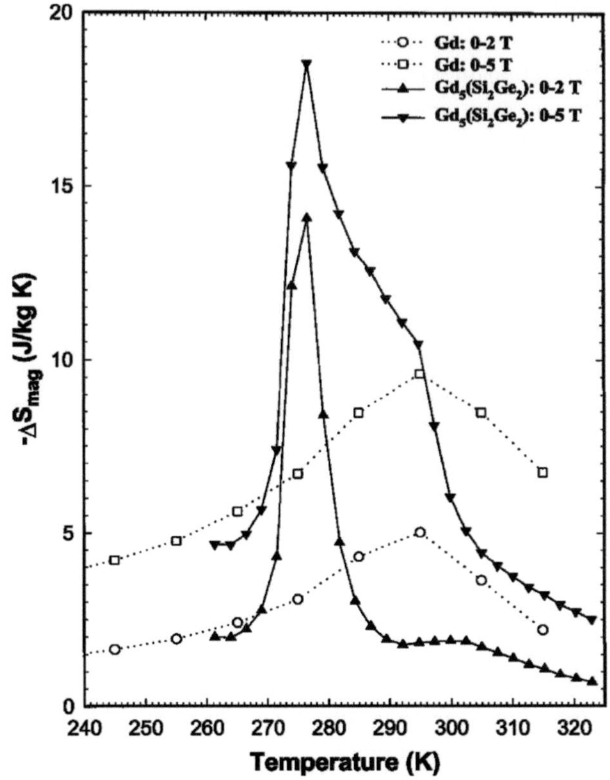

Fig. 4.18 Entropy change ΔS of the $Gd_5Si_2Ge_2$ for a magnetic field change from 0 to 2 T and 0 to 5 T, compared to that of pure Gd as determined from magnetisation measurements [10]

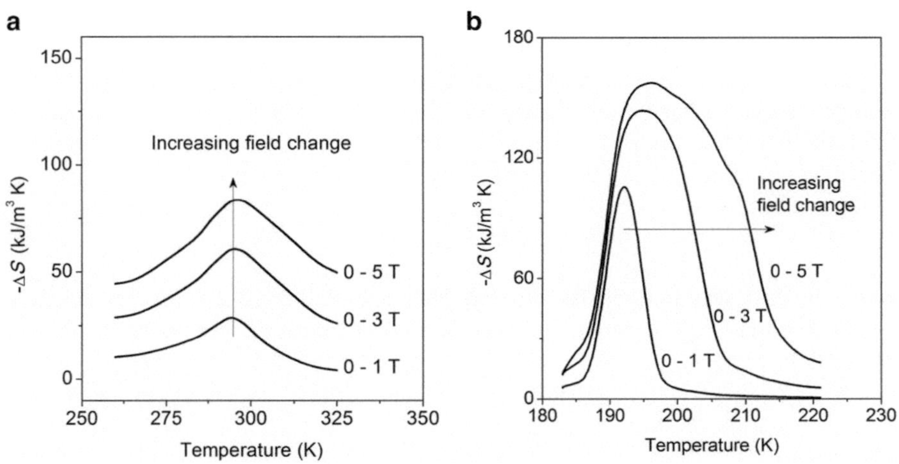

Fig. 4.19 Entropy change ΔS of the polycrystalline Gd (**a**) and $LaFe_{11.6}Si_{1.4}$ (**b**) for a magnetic field change from 0 to 2 T, 0 to 3 T and 0 to 5 T

a term in H^n, ΔS_{max} contains an extra term independent of the field. Based on this observation and by using the Landau expansion, Lyubina et al. [23] obtained the following law describing the field dependence of the maximum entropy change ΔS_{max} in the complete field range:

$$(-\Delta S)_{max} = A(H + H_0)^{2/3} - AH_0^{2/3} + BH^{4/3}, \quad (4.29)$$

where A and B are intrinsic material constants, both being combinations of Landau's coefficients, and H_0 is an extrinsic parameter proportional to the width of the distribution of Curie temperatures that represents the material's homogeneity and/or purity. This law was shown to hold for various materials undergoing a second-order transition [23] and also be valid for the maximum adiabatic temperature change [22]:

$$\Delta T_{ad}^{max} = A'(H + H_0)^{2/3} - A'H_0^{2/3} + B'H^{4/3} \quad (4.30)$$

where A' and B' are combinations of Landau's coefficients, different to A and B.

In materials undergoing the first-order transition, the biggest change of ΔS_{max} takes place in weak to moderate magnetic fields. In higher magnetic fields, the peak height is hardly changing (Fig. 4.19b). The critical field of the transition H_{mi} is temperature dependent (Fig. 4.14), and, thus, the effect of the magnetic field is to asymmetrically broaden the ΔS peak. Despite this different behaviour and the fact that Eq. (4.29) is strictly valid for the second-order transition only, ΔS_{max} in some materials undergoing first-order transitions was shown to follow the same field dependence (Fig. 4.20); the reason for this behaviour could be the non-uniform first-order transition observed in real materials. The applicability of Eq. (4.29) to some materials undergoing the non-uniform first-order transition is corroborated by the observation by Belo et al. [82] of the following scaling behaviour:

$$(-\Delta S)_{max} \sim T_c^{-2/3}, \quad (4.31)$$

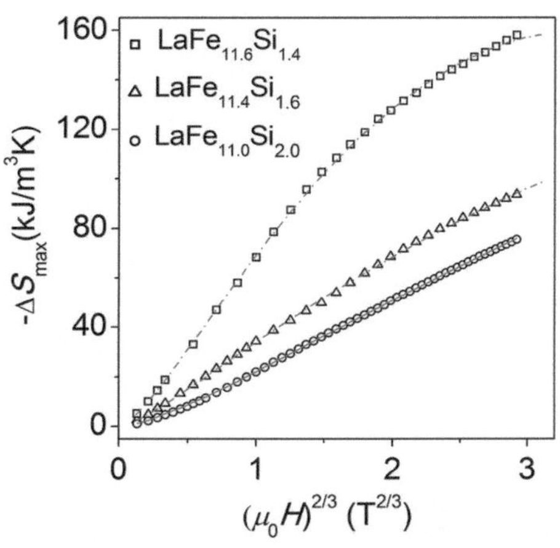

Fig. 4.20 Magnetic field dependence of the maximum entropy change ΔS_{max} for melt-spun $LaFe_{13-x}Si_x$ alloys undergoing the first-order metamagnetic transition for x = 1.4 and 1.6 and second-order transition for x = 2.0. The lines are fits to Eq. (4.29) [23]

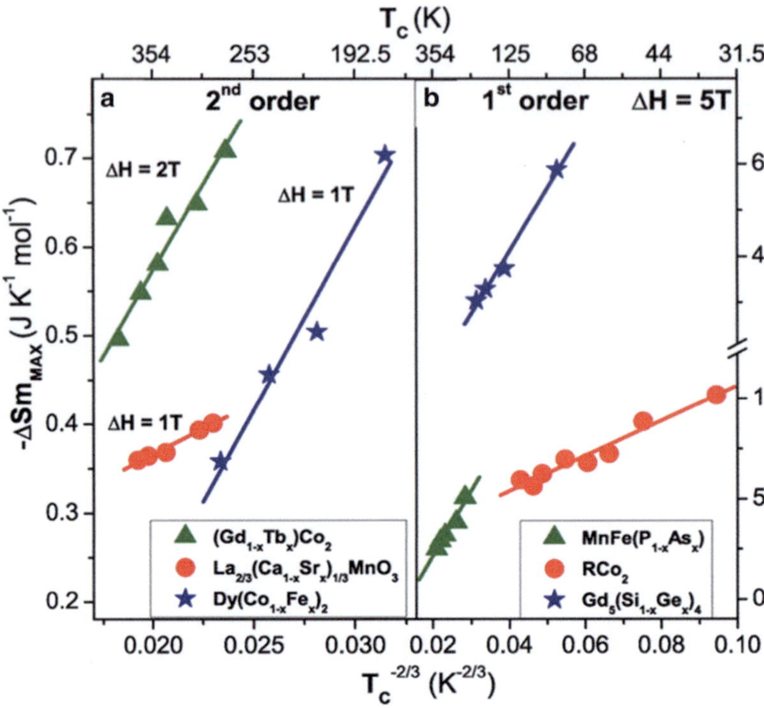

Fig. 4.21 Maximum entropy change as a function of $T_c^{-2/3}$ for materials undergoing (**a**) second-order and (**b**) first-order transition. The lines are linear fits to the experimental data expressed by symbols [82]

which was surprisingly found to be valid not only for materials undergoing the second-order transition but also for a number of materials with first-order transitions (Fig. 4.21).

The knowledge of both entropy change ΔS and adiabatic temperature change ΔT_{ad} is required to assess the suitability of the particular material for magnetic refrigeration. In general, the entropy change ΔS gives a measure of the cooling power, while sufficient ΔT_{ad} is crucial to make it possible to transfer the heat from the magnetic refrigerant to the heat transfer fluid. In magnetic refrigerators employing materials with the giant MCE, a higher cooling power can potentially be achieved compared to the refrigerators using second-order transition materials.

There have been several attempts to introduce a single parameter that could serve as a figure of merit for magnetocaloric materials. Wood and Potter [83] introduced the concept of magnetic refrigerant capacity:

$$\mathrm{RC} = \Delta S_{\mathrm{cold}} \Delta T_{\mathrm{span}} \tag{4.32}$$

where the entropy change at the cold-side ΔS_{cold} is equal to the entropy change at the hot side, $\Delta S_{\mathrm{cold}} = \Delta S_{\mathrm{hot}} = \Delta S\big|_{T_{\mathrm{cold}}}$. However, the authors assumed a constant temperature of the magnetic refrigerant across the material bed, which is not the case in the AMR, and neglected the necessary condition for the material to possess sufficient ΔT_{ad} to be able to move heat away from the regenerator.

Another frequently reported quantity is the relative cooling power ($RCP_{\Delta S}$), which is defined as an area under the $\Delta S(T)$ curve or a product of ΔS_{max} and the full width at half maximum (FWHM) of the ΔS curve [84, 85]. The latter is equivalent to RC in Eq. (4.32) multiplied by two. The relative cooling power can also be defined for the adiabatic temperature change ($RCP_{\Delta T}$) as the product of its maximum ΔT_{ad}^{max} and FWHM of the $\Delta T_{ad}(T)$ curve [84]. From the thermodynamic considerations, it can be obtained that for a given magnetic field the RCP of a refrigerant undergoing a first-order phase transition is determined solely by the magnetisation difference [85]:

$$\mathrm{RCP} = H(M_2 - M_1). \tag{4.33}$$

Consequently, RCP is maximised by choosing materials with a large magnetisation jump, $M_2 - M_1$, i.e. materials whose energy profiles regarded as a function of the magnetic moment have two minima situated as far apart as possible [86].

The knowledge of RC or $RCP_{\Delta S}$ alone is insufficient for the magnetic refrigerant performance characterisation: if a material has a large ΔS, it will not necessarily exhibit a large ΔT_{ad} and vice versa; this situation is illustrated in Fig. 4.22. This fact

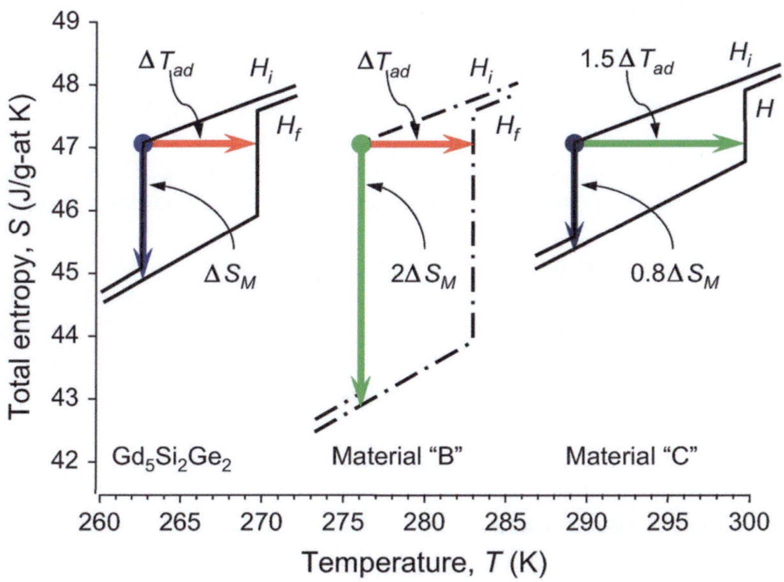

Fig. 4.22 Schematic S–T diagrams of three different materials exhibiting the giant MCE in the vicinity of the corresponding phase transitions for a small ΔH: $Gd_5Si_2Ge_2$ and hypothetical materials "B" and "C". Material "B" has ΔS twice that of $Gd_5Si_2Ge_2$, but the same ΔT_{ad}. Material "C" has ΔS smaller than that of $Gd_5Si_2Ge_2$ but a greater ΔT_{ad} [87]

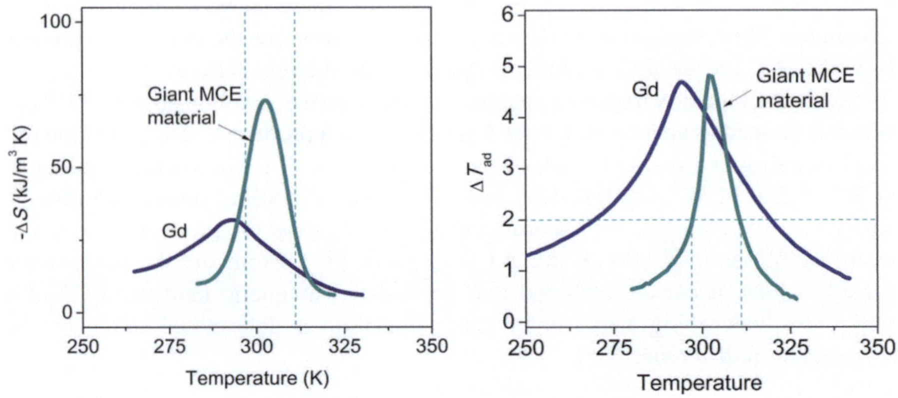

Fig. 4.23 Determination of $RCP_{\Delta S}$ and RC from the $\Delta S(T)$ diagram in the temperature range, where the corresponding $\Delta T_{ad}(T) \geq 2$ K

leads to erroneous statements that can be found in the literature, that the performance (evaluated by the calculation of $RCP_{\Delta S}$) of materials with a very broad but very moderate ΔS is the same as that of giant MCE materials. A broad ΔS varying over the temperature range of about 100 K is, for example, observed in amorphous and nanocomposite materials [88]. ΔT_{ad} in such materials is usually very low (<1 K in a field change from 0 to 2 T) eliminating the possibility of performing cooling with these materials [89].

The adiabatic temperature change a material should possess is determined by the desired device specifications. Using 1D numerical AMR model, Engelbrecht and Bahl [89] showed that the cooling power is almost brought to nothing when ΔT_{ad} is reduced below 2 K. By setting the temperature range, within which the RCP or RC is determined, to that where ΔT_{ad} is above 2 K (Fig. 4.23), one can get an $RCP_{\Delta S}$ value that can be useful for an engineer. This, however, requires the knowledge of ΔT_{ad}.

If the material type or the phase composition in a given material varies, it is more meaningful to compare the entropy change ΔS values normalised per sample volume, J/m^3 K, instead of those normalised per weight, J/kg K. Also for applications, where the magnetic field is produced, e.g. by permanent magnets, it is desirable to have the largest entropy change in the smallest volume [17, 85].

4.4.2 Curie Temperature

The advantage of materials with the first-order transition is the giant MCE, i.e. a large (giant) entropy change ΔS. However, the entropy change is concentrated in a narrow temperature region (cf. Figs. 4.18 and 4.23), by this limiting the operating temperature span of a refrigerator $\Delta T_{span} = T_{hot} - T_{cold}$. In order to use the

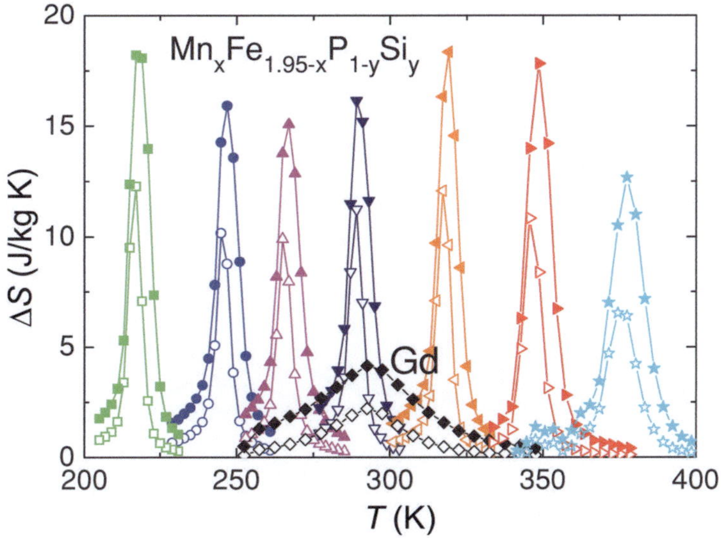

Fig. 4.24 Entropy change ΔS for a field change of 0-1 T (open symbols) and 0-2 T (solid symbols) for the $Mn_xFe_{1.95-x}P_{1-y}Si_y$ compounds with x = 1.34, 1.32, 1.30, 1.28, 1.24, 0.66, 0.66 and y = 0.46, 0.48, 0.50, 0.52, 0.54, 0.34, 0.37, from left to right, respectively. The data of Gd metal for a field change of 0-1 T (open diamond) and of 0-2 T (solid diamond) are included [77]

advantage of the first-order transition materials, their large ΔS, a stack of materials with different Curie temperatures should be used in the regenerator. Such a layered bed containing giant MCE materials allows to cover a wide temperature range and at the same time provides the cooling power significantly larger than that of Gd (see Sect. 4.3). In order to be able to produce such a layered bed, the Curie temperature of the refrigerant materials should be tunable. Usually this is achieved by the composition variation. In terms of retaining large entropy change, it is advantageous to select substitution or interstitial elements so as to retain the first-order transition. In terms of economic merits, the substitution elements should be preferably selected from the group of non-critical elements.

An example of adjusting the concentration by simultaneous variation of the Fe:Mn and P:Si ratios in the $Mn_xFe_{1.95-x}P_{1-y}Si_y$ compounds was provided by Dung et al. [77]. With this stoichiometry variation, Dung et al. have been able to achieve constant ΔS in a very large temperature range of 200 K (Fig. 4.24).

Sharp ΔS peaks regarded as a function of temperature in the first-order transition materials cause uneven distribution of the MCE in the bed and compromise the performance of the refrigerator [89]. The example in Fig. 4.25 shows that gaps in the performance can be avoided by fine-tuning of the Curie temperature by varying Mn concentration and full hydrogenation of the $La(Fe,Si)_{13}$ compounds [90]. Ideally, a large and constant ΔS is desired in the whole ΔT_{span}.

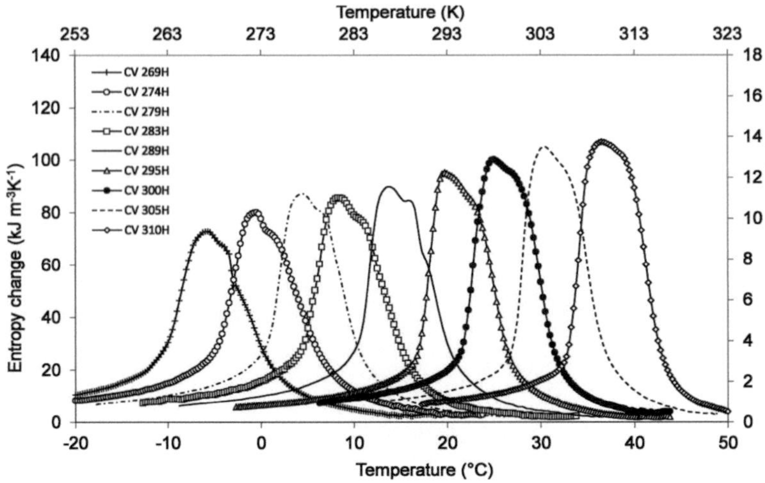

Fig. 4.25 Magnetic entropy change for a field change of 1.6 T in La(Fe,Mn,Si)$_{13}$H$_z$ alloys with various Mn contents and maximum possible amount of hydrogen (commercial name CALORIVAC® by Vacuumschmmelze); Mn content is decreasing from left to right (data courtesy of Barcza and Katter [90])

4.4.3 Hysteresis

In the magnetic cooling, reducing magnetic and thermal hysteresis is critical for refrigeration cycle efficiency. The giant MCE observed at first-order transitions comes at a cost of a hysteresis. There are two sources of the hysteresis in the materials undergoing the first-order transition: intrinsic and extrinsic.

The intrinsic hysteresis in first-order transition materials is governed by the thermodynamics and in particular by the existence of the latent heat at the transformation [25]. In the free energy vs. magnetisation curves F(M) of these materials, at least two minima should be present [86, 91]. Overcoming the energy barrier U between the two minima at T close to 300 K is a thermally activated process (Fig. 4.26), whose characteristic time constant is expected to obey the Arrhenius law. If at the transition point (point where the free energies of the states 1 and 2 are equal, Fig. 4.2) the latent heat given by Eq. (4.13) is $L < 0$, then heating will shift the equilibrium to the direction of endothermic process, while cooling will shift the equilibrium to the direction of exothermic process. Thus, the thermal hysteresis arises; increasing the heating/cooling rate will lead to the increase of the hysteresis. On the other hand, if we allow the system to transform in an infinite amount of time (quasistatically), the transition will take place at T_c.

At quasistatic conditions, the transition in the field would occur at a certain critical field H_m, such that the free energies of both phases are equal (Fig. 4.26a). In a non-quasistatic regime, the transition takes place at a higher field, $H_m + \Delta H'/2$ [86]. The role of the extra field $\Delta H'$ is to lower the barrier in the direction of the transition

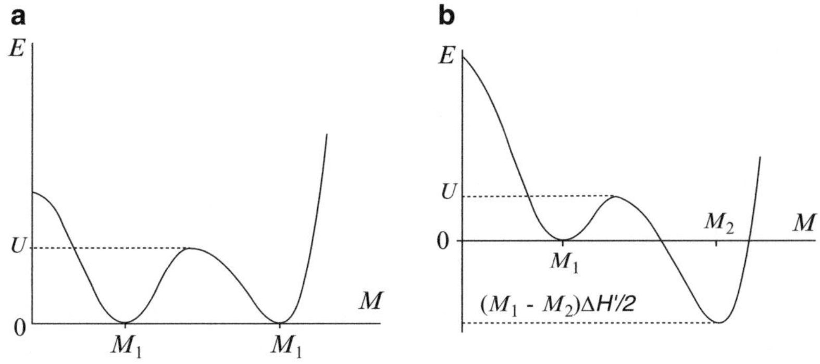

Fig. 4.26 Energy-vs.-magnetisation dependence of a hypothetical refrigerant undergoing a first-order phase transition. (**a**) The transition takes place quasistatically; the energies of the two phases are equal. (**b**) The transition $1 \rightarrow 2$ proceeds non-quasistatically in the presence of an extra magnetic field $\Delta H'/2$ which favours the higher magnetisation and lowers the barrier [86]

(from left to right in Fig. 4.26b), so that the barrier can be overcome. The energy difference between the final and the initial states, $(M_2 - M_1)\Delta H'/2$, will be dissipated as heat. Similarly, the inverse transition $2 \rightarrow 1$ will take place at $H_m - \Delta H'/2$ on reducing field and will be accompanied by dissipation of the same energy. The total amount of heat released in one cycle equals the area of the hysteresis loop, $(M_2 - M_1)\Delta H'$. This amount should be deducted from the RCP in Eq. (4.33). Obviously, the broader the hysteresis, the less efficient the operation of the refrigerator. When the hysteresis width $\Delta H'$ becomes comparable to the available magnetic field H, the cooling will cease.

Thus, for the applications it is favourable to choose first-order type materials whose free energy profile F(M) has characteristic low energy barriers, as those observed, e.g. in $La(Fe,Si)_{13}$-type compounds [86, 92].

The extrinsic hysteresis in materials with the first-order transition is caused by the peculiarities of the microstructure, e.g. strain, crystallite size and inhomogeneities, and by the kinetics of the transformation. The existence of the extrinsic hysteresis is a phenomenon common to most magnetic materials.

A common feature of the first-order transitions is the change of the exchange interaction during the transition and a concomitant volume and/or crystal symmetry change [26, 93]. The exchange interaction is a function of the interatomic distance, and any lattice deformation results in the appearance of the spontaneous magnetisation [26, 91, 93]. This leads to a change of the sample dimensions and the appearance of strains. As a result a new effective value of the Curie temperature is observed, different from that of the rigid system, i.e. a hysteresis appears [93].

The lattice constants of the magnetocaloric compounds vary substantially during the first-order magnetic or magnetostructural transition. In $La(Fe,Si)_{13}$-type compounds, an isotropic lattice expansion of about 1.0–1.4 %, depending on the temperature, composition and applied field strength, is observed during the transition from the para- to ferromagnetic state [29, 94]. In $(Mn,Fe)_2(P,Ge)$- and

$Gd_5(Si,Ge)_4$-type alloys, the overall volume expansion at the PM–FM transition is below 0.1 and 0.4 %, respectively. However, these alloys exhibit a large anisotropic change of the lattice constants at the transition: 1.5 % contraction along the a-axis and 2.8 % elongation along the c-axis in $MnFeP_{0.85}Ge_{0.15}$ [95] and 0.9 % contraction along the a-axis and 0.2 % elongation along the c-axis in $Gd_5Si_{1.8}Ge_{2.2}$ [96].

The majority of the magnetocaloric materials are polycrystalline. The large volume change as well as the anisotropic expansion at the first-order transition has very significant implications on the MCE and the hysteresis: both are greatly influenced by internal constraints imposed by the grain boundaries [14]. The increase of the lattice constants, i.e. the volume expansion or linear magnetostriction of each individual grain, is constrained by neighbouring grains in the polycrystalline alloys. Consider the case, when the dimensions (volume) of the FM phase are larger than those of the PM phase. When going from the PM to the FM state, the constrained volume expansion/elongation of the crystallites will result in the effective reduction of T_c, whereas during the reverse FM to PM transformation, no such constraints exist and the transition point is closer to the "intrinsic" T_c value for a hypothetical non-compressible lattice [93]. The same is true for the corresponding first-order transformation in the magnetic field, which also occurs with the concomitant change of the lattice constants.

Lyubina et al. [14] showed the importance of microstructure design for the control of the hysteresis of materials undergoing the first-order transition. Drastic reduction of the hysteresis can be achieved by reduction of the internal constraints through the introduction of porosity (cf. the magnetisation curves and ΔT_{ad} of the bulk and porous $LaFe_{11.6}Si_{1.4}$ in Fig. 4.27). The larger ΔT_{ad} observed during the PM to FM transition in the bulk alloy (Fig. 4.27c) is due to the so-called "undercooling" effect. On cooling,

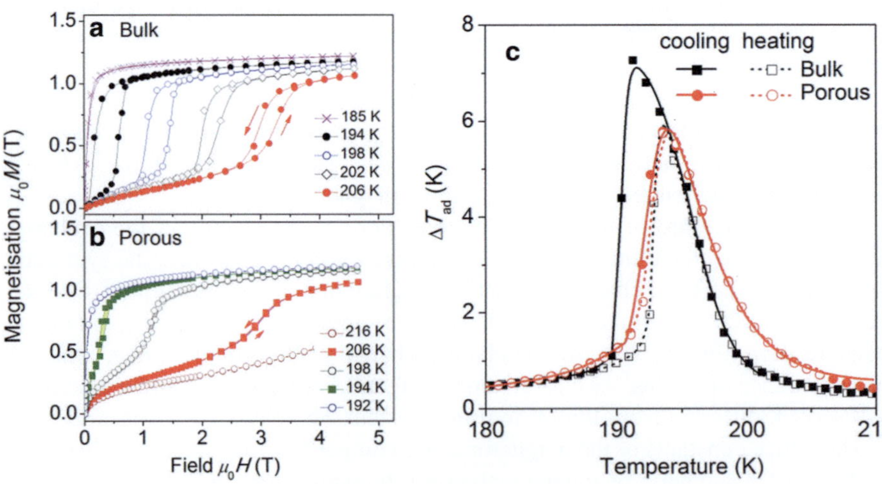

Fig. 4.27 Magnetisation isotherms of bulk $LaFe_{11.6}Si_{1.4}$ (**a**) and porous $LaFe_{11.6}Si_{1.4}$ (**b**) and the adiabatic temperature change of bulk and porous $LaFe_{11.6}Si_{1.4}$ in the vicinity of the Curie temperature. The bulk alloy was prepared by induction melting followed by annealing at 1150 °C/7 days. The porous alloy was prepared by pulverising the bulk alloy followed by compaction at 730 °C/10 min (adapted from [29])

the PM to FM transition is hindered until the strain barrier experienced as a result of hindered volume expansion is overcome and an abrupt, burst-like transformation can proceed. It is also apparent from Fig. 4.27c that the adiabatic temperature change for the porous alloys resembles to a great extent ΔT_{ad} observed in the bulk alloy on heating, i.e. during the reverse FM to PM transformation, where the lattice volume contracts and the transition is not hindered by the neighbouring grains. The hysteresis reduction in porous alloys comes, however, at a cost of somewhat reduced ΔS_{max} and ΔT_{ad}^{max} : the maximum adiabatic temperature change and the volumetric entropy change are reduced by about 20 and 30 %, respectively, as compared to the bulk alloy [15, 29]. This MCE reduction can to some extent be circumvented by adjusting the degree of porosity or by the introduction of a matrix material. The matrix material (e.g. copper, silver and their alloys, polymers) allows on the one hand obtaining a higher density and on the other hand accommodating the strain during the expansion/contraction of the material particles [15].

The actual cooling power of the magnetic refrigerator is a product of the RCP (Eq. 4.33) and the operation frequency f of the device. Obviously, an increase in operation frequency means that the magnetocaloric material is magnetised and demagnetised more often per unit time. This can theoretically increase the cooling powder and would allow to minimise the volume of the magnetic refrigerants and permanent magnets and is therefore highly desired in the magnetic cooling applications. The upper limit of the frequency f at which magnetic refrigerators can operate was shown by Kuz'min [97] to be set by the balance between thermal conductivity and viscous friction and is in the order of 100 Hz. As far as magnetic refrigerants are concerned, frequency scaling can be a very challenging task. Current magnetic refrigerator devices operate at frequencies in the order of 1 Hz [64–66]. Low magnetic hysteresis and fast dynamics are essential for achieving higher operation frequencies.

Generally, magnetisation vs. field loops of first-order transition materials show "flaring" as the field rate is increased resulting in increased magnetic hysteresis (Fig. 4.28, [34]). One of the reasons for the flaring of the M(H) loops is the extrinsic thermal effect [34]: once the PM to FM transition is induced at some part of the material, the increase in temperature (the magnetocaloric effect) from the nucleated region heats the surrounding regions and therefore increases their critical fields. Consequently, increasingly larger fields are required for the remaining PM volumes to reach their critical field, as more of the material transforms and the whole system heats. On the reverse, FM to PM transition, faster field rates keep the sample out of thermal equilibrium longer if the heat has no easy means of transfer into and out of the sample. Thus, a poor thermal conductance at the interface between material and thermal bath results in an increased hysteresis. Improving the thermal linkage greatly reduces the effects of the field rate [98]. The thermal linkage between material and thermal bath also improves when small material fragments are used [99].

The kinetics of the first-order transitions was studied by Yako et al. [100], Fujita et al. [101] and Fujita and Yako [102]. The PM to FM transition rate shows a sigmoid behaviour with temperature and depends on the cooling rate. Temperature

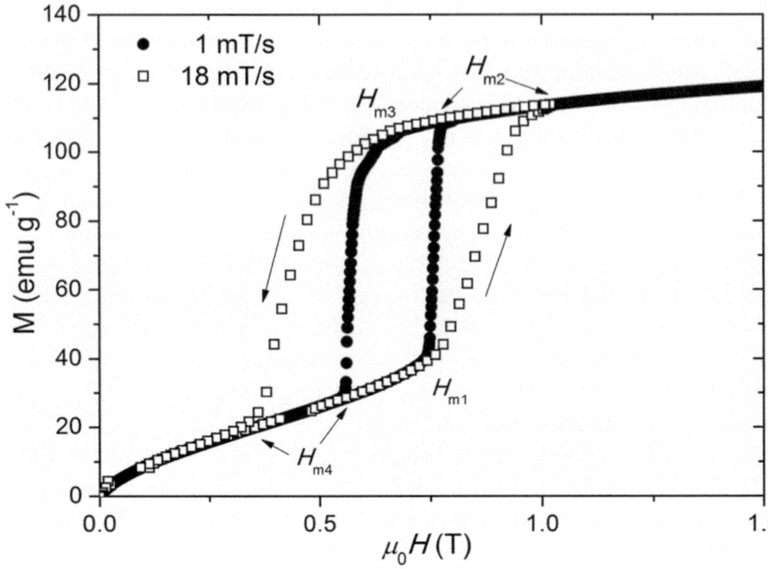

Fig. 4.28 Magnetisation as a function of applied field for a $LaFe_{11.6}Si_{1.4}$ piece with dimensions of about $1.7 \times 1.1 \times 0.40\ mm^3$at 194.5 K, 3.5 K above T_c as the material transforms from the PM (low field) to FM (high field) state and back at different field sweep rates. The transition start fields H_{m1} and H_{m3} depend only slightly on the field, whereas transition finish fields H_{m2} and H_{m4} show a strong dependence on the field sweep rates (adapted from [34])

dependence of the transition rate upon cooling in the $La(Fe,Si)_{13}$-type alloys was shown to exhibit the Kolmogorov–Johnson–Mehl–Avrami (KJMA) kinetics with the Avrami exponent of about 2.2 indicating a 2D-like growth. The kinetics of the transformation was found to be governed by the generation and dissipation of the latent heat at the interface between the FM and PM regions. Moreover, the nucleation and growth processes were shown to be influenced by the demagnetising fields [34, 100, 101].

Time dependence and field rate effects of the MCE were examined by Lovell et al. [34], who showed that thin plates of magnetocaloric materials are required in order to mitigate the extrinsic hysteresis created in the presence of rapidly changing magnetic field and to shorten the time required to complete the magnetic transition, thereby allowing an increased refrigeration operation frequency f.

4.4.4 Thermal Transport Properties

In the active magnetic regenerator, two major types of the magnetic refrigerant beds can be distinguished: a sphere packed bed and a parallel-channel AMR. Sphere packed beds are conventionally fabricated from spherical particles or powders with a diameter of about 200–500 μm; their fabrication is uncomplicated, and they can provide superior heat transfer properties [62, 66]. However, the pressure drop in the heat transfer fluid across the regenerator bed can be a limiting factor for this geometry, which in turn translates into a low operation frequency, typically $f \lesssim 4\,\mathrm{Hz}$. Parallel channel AMR exhibits good heat transfer characteristics while maintaining a small pressure drop [103, 104] provided the channel size and channel spacing are uniform [105]. Gd plates with the thickness of about 100 μm are required to obtain heat transfer characteristics similar to those obtained when using spheres [103, 104]. For first-order transition materials, plate thickness below 100 μm can be advantageous in terms of reduction of the hysteresis and improving thermal transport characteristics (Sect. 4.4.3).

In the AMR, the challenge is to achieve a large thermal gradient, i.e. large difference between T_{cold} and T_{hot} (see Chap. 4.3), across a short length that would allow to reduce the size of the magnets and consequently the size of the complete AMR. On the other hand, an efficient regenerator should be able to transfer heat from the magnetic refrigerant interior to the heat transfer fluid. This means that the magnetic refrigerant material should have a very high thermal conductivity transverse to the direction of the flow and very low thermal conductivity in the direction of the flow (axial direction) [106]. The thermal conductivity κ of known magnetocaloric materials does not show the required anisotropy and lies in the range between approximately 1 and 16 W/m K near room temperature (Table 4.1). AMR modelling by Nielsen and Engelbrecht [106] showed that if the AMR device is operated at a high frequency, the sensitivity of the AMR performance to variations in the thermal conductivity is increased.

Ideally, the thermal conductivity of the magnetic refrigerant should be engineered in a way to achieve, if required, an anisotropic distribution of the thermal conductivity. The decrease of κ in the axial direction of the regenerator may be obtained by the use of specific geometries [113]. Engineering magnetic refrigerant materials to achieve defined thermal transport properties is another approach that

Table 4.1 Thermal conductivity (κ) of different magnetocaloric materials at 300 K

Material	κ (W/K m)	Reference
Gd (single crystal)	10.9	[107]
Gd (polycrystalline, purity 99.7 %)	8.0	[108]
$Gd_5Si_2Ge_2$	5.2	[109]
$LaFe_{11.6}Si_{1.4}$ (bulk)	9.7	[15]
$LaFe_{11.6}Si_{1.4}$ (porous)	0.9	[15]
$LaFe_{11.44}Si_{1.56}H_{1.0}$ (bulk)	9.1	[109]
$LaFe_{11.6}Si_{1.4}H_y$ (porous)	2.0	[15]
$LaFe_{11.6}Si_{1.4}H_y$/3 wt.% Cu (porous)	2.7	[15]
$LaFe_{11.6}Si_{1.4}H_y$/4 wt.% Cu at surface (porous)	2.9	[15]
$LaFe_{11.6}Si_{1.4}$/5 wt.% Ag epoxy	5.0	[110]
$La_{0.67}Ca_{0.33}MnO_3$ (bulk)	2.0	[111]
$La_{0.67}Ca_{0.33}MnO_3$/Ag	2.5	[111]
Ni_2MnGa	15.6	[112]

was demonstrated by Lyubina et al. [15] and Turcaud et al. [111]. In composites consisting of porous first-order transition $La(Fe,Si)_{13}$ and high thermal conductivity Cu prepared by electroless copper plating, the thermal conductivity can be improved by about 300 %, compared to the base material. The Cu content is as low as 4 wt.%, and, thus, the magnitude of the magnetocaloric effect in the composites remains high [15]. Also in second-order transition materials, such as low thermal conductivity $La_{0.67}Ca_{0.33}MnO_3$, silver impregnation can be used to achieve a threefold increase in the thermal conductivity [111].

4.4.5 Mechanical Properties

The majority of magnetocaloric materials (with the exception of Gd and manganites) are based on intermetallic compounds. The substantial volume change or linear magnetostriction at the transition combined with a relatively high brittleness of the intermetallic compounds is a severe shortcoming for magnetic cooling applications, where the magnetic refrigerant is exposed to a cyclic application and removal of the magnetic field and cyclic temperature variation. Over a period of 10 years, the magnetic refrigerant can experience $\sim 10^7$–10^8 magnetisation–demagnetisation cycles. Sphere beds with first-order transition materials can change their characteristics over time due to pulverising and clog the regenerator bed [3]. During cycling of first-order transition material plates, microcrack formation results in the reduction of the MCE and eventually can lead to the mechanical failure (Fig. 4.29, [14]).

In materials with the second-order transition, the volume change at the transition is lower (volume change of 0.03 % is observed in $LaFe_{10.7}Co_{1.3}Si$ in a magnetic field change of 0.6 T), and no degradation is apparent over 10^6 cycles [114]. The mechanical stability is correlated to the size of the lattice parameter discontinuities at the transition. Thus, materials with a weakly pronounced first-order transition and a correspondingly lower volume change, such as a $MnFe_{0.95}P_{0.582}B_{0.078}Si_{0.34}$ alloy reported by Guillou et al. [115], show an enhanced mechanical stability.

One of the ways to provide the mechanical stability in the materials with the first-order transition is to allow the crystallites to change their dimensions. Porous architectures allow this situation and the MCE is stable over hundreds of cycles [14]; the measurement has been stopped at that point and the sample was intact. However, the thermal conductivity of porous materials can reduce by a factor of 10 [15]. Therefore, a composite material, where the magnetocaloric material particles are embedded into an elastic matrix with moderate to high thermal conductivity, can further improve the mechanical stability of the magnetic refrigerants with the first-order transition, while providing good thermal transport characteristics [15, 110, 116]. Notably, when the composites are prepared by cold pressing, the formation of cracks and comminution of the magnetocaloric material particles can reduce the thermal conductivity and the MCE [15, 110], and, thus, the compaction pressure should be selected so as to avoid these effects.

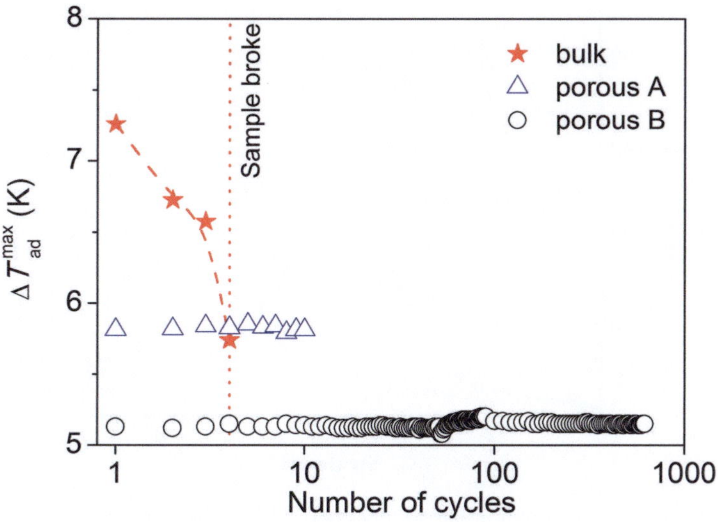

Fig. 4.29 The maximum adiabatic temperature change ΔT_{ad}^{max} vs. number of temperature–field cycles in $LaFe_{11.6}Si_{1.4}$. The bulk plate loses its mechanical integrity after the 4th cycle. ΔT_{ad}^{max} remains constant in the porous materials A and B. Material B has a higher amount of the secondary phase (α-Fe) compared to material B and, thus, lower ΔT_{ad}^{max} values [14]

Poor mechanical stability complicates shaping of magnetocaloric materials into thin plates. For instance, in materials with T_c around room temperature, such as $La(Fe_{0.86}Si_{0.09}Co_{0.05})_{13}$ [117], the conventional machining by grinding and wire cutting induces local temperature gradients, even if the area to be cut is well cooled. The local gradients result in the situation that some parts of the material experience the PM ↔ FM transitions leading to the formation of microcracks and fracture of the parts.

Depending on the specific material type, several strategies can be used to overcome the problem. In the case of the $La(Fe,Si)_{13}$ material family, either a so-called thermally induced decomposition and recombination (TDR) process [117] or a subsequent hydrogenation of the base material with T_c well below room temperature can be used. In the TDR process, $La(Fe,Co,Si)_{13}$ materials are heated to about 1073 K, where a decomposition of the $La(Fe,Co,Si)_{13}$ phase into predominantly α-Fe and LaFeSi occurs. As a result, the compressive and bending strength increases from 658 to 1176 N/mm^2 and from 150 N/mm^2 to 260 N/mm^2, respectively [117]. The increase of the mechanical strength and the absence of the $La(Fe,Co,Si)_{13}$ phase undergoing the PM ↔ FM transitions make the machining possible. After machining of the parts, the $La(Fe,Co,Si)_{13}$ phase can be restored by a heat treatment at about 1323 K. Another strategy is the adjustment of the Curie temperature by hydriding of the base $La(Fe,Si)_{13}$ or $La(Fe,Ni,Si)_{13}$ compounds with T_c well below room temperature after shaping those into a final form [14, 118, 119].

In other magnetocaloric materials, where the post-shaping T_c adjustment cannot be made by interstitial insertion of hydrogen (examples are manganites and $Gd_5(Si,Ge)_4$- and $(Mn,Fe)_2(P,Ge)$-type alloys), powder processing routes can

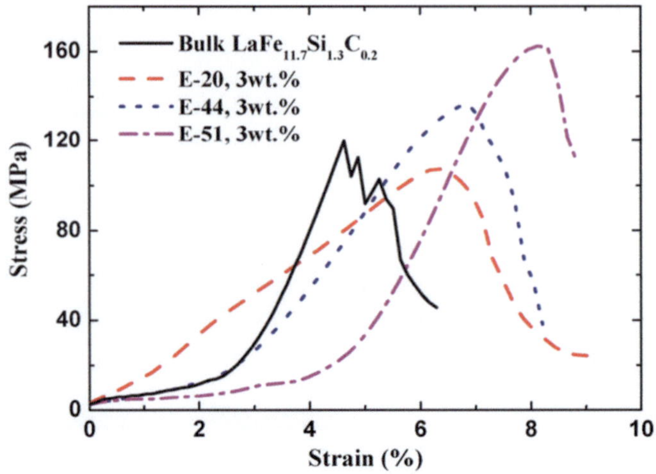

Fig. 4.30 Compressive stress–strain curves for $LaFe_{11.7}Si_{1.3}C_{0.2}H_{1.8}$ bonded using different commercial epoxy resins (E-20, E-44, E-51) in comparison with bulk $LaFe_{11.7}Si_{1.3}C_{0.2}$ compound [116]

be employed. These can include polymer bonding of the powders [110, 116], electroless coating of powders followed by pressing [15], additive manufacturing [114, 120] and tape casting [121]. As an example, Fig. 4.30 shows how the mechanical properties can be improved in epoxy-bonded magnetocaloric materials of the $La(Fe,Si)_{13}$ family as compared to the bulk alloy; the highest compressive stress of about 162 MPa and a maximum strain of about 8 % were demonstrated in the bonded materials with the volumetric entropy change ΔS being about 80 % of the bulk alloy ΔS value [116].

4.4.6 *Chemical Stability*

In the majority of the existing AMR prototypes, the magnetic refrigerant is in contact with a heat transfer fluid, usually water with antifreeze and corrosion inhibitors [66, 122]. The addition of the anti-corrosion agents allows to inhibit corrosion, but only for a limited period of time [122]. Deaeration of the heat transfer fluid was shown by Fujieda et al. [123] to be effective in corrosion reduction in $La(Fe_{0.88}Si_{0.12})_{13}$; the effect is only observable for a very low dissolved oxygen content of 0.1 ppm. With the exception of ceramic materials (manganites), which have a high corrosion resistance, surface protection of the magnetocaloric materials by coating [122, 124] or via formation of passivation layers [125] is the most effective means for ensuring the chemical stability and avoiding deterioration of the refrigerant performance.

Other ageing effects that may arise during the use of the magnetic refrigerant should also be taken into account. A prominent example in this respect is hydrogenated $La(Fe,Si)_{13}H_y$. Heating of $La(Fe,Si)_{13}H_y$ hydrides in hydrogen-free atmosphere to approximately above 150 °C leads to a partial or complete dehydrogenation, and as a result the Curie temperature will be lowered that can compromise the AMR performance. Partially hydrided ternary $La(Fe,Si)_{13}$ decompose into a fraction with a higher T_c and a second fraction with a lower T_c compared to the starting material when held at the Curie temperature for several days [118]. Such degrading material cannot be used in applications, which require constant properties over a long period of time. However, by fully hydriding manganese containing $La(Fe,Mn,Si)_{13}$ alloys, stable alloys with adjustable Curie temperatures around room temperature can be produced. Thus, it is crucial to ensure the stability of the magnetocaloric materials to ensure reliable operation of the magnetic refrigerators or heat pumps.

4.5 Materials

In this section, an overview of the origin of the magnetocaloric effect and properties of the major classes of magnetocaloric materials is given. It is not the aim to include all known materials exhibiting the magnetocaloric effect in this section (clearly all magnetically ordered materials do exhibit a MCE), but to provide a brief overview of materials that are discussed in relation to the magnetic heat pumping applications near room temperature. The reader is referred to works by Brück et al. [16, 19], Franco et al. [126], Gschneidner et al. [17] and Tishin and Spichkin [18] for a review on materials not discussed in the present chapter.

4.5.1 Gadolinium

Although it is unlikely that gadolinium will be used for magnetic heat pumping due to its high cost, Gd is often used as a benchmark material against which the performance of other magnetic refrigerant materials is compared. The MCE in Gd is due to the second-order transition from the para- to ferromagnetic state around room temperature. Pure single-crystal Gd orders ferromagnetically at 294(1) K [38, 127]; the Curie temperature decreases with an increasing amount of impurities and inhomogeneity [23, 38]. The magnetic moment of Gd is parallel to the c-axis between 294 and 232 K; below 232 K a spin reorientation takes place, where the moment starts to form a cone with the maximum deviation from the c-axis of 65° around 180 K and moves back to within 32° at lower temperatures [127]. The peak temperature T_{max} of the entropy change ΔS essentially coincides with T_c in pure single crystal; depending on the purity and homogeneity, the position of the ΔS peak of polycrystalline Gd may lie as far as 4 K below the Curie temperature of a single crystal ([123], Table 4.2). Both the entropy change and the adiabatic temperature change depend on the purity: the purer and the more homogenous is Gd, the larger is the MCE (Table 4.2). Hysteresis is absent; see Fig. 4.15a.

Table 4.2 Properties of Gd: the Curie temperature (T_c), the ΔS peak temperature in the lowest provided field change (T_{max}), the maximum entropy change (ΔS_{max}) and the maximum adiabatic temperature change (ΔT_{ad}^{max}) in a field change $\mu_0\Delta H$. The values of ΔS_{max} are calculated from gravimetric to volumetric units or vice versa using the density of 7.89 g/cm^3

	T_c (K)	T_{max} (K)	$\mu_0\Delta H$ (T)	$-\Delta S_{max}$ (J/kg K)	$-\Delta S_{max}$ (kJ/m^3 K)	ΔT_{ad}^{max} (K)	Reference
Single crystal	295	295	1	4.0	31.8	3.1[a]	[23]; Lyubina et al., unpublished data[a]; [41][b]
			2	6.2	49.4	5.2[a]	
			5	10.8	85.5	12.0[b]	
Polycrystalline, purity 99.7 %		294	1	3.2	25.5	2.9	[23], Fig. 2.15
			2	6.1	48.0	5.2	
			5	10.9	86.2		
Polycrystalline, purity 97.7 %		291	1	3.2	25.2	2.9[a]	[23]; Lyubina et al., unpublished data[a]
			2	5.2	41.4	4.9[a]	
			5	9.8	77.5		
Nanocrystalline, purity 99.8 %		290	2	3.4	26.8		[128]
			5	6.9	54.4		

The magnetocaloric effect of nanocrystalline Gd with grain size of about 15 nm synthesised by the inert gas condensation technique shows a 1.5-fold reduction in the entropy change compared to the single-crystal Gd ([128], Table 4.2). The reduction was attributed by Mathew et al. [128] to the incomplete saturation due to large interfacial anisotropy fields. Severe plastic deformation allows obtaining thin foils of relatively ductile materials. Taskaev et al. [129] reported a 3- to 3.5-fold decrease in the MCE as a result of strong magnetisation reduction in Gd foils subjected to severe plastic deformation; this reduction can to some extent be recovered by a subsequent annealing.

For increasing operation temperature span ΔT_{span}, materials with different Curie temperatures are required for the construction of graded regenerator beds (Sect. 4.3). The variation of T_c can be achieved by forming solid solutions of Gd with, e.g. Tb, Dy and Nd ([17] and references therein).

4.5.2 $Gd_5(Si_xGe_{1-x})_4$ Alloys

Intermetallic alloys of Gd and, in particular, second-order transition Gd_5Si_4 with a Curie temperature of 336 K were indicated by Brown as possible magnetic refrigerants in his seminal paper in 1976 [8]. The ternary intermetallic compounds $Gd_5(Si_xGe_{1-x})_4$ with Curie temperatures in the range between 295 and 335 K were discovered in the Gd–Si–Ge system by Holtzberg et al. [11]. In a study of the magnetic properties of Gd-based materials with Curie temperatures between 250 and 350 K, Pecharsky and Gschneidner [10] discovered an extraordinarily large (giant) magnetocaloric effect in $Gd_5Si_2Ge_2$. Holtzberg et al. [11] did not investigate the magnetocaloric effect in the $Gd_5(Si_xGe_{1-x})_4$ compounds; the magnetisation variation with temperature of the nominally the same stoichiometry compound,

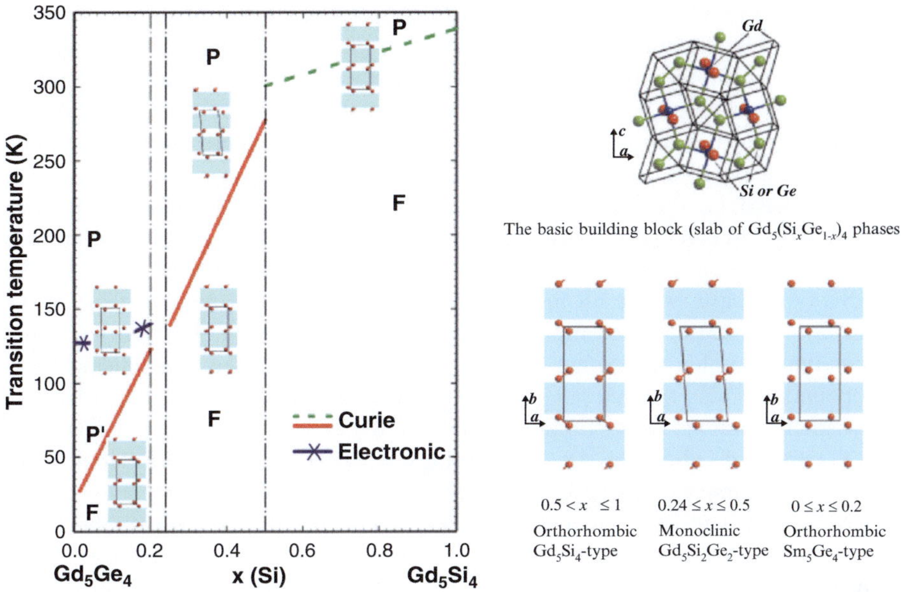

Fig. 4.31 Magnetic and crystallographic phases in the $Gd_5(Si_xGe_{1-x})_4$ system in zero magnetic field. Right: the dashed line corresponds to a second-order para- to ferromagnetic $P \leftrightarrow F$ phase transition, when $0.5 < x \le 1$. The cross line is a second-order paramagnetic to electronic phase transition $P \leftrightarrow P'$, when $0 \le x \le 0.2$. The solid line is a first-order ferromagnetic to paramagnetic phase transitions, $F \leftrightarrow P'$ and $F \leftrightarrow P$, when $0 \le x \le 0.2$ and $0.24 \le x \le 0.5$, respectively. Left: the corners of all cubes and trigonal prisms traced in the slab of $Gd_5(Si_xGe_{1-x})_4$ are occupied by Gd atoms. The remaining Gd atoms are blue and Si (Ge) atoms located inside the slab are shown in green. The red circles in the three schematic crystallographic phases below represent exterior Si and/or Ge atoms (depending on the composition) responsible for bonding between the slabs. The line between these circles indicates the Si(Ge)–Si(Ge) distances consistent with covalently bonded pairs of atoms (adapted from [12])

$Gd_5Si_2Ge_2$, revealed the second-order transition from the para- to ferromagnetic state. The reason for the discrepancy could be a very narrow composition region, where the transition type abruptly changes (Fig. 4.31); a slight deviation from the exact $Gd_5Si_2Ge_2$ stoichiometry to the Si-rich side of the phase diagram leads to a change of the transition from first to second order.

In the magnetic phase diagram of $Gd_5(Si_xGe_{1-x})_4$, three composition regions can be distinguished (Fig. 4.31, [12, 31]). These arise as a result of the change in the bonding between the subnanometre-thick slabs (Fig. 4.31) in the $Gd_5(Si_xGe_{1-x})_4$ crystals with the change of the composition x. In the orthorhombic Gd_5Si_4-type structure ($0.5 < x \le 1$), all of the Si(Ge)–Si(Ge) covalent bonds exist between the slabs. The transition in this composition range is the purely magnetic second-order phase transition, occurring without the crystal structure change. In the intermediate phase region $0.24 \le x \le 0.5$, one half of the inter-slab Si(Ge)–Si(Ge) bonds is broken and the room temperature crystal structure of the $Gd_5(Si_xGe_{1-x})_4$ alloys is monoclinically distorted and is composed of alternating strongly and weakly interacting slabs. On cooling, the $Gd_5(Si_xGe_{1-x})_4$ ($0.24 \leq x \leq 0.5$) alloys undergo a first-order magnetostructural transformation from the monoclinic paramagnetic phase to the orthorhombic ferromagnetic phase with strong interactions between all slabs (Fig. 4.31). The room temperature crystal structure of the Gd_5Ge_4-based alloys ($0 \leq x \leq 0.2$) restores the orthorhombic symmetry, where the layers remain essentially the same, except for the composition (Si/Ge ratio), but are no longer interconnected, i.e. there exist no inter-slab Si(Ge)–Si(Ge) bonds. On cooling, they first undergo an electronic transition without a change of the crystal structure and then a ferromagnetic ordering coupled with a structural transformation to the Gd_5Si_4-type phase, where all the slabs are connected to one another (Fig. 4.31). Paudyal et al. [130] studied the transition in $Gd_5Si_2Ge_2$ using electronic structure calculations coupled with magnetothermodynamic models; the magnetism of this compound was also theoretically studied by de Oliveira and von Ranke [131].

Magnetisation vs. magnetic field M(H) and the entropy change vs. temperature $\Delta S(T)$ curves of $Gd_5Si_2Ge_2$ are given in Figs. 4.10 and 4.18. The behaviour of the M(H) below approximately 272 K is typical of a soft ferromagnet and above 309 K it has the characteristics of a paramagnet. Between 278 and 300 K, the M(H) isotherms display an S-shape, which is due to the coupled magnetostructural transition discussed above [10, 31]. The magnetostructural first-order transition in the composition range $x \leq 0.5$ is responsible for the appearance of the giant magnetocaloric effect. Of importance for applications is the fact that this transition can be induced not only by the temperature but also by the magnetic field and, due to the large linear magnetostriction (Table 4.3), also by pressure [132]. The transition in the field is accompanied by a significant hysteresis of $\sim$1 T; thermal hysteresis is about 2–5 K.

The variation of the transition temperature in $Gd_5(Si_xGe_{1-x})_4$ is achieved by changing the stoichiometry x. Obviously, the increase of x above 2 leads to the change of the transition from first to second order, and the MCE is decreased (Table 4.3, Fig. 4.31). Likewise, the addition of a fourth element Z being Fe, Cu, Ga, Mn or Al leads to a shift of the transition temperature of $Gd_5Ge_{1.9}Si_2Z_{0.1}$ to a somewhat higher temperatures and the change of the transition from first- to second-order type [133, 134]. Substitution of small amounts of Sn [135] and Nb [136] was found to increase T_c and retain the first-order character of the transition. Substitution of other rare earths (Dy, Tb, Pr) for Gd generally reduces T_c and MCE [17].

When a commercial-grade Gd metal is used to prepare $Gd_5Si_2Ge_2$, the MCE $|\Delta S_{max}| \approx 46$ kJ/m^3 K is significantly inferior to the MCE ($|\Delta S_{max}| \approx 205$ kJ/m^3 K) obtained in appropriately heat-treated materials prepared from high-purity Gd metal [31, 137]. The enhancement of the MCE after a homogenisation heat treatment at about 1570 K was ascribed to the elimination of foreign phases and a partial ordering of the crystal structure via a redistribution of Si and Ge atoms among different crystallographic sites (Table 4.3, [31, 96]). When using a commercial-grade Gd (purity between 90 and 98 at.%), a complex process with dynamic vacuum and high-temperature heat treatment is necessary to obtain a large entropy change in the range of 131–160 kJ/m^3 K [138, 139]. Generally, the high cost of the constituent elements and large hysteresis limit the attractiveness of $Gd_5(Si_xGe_{1-x})_4$-type materials.

Table 4.3 Properties of Gd–Si–Ge alloys at the transition from a high- to low-temperature phase: the ΔS peak temperature in the lowest provided field change (T_{max}), the maximum entropy change (ΔS_{max}) and the maximum adiabatic temperature change (ΔT_{ad}^{max}) in a field change $\mu_0 \Delta H$, the change of the lattice cell parameters (Δa/a and Δc/c) and volume (ΔV/V), the maximum thermal hysteresis ΔT_{hys} and maximum field hysteresis ΔH_{hys}. The values of ΔS_{max} are calculated from gravimetric to volumetric units or vice versa using the density of 5.7 g/cm^3

Material	T_{max} (K)	$\mu_0\Delta H$ (T)	$-\Delta S_{max}$ (J/kg K)	$-\Delta S_{max}$ (kJ/m^3 K)	ΔT_{ad}^{max} (K)	Lattice change (%) Δa/a	Δc/c	ΔV/V	ΔH_{hys} (T)	ΔT_{hys} (K)	Reference
$Gd_5Si_2Ge_2$ (arc-melted)	277	1	6.5	37					1	2–5	[10], [31]
		2	15	85	5						
		5	20	114	11						
$Gd_5Si_2Ge_2$ (homogenised at 1570 K/1 h)	272	1	12	68		−0.9[a]	0.2[a]	0.4[a]			[31], [96][a]
		2	27	154	7						
		5	36.4	207	17						
Gd_5Si_4	326	2			4.3[b]						[84], [17][b]
		5		61.7	7.0[b]						
$Gd_5Si_3Ge_1$	323	2			4.2[b]						[84], [17][b]
		5	11.4	65.3	6.8[b]						
Gd_5Ge_4	20	5		128							[84], [17][b]
$Gd_5Si_1Ge_3$	140	2			4.2[b]						[84], [17][b]
		5		538	12[b]						

4.5.3 $LaFe_{13-x}Si_x$ Alloys

In the concentration range $x \leq 2.5$, the $LaFe_{13-x}Si_x$ compounds crystallise in the cubic $NaZn_{13}$-type structure (space group $Fm\overline{3}c$). The binary $LaFe_{13}$ phase does not exist; therefore, the addition of Si, Al or Co is required for the stabilisation of the $NaZn_{13}$-type phase [140, 141]. The La atoms fully occupy the 8a sites, and all the 8b sites are solely occupied by Fe; the Si atoms and the Fe atoms are distributed among the 96i sites [29].

The giant MCE accompanied by a very low thermal and field hysteresis was reported by Hu et al. [142] in $LaFe_{11.4}Si_{1.6}$. The compound with this stoichiometry is on the verge of the changing transition type from first to second order, which is reflected by a weakly pronounced S-shape of the M(H) curves and essentially absent field hysteresis (Fig. 4.32). Above $x = 1.6$ the transition is second order. In $LaFe_{13-x}Si_x$ with $x \leq 1.6$, the giant MCE is associated with the thermally induced first-order phase transition from the paramagnetic to the ferromagnetic state at

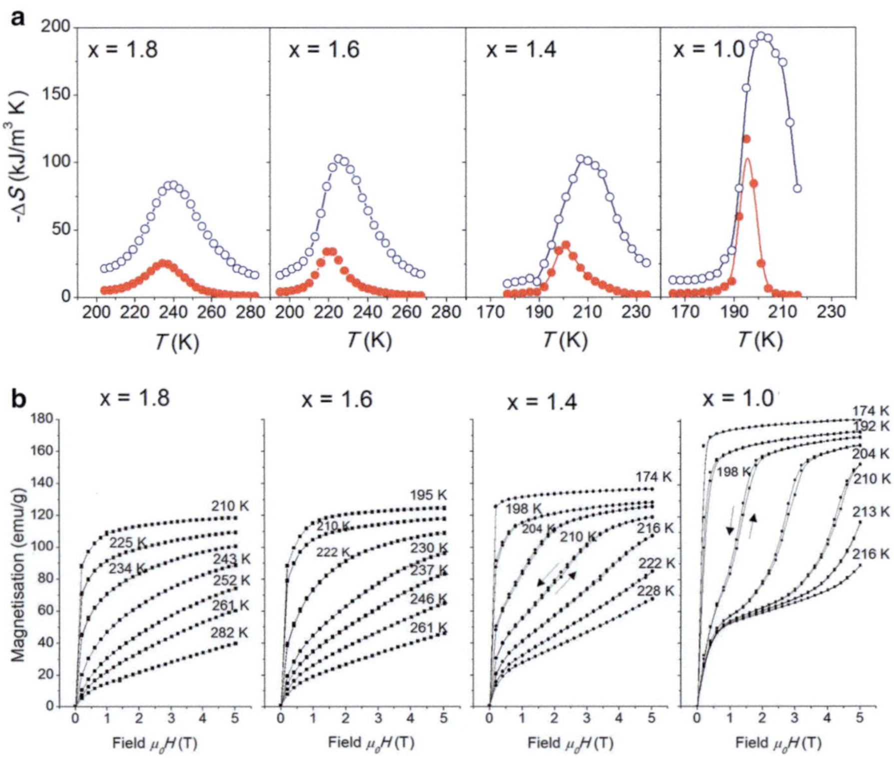

Fig. 4.32 (**a**) Temperature dependence of the entropy change ΔS of melt-spun $LaFe_{13-x}Si_x$ ribbons in a field change of 0–2 T (closed symbols) and 0–5 T (open symbols). (**b**) The isothermal magnetisation of melt-spun $LaFe_{13-x}Si_x$. Magnetisation isotherms are shown for increasing and decreasing magnetic field in the vicinity of T_c (adapted from [85])

temperatures around 200 K [143]. Additionally, above the Curie temperature, an itinerant-electron metamagnetic (IEM) transition is induced by an external magnetic field; the magnetisation isotherms at the IEM transition have a characteristic S-shape (x = 1.0 in Fig. 4.32b). The IEM transition occurs due to the change of the density of states at the Fermi level under the action of the magnetic field. Both thermal and field transitions to the ferromagnetic state are accompanied by a sizable volume change of about 1.0–1.4 % (Table 4.4, [29, 94]).

Electronic structure calculations suggest that $LaFe_{13-x}Si_x$ owes its excellent magnetocaloric properties to the exceptional flatness of its total energy dependence on the magnetic moment (Fig. 4.33a, [86]). The presence of several shallow minima and maxima in the energy profile that are very sensitive to the lattice parameter implies that transition to the ferromagnetic state involves a series of consecutive transitions; these have been observed experimentally in the hydride of the $LaFe_{13-x}Si_x$ alloy [92]. Thus, the $LaFe_{13-x}Si_x$ materials have a very low intrinsic hysteresis.

In $LaFe_{13-x}Si_x$, the Curie temperature increases linearly with increasing Si content (Table 4.4, [94]). The maximum T_c of about 250 K is achieved for x = 2.5, which is below the room temperature and the transition becomes second order. The addition of Co can be used to raise T_c to the room temperature and above; however, also in this case the transition changes from first to second order, and the entropy change is reduced significantly [85].

Hydrogen insertion into the $LaFe_{13-x}Si_x$ phase can be used to adjust the Curie temperature. The increase of the hydrogen concentration y from 0 to about 2.3 in the $La(Fe,Si)_{13}H_y$ hydrides enables a linear increase of T_c to about 350 K [85]. The first-order type of the phase transition is preserved, and as a consequence the MCE magnitude in the $La(Fe,Si)_{13}H_y$ hydrides remains large in contrast to the situation where the increase of T_c is achieved by Co substitution (Table 4.4). Hydrogen is incorporated into the interstitial 24d sites of the $LaFe_{13-x}Si_x$ crystal structure with the maximum occupancy fraction of about 50 %; the crystal symmetry is not modified, but a lattice expansion is observed [29, 145]. The absorption of hydrogen does not only increase the lattice constants of $LaFe_{13-x}Si_x$ but also modifies the electronic structure of the material, as observed by the study of the thermopower in these alloys [146]. Hydriding can be performed by a high-temperature gas process [118, 145] or by an alternative cost-effective room temperature electrolytic process [119].

However, partially hydrogenated $La(Fe,Si)_{13}H_y$-based alloys are not stable when used at temperatures close to their Curie temperature but decompose into hydrogen-rich and hydrogen-poor fractions with deteriorated MCE [118]. This behaviour makes the material less suitable for applications. Thus, it is more appropriate to first adjust the Curie temperature of the parent alloy, e.g. by partial substitution of Mn for Fe [118, 147] or Ce for La [148], and subsequently subject the alloys to full hydrogenation. The substitution of Ce for La is also attractive because of the substantially higher availability of Ce as compared to other rare-earth elements. The working temperature range of the $La(Fe,Si)_{13}H_y$ hydrides is limited to below 150 °C (see Sect. 4.4.6).

Table 4.4 Properties of $LaFe_{13-x}Si_x$-type alloys at the transition from a high- to low-temperature phase: the ΔS peak temperature in the lowest provided field change (T_{max}), the maximum entropy change (ΔS_{max}) and the maximum adiabatic temperature change (ΔT_{ad}^{max}) in a field change $\mu_0\Delta H$, the isotropic volume change (ΔV/V), the maximum thermal hysteresis ΔT_{hys} and maximum field hysteresis ΔH_{hys}. The values of ΔS_{max} are calculated from gravimetric or volumetric units to vice versa using the density of 6.8 g/cm^3

Material	T_{max} (K)	$\mu_0\Delta H$ (T)	$-\Delta S_{max}$ (J/kg K)	$-\Delta S_{max}$ (kJ/m^3 K)	ΔT_{ad}^{max} (K)	ΔV/V (%)	ΔH_{hys} (T)	ΔT_{hys} (K)	Reference
$LaFe_{12.0}Si_{1.0}$ (melt-spun)	196	2 5	17 29	118 195			0.07	0.7	[85]
$LaFe_{11.4}Si_{1.6}$ (melt-spun)	220	2 5	5 15	35 102			0.02		[85]
$LaFe_{11.2}Si_{1.8}$ (melt-spun)	234	2 5	4 12	25 83			0		[85]
$LaFe_{11.8}Si_{1.2}$	198	5	30	204		1.35			[85]
$LaFe_{11.4}Co_{0.4}Si_{1.2}$	244	5	19	132					[85]
$LaFe_{11.0}Co_{0.8}Si_{1.2}$	290	5	14	93					[85]
$LaFe_{11.6}Si_{1.4}H_{0.8}$	240	5	17	114					[85]
$LaFe_{11.6}Si_{1.4}H_{1.2}$	273	5	17	118					[85]
$LaFe_{11.6}Si_{1.4}H_{1.6}$	333	5	21	140			0.1	<0.5	[92]
$LaFe_{11.6}Si_{1.4}H_{2.3}$	342	5	18	120					[85]
$LaFe_{11.35}Mn_{0.39}Si_{1.26}H_{1.53}$	282	1.6	11	75					[118]
$LaFe_{11.35}Mn_{0.39}Si_{1.26}H_{1.53}$	289	1.6	13	88					[118]
$LaFe_{11.35}Mn_{0.39}Si_{1.26}H_{1.53}$	295	1.6	12	82					[118]
$LaFe_{11.6}Si_{1.4}$ (bulk)	196	1 2 5	22 24 26	147[a] 160 180	4.2 7.8	1.09[b]	0.4	2.3	[85]; Lyubina, unpublished data[a]; [29][b]
$LaFe_{11.6}Si_{1.4}$ (melt-spun)	201	1 2 5	10 16 22	68[a] 107 151	2.2[a] 4.1[b]		0.03	1.0	[92]; Lyubina, unpublished data[a]; [29][b]
$LaFe_{11.6}Si_{1.4}$ (porous)	196	1 2 5	11 16 18	76[a] 109 125[a]	3.3 5.8		0.09	0.5	[29]; Lyubina, unpublished data[a]
$LaFe_{11.6}Si_{1.4}$ (grain size 70 nm)	200	2 5	8 14	54 96	2.2		0.03		[29]

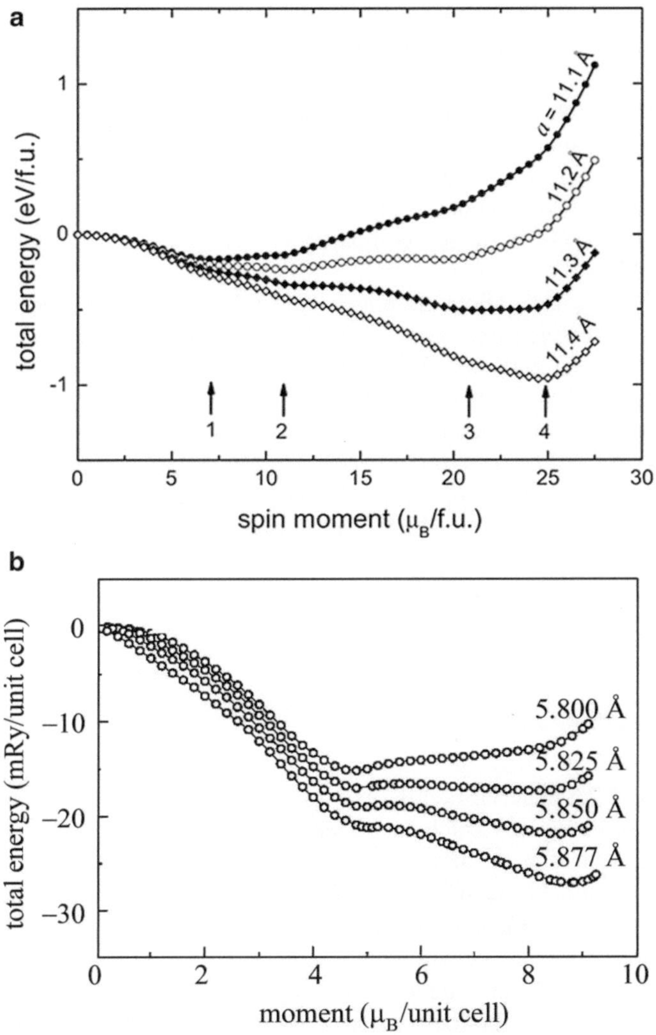

Fig. 4.33 Calculated values of the total energy vs. magnetic moment curves at the indicated values of lattice parameters for (**a**) $LaFe_{12}Si_1$ [86] and (**b**) Fe_2P (adapted from [144]). The arrows mark the positions of minima

As discussed in Sect. 4.4.3, substantial hysteresis can be observed in materials with the first-order transition, which is of extrinsic, microstructure-driven nature [14]. Therefore, tailoring of the microstructure can be an effective means in improving the material performance without changing the order of the transition. In the family of the $LaFe_{13-x}Si_x$ materials, bulk alloys show a thermal and field hysteresis of 2.3 K and 0.4 T, respectively (Table 4.4, [14, 29]). The reduction of grain constraints by creating the appropriate microstructure is crucial in tailoring the transition and the hysteresis [14]. In porous and melt-spun $LaFe_{13-x}Si_x$, the hysteresis is reduced drastically, and the values of the entropy change normalised to the mass density are at the high level (Table 4.4). In the design of the microstructure, further effects, such as particle surface oxidation [15] and random anisotropy [29], should be taken into account, as those can have a negative effect on the MCE. The intergrain exchange coupling between the crystallites with nanometre dimensions brings about a substantial reduction in the MCE (Table 4.4): in $LaFe_{13-x}Si_x$ with the crystallite size of about 70 and 40 nm, the maximum entropy change ΔS_{max} is about 40 and 60 % lower compared to the microcrystalline counterpart, respectively [29].

The $La(Fe,Si)_{13}$-type alloys do not contain critical elements; they can be produced by cost-effective techniques (e.g. melt spinning). When the microstructure of $La(Fe,Si)_{13}$-type materials is designed appropriately, a large MCE accompanied by a very low hysteresis can be obtained (Table 4.4), which makes these alloys very attractive for magnetic heat pumping. The reported specific cooling power of $La(Fe,Si)_{13}H_y$-type hydrides (≈450 W/kg) arranged in a layered bed exceeds significantly that of Gd (≈170 W/kg). Furthermore, values as high as 1300 W/kg were reported for layered $La(Fe,Si)_{13}H_y$ beds (the values are compiled from [66, 70]). It should be noted, however, that the achievable specific cooling power depends on the particular magnetic cooling device design.

4.5.4 $(Mn,Fe)_2(P,X)$ Compounds

The $(Mn,Fe)_2(P,X)$ material family (X = As, Si, Ge, B) is derived from the prototypical Fe_2P compound, which has been known since the 1980s to exhibit a moderate latent heat at the first-order transition from the para- to ferromagnetic state at a Curie temperature of 217 K [149]. In the hexagonal Fe_2P compound (space group $P\overline{6}2m$), the Fe atoms occupy two non-equivalent Wyckoff positions 3f and 3g. Using band structure calculations, Yamada and Terao [144] showed that Fe(3g) is ferromagnetic both at smaller and larger lattice constants and Fe(3f) is non-magnetic at smaller lattice constants, but a small moment of about 0.3 μ_B is induced on Fe(3f) due to the hybridisation between the orbitals on Fe(3g) and Fe(3f). In other words, Fe atoms at the 3f site exhibit a metamagnetic behaviour, while the Fe atoms at the 3g site are ferromagnetic. Yamada and Terao [144] ascribed the first-order transition in Fe_2P to the metastable state of the Fe moment at the 3f site. With increasing temperature below the Curie temperature, the ferromagnetic moment of Fe(3g) decreases and so does the effective exchange field created by Fe(3g) on Fe atoms at the 3f site. Eventually, the magnetic moment of Fe(3f) collapses at the critical field of the metamagnetic transition or in zero field at T_c. Simultaneously, Fe(3g) also loses its moment.

The itinerant metamagnetic transition in Fe_2P-type materials occurs without the change of the crystal symmetry; both ferro- and paramagnetic phases are hexagonal. A significant linear magnetostriction (anisotropic volume change) is observed at the transition (Table 4.5). The total energy vs. magnetic moment of Fe_2P (and apparently that of related MnFeP and $MnFeP_{0.7}As_{0.3}$ compounds) shows very low barriers between the para- and ferromagnetic states (Fig. 4.33b), thus implying that a low intrinsic hysteresis can be obtained for some compositions in this material class.

The addition of As allows to increase T_c to the room temperature range at the same time retaining the first-order character of the transition [156]. A giant magnetocaloric effect was reported by Tegus et al. [151] in $MnFeP_{0.45}As_{0.55}$ ($|\Delta S_{max}| = 95$ kJ/m^3 K in a field change of 0–2 T near 303 K). In correspondence with the expected low intrinsic hysteresis in this material class, the first-order transition in $MnFeP_{1-x}As_x$ is accompanied by a low thermal and field hysteresis (Table 4.5). Despite the excellent magnetic properties, the poisonous nature of As prohibits the use of these materials in domestic appliances.

In the search for alternatives for arsenic, Ge and Si were found to be suitable substitution elements that do not change the transition type and allow the T_c variation (Fig. 4.24). In particular, the same mechanism discussed above for Fe_2P was found to be responsible for the metamagnetic transition and giant MCE in MnFe(P,Si) [77]. The calculations show that layers occupied by Mn (3g) are strongly magnetic; the disappearance of the magnetic moments in weak itinerant Fe layers (3f) is ascribed to a change from non-bonding electron density at the Fe site below T_c into a distribution, which is hybridised with Si/P above T_c. This change in hybridisation causes the distinct change in the lattice parameters at T_c (Table 4.5). However, the substitution with Ge and Si leads to a strong increase of the thermal and field hysteresis to about 20–40 K and 2.5 T, respectively (Table 4.5, [153, 154]).

Substitution of Co and Ni for Fe in MnFe(P,Si) allows to reduce the hysteresis, albeit at a price of weakening the first-order transition and reducing the MCE (Table 4.5, [153]). Unusually, the Co substitution leads to the reduction of the Curie temperature. B substitution leads to the disappearance of the first-order magnetic transition, with the rate being more pronounced than observed when varying the Mn/Fe and P/Si ratio or As and Ge substitution [155, 157]. However, for some B-substituted compounds, e.g. $MnFe_{0.95}P_{0.582}Si_{0.34}B_{0.078}$, it is possible to obtain high MCE and low hysteresis (Table 4.5, [115]).

It is clear from Table 4.5 that the large hysteresis is always accompanied by a large lattice parameter and lattice volume change. In the $(Mn,Fe)_2(P,X)$-type alloys, the large discontinuity in the lattice parameters/large anisotropic volume change at the transition leads to the observation of the so-called virgin effect. The effect consists in the observation of the Curie temperature on first cooling through the PM–FM transition at a much lower temperature than that on subsequent cooling; the difference can reach 20–50 K [16, 19, 115]. The effect can be traced back to the formation of cracks leading eventually to the pulverisation of bulk materials. In the weak first-order transition $MnFe_{0.95}P_{0.582}Si_{0.34}B_{0.078}$, $\Delta V/V = 0.09$ % is very low, and additionally to the low hysteresis, the material was found to be mechanically stable on cycling [115]. The $(Mn,Fe)_2(P,X)$-type alloys with the large MCE, low

Table 4.5 Properties of $(Mn,Fe)_2(P,X)$-type alloys at the transition from a high- to low-temperature phase: the ΔS peak temperature in the lowest provided field change (T_{max}), the maximum entropy change (ΔS_{max}) and the maximum adiabatic temperature change (ΔT_{ad}^{max}) in a field change $\mu_0\Delta H$, the change of the lattice cell parameters ($\Delta a/a$ and $\Delta c/c$) and volume ($\Delta V/V$), the maximum thermal hysteresis ΔT_{hys} and maximum field hysteresis ΔH_{hys}. The values of ΔS_{max} are calculated from gravimetric to volumetric units or vice versa using the density of 6.3 g/cm^3

Material	T_{max} (K)	$\mu_0\Delta H$ (T)	$-\Delta S_{max}$ (J/kg K)	$-\Delta S_{max}$ (kJ/m^3 K)	ΔT_{ad}^{max} (K)	Lattice change (%)			ΔH_{hys} (T)	ΔT_{hys} (K)	Reference
						$\Delta a/a$	$\Delta c/c$	$\Delta V/V$			
$MnFeP_{0.45}As_{0.55}$	303	1 2 5	13 15 18[a]	82 95 113	2.8					7.0, 1.0[a]	[150], [151][a]
$MnFeP_{0.47}As_{0.53}$	300	1 2	8 13	50 82					≈0.1[b]	1.7	[150], [152][b]
$Mn_{1.1}Fe_{0.9}P_{0.47}As_{0.53}$	294	1 2	11 20	69 126	2.9					2.0	[150]
$MnFeP_{0.78}Ge_{0.22}$	323	2	30	189					2.5	20	[153]
$MnFe_{0.94}Co_{0.06}P_{0.78}Ge_{0.22}$	294	2	20	126						8	[153]
$MnFe_{0.88}Co_{0.12}P_{0.78}Ge_{0.22}$	260	2	17	107						4	[153]
$MnFe_{0.88}Co_{0.12}P_{0.74}Ge_{0.26}$	300	2	11	69						2	[153]
$MnFeP_{0.55}Si_{0.3}Ge_{0.15}$	292	2	6	38						9	[154]
$MnFeP_{0.59}Si_{0.3}Ge_{0.11}$	288	2	14	88						18	[154]
$MnFeP_{0.59}Si_{0.34}Ge_{0.07}$	280	2	16	101						27	[154]
$MnFe_{0.95}P_{0.55}Si_{0.45}$	282	1	10	63		−1.74	3.73	0.15		36	[115]
$Mn_{1.25}Fe_{0.7}P_{0.5}Si_{0.5}$	274	1	9	57		−0.79	1.86	0.26	≈0.05[b]	3	[115]; [152][b]
$MnFe_{0.95}P_{0.582}Si_{0.34}B_{0.078}$	289	1	9	57		−0.73	1.57	0.09		2.5	[115]
$MnFe_{0.95}P_{0.595}Si_{0.3}B_{0.075}$	281	1 2	9.8 14.5	62 91	2.7			≤0.05		1.6	[155]

hysteresis and not containing poisonous (As) and expensive (Ge) elements are very attractive for magnetic cooling applications.

4.5.5 Heusler Alloys

Heusler materials are intermetallic compounds with a general formula X_2YZ, where X atoms occupy the 8c Wyckoff positions and Y and Z atoms the 4a and 4b positions of the space group $Fm\overline{3}m$, respectively. X atoms are transition elements, Y atoms are usually transition elements, and in some instances rare-earth or alkaline rare-earth elements and Z atoms are IIIA–VA group elements. Several magnetic field-driven effects are observed in magnetic Heusler alloys including magnetic shape memory effect [158], giant magnetoresistance, exchange bias and giant magnetocaloric effect (see reviews [159, 160] and references therein). All these effects rely on the presence of the martensitic phase in the Heusler alloys.

The magnetocaloric effect in Heusler alloys was first reported in Ni_2MnGa [161]. The martensitic transformation at around 200 K in stoichiometric Ni_2MnGa and close to stoichiometric alloys is of first order, and the application of a magnetic field close to this transition leads to a large MCE [161, 162]. Both austenite and martensite phases in Ni_2MnGa-type alloys are ferromagnetic and the magnetisation jump is mainly controlled by the magnetostructural coupling at the mesoscopic scale [162]. The entropy change is largest ($\Delta S_{max} \approx -160$ kJ/m^3 K in a field change of 5 T) when the austenite Curie temperature nearly coincides with the austenite–martensite transformation temperature [159].

In Ni_2MnZ-based Heusler alloys with Z = Sn, In and Sb, a giant MCE is also observed at the martensite transformation [163]. In these Heusler alloys, a so-called inverse magnetocaloric effect at the austenite–martensite transition is observed, i.e. the entropy change ΔS is positive and is about 56 kJ/m^3 K and 144 kJ/m^3 K in Ni_2MnSn in a field change of 2 and 5 T, respectively. The sign of ΔS is determined by the sign of the magnetisation derivative $(\partial M/\partial T)_H$ in Eq. (4.22). The positive $(\partial M/\partial T)_H$ is a consequence of a higher magnetisation of the high-temperature ferromagnetic austenite phase compared to the magnetisation of the low-temperature martensite [163–165]. The first-order transformation in the Heusler alloys is accompanied by a large temperature hysteresis (about 10–20 K) and field hysteresis (about 1–2 T). Due to the large hysteresis, the adiabatic temperature change upon cooling and heating differs significantly (Table 4.6, [167]) making the use of this material type in the AMR cycle inefficient.

4.5.6 Manganites

Ferromagnetic manganese oxides with the perovskite structure, so-called manganites, with the general formula $R_{1-x}M_xMnO_3$, where R = La, Pr, Nd, Sm, Eu, Gd,

Ho, Tb, Y and M = Ca, Sr, Ba, Pb, Na^{1+}, K^{1+}, Ag^{1+}, are another class of materials considered for magnetic cooling applications; their properties have been reviewed by Phan and Yu [168]. The application field of manganites ranges from colossal magnetoresistance to spintronics [169]. In terms of the magnetic cooling applications, most interesting manganites are those from the family of (La,Sr,R)MnO_3, where R = Pr, Ca [121, 170].

The fact that the materials are oxides makes it possible to use cost-effective synthesis and shaping techniques that are usually not suitable for the preparation of metal alloys. Recently, Bahl et al. [121] reported the preparation of thin and flat ceramic plates of $La_{0.67}Ca_{\approx 0.29}Sr_{\approx 0.04}Mn_{1.05}O_3$ using spray pyrolysis followed by short-time annealing at about 1000 °C and tape casting. Next to the low cost, the advantage of the manganites is the ease of the Curie temperature tuning near room temperature that is achieved simply by varying the La:Ca:Sr (R:M) ratio, the absence of hysteresis and excellent chemical stability. However, due to the second-order transition at the Curie temperature, the MCE in the majority of the manganites is low (Table 4.7, [168]). An exception is $La_{1-x}Ca_xMnO_3$ ($x = 0.25$–0.50) that was reported to show a first-order transition due to the simultaneous paramagnetic insulator to ferromagnetic metal transition accompanied by a large volume change [175]. Although the entropy change value is improved, the adiabatic temperature change remains rather low (Table 4.6, [174])

Although the gravimetric entropy change in manganites is close to that in Gd [172, 173], the low density of manganites leads to moderate values of the volumetric entropy change (cf. Tables 4.2 and 4.7). Due to a high heat capacity (cf. Eq. 4.27), the adiabatic temperature change ΔT_{ad}^{max} of manganites is below 2 K in a field change of 2 T and is noticeably lower than that in Gd (cf. Tables 4.2 and 4.7, [121]). The often reported high RCP or RC values for manganites must be taken with care, since if one takes into account the low ΔT_{ad} and low volumetric ΔS of manganites, the cooling capacity of manganites ($RCP_{\Delta S}$) in the temperature range where ΔT_{ad} is above 2 K can be negligibly small (see Sect. 4.4.1). However, it must be noted that the achievable cooling power in an AMR depends strongly on its particular design and cooling can also be performed using manganite materials with the low MCE, albeit with a low specific cooling power of 35 W/kg [121]. Bahl et al. [121] reported that the highest specific cooling power obtained in the same device for Gd plates was 16 W/kg, i.e. lower than for manganites.

4.5.7 Amorphous and Nanocomposite Materials

A wide range of amorphous materials and nanocomposites containing transition metal or rare-earth nanocrystals dispersed in an amorphous matrix exhibit a second-order magnetic phase transition in a very broad temperature range of ~100 K. These alloys include exemplarily Fe-based amorphous alloys and nanocomposites, e.g. Fe–Zr [176], FeCoSiB, Finemet-type FeSiBCuNb [126, 177], $Fe_{75}Nb_{10}B_{15}$

Table 4.6 Properties of Heusler alloys at the transition from a high- to a low-temperature phase: the ΔS peak temperature in the lowest provided field change (T_{max}), the maximum entropy change (ΔS_{max}) and the maximum adiabatic temperature change (ΔT_{ad}^{max}) in a field change $\mu_0\Delta H$, the change of the lattice cell parameters ($\Delta a/a$ and $\Delta c/c$) and volume ($\Delta V/V$), the maximum thermal hysteresis ΔT_{hys} and maximum field hysteresis ΔH_{hys}. The values of ΔS_{max} are calculated from gravimetric to volumetric units or vice versa using the density of 8 g/cm^3

Material	T_{max} (K)	$\mu_0\Delta H$ (T)	$-\Delta S_{max}$ (J/kg K)	$-\Delta S_{max}$ (kJ/m^3 K)	ΔT_{ad}^{max} (K)	ΔH_{hys} (T)	ΔT_{hys} (K)	Reference
$Ni_{52.9}Mn_{22.4}Ga_{24.7}$	305	5	8.6	69				[166]
$Ni_{55.2}Mn_{18.6}Ga_{26.2}$	315	5	20.4	163				[166]
$Ni_{52.7}Mn_{23.9}Ga_{23.4}$	338	5	15.6	125				[166]
$Ni_{50}Mn_{34}In_{16}$	225	1	−2.4	−19		1.6	30[a]	[165], cooling[a], and heating[b] data (Lyubina, unpublished)
		2	−4.6	−37[a]	−0.5[a] −1.3[b]			
		5	−9.4	−75				
$Ni_{50}Mn_{37}Sn_{13}$	295	1	−3.3	−26				[163]
		2	−6.9	−55				
		5	−18.0	−144				
$Ni_{50}Mn_{36}Co_1Sn_{13}$	283	1			−0.4		16	[167]
		2	−13.8	−110	−1.25			

powders [178] and rare-earth-based composites and glasses, e.g. $Gd_{60}Al_{10}Mn_{30}$ nanocomposite [88], $Tb_{60}Ni_{30}Al_{10}$ and other Nd-, Tb-, Dy-, Ho-, Er- and Gd-based metallic glasses ([179] and references therein). The Curie temperature of the rare-earth-based alloys is below 200 K, whereas T_c of the Fe-based amorphous materials and nanocomposites can be adjusted by composition variation and addition of alloying elements near room temperature.

The close to 300 K and adjustable transition temperatures, low cost, low hysteresis, high corrosion resistance and good mechanical properties guide one to consideration of Fe-based materials as potential magnetic refrigerants. However, the MCE in the Fe-based amorphous and nanocomposite materials is very low. Typical $|\Delta S|$ values are in the range between 3.4 and 13.6 kJ/m^3 K for a field change of 1.5 T (Table 4.8, [126]). Taking into account the low adiabatic temperature change of 1.4 K in a field change of 1.8 T [182], only a low specific cooling power can be expected for this material type.

4.5.8 Materials with Transitions Between Magnetically Ordered States

4.5.8.1 Fe–Rh Alloys

The magnetic properties of the Fe–Rh system (space group $Pm\overline{3}m$) were first investigated in the 1930s [183, 184]. It was observed that when the rhodium concentration reaches 50 %, a phase transition around 350 K at which the bulk magnetisation disappears takes place. Later it was demonstrated that the observed phase transition is a first-order transition from the high-temperature ferromagnetic to the low-temperature antiferromagnetic (AF) state [185]. The first observation of the giant magnetocaloric effect ($\Delta T_{ad}^{max} \approx -6.8$ and –12.9 K in a field change of 1.25 and 1.95 T, respectively) was in a slightly off-stoichiometric $Fe_{49}Rh_{51}$ around 313 K [13]. The entropy change saturates in a field of approximately 1 T to a value of $\Delta S_{max} \approx 123$ kJ/m^3 K [186]. The MCE is inverse and is accompanied by a large hysteresis of about 10 K (Table 4.9). The large hysteresis is the reason for the observation of the large MCE only at the first cycle [13, 190]. After the first magnetisation–demagnetisation cycle, some part of the sample transforms to the FM state and does not return to the AF on subsequent field removal. Thus, the subsequent application of the magnetic field induces a much lower MCE [190]. The large hysteresis and the prohibitively high price of Rh make this alloy system rather unsuitable for magnetic heat pumping applications.

4.5.8.2 Spin Reorientation Transition in $Tb_2Fe_{14}B$ and $NdCo_5$

An abrupt change of the magnetisation is required for the MCE maximisation (see Eq. 4.22 and Sect. 4.4.1). A large MCE can be also obtained at a first-order spin

Table 4.7 Properties of manganites at the transition from a high- to a low-temperature phase: the ΔS peak temperature in the lowest provided field change (T_{max}), the maximum entropy change (ΔS_{max}) and the maximum adiabatic temperature change (ΔT_{ad}^{max}) in a field change $\mu_0 \Delta H$, the change of the lattice volume ($\Delta V/V$) and the maximum thermal hysteresis ΔT_{hys}. The values of ΔS_{max} are calculated from gravimetric to volumetric units or vice versa using the density of 6.2 g/cm^3

Material	T_{max} (K)	$\mu_0 \Delta H$ (T)	$-\Delta S_{max}$ (J/kg K)	$-\Delta S_{max}$ (kJ/m^3 K)	ΔT_{ad}^{max} (K)	$\Delta V/V$ (%)	ΔT_{hys} (K)	Reference
$La_{0.67}Sr_{0.33}MnO_3$	370	1	1.6	9.9	1.0			[171]
		2	2.7	16.7	1.6			
		5	5.15	31.9	3.3			
$La_{0.67}(Ca_{0.95}Sr_{0.05})_{0.33}MnO_3$	275	1	3.26[a]	20.2[a]				[172][a], [173][b]
		5	10.5[b]	65.1[b]				
$La_{0.67}(Ca_{0.887}Sr_{0.113})_{0.33}Mn_{1.05}O_3$	275	1	3.7	22.9	1.3			[121]
$La_{0.67}Ca_{0.33}MnO_3$	268	1			1.3	≈0.1[c]	5	[174], [175][c]
		2	6.9	42.8	2.4			

Table 4.8 Properties of amorphous and nanocomposite materials at the transition from a high- to a low-temperature phase: the ΔS peak temperature in the lowest provided field change (T_{max}), the maximum entropy change (ΔS_{max}) and the maximum adiabatic temperature change (ΔT_{ad}^{max}) in a field change $\mu_0\Delta H$, the change of the lattice cell parameters (Δa/a and Δc/c) and volume (ΔV/V), the maximum thermal hysteresis ΔT_{hys} and maximum field hysteresis ΔH_{hys}. The values of ΔS_{max} are calculated from gravimetric to volumetric units or vice versa using the density of 6.8 g/cm^3

Material	T_{max} (K)	$\mu_0\Delta H$ (T)	$-\Delta S_{max}$ (J/kg K)	$-\Delta S_{max}$ (kJ/m^3 K)	Reference
$Fe_{90}Zr_{10}$	230	1.4	1.4	9.5	[176]
		5	4.0	27	
$Fe_{84}Nb_7B_9$	299	1.5	1.44	9.8	[180]
$Fe_{88}Zr_7B_4Cu_1$	295	1.5	1.32	9.0	[181]
$Gd_{60}Al_{40}Mn_{30}$	171	5	3.3	22	[88]
$Tb_{60}Ni_{30}Al_{10}$	48	2	2.0	14	[179]
		4.6	6.5	44	

reorientation transition, which is accompanied by a large magnetocrystalline anisotropy constant(s) variation. MCE due to changes of the magnetic anisotropy energy during rotation of the spontaneous magnetisation was first observed by Akulov and Kirensky [191] in a Ni single crystal. The entropy is obtained from the free energy of the magnetic anisotropy as [192]

$$S_{\text{anis}} = -\left(\frac{\partial F_{\text{anis}}}{\partial T}\right)_H \tag{4.34}$$

and the corresponding infinitesimal adiabatic temperature change for a uniaxial crystal is

$$dT_{\text{ad}}^{\text{anis}} = -\frac{T}{C_{p,H}} dS_{\text{anis}} = \frac{T}{C_{p,H}} \frac{\partial K_1}{\partial T} d\sin^2\theta(H) \tag{4.35}$$

where K_1 is the uniaxial anisotropy constant, $C_{p,H}$ is the heat capacity at constant pressure and field and $\theta(H)$ is the angle between the magnetisation and the c-axis of the crystal in the absence of a magnetic field. The MCE due to the change of the anisotropy can be utilised in a special design of the magnetic heat pumping device, where differently to the usual process in which one applies and removes a magnetic field, a rotation of the magnetic material leading to a reorientation of the spins is performed.

In the family of $RE_2Fe_{14}B$ compounds (space group $P4_2/mnm$), of which $Nd_2Fe_{14}B$ is the most prominent example, as it forms the basis of the high-performance Nd–Fe–B magnets, there exist compounds with heavy rare earths (RE = Er, Tm) that show a spin reorientation of the first-order type at temperature T_{SR} near room temperature [27]. The spin reorientation transition in the compounds with RE = Er, Tm is between the “easy axis”, i.e. the magnetisation parallel to the c-axis above T_{SR} and the easy plane, where the magnetisation is perpendicular to the c-axis below T_{SR}. The spin reorientation temperature T_{SR} is 323 K in $Er_2Fe_{14}B$ and 310 K in $Tm_2Fe_{14}B$. The application of the magnetic field parallel to the c-axis

Table 4.9 Properties of materials with transitions between magnetically ordered states at the transition from a high- to a low-temperature phase: the ΔS peak temperature in the lowest provided field change (T_{max}), the maximum entropy change (ΔS_{max}) and the maximum adiabatic temperature change (ΔT_{ad}^{max}) in a field change $\mu_0 \Delta H$, the change of the volume ($\Delta V/V$), the maximum thermal hysteresis ΔT_{hys} and maximum field hysteresis ΔH_{hys}

Material	T_{max} (K)	$\mu_0 \Delta H$ (T)	$-\Delta S_{max}$ (J/kg K)	$-\Delta S_{max}$ (kJ/m^3 K)	ΔT_{ad}^{max} (K)	Density (g/cm^3)	$\Delta V/V$ (%)	ΔT_{hys} (K)	Reference
Fe–Rh	313–320	1	−12.5[a]	−123.2	≈−5.6	9.86	0.27[b]	10	[13], [186][a], [187][b]
		2	−12.5[a]	−123.2	≈−13				
		5	−12.5[a]	−123.2					
$Er_2Fe_{14}B$ (**H** ‖ *c*-axis)	323	1	0.8	6.6	−0.5[c]	8.22		0.2[d]	[188], [60][c], [27][d]
		2	0.8	6.6	−0.5[c]				
		5	0.8	6.6	−0.5[c]				
$NdCo_5$ (**H** ‖ *c*-axis)	290	1.3	−1.9	15.9	−1.6	8.38			[189]

of the $Er_2Fe_{14}B$ single crystal results in the cooling with the maximum adiabatic temperature change being about − 0.5 K at $T < T_{SR}$, whereas the application of the magnetic field perpendicular to the c-axis results in a positive ΔT_{ad} with the maximum of 0.5 K at $T > T_{SR}$; in both cases ΔT_{ad} saturates in a field of about 0.5–0.7 T [60]. The entropy change at the spin reorientation transition is about 6.6 kJ/m^3 K; this saturation value is related to the temperature derivative of the magnetic anisotropy [188]. The entropy change saturates in very low magnetic fields ~0.01 T, and further increase of the magnetic field acts to broaden the temperature range with the constant ΔS.

Such spin reorientation process is attractive for magnetic heat pumping, since the MCE can be induced by the rotation of a magnetic field rather than by the variation of its amplitude. Instead of the single crystals that are grown using expensive techniques, oriented polycrystalline powders fixed by a (polymer) binder can be used. A disadvantage is the necessity of (heavy) rare-earth element use in order to obtain a large derivative $\partial K_1/\partial T$. Moreover, the magnitude of the MCE is low. However, a giant MCE due to the spin reorientation transition has recently been observed in a single-crystal $NdCo_5$ by Nikitin et al. [189]. $NdCo_5$ possesses high magnetocrystalline anisotropy and exhibits a spin reorientation transition between "easy plane" and "easy cone" at 250 K and between "easy cone" and "easy axis" at 290 K. By rotating the $NdCo_5$ single crystal in the magnetic field of 1.3 T, the adiabatic temperature change of 1.6 K and entropy change of 15.9 kJ/m^3 K can be obtained; these values are about 50 % lower than in Gd (cf. Tables 4.2 and 4.9).

4.6 Summary

The principle of the magnetic heat pumping near room temperature utilising a magnetic phase transition has been known for four decades [8]. The last two decades have been marked by a rapid growth in the research field of magnetocaloric materials, which brought the understanding of the importance of a number of properties a material should possess to be useful for the magnetic heat pumping. A large number of prototypes existing nowadays substantiate the fact that it is still not clear which design is most favourable in terms of maximising the efficiency and minimising losses, and materials with drastically improved performance are sought after.

The intrinsic magnetic properties of the magnetocaloric materials and their variation are relatively well understood. A number of questions, nevertheless, remain open and further improvements are desired. Further reduction of the magnetic field (<<1 T) required for obtaining as large MCE as possible is desired, as this would eliminate the main obstacle for the widespread application being the high cost of the magnetic field (permanent magnets). In materials with the first-order transition, the minimisation of the H_0 term in Eq. (4.29) and (4.30) is required for further maximising the MCE; this can be achieved in materials consisting of fine, preferably single-crystalline particles with a uniform size distribution. In terms of the minimisation of the cooling system losses, reduction of the hysteresis and thermal transport management remain a challenge. Providing mechanical and chemical stability is crucial for the reliable device operation. Shaping the magnetocaloric materials into an appropriate form using industrial-scale relevant techniques is of utmost importance. Along with the possibility of the discovery of new materials, further improvements of solid magnetic refrigerant materials are expected to be achievable by creating specific material architectures. In this effort, the understanding of the magnetic phase transitions can be used to tailor and improve the performance of magnetocaloric materials.

Acknowledgement I would like to thank Lesley F. Cohen for her support and encouragement and fruitful discussions and Ullrich Hannemann for fruitful discussions and critical reading of the manuscript.

References

1. European Commission, Energy Efficiency, Buildings (2015). http://ec.europa.eu/energy/en/topics/energy-efficiency/buildings. Accessed 30 Apr 2015
2. Goetzler, W., Goffri, S., Jasinski, S., Legett, R., Lisle, H., Marantan, A., Millard, M., Pinault, D., Westphalen, D., Zogg, R.: Energy Savings Potential and R&D Opportunities for Commercial Refrigeration, U.S. Department of Energy, Energy Efficiency and Renewable Energy Building Technologies Program, 23 Sept 2009
3. Gschneidner Jr., K.A., Pecharsky, V.K.: Int. J. Refrigeration **31**, 945 (2008)
4. Zimm, C., Jastrab, A., Sternberg, A., Pecharsky, V., Gschneidner Jr., K., Osborne, M., Anderson, I.: Adv. Cryog. Eng. **43**, 1759 (1998)
5. Weiss, P., Piccard, A.: J. Phys. (Paris) **7**, 103 (1917). 5th Ser
6. Giauque, W.F., MacDougall, D.P.: Phys. Rev. **43**, 768 (1933)
7. Smith, A.: Eur. Phys. J. H **38**, 507 (2013)
8. Brown, G.V.: J. Appl. Phys. **47**, 3673 (1976)
9. Cooling post, Debut for magnetic refrigeration wine cooler. (2015). http://www.coolingpost.com/world-news/debut-for-magnetic-refrigeration-wine-cooler/. Accessed 30 Apr 2015
10. Pecharsky, V.K., Gschneidner Jr., K.A.: Phys. Rev. Lett. **78**, 4494 (1997)
11. Holtzberg, F., Gambino, R.J., McGuire, T.R.: J. Phys. Chem. Solids **28**, 2283 (1967)
12. Pecharsky, V.K., Gschneidner Jr., K.A.: Adv. Mater. **13**, 683 (2001)
13. Nikitin, S.A., Myalikgulyev, G., Tishin, A.M., Annaorazov, M.P., Asatryan, K.A., Tyurin, A. L.: Phys. Lett. A **148**, 363 (1990)
14. Lyubina, J., Schäfer, R., Martin, N., Schultz, L., Gutfleisch, O.: Adv. Mater. **22**, 3735 (2010)
15. Lyubina, J., Hannemann, U., Cohen, L.F., Ryan, M.P.: Adv. Energy Mater. **2**, 1323 (2012)
16. Brück, E.: In: Buschow K.H.J. (ed.) Handbook of Magnetic Materials, vol. 17, p. 235. Elsevier B.V., (2008)
17. Gschneidner Jr., K.A., Pecharsky, V.K., Tsokol, A.O.: Rep. Prog. Phys. **68**, 1479 (2005)
18. Tishin, A.M., Spichkin, Y.I.: The Magnetocaloric Effect and Its Applications. Institute of Physics Publishing, Bristol (2003)
19. Brück, E., Tegus, O., Cam Thanh, D.T., Trung, N.T., Buschow, K.H.J.: Int. J. Refrigeration **31**, 763 (2008)
20. Ashcroft, N.V., Mermin, N.D.: Solid State Physics, p. 464. Saunders College Publishing, Philadelphia (1976)

21. Wohlfarth, E.P.: In: Wohlfarth, E.P. (ed.) Handbook of Magnetic Materials, vol. 1, p. 47. North-Holland Publishing Company, Amsterdam (1980)
22. Kuz'min, M.D., Skokov, K.P., Karpenkov, D.Y., Moore, J.D., Richter, M., Gutfleisch, O.: Appl. Phys. Lett. **99**, 012501 (2011)
23. Lyubina, J., Kuz'min, M.D., Nenkov, K., Gutfleisch, O., Richter, M., Schlagel, D.L., Lograsso, T.A., Gschneidner Jr., K.A.: Phys. Rev. B **83**, 012403 (2011)
24. Oesterreicher, H., Parker, F.T.: J. Appl. Phys. **55**, 4334 (1984)
25. Landau, L.D., Lifshitz, E.M.: Statistical Physics, 3rd edn. Elsevier, Oxford (1980)
26. Grazhdankina, N.P.: Sov. Phys. Usp. **11**, 727 (1969)
27. Pique, C., Burriel, R., Bartolome, J.: J. Magn. Magn. Mater. **154**, 71 (1996)
28. Tu, P., Heeger, A.J., Kouvel, J.S., Comly, J.B.: J. Appl. Phys. **40**, 1368 (1969)
29. Lyubina, J.: J. Appl. Phys. **109**, 07A902 (2011)
30. Pecharsky, V.K., Gschneidner Jr., K.A., Fort, D.: Scripta Mater. **35**, 843 (1996)
31. Pecharsky, A.O., Gschneidner Jr., K.A., Pecharsky, V.K.: J. Appl. Phys. **93**, 4722 (2003)
32. Morrison, K., Bratko, M., Turcaud, J., Berenov, A., Caplin, A.D., Cohen, L.F.: Rev. Sci. Instr. **83**, 033901 (2012)
33. Morrison, K., Moore, J.D., Sandeman, K.G., Caplin, A.D., Cohen, L.F.: Phys. Rev. B **79**, 134408 (2009)
34. Lovell, E., Pereira, A.M., Caplin, A.D., Lyubina, J., Cohen, L.F.: Adv. Energy Mater. **5**, 1401639 (2015)
35. Roy, S.B., Perkins, G.K., Chattopadhyay, M.K., Nigam, A.K., Sokhey, K.J.S., Chaddah, P., Caplin, A.D., Cohen, L.F.: Phys. Rev. Lett. **92**, 147203 (2004)
36. Roy, S.B., Chattopadhyay, M.K., Chaddah, P., Moore, J.D., Perkins, G.K., Cohen, L.F., Gschneidner Jr., K.A., Pecharsky, V.K.: Phys. Rev. B **74**, 012403 (2006)
37. Belov, K.P.: Magnetic Transitions. Consultants Bureau, New York (1961)
38. Dan'kov, S.Y., Tishin, A.M., Pecharsky, V.K., Gschneidner Jr., K.A.: Phys. Rev. B **57**, 3478 (1998)
39. Reif, F.: Fundamentals of Statistical and Thermal Physics. McGraw-Hill International Editions, Singapore (1985)
40. Fliessbach, T.: Statistische Physik, 2nd edn. Spektrum Akademischer Verlag, Heidelberg (1995)
41. Pecharsky, V.K., Gschneidner Jr., K.A.: J. Appl. Phys. **86**, 565 (1999)
42. Pecharsky, V.K., Gschneidner Jr., K.A.: J. Appl. Phys. **86**, 6315 (1999)
43. Morrison, K., Lyubina, J., Moore, J.D., Sandeman, K.G., Gutfleisch, O., Cohen, L.F., Caplin, A.D.: Phil. Magazine **92**, 292 (2012)
44. Basso, V., Küpferling, M., Sasso, C.P., Giudici, L.: Rev. Sci. Instrum. **79**, 063907 (2008)
45. Basso, V., Küpferling, M., Sasso, C.P.: Rev. Sci. Instrum. **81**, 113904 (2010)
46. Chernyshov, A.S., Tsokol, A.O., Tishin, A.M., Gschneidner Jr., K.A., Pecharsky, V.K.: Phys. Rev B **71**, 184410 (2005)
47. Morrison, K., Sandeman, K.G., Cohen, L.F., Sasso, C.P., Basso, V., Barcza, A., Katter, M., Moore, J.D., Skokov, K.P., Gutfleisch, O.: Int. J. Refrigeration **35**, 1528 (2012)
48. Bratko, M., Morrison, K., de Campos, A., Gama, S., Cohen, L.F., Sandeman, K.G.: Appl. Phys. Lett. **100**, 252409 (2012)
49. Caron, L., Ou, Z.Q., Nguyen, T.T., Cam Thanh, D.T., Tegus, O., Brück, E.: J. Magn. Magn. Mater. **321**, 3559 (2009)
50. Manosa, L., Planes, A., Moya, X.: Adv. Mater. **21**, 3725 (2009)
51. Morrison, K., Lyubina, J., Moore, J.D., Caplin, A.D., Sandeman, K.G., Gutfleisch, O., Cohen, L.F.: J. Phys. D Appl. Phys. **43**, 132001 (2010)
52. Giguère, A., Foldeaki, M., Ravi Gopal, B., Chahine, R., Bose, T.K., Frydman, A., Barclay, J. A.: Phys. Rev. Lett. **83**, 2262 (1999)
53. Casanova, F., Batlle, X., Labarta, A., Marcos, J., Manosa, L., Planes, A.: Phys. Rev. B **66**, 100401(R) (2002)
54. Fujita, A., Yako, H., Kano, M.: J. Appl. Phys. **113**, 17A924 (2013)

55. Smith, A., Bahl, C.R.H., Bjørk, R., Engelbrecht, K., Nielsen, K.K.: N. Pryds. **2**, 1288 (2012)
56. Christensen, D.V., Bjørk, R., Nielsen, K.K., Bahl, C.R.H., Smith, A., Clausen, S.: J. Appl. Phys. **108**, 063913 (2010)
57. Otowski, W., Glorieux, C., Hofman, R., Thoen, J.: Thermochimica Acta **218**, 123 (1993)
58. Bjørk, R., Bahl, C.R.H., Katter, M.: J. Magnet. Magnet. Mater. **322**, 3882 (2010)
59. Lyubina, J., Gutfleisch, O., Kuz'min, M., Richter, M.: In: Proceedings of the 3rd International Conference on Magnetic Refrigeration at Room Temperature, Des Moines, IA, USA, May 11–15, 2009, p. 49
60. Skokov, K.P., Koshkid'ko, Yu.S., Pastushenkov, Yu.G., Lyubina, J., Gutfleisch, O.: In: Proceedings of the 3rd International Conference on Magnetic Refrigeration at Room Temperature, Des Moines, IA, USA, May 11–15, 2009, p. 161
61. Barclay, J.A.: J. Appl. Phys. **53**, 2887 (1982)
62. Zimm, C., Boeder, A., Chell, J., Sternberg, A., Fujita, A., Fujieda, S., Fukamichi, K.: Int. J. Refrigeration **29**, 1302 (2006)
63. Rowe, A.: Int. J. Refrigeration **34**, 168 (2011)
64. Tura, A., Rowe, A.: Int. J. Refrigeration **34**, 628 (2011)
65. Engelbrecht, K., Eriksen, D., Bahl, C.R.H., Bjørk, R., Geyti, J., Lozano, J.A., Nielsen, K.K., Saxild, F., Smith, A., Pryds, N.: Int. J. Refrigeration **35**, 1498 (2012)
66. Jacobs, S., Auringer, J., Boeder, A., Chell, J., Komorowski, L., Leonard, J., Russek, S., Zimm, C.: Int. J. Refrigeration **37**, 84 (2014)
67. Barclay, J.A., Steyert W.A.: U.S. Patent 4332135 (1982)
68. Bjørk, R., Bahl, C.R.H., Smith, A., Pryds, N.: Int. J. Refrigeration **33**, 437 (2010)
69. Yu, B., Liu, M., Egolf, P.W., Kitanovski, A.: Int. J. Refrigeration **22**, 1029 (2010)
70. Russek, S., Auringer, J., Boeder, A., Chell, J., Jacobs, S., Zimm, C.: In: Proceedings of 4th International Conference on Magnetic Refrigeration at Room Temperature, Baotou, China, 23–28 Aug. (2010) p. 339
71. WIPO Patentscope, Search World Intellectual Property Organization (2015). https://patentscope.wipo.int. Accessed 30 Apr 2015
72. Stefan, J.: Ann. Phys. **274**, 427 (1889)
73. Edison, T.A.: U.S. Patent 476983 (1892)
74. Tesla, N.: U.S. Patent 428057 (1890)
75. Kirol, L.D., Mills, J.I.: J. Appl. Phys. **56**, 824 (1984)
76. Solomon, D.: Energy Convers. Mgmt. **31**, 157 (1991)
77. Dung, N.H., Ou, Z.Q., Caron, L., Zhang, L., Thanh, D.T.C., de Wijs, G.A., de Groot, R.A., Buschow, K.H.J., Brück, E.: Adv. En. Mater. **1**, 1215 (2011)
78. Vuarnoz, D., Kitanovski, A., Gonin, C., Borgeaud, Y., Delessert, M., Meinen, M., Egolf, P. W.: Appl. Energ. **100**, 229 (2012)
79. Tishin, A.M.: Cryogenics **30**, 127 (1990)
80. Franco, V., Blazquez, J.S., Conde, A.: Appl. Phys. Lett. **89**, 222512 (2006)
81. Dong, Q., Zhang, H., Shen, J., Sun, J., Shen, B.: J. Magn. Magn. Mater. **319**, 56 (2007)
82. Belo, J.H., Amaral, J.S., Pereira, A.M., Amaral, V.S., Araújo, J.P.: Appl. Phys. Lett. **100**, 242407 (2012)
83. Wood, M.E., Potter, W.H.: Cryogenics **25**, 667 (1985)
84. Gschneidner Jr., K.A., Pecharsky, V.K.: Annu. Rev. Mater. Sci. **30**, 387 (2000)
85. Lyubina, J., Gutfleisch, O., Kuz'min, M., Richter, M.: J. Magn. Magn. Mater. **321**, 3571 (2009)
86. Kuz'min, M.D., Richter, M.: Phys. Rev. B **76**, 092401 (2007)
87. Pecharsky, V.K., Gschneidner Jr., K.A.: Int. J. Refrigeration **29**, 1239 (2006)
88. Gorsse, S., Chevalier, B., Orveillon, G.: Appl. Phys. Lett. **92**, 122501 (2008)
89. Engelbrecht, K., Bahl, C.R.H.: J. Appl. Phys. **108**, 123918 (2010)
90. Barcza, A., Katter, M.: Private communication (2015) In: Proceedings of the magnetic materials for the 21st century seminar, Hanau, 22–23 Oct 2012
91. Yamada, H.: Phys. Rev. B **47**, 11211 (1993)

92. Lyubina, J., Nenkov, K., Schultz, L., Gutfleisch, O.: Phys. Rev. Lett. **101**, 177203 (2008)
93. Bean, C.P., Rodbell, D.S.: Phys. Rev. **126**, 104 (1962)
94. Fujita, A., Fujieda, S., Fukamichi, K., Mitamura, H., Goto, T.: Phys. Rev. B **65**, 014410 (2001)
95. Yabuta, H., Umeo, K., Takabatake, T., Koyama, K., Watanabe, K.: J. Phys. Soc. Jpn. **75**, 113707 (2006)
96. Morellon, L., Algarabel, P.A., Ibarra, M.R., Blasco, J., Garcia-Landa, B., Arnold, Z., Albertini, F.: Phys. Rev. B **58**, R14721 (1998)
97. Kuz'min, M.D.: Appl. Phys. Lett. **90**, 251916 (2007)
98. Kuepferling, M., Sasso, C.P., Basso, V.: EPJ Web. Conf. **40**, 06010 (2013)
99. Moore, J.D., Morrison, K., Sandeman, K.G., Katter, M., Cohen, L.F.: Appl. Phys. Lett. **95**, 252504 (2009)
100. Yako, H., Fujieda, S., Fujita, A., Fukamichi, K.: IEEE Trans. Magn. **47**, 2482 (2011)
101. Fujita, A., Kondo, T., Kano, M., Yako, H.: Appl. Phys. Lett. **102**, 041913 (2013)
102. Fujita, A., Yako, H.: J. Alloys Comp. **577**, S48 (2013)
103. Tura, A., Nielsen, K.K., Rowe, A.: Int. J. Refrigeration **35**, 1518 (2012)
104. Tusek, J., Kitanovski, A., Poredos, A.: Int. J. Refrigeration **36**, 1456 (2013)
105. Nielsen, K.K., Engelbrecht, K., Christensen, D.V., Jensen, J.B., Smith, A., Bahl, C.R.H.: Appl. Therm. Eng. **40**, 236 (2012)
106. Nielsen, K.K., Engelbrecht, K.: J. Phys. D Appl. Phys. **45**, 145001 (2012)
107. Hannemann, U., Lyubina, J., Gschneidner, K.A., Jr., Cohen, L.F.: Unpublished data. (2011).
108. Jacobsson, P., Sundqvist, B.: Phys. Rev. B **40**, 9541 (1989)
109. Fujieda, S., Hasegawa, Y., Fujita, A., Fukamichi, K.: J. Appl. Phys. **95**, 2429 (2004)
110. Skokov, K.P., Karpenkov, D.Y., Kuz'min, M.D., Radulov, I.A., Gottschall, T., Kaeswurm, B., Fries, M., Gutfleisch, O.: J. Appl. Phys. **115**, 17A941 (2014)
111. Turcaud, J.A., Morrison, K., Berenov, A., Alford, N.M.N., Sandeman, K.G., Cohen, L.F.: Scripta Mater. **68**, 510 (2013)
112. Abhyankar, A.C., Yu, Y.T., Kuo, Y.K., Huang, G.W., Lu, C.S.: Intermetallics **18**, 2090 (2010)
113. Oezcan, S., Burdett, P., Wilson, N.: U.S. Patent 2014/0216057A1 (2014)
114. Moore, J.D., Klemm, D., Lindackers, D., Grasemann, S., Träger, R., Eckert, J., Löber, L., Scudino, S., Katter, M., Barcza, A., Skokov, K.P., Gutfleisch, O.: J. Appl. Phys. **114**, 043907 (2013)
115. Guillou, F., Yibole, H., van Dijk, N.H., Zhang, L., Hardy, V., Brück, E.: J. Alloys Comp. **617**, 569 (2014)
116. Zhang, H., Sun, Y., Li, Y., Wu, Y., Long, Y., Shen, J., Hu, F., Sun, J., Shen, B.: J. Appl. Phys. **117**, 063902 (2015)
117. Katter, M., Zellmann, V., Reppel, G.W., Uestuener, K.: In: Proceedings of the 3rd International Conference on Magnetic Refrigeration at Room Temperature, Des Moines, IA, USA, May 11–15, 2009, p. 83
118. Barcza, A., Katter, M., Zellmann, V., Russek, S., Jacobs, S., Zimm, C.: IEEE Trans. Magn. **47**, 3391 (2011)
119. Lyubina, J., Hannemann, U., Ryan, M.P., Cohen, L.F.: Adv. Mater. **24**, 2042 (2012)
120. Seeler, F., Degen, G., Reesink, B., Kaczun, J.: U.S. Patent US20120033002A1 (2012)
121. Bahl, C.R.H., Velázquez, D., Nielsen, K.K., Engelbrecht, K., Andersen, K.B., Bulatova, R., Pryds, N.: Appl. Phys. Lett. **100**, 121905 (2012)
122. Engelbrecht, K., Bahl, C.R.H., Nielsen, K.K.: Int. J. Refrigeration **34**, 1132 (2011)
123. Fujieda, S., Fukamichi, K., Suzuki, S.: J. Alloys Comp. **600**, 67 (2014)
124. Tian, N., Zhang, N.N., You, C.Y., Gao, B., He, J.: J. Appl. Phys. **113**, 103909 (2013)
125. Chennabasappa, M., Chevalier, B., Lahaye, M., Labrugere, C., Toulemonde, O.: J. Alloys Comp. **584**, 34 (2014)
126. Franco, V., Blazquez, J.S., Ingale, B., Conde, A.: Annu. Rev. Mater. Res. **42**, 305 (2012)
127. Cable, J.W., Wollan, E.O.: Phys. Rev. **165**, 733 (1968)

128. Mathew, S.P., Kaul, S.N., Nigam, A.K., Probst, A.-C., Birringer, R.J.: Physics: Conf. Series **200**, 072047 (2010)
129. Taskaev, S.V., Buchelnikov, V.D., Pellenen, A.P., Kuz'min, M.D., Skokov, K.P., Karpenkov, D.Y., Bataev, D.S., Gutfleisch, O.: J. Appl. Phys. **113**, 17A933 (2013)
130. Paudyal, D., Pecharsky, V.K., Gschneidner Jr., K.A., Harmon, B.N.: Phys. Rev. B **73**, 144406 (2006)
131. de Oliveira, N.A., von Ranke, P.J.: Phys. Rep. **489**, 89 (2010)
132. Tseng, Y.C., Haskel, D., Lang, J.C., Pecharsky, V.K., Gschneidner Jr., K.A.: J. Appl. Phys. **103**, 07B301 (2008)
133. Provenzano, V., Shapiro, A.J., Shull, R.D.: Nature **429**, 853 (2004)
134. Shull, R.D., Provenzano, V., Shapiro, A.J., Fu, A., Lufaso, M.W., Karapetrova, J., Kletetschka, G., Mikula, V.: J. Appl. Phys. **99**, 08K908 (2006)
135. Li, J.Q., Sun, W.A., Jian, Y.X., Zhuang, Y.H., Huang, W.D., Liang, J.K.: J. Appl. Phys. **100**, 073904 (2006)
136. Prabahar, K., Raj Kumar, D.M., Manivel Raja, M., Palit, M., Chandrasekaran, V.: Mater. Sci. Eng. B **172**, 294 (2010)
137. Gschneidner Jr., K.A., Pecharsky, V.K.: J. Appl. Phys. **85**, 5365 (1999)
138. Fu, H., Chen, Y., Tu, M., Zhang, T.: Acta Mater. **53**, 2377 (2005)
139. Gschneidner Jr., K.A., Pecharsky, A.O., Pecharsky, V.K., Lograsso, T.A., Schlagel, D.L.: In: Bautista, R.G., Mishra, B. (eds.) Rare Earth and Actinides: Science, Technology and Applications IV, p. 63. TMS, Warrendale, PA (2000)
140. Helmholdt, B., Palstra, T.T.M., Nieuwenhuys, G.J., Mydosh, J.A., van der Kraan, A.M., Buschow, K.H.J.: Phys. Rev. B **34**, 169 (1986)
141. Palstra, T.T.M., Mydosh, J.A., Nieuwenhuys, G.J., van der Kraan, A.M., Buschow, K.H.J.: J. Magn. Magn. Mater. **36**, 290 (1983)
142. Hu, F.X., Shen, B.G., Sun, J.R., Cheng, Z.H.: Appl. Phys. Lett. **78**, 3675 (2001)
143. Fujita, A., Fujieda, S., Hasegawa, Y., Fukamichi, K.: Phys. Rev. B **67**, 104416 (2003)
144. Yamada, H., Terao, K.: Phase Trans. A Multinational J. **75**, 231 (2002)
145. Fujieda, S., Fujita, A., Fukamichi, K., Yamaguchi, Y., Ohoyama, K.: J. Phys. Soc. Jpn. **77**, 074722 (2008)
146. Hannemann, U., Lyubina, J., Ryan, M.P., Alford, N.M., Cohen, L.F.: Europhys. Lett. **100**, 57009 (2012)
147. Wang, C., Long, Y., Ma, T., Fu, B., Ye, R., Chang, Y., Hu, F., Shen, J.: J. Appl. Phys. **109**, 07A910 (2011)
148. Fujita, A., Fujieda, S., Fukamichi, K.: IEEE Trans. Magn. **45**, 2620 (2009)
149. Beckman, O., Lundgren, L., Nordblad, P., Svedlindh, P., Törne, A., Andersson, Y., Rundqvist, S.: Phys. Scr. **25**, 679 (1982)
150. Brück, E., Ilyn, M., Tishin, A.M., Tegus, O.: J. Magn. Magn. Mater. **290–291**, 8 (2005)
151. Tegus, O., Brück, E., Buschow, K.H.J., de Boer, F.R.: Nature **415**, 150 (2002)
152. Brück, E., Guillou, F., Caron, L., Yibole, H., Miao, X.F.: Presented at Moscow International Symposium on Magnetism MISM-2014, Moscow, Russia, 29 June–3 July, (2014)
153. Brück, E., Trung, N.T., Ou, Z.Q., Buschow, K.H.J.: Scripta Mater. **67**, 590 (2012)
154. Cam Thanh, D.T., Brück, E., Tegus, O., Klaasse, J.C.P., Buschow, K.H.J.: J. Magn. Magn. Mater. **310**, e1012 (2007)
155. Guillou, F., Porcari, G., Yibole, H., van Dijk, N., Brück, E.: Adv. Mater. **26**, 2671 (2014)
156. Zach, R., Guillot, M., Fruchart, R.: J. Magn. Magn. Mater. **89**, 221 (1990)
157. Guillou, F., Yibole, H., van Dijk, N.H., Brück, E.: J. Alloys Comp. **632**, 717 (2015)
158. Ullakko, K., Huang, J.K., Kantner, C., O'Handley, R.C.: Appl. Phys. Lett. **69**, 1966 (1996)
159. Acet, M., Manosa, L., Planes, A.: In: Buschow K.H.J. (ed.) Handbook of Magnetic Materials, vol. 19, p. 231. Elsevier B.V., (2011)
160. Manosa, L., Planes, A., Acet, M.: J. Mater. Chem. A **1**, 4925 (2013)
161. Hu, F.X., Shen, B.G., Sun, J.R.: Appl. Phys. Lett. **76**, 3460 (2000)

162. Marcos, J., Planes, A., Manosa, L., Casanova, F., Batlle, X., Labarta, A., Martinez, B.: Phys. Rev. B **66**, 224413 (2002)
163. Krenke, T., Duman, E., Acet, M., Wassermann, E.F., Moya, X., Manosa, L., Planes, A.: Nature Mater. **4**, 450 (2005)
164. Liu, J., Lyubina, J., Gutfleisch, O.: In: 3rd IIF-IIR International Conference on Magnetic Refrigeration at Room Temperature, Des Moines, IA, USA, May 11–15, 2009, p. 27
165. Manosa, L., Moya, X., Planes, A., Gutfleisch, O., Lyubina, J., Barrio, M., Tamarit, J.-L., Aksoy, S., Krenke, T., Acet, M.: Appl. Phys. Lett. **92**, 012515 (2008)
166. Zhou, X., Li, W., Kunkel, H.P., Williams, G., Zhang, S.: J. Appl. Phys. **97**, 10M515 (2005)
167. Khovaylo, V.V., Skokov, K.P., Gutfleisch, O., Miki, H., Takagi, T., Kanomata, T., Koledov, V.V., Shavrov, V.G., Wang, G., Palacios, E., Bartolomé, J., Burriel, R.: Phys. Rev. B **81**, 214406 (2010)
168. Phan, M.-H., Yu, S.-C.: J. Magn. Magn. Mater. **308**, 325 (2007)
169. Coey, J.M.D., Viret, M., von Molnar, S.: Adv. Phys. **58**, 571 (2009)
170. Zhang, Y., Lampen, P.J., Phan, T.-L., Yu, S.-C., Srikanth, H., Phan, M.-H.: J. Appl. Phys. **111**, 063918 (2012)
171. Rostamnejadi, A., Venkatesan, M., Kameli, P., Salamati, H., Coey, J.M.D.: J. Magn. Magn. Mater. **323**, 2214 (2011)
172. Mira, J., Rivas, J., Hueso, L.E., Rivadulla, F., Lopez Quintela, M.A.: J. Appl. Phys. **91**, 8903 (2002)
173. Phan, M.H., Yu, S.C., Hur, N.H.: Appl. Phys. Lett. **86**, 072504 (2005)
174. Lin, G.C., Wei, Q., Zhang, J.X.: J. Magn. Magn. Mater. **300**, 392 (2006)
175. Radaelli, P.G., Cox, D.E., Marezio, M., Cheong, S.-W., Schiffer, P.E., Ramirez, A.P.: Phys. Rev. Lett. **75**, 4488 (1995)
176. Maeda, H., Sato, M., Uehara, M.: J. Jpn. Inst. Met. **47**, 688 (1983)
177. Franco, V., Blazquez, J.S., Conde, C.F., Conde, A.: Appl. Phys. Lett. **88**, 042505 (2006)
178. Ipus, J.J., Blázquez, J.S., Franco, V., Conde, A., Kiss, L.F.: J. Appl. Phys. **105**, 123922 (2009)
179. Gorsse, S., Mayer, C., Chevalier, B.: J. Appl. Phys. **109**, 033914 (2011)
180. Min, S.G., Kim, K.S., Yu, S.C., Lee, K.W.: Mater. Sci. Eng. A **449–451**, 423 (2007)
181. Caballero-Flores, R., Franco, V., Conde, A., Knipling, K.E., Willard, M.A.: Appl. Phys. Lett. **96**, 182506 (2010)
182. Law, J.Y., Franco, V., Ramanujan, R.V.: J. Appl. Phys. **110**, 023907 (2011)
183. Fallot, M.: Ann. Phys. **10**, 291 (1938)
184. Shirane, G., Chen, C.W., Flinn, P.A., Nathans, R.: J. Appl. Phys. **34**, 1044 (1963)
185. Kouvel, J.S., Hartelius, C.C.: J. Appl. Phys. Suppl. **33**, 1343 (1962)
186. Stern-Taulats, E., Planes, A., Lloveras, P., Barrio, M., Tamarit, J.L., Pramanick, S., Majumdar, S., Frontera, C., Manosa, L.: Phys. Rev. B **89**, 214105 (2014)
187. Annaorazov, M.P., Nikitin, S.A., Tyurin, A.L., Asatryan, K.A., Dovletov, A.K.: J. Appl. Phys. **79**, 1689 (1996)
188. Basso, V., Sasso, C.P., Küpferling, M., Skokov, K.P., Gutfleisch, O.: J. Appl. Phys. **109**, 083910 (2011)
189. Nikitin, S.A., Skokov, K.P., Koshkid'ko, Y.S., Pastushenkov, Y.G., Ivanova, T.I.: Phys. Rev. Lett. **105**, 137205 (2010)
190. Manekar, M., Roy, S.B.: J. Phys. D Appl. Phys. **41**, 192004 (2008)
191. Akulov, N.S., Kirensky, L.V.: J. Phys. USSR **9**, 31 (1940)
192. Vonsovsky, S.V.: Magnetism. Nauka, Moscow (1971)

Chapter 5
Above Room Temperature Ferromagnetism in Dilute Magnetic Oxide Semiconductors

A.S. Semisalova, A. Orlov, A. Smekhova, E. Gan'shina, N. Perov, W. Anwand, K. Potzger, E. Lähderanta, and A. Granovsky

"Do not give up hope, maestro!"

Bulat Okudjava

Abbreviations

AHE	Anomalous Hall effect
DMO	Dilute magnetic oxides
DMS	Dilute magnetic semiconductors
EDX	Energy dispersive X-ray analysis
MO	Magneto-optical
PAS	Positron annihilation spectroscopy
RF	Radio frequency

A.S. Semisalova
Faculty of Physics, Lomonosov Moscow State University, Moscow 119991, Russia

Lappeenranta University of Technology, Lappeenranta 53851, Finland

Helmholtz-Zentrum Dresden - Rossendorf, Dresden 01328, Germany

A. Orlov
Federal State Research and Design Institute of Rare Metal Industry, Moscow 119017, Russia

A. Smekhova
Faculty of Physics, Lomonosov Moscow State University, Moscow 119991, Russia

Fakultät für Physik, Experimentalphysik, Universität Duisburg-Essen, Duisburg 47048, Germany

E. Gan'shina • N. Perov • A. Granovsky (✉)
Faculty of Physics, Lomonosov Moscow State University, Moscow 119991, Russia
e-mail: granov@magn.ru

W. Anwand • K. Potzger
Helmholtz-Zentrum Dresden - Rossendorf, Dresden 01328, Germany

E. Lähderanta
Lappeenranta University of Technology, Lappeenranta 53851, Finland

A. Zhukov (ed.), *Novel Functional Magnetic Materials*, Springer Series in Materials Science 231, DOI 10.1007/978-3-319-26106-5_5

SQUID	Superconducting quantum interference device
TKE	Transversal Kerr effect
TM	Transition metal
XANES	X-ray absorption near-edge structure
XMCD	X-ray magnetic circular dichroism
XRD	X-ray diffraction
ZFC/FC	Zero field cooled/field cooled

5.1 Introduction

Ferromagnetic semiconductors with application potential in modern computing should keep both ferromagnetic and semiconducting properties at room temperature. To develop such multifunctional materials is one of the most important issues of modern material science. The utilization of spin functionality hand in hand with electrical charge-based electronics opens the wide field of phenomena combining brand new physics and extensive potential for applications in the next-generation logic devices and storage, spintronics, magnetophotonics, etc. Such materials can make a revolutionary step in technology. In spite of enormous efforts over the last decades to the creation, understanding and manipulation of ferromagnetism in doped semiconducting materials, the problem is far from being solved. Moreover, there are serious doubts that it is possible to create robust and homogeneous above room temperature ferromagnetic semiconductors with simultaneously high spin polarization, mobility of current carriers, and spontaneous magnetization.

The initial approach for creating semiconductors which are ferromagnetic above room temperature was straightforward: Some amount of 3*d* or 4*f* magnetic ions have been incorporated into a para- or diamagnetic semiconductor or oxide with the hope that exchange interaction between them is strong enough to provide a long-range ferromagnetic order. Such materials were called dilute magnetic semiconductors (DMS), in the case of the host material being an oxide, also dilute magnetic oxides (DMO). In the case of DMS, it is very difficult, if possible at all, to reach a high Curie temperature [1–3]; at least up to now, the record value for the thoroughly studied GaAs:Mn system does not exceed 192 K [4, 5]. At a small volume fraction of transition metal (TM) impurities, they are far from each other to provide high Curie temperature, but if their concentration is large, there is no way to avoid their aggregation or non-substitutional positions in lattice. Moreover, the semiconducting properties might be lost. In the case of DMO, the Curie temperature can be very high. For example, in $TiO_{2-\delta}$:Co, it can exceed 800 K [6]. The persistence of spontaneous magnetization above room temperature has been found also in materials which do not contain TM ions [7]. This phenomenon was called as "d^0-ferromagnetism", and its discovery triggered a great interest to find the actual origin of magnetic moments and ferromagnetic exchange in DMS and DMO materials. There are also two other routes to develop above room temperature ferromagnetic semiconductors, namely, to turn ferromagnetic or ferrimagnetic oxides, showing spontaneous magnetization up to 800 K, into good semiconductors by carrier doping [1] or by tuning the composition or microstructure of concentrated paramagnetic "prophase" alloys, as it has been shown recently for $Si_{50-x}Mn_{50+x}$ alloys [8].

Magnetic moments in DMS and DMO are usually associated with TM impurities and related to hybridization with their non-TM neighbors. However, long-range ferromagnetic order in DMS and DMO can be of defect-induced origin associated with vacancies, F-centers, dangling bonds, and any unpaired electrons, located inside crystallites, intercrystalline grains, or at the surface and interfaces [9–11]. Numerous models of exchange interaction in DMS and DMO were suggested in literature starting with pioneering work of T. Dietl [12]: carrier-mediated ferromagnetism with Zener or RKKY-type exchange [1, 12], ferromagnetic superexchange [1, 13], double exchange [1], kinematic exchange [14], percolating magnetic polarons [15], charge-transfer magnetism, and Stoner-type magnetism of impurity bands [9]. We refer the readers to the original papers and a recent review [1].

The aggregation of TM impurities in DMS or DMO, as well as a possible presence of contamination or a very small amount of uncontrolled parasitic ferromagnetic phases (handling effect [16]), is a very serious obstacle to detect an intrinsic ferromagnetism. In the presence of a small amount of clusters, a hysteresis loop cannot be considered as an evidence of intrinsic ferromagnetism. Therefore, in the case of DMS, DMO, and defect-induced magnetism, a comprehensive study of samples using different modern methods of investigation is of great importance.

Another problem is a very high sensitivity of the magnetic properties of the samples to tiny details of technology and thermal treatment. The magnetic properties of samples of nominally the same composition obtained by different groups even by the same method may drastically differ. Moreover, sometimes the samples are inhomogeneous due to distribution of defects and embedded TM ions. Therefore, a comprehensive study using different methods should be done for the same sample, if possible.

The GaAs:Mn compound is the most extensively and successfully investigated DMS among Mn-doped III–V, II–VI, IV–VI, V_2–VI_3, I–II–V, and elemental group IV semiconductors [1]. The comprehensive research on this system has demonstrated outstanding low-temperature functionalities [1]. The early investigations of DMO were summarized in the brief reviews [17, 18] and collective monograph [19], but the obtained results are controversial and the theoretical models meet serious difficulties while explaining available experimental data. In this chapter we review recent results from experiments on the magnetic properties of several DMO systems focusing on TM-doped TiO_2.

The wide band gap semiconductor oxide TiO_2 is especially interesting as a host for DMO due to well-known special properties of this material [17–19]. TiO_2:Co is undoubtedly the most explored DMO which is, probably, because Co-doped titania was the first oxide where above room temperature ferromagnetism was reported [6].

TiO_2 has three basic polymorphs, namely, rutile, anatase, and brookite. The rutile and anatase phases, which have been explored in the context of DMO studies, have indirect band gaps of 3.0 and 3.2 eV, respectively. They are n-type conductors with shallow donor levels [18]. Doped, for example, with Co, V, or Fe, TiO_2

exhibits a ferromagnetic behavior up to 800 K and higher temperatures. The carrier-mediated ferromagnetism in TiO_2:Co is the most popular and believable point of view, but it fails in explaining ferromagnetism in dielectric $TiO_{2-\delta}$:Co [20] and extra large magnetic moment per Co impurity at low doping [21].

The case of doping with V was studied more rarely in comparison with Co doping because it was widely supposed that the mechanism of long-range ferromagnetic order in $TiO_{2-\delta}$:V is the same as in $TiO_{2-\delta}$:Co [19]. However, magnetic properties of these two systems are quite different, as it is shown below (Sect. 5.3 and 5.5). In the case of $TiO_{2-\delta}$:V besides magnetic, structural, and magneto-optical (MO) measurements, we used positron annihilation spectroscopy (PAS). PAS is one of the effective methods for nondestructive studies of defects inside thin films and could be used for either conductive or dielectric samples. Since this method is relatively new and was not widely used for DMO, we describe the physics of the PAS experiment and recent results in Sect. 5.4.

Since the possibility of extrinsic ferromagnetism (clusters formation) cannot be excluded at high TM concentration in $TiO_{2-\delta}$, the comparison of magnetic properties of Co- and V-doped $TiO_{2-\delta}$ films at extra low level of doping and produced by the same method under identical conditions is of special interest (Sect. 5.5).

The recent experimental data on another DMO, TM-doped ZnO, is presented in Sect. 5.6.

5.2 Above Room Temperature Ferromagnetism in $TiO_{2-\delta}$:Co

The above room temperature ferromagnetism in DMO for the first time was experimentally discovered in the anatase form of doped $Ti_{1-x}Co_xO_{2-\delta}$ [6] and soon in other doped oxide semiconductor matrixes, i.e., ZnO [22] and SnO_2 [23]. The state of investigations in this area up to 2006 year was discussed in the collective monograph [19]. To that time the ferromagnetic phase arising with the Curie temperature $T_C > 400$ K was observed in the films of titanium oxide doped with TM impurities V, Cr, Mn, Fe, Co, and Ni [24]. In the case of substrate temperature of 650 °C, authors [24] obtained high values of saturation magnetization at 300 K for all the films of $TiO_{2-\delta}$:TM with the highest magnetization—up to 4.2 μ_B per an impurity atom—for $TiO_{2-\delta}$:V (Fig. 5.1).

The ferromagnetic phase with lower magnetization was also observed in the oxides SnO_2 [23, 25] and In_2O_3 [24, 26] at doping with TM impurities. Room temperature ferromagnetism was reported for the undoped oxide films TiO_2, In_2O_3, and HfO_2 as well; it was ascertained that ferromagnetic order originates from crystal structure defects and oxygen vacancies in the films [7, 27, 28].

Common techniques for DMO thin film growth are magnetron sputtering and pulsed laser deposition (PLD). For both methods, either metallic or ceramic targets can be used where the desired stoichiometry in the film is adjusted by the partial pressure during deposition. Also molecular beam epitaxy or laser molecular beam

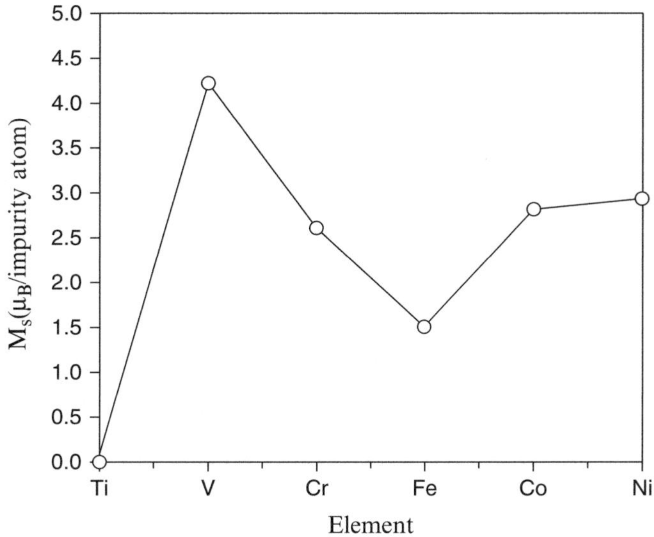

Fig. 5.1 Saturation magnetization vs. elements for TiO_2:TM films grown on $LaAlO_3$ substrates [24]

epitaxy is used for the deposition of the magnetic oxides. For creation of a material with high-quality crystal structure, the film deposition is carried out onto substrates heated up to high temperatures, as a rule in the range of 500–750 °C. Thin polished dielectric $SrTiO_3$ or $LaAlO_3$ plates with (001) orientation are used as substrates for deposition. These materials have cubic or three-wedged crystal structure, correspondingly, with the lattice parameters of 0.390 and 0.379 nm, which match with TiO_2 crystal lattice parameters.

The structural investigation of the $Ti_{1-x}Co_xO_{2-\delta}$ films with x from 3 up to 8 at.% [29, 30] showed that the films fabricated at a low partial oxygen pressure consist mostly of amorphous metallic phase. With increasing oxygen content in the gas mixture, the polycrystalline films grow firstly with a structure of cubic oxide TiO and secondly with tetragonal dioxide TiO_2—anatase and rutile (or mixed phases with a different oxygen deficiency). Single-phase rutile films can be created only at high oxygen content and a substrate temperature above 700 °C. The phases of anatase and rutile are indirect band gap semiconductors, while the monoxide TiO is ionic crystal with metallic conductivity [31].

The inevitability of magnetic clusters formation during the deposition of films based on Ti with 3*d* TM and O_2 is dictated by the thermodynamic phase diagrams. Solubility of Fe or Co in a Ti matrix is not higher than 0.2 and 0.3 at.%, correspondingly. The details of the corresponding ternary phase diagrams including oxygen are unknown, but certain data reveal that the character of metal-metal interaction changes insignificantly in the presence of oxygen. Thus, when deposition of titanium dioxide with the high TM content is used, the created films are always metastable and can contain TM clusters with TM ions in metallic state. A large number of investigations on TiO_2:TM films have revealed the presence of magnetic clusters, namely, extrinsic ferromagnetism.

For example [29, 30, 32, 33], it was shown that ferromagnetism in the Co-doped anatase and rutile phases as well as in the cubic TiO_2 phase in a charge carrier range from 1.10^{18} up to 3.10^{22} cm^{-3} originates from inclusions of metallic Co or Co-enriched phase. Later, the charge carrier concentration limitations for ferromagnetic phase forming in the films of $TiO_{2-\delta}$:Co (3–10 %)—from 7.10^{18} up to 4.10^{22} cm^{-3}—was established in [34]. The investigations were carried out with atomic force and magnetic force microscopy and revealed well-defined clusters with sizes ranging from 20 to 60 nm in $TiO_{2-\delta}$:3%Co. The authors [34] also observed soft X-ray magnetic circular dichroism (XMCD) spectral line shape nearly identical to that of metallic Co. In addition, a low magnetic moment was reported at a Co concentration less than 5 %, where no extrinsic ferromagnetic clusters have been observed, while a high magnetic moment was reported for a Co content higher than 5 % in the presence of Co clusters [33].

In order to obtain an intrinsic ferromagnetic dielectric TiO_2 without any magnetic clusters, Griffin et al. [20] have grown epitaxial dielectric paramagnetic films of anatase TiO_2:Co by magnetron sputtering from a ceramic oxide $Ti_{1-x}Co_xO_2$ target. After the low-temperature annealing in vacuum, the films kept the highly insulating state and simultaneously became intrinsically ferromagnetic at room temperature without magnetic clusters in a matrix. These films kept the dielectric state because localized oxygen vacancies were bound to Co^{2+} dopant ions, and the ferromagnetic ordering was provided with bound magnetic polarons. This result clearly shows that the charge carrier concentration is not the only parameter determining the room temperature ferromagnetism and rules out the carrier-mediated ferromagnetism of RKKY-type as the only one mechanism for long-range ferromagnetic order in $TiO_{2-\delta}$:Co.

To avoid clustering it is necessary to study samples containing a small volume fraction of TM ions. At a small ferromagnetic impurity concentration in DMO, "giant magnetic moments" per impurity atom were observed. Such high magnetization was reported, for example, for Co-doped of SnO_2 [23] and CeO_2 [35], as well as in a high dilute ferromagnetic semiconductor $TiO_{2-\delta}$:Co [30]. This behavior might be caused by unquenched orbital magnetic moments, by the crystal lattice polarization, and by defect-induced magnetism. However, we cannot completely exclude the influence of parasitic phases.

Several attempts to determine the magnetic moment at Co sites in $TiO_{2-\delta}$:Co by element-selective spectroscopic techniques such as X-ray absorption near-edge structure (XANES) and X-ray magnetic circular dichroism (XMCD) gave controversial results (see [36, 37] and references therein). As a rule, the data obtained by XMCD and magnetic measurements do not correspond to each other. It also supports the contribution of defect-induced magnetism to total magnetization. For example, we performed XANES and XMCD measurements at the Co $L_{2,3}$ absorption edges (soft X-ray region) for the $Ti_{0.92}Co_{0.08}O_{2-\delta}$ thin film [36]; the results of macroscopic and local-probe measurements are shown in Figs. 5.2 and 5.3. The element-selective magnetization curve recorded by XMCD at the Co L_3 absorption

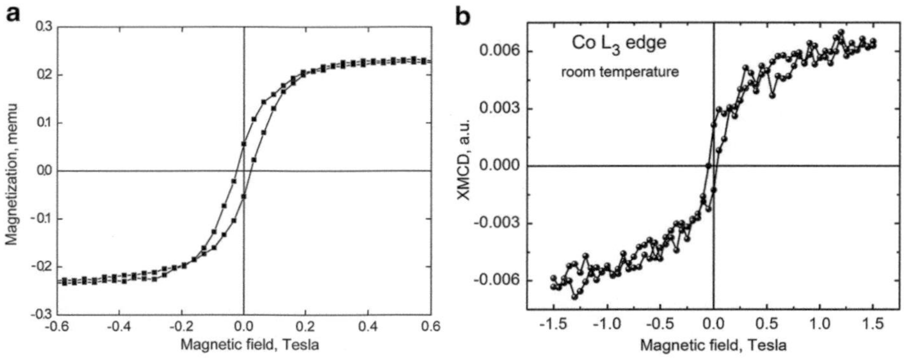

Fig. 5.2 (**a**) Magnetic hysteresis loop of the $Ti_{0.92}Co_{0.08}O_{2-\delta}$ film ($\rho = 3\ \Omega$ cm) annealed in vacuum for 2 h and (**b**) the element-selective magnetization loop recorded at the Co L_3 absorption edge at room temperature for the same film ($\rho = 10\ \Omega$ cm) after annealing for 1 h [36]

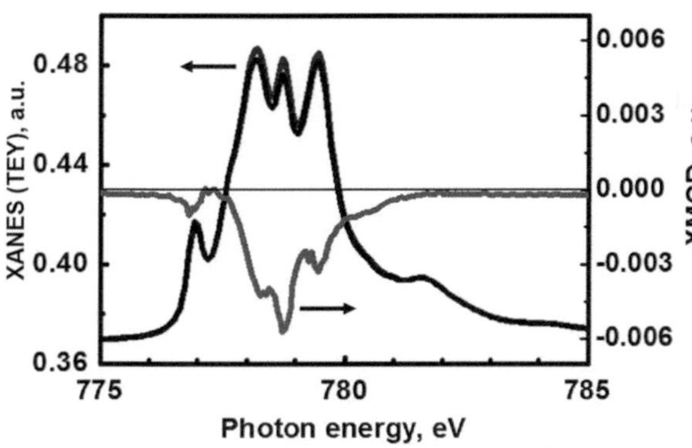

Fig. 5.3 X-ray absorption near-edge structure and X-ray magnetic circular dichroism spectra of the $Ti_{0.92}Co_{0.08}O_{2-\delta}$ film annealed for 1 h [36]

edge clearly exhibited ferromagnetic behavior at room temperature. It directly indicates on ferromagnetic exchange between the Co ions. But nevertheless it is difficult to claim about intrinsic ferromagnetism in this case because of too large volume fraction of Co in the sample (8 %).

Magneto-optical (MO) spectroscopy in the range of visible light is an effective tool to study the electronic and magnetic structure of homogeneous and inhomogeneous ferromagnetic materials. When the magnetization is weak, it can be difficult to rule out the existence of parasitic phases or ferromagnetic clusters from magnetic measurements and from X-ray diffraction data. Since all MO effects are due to the influence of spin–orbit and exchange interaction on interband and intraband optical transitions, MO spectra are extremely sensitive to the type and concentration of magnetic and nonmagnetic impurities, additional phases, magnetic and crystalline microstructure, band structure, concentration, and polarization of charge carriers. Besides, various MO effects measured in different spectral ranges complement each other nicely. For instance, if magnetic circular dichroism is proportional to the energy derivative of the absorption coefficient spectrum and therefore is large close to the absorption threshold, transversal Kerr effect (TKE) is not small in a wider spectral range. Magnetic circular dichroism spectra at visible band were systematically examined in rutile and anatase $Ti_{1-x}Co_xO_{2-\delta}$ epitaxial thin films [38–40] with x > 1 at.%. The obtained data indicates intrinsic ferromagnetism in the studied samples in the limited range of x. Using magnetron sputtering, we succeeded to fabricate films with x < 1 at.% which showed above room temperature ferromagnetism; the obtained TKE spectra clearly show the intrinsic ferromagnetic response (see details in [41]). A highly sensitive MO setup has allowed us to find out a signal and to resolve spectral lines. It should be emphasized that at so low doping level, obviously smaller than a threshold of solubility Co in $TiO_{2-\delta}$, the Co clusters formation is rather improbable. Moreover, even if they present, the size of Co clusters should be much smaller than 10 nm, and they locate rather far from each other. Thus, Co clusters can provide only superparamagnetic behavior but not long-range ferromagnetic order. Fig. 5.4 shows examples of the TKE spectra. It should be emphasized that the overall shape of the TKE spectra for examined samples of oxides with x > 1 % and x < 1 % has nothing in common with the TKE spectra for bulk Co or Co nanoclusters in nonmagnetic matrix [41]. With the reduction of the concentration of Co, the TKE amplitude decreases approximately in the same manner as magnetization and the TKE spectrum becomes more structured. Among all presented curves, the spectrum of anatase doped with 0.4 % Co has the most fine structure (Fig. 5.4). What is the origin of this fine structure at x = 0.4 %? As the measured MO spectrum entirely is located in the range of a relative transparency (the width of a band gap for anatase and rutile TiO_2 is 3.2 and 3.03 eV, correspondingly), it seems reasonable to relate the features found with a presence of impurities or defects. Spectral positions of intraionic optical transitions in the most stable impurity Co^{2+} in the octahedral complex of oxygen are 2.5, 1.8, and 0.9 eV and in the tetrahedral complex of oxygen are 2.0 – 1.9, 0.9 – 0.8, and 0.5 eV [42]. These six values are very close to the experimental observation of the peak structure for x = 0.4 %. The exact positions of these lines depend on the value of crystalline fields. So the fine structure of the TKE spectra can be associated with intraionic transitions in Co^{2+} ions, located in various coordinate surrounding.

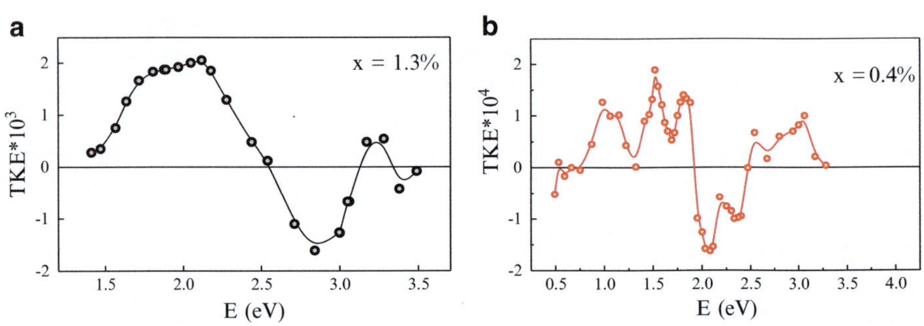

Fig. 5.4 The TKE spectra of $Ti_{1-x}Co_xO_{2-\delta}$ anatase films with different x: (**a**) x = 1.3 %; (**b**) x = 0.4 %. From [41]

It proves the intrinsic ferromagnetism in $Ti_{1-x}Co_xO_{2-\delta}$ at low-level doping, and it looks like Co^{2+} ions built-in oxygen octahedrons at least partly are responsible for ferromagnetic order.

To summarize this paragraph, we conclude that above room temperature ferromagnetism in $TiO_{2-\delta}$:Co is mostly connected with long-range magnetic order of magnetic moments of embedded Co ions but also partly with magnetic polarization of surrounding Co ions and defects. There is no universal mechanism of ferromagnetic exchange in this DMO system. It strongly depends on Co volume fraction, charge carrier concentration, and technological conditions of film synthesis. At very low doping level (<1 %), ferromagnetism in $TiO_{2-\delta}$:Co is intrinsic, but as a rule Co clusters are also responsible for magnetic properties at $x > 1$ % Co.

5.3 Above Room Temperature Ferromagnetism in $TiO_{2-\delta}$:V

Titanium oxides doped with vanadium are promising candidates for intrinsic ferromagnetic semiconductors with a high magnetization because of the high solubility of vanadium. The complete ternary phase diagram of this system is unknown, but the analysis of binary Ti-V diagram and available data on the solubility of vanadium in titanium in the presence of oxide phase [43] show that it is possible to fabricate single-phase thin films $TiO_{2-\delta}$:V with V content up to 3 at.% by deposition (solubility limit at room temperature) and to obtain the single-phase samples with V concentration at about 18 % by ultrafast quenching (eutectic point at 675 °C). Besides, metallic vanadium clusters are paramagnetic and therefore, even if they appear in the matrix, they do not play role in ferromagnetic response.

Pioneer works on synthesis of $TiO_{2-\delta}$:V ferromagnetic films were accomplished by N.H. Hong et al. [24]. The films of ferromagnetic dielectric TiO_2:5%V were obtained by the laser ablation technique on $LaAlO_3$ substrates. The abnormal high magnetic moment of 4.23 μ_B per V atom was explained by the contribution of unquenched orbital moments at low magnetic impurity content. It cannot be excluded that this high value was not connected with vanadium at all because it is difficult to avoid noncontrolled impurities in the laser ablation technique. For example, only slight room temperature magnetization has been found in the TiO_2:V powder samples [44] and nanocrystals [45].

In [46] the structure, magnetic and MO properties of $TiO_{2-\delta}$:V system were examined at different V content and in a wide resistivity range spreading from a deep dielectric down to a degenerate semiconductor. The thin films $Ti_{1-x}V_xO_{2-\delta}$ on substrates of either $LaAlO_3$ or rutile TiO_2 having (001) orientation were grown by radio frequency (RF) magnetron sputtering of metallic alloy targets in the argon–oxygen atmosphere. The concentration of V impurities (x) in the films of 3, 10, and 18 at.% has been defined by the composition of targets and checked by local energy dispersive X-ray spectroscopy (EDX). The oxygen partial pressure in plasma was in the range of 2×10^{-6}–2×10^{-4} Torr. The substrate temperatures were 300, 500, and 650 °C; the growth rate 0.05–0.09 nm/s; and the films thickness 0.1–0.2 μm. The surface topography of the films was recorded with atomic force microscope AFM 2000. XRD analysis was carried out at the D8 Discover diffractometer (Bruker-AXS), and X-ray photoelectron spectroscopy study was performed using the Axis Ultra (Kratos) spectrometer. XANES investigations were carried out at the source ID12 of the European Synchrotron Radiation Facility. Magnetic measurements were performed using the vibrating sample magnetometer (VSM, LakeShore 7400 System) having a sensitivity of 1×10^{-6} emu. Magneto-optical spectra were measured in the transversal Kerr effect geometry in the 1.4–3.2 eV energy range with the applied magnetic field of 3 kOe.

The study of the chemical state of the V impurities by X-ray photoelectron spectroscopy (XPS) has shown that in the films with 3 at.% V all impurities are in the oxidized state (Fig. 5.5). The binding energy of the vanadium peak $2p^{3/2}$ was found to be shifted comparatively to the value of 513.1 eV which related to the metal vanadium and slipped from 515.0 eV up to 516.3 eV when the resistivity changed from 10^{-2} up to 10^{4} Ω cm. It means the oxidation degree also increases correspondingly to the valence between +3 and +4. The character of bulk sensitive XANES spectrum at the vanadium K-edge in the film with 3 % V (Fig. 5.6) also confirms the ionic (nonmetallic) state of V atoms [46].

Depending on the substrate types, the deposition temperature, and the oxygen content in the argon–oxygen atmosphere, the grown films revealed the structure either anatase or rutile with the (001) crystallography orientation. The films' resistivity changed in the range of 10^{-3}–10^{6} Ω cm depending on the oxygen content and the deposition rate. All grown films were found to have an electron type of conductivity. TiO_2:V films revealed a room temperature magnetization in the whole range of a studied resistivity (from a deep dielectric down to a degenerate

Fig. 5.5 X-ray photoelectron V 2p spectra of the TiO_2:3 % V films with the resistivity of (*1*) 10^{-2} and (*2*) 10^{4} Ω cm (from [46])

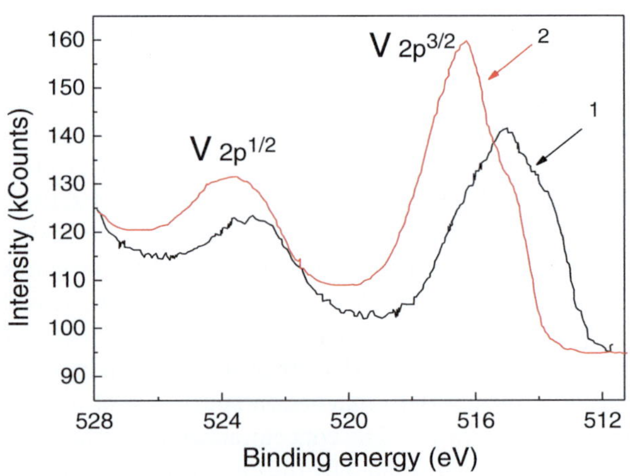

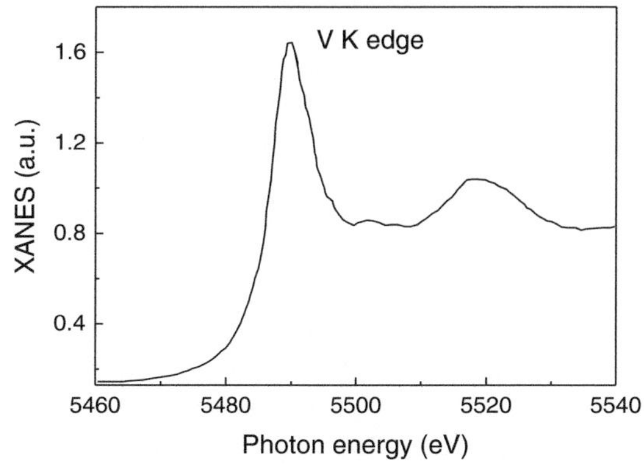

Fig. 5.6 X-ray absorption near-edge structure (XANES) of the TiO_2:3 % V film (from [46])

semiconductor). It is contrary to the TiO_2:Co system where the as-grown dielectric films did not reveal any ferromagnetic ordering [20, 30]. For degenerate semiconductor TiO_2:V and for deep insulator, the magnetization was observed to be reduced. The deposition of films at 650 °C ensures the higher magnetization comparatively to the films grown at lower temperatures that agrees with the observation in [6]. The magnetization of the films grown at rutile substrates was always slightly smaller.

The maximum magnetization was obtained for the films with 3 at.% V, and it was equal to 42 emu cm^{-3} which corresponds to 4.8 μ_B per V atom. This value is slightly higher than observed one in [24] (Fig.5.1) for the films with 5 % V (4.23 μ_B/at.); it considerably exceeds the value of the magnetic moment 3 μ_B for a single V atom. It is interesting to note that for high-quality anatase films fabricated by pulsed laser deposition with 6 % V, the estimated value for magnetic moment per V atom was only 1.0 μ_B [47]. It should be also underlined that in the case of TiO_2:V, the magnetization is very sensitive to the fabrication conditions, and so large value as 4.8 μ_B per V atom is the maximum that has been achieved. More often magnetic moment per V atom for 1 – 6 at.% V doping was about 1.0 μ_B (see below and [47]) or sometimes even smaller.

According to the calculations in [48], V^{3+} ions located on substitutional lattice sites lead to a ferromagnetic state of n-type titanium oxide. The authors of [49] have shown that doping of TiO_2 with V leads to the formation of deep levels in the semiconductor band gap and these impurity levels are responsible for the appearance of a magnetic moment in the material. It also has been calculated [50] that the total magnetic moment in TiO_2:V with oxygen vacancies can reach up to 2.47 μ_B/at. and 3.0 μ_B/at. in the structures of anatase or rutile, respectively.

In a simplified picture, the origin of ferromagnetism in TiO_2:V is the following: at low V content, the magnetic moments located at V sites are enhanced due to unquenched orbital moments. The vanadium magnetic moments percolate into a ferromagnetic network due to exchange interaction through magnetic polarons or/and conduction electron (Zener-type exchange or carrier-mediated ferromagnetism). The concentration of current carriers depends on the number of oxygen vacancies. The magnetization decreases at large V content (in the films with 18 % V as well as with 10 % V) because most of the V ions are close to each other and therefore aggregate into paramagnetic clusters. This scenario is very similar to that for TiO_2: Co but there is serious doubt that the origin of ferromagnetic ordering is the same. For example, in contrast to TiO_2:Co, we did not succeed to obtain a ferromagnetic XMCD signal from V. We could not detect an anomalous Hall effect (AHE) and, as a rule, the MO signal in TiO_2:V is much smaller than in TiO_2: Co (see below in 5.5 and [46, 51, 52]). Therefore, a scenario based on defect-induced magnetism is also possible. It is worth to notice that for the thin films fabricated by the same method without V, the magnetization is insignificant. It means that if the V atoms do not bear magnetic moment themselves, they induce defects with magnetic moments or Stoner-type ferromagnetism in the impurity band [53]. Therefore to study defects in TiO_2:V is of a primary importance.

5.4 Positron Annihilation Spectroscopy of Defects in $TiO_{2-\delta}$:V(1 ÷ 3 %)

Positron annihilation spectroscopy (PAS) is one of the effective methods for nondestructive studies of open-volume defects in solids and liquids and could also be applied to conductive and dielectric thin films. Mostly it is exploited for studies of vacancies and porous structures with porous sizes around 0.3÷30 nm in thin layers with the help of low-energy positron implantation from a table-top beam in laboratory conditions. Depth profiling with PAS is an ideal method for estimations of inhomogeneities and damages buried under the surface or under the diffusion layer [54, 55].

The main idea of different PAS methods could be formulated as follows: when a positron (e^+) is in contact with a dense environment, it significantly slows down and annihilates with an electron; as a result an emission of two photons occurs whose energy, momentum, and a time of emission can be accurately measured. Since the positron annihilation rate depends on the electron wave function density in the matter, the annihilation process is sensitive to the electronic structure of the sample studied. Thus, the precision measurements of annihilation gamma rays parameters reflect characteristic parameters of the electronic structure of the test substance. The technology based on these measurements has become a unique tool for nanomaterials studies, rapidly progressing in recent years. This technique significantly complements such well-known and widely used methods as X-ray analysis based on synchrotron radiation, neutron scattering, etc. The PAS occupies a unique niche among the methods used to determine the type and size of defects and clearly dominates the methods used to determine their concentration.

5.4.1 Physics Behind the Experiment

Probing the sample with the help of positrons is the easiest method of studying the matter with antiparticles ("antimatter probe"). The positron is the antiparticle to the electron, so it has the same mass but opposite charge. When a positron enters the matter, it quickly (within ~10^{-12}s) slows to a thermal velocity (~$0.5 \div 1 \times 10^{-3}$ eV) through inelastic collisions. Further, thermal positrons diffuse into the material (at times ~ 10^{-10}s), repelling on the positive potential of protons in the atom nuclei and attracting by a negative potential of areas with missing kernels, i.e., by structural defects. In these defect regions, the annihilation with electrons may occur (~$10^{-10} \div 10^{-7}$s) (see. Fig. 5.7).

When a positron collides with an electron, the masses of two particles are converted into the energy of formed gamma rays (i.e., in the electromagnetic radiation of high frequency), which is determined by the well-known Einstein's formula $E = mc^2$. Taken into account electron and positron masses, the energy of the formed gamma rays is equal to 511 keV; moreover, these two gamma photons are always emitted in nearly opposite directions (due to conservation of momentum) and can be detected using one or two detectors. In the latter case detectors are facing each other and perform the measurements in a coincidence mode. Positron and electron can also form a positron-electron pair called positronium (Ps), which is an analogue of a hydrogen atom where a proton is replaced by a positron. The lifetime of Ps depends on the relative orientation of positron and electron spins: in the case of antiparallel spins, the lifetime is short (125 ps), whereas for Ps with parallel spins, the lifetime is prolongated up to 142 ns.

Positron sources, used for PAS, are mainly β^+ emitters, like ^{22}Na, or accelerator-based positron sources, where the positrons are created via bremsstrahlung and pair creation. PAS methods differ mainly in the way of the gamma rays detection and subsequent analysis of the resulting spectrum. Nowadays the distinguished methods are Doppler broadening spectroscopy (DBS), angular correlation of annihilation radiation (ACAR), Ps time-of-flight (Ps-TOF) method, 3γ annihilation spectroscopy (rate 3γ/2γ), and positron annihilation lifetime spectroscopy (PALS). In the following, the further explanations will focus on the DBS. The so-called Doppler

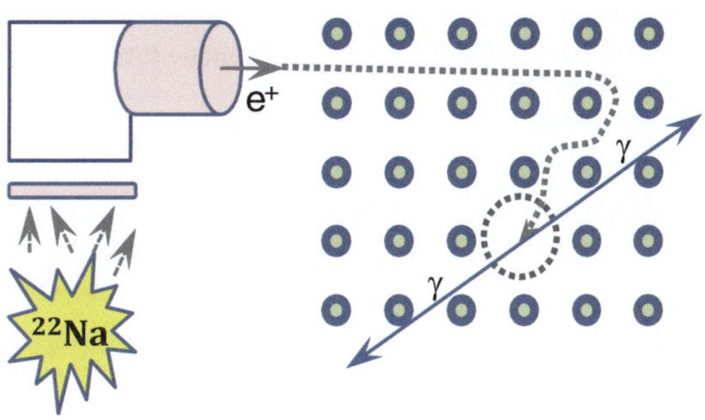

Fig. 5.7 Positron's thermalization (energy loss down to thermal energies) and diffusion in the matter and scheme of annihilation in negatively charged vacancy

broadening of the 511 keV annihilation line reflects the distribution of the momenta of the electrons which contribute to the annihilation.

In areas without defects, the positron annihilation occurs predominantly with a core electron which has a high momentum, while in areas with defects, positrons annihilate mostly with valence electrons having a low momentum. Since positrons, slowed down to thermal energies, are practically in the rest relative to the electron, the energy broadening of the annihilation line (Doppler broadening) is caused mainly by the momenta of the electrons involved in the annihilation. Characteristic shifts energies lie in the order of few keV.

Positrons can be successfully applied to studies of defects in the size range from an atomic size to a nm scale (10^{-7} m) and can identify and locate vacancies or voids in metals and dielectrics. At room temperatures the probability of positron annihilation in negatively charged vacancies is by an order of magnitude larger than in the neutral vacancy, while in positively charged vacancies it is 1.5 orders of magnitude smaller than in the neutral one. At low temperatures, the difference in such an effective coefficient of "attraction" is much larger [56]. Therefore, PAS measurements carried out at room temperature are mainly restricted to the investigation of negatively charged defects which in the case of DMS and DMO are of primary importance.

5.4.2 *Doppler Broadening Spectroscopy (DBS)*

For the investigations of thin layers on top of a solid substrate, a mono-energetic slow positron beam can be used. Thereby positrons are implanted into the surface with different incident energies. Depending on their incident energy, positrons are stopped in varying depth below the surface and annihilate there with electrons after slowing down to a thermal equilibrium. In this way, a depth-dependent defect characterization becomes possible up to a depth of some micrometers from the surface. A scheme of a mono-energetic slow positron beam is shown in Fig. 5.8. Positrons from a ^{22}Na source are moderated with a tungsten foil to an energy of

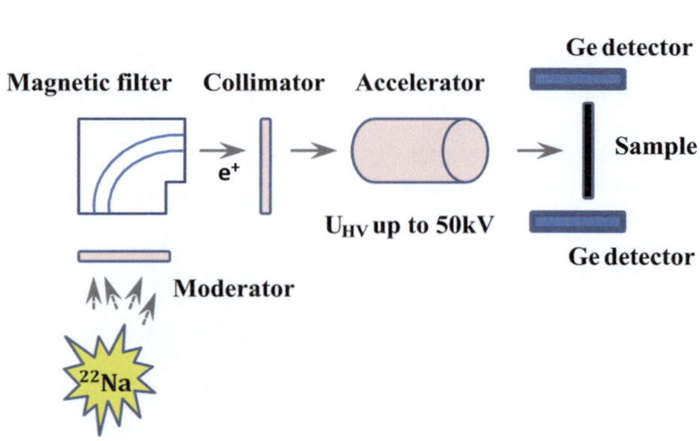

Fig. 5.8 Scheme of a slow positron beam setup for coincidence DBS measurements

2–3 eV due to the negative positron work function of tungsten. The moderated positrons are pre-accelerated with about 30 V and magnetically guided through a bent tube and in particular cases also through a collimator to the main accelerator where they can be accelerated to higher energies. The bent tube is arranged in order to separate the slow positrons from the fast ones. Finally, the positrons hit the sample and annihilate in a depth predominated from their incident energy. The energy of the annihilation radiation is measured using Ge detectors with a high energy resolution.

During the annihilation process, the center of mass of an electron and a positron is moving, so the photo-peak associated to the resulting gamma rays will get a Doppler broadening of the central peak at 511 keV. The energy of annihilation photons registered in the experiment is mainly influenced by the momentum of an involved electron. The direction of the electron momentum can coincide with the propagation direction of the annihilation photon and can be against it or be somewhere in between followed by increased or decreased photon energies. This causes the Doppler broadening of the annihilation line.

In the DBS method, one of the characteristic parameters of the Doppler spectrum is a so-called parameter S, defined as the ratio of counts (i.e., area) in the central area around the photo-peak to the total peak area (see Fig. 5.9). Parameter S has a simple link to the Doppler broadening: if the annihilation peak is narrow, what happens if slow positrons annihilate mainly with low momentum valence electrons, the parameter S is increasing. So, the parameter S represents a measure for the positron annihilation in open volumes (vacancies and their clusters, pores, voids but also grain boundaries and dislocations), where annihilation takes place preferentially with valence electrons due to missing atoms. Similarly, it is possible to obtain another characteristic parameter of the annihilation—the parameter W defined at the wings of the photo-peak area. It can be used to study the contribution from core electrons (electrons located in the inner shell of atoms). The W parameter is defined as the ratio of counts in the two wings of the annihilation spectrum to the total number of counts in the peak. For this reason, the W parameter is sensitive to the chemical environment of the annihilating zone.

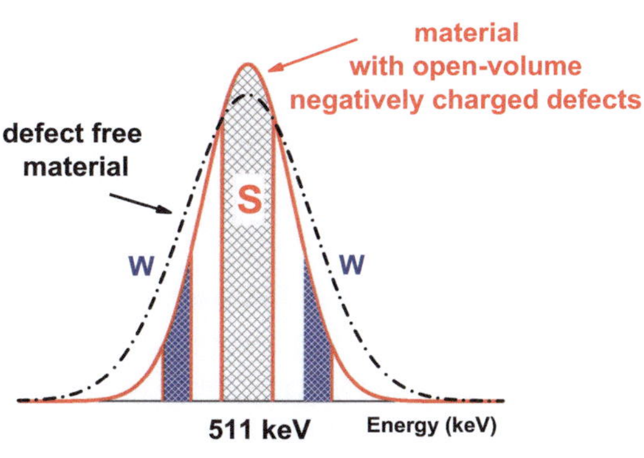

Fig. 5.9 Scheme of registered photo-peak in the DBS method with regions for S and W parameters

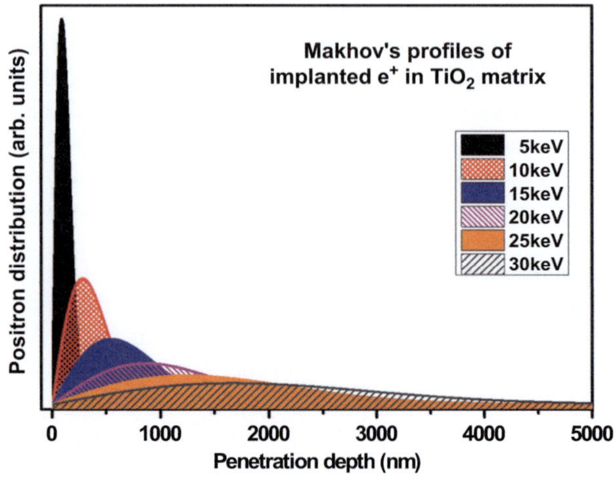

Fig. 5.10 Depth profiles of mono-energetic e^+ in TiO_2 matrix depending on the initial energy

By the DBS method various types of defects in crystals, metals, semiconductors and polymers, as well as features in chemical structure and biological fluids systems, can be studied.

Using simple parameters such as S and W, comprehensive information can be obtained about the properties of material areas with a low electron density. S and W parameters are usually measured as a function of the positron beam energy E. Assuming a Makhovian profile for the positron implantation in a solid, the mean penetration depth of the positrons with a certain energy E can be calculated according to

$$z(\text{nm}) = \frac{A.E^n}{\rho}$$

where $A = 36$ μg $cm^{-2}(keV)^{-n}$ and $n = 1.62$ are empirical constants [57], ρ is a material density in [g/cm^3], and E is the energy of positrons in [keV]. Illustration of this formula is shown in Fig. 5.10.

Furthermore, it is possible to calculate a depth-dependent defect profile S(d) with the depth d in [nm] from the experimentally measured S(E) curve using the software package VEPFit [58]. In addition, with the knowledge of S_{max} (S parameter of saturation trapping) and $S_{undamaged}$ (S parameter of the defect-free reference), conclusions about the defect concentration can be drawn.

5.4.3 Experimental Results

The results of depth-resolving DBS experiments presented in this chapter have been obtained at the Slow Positron System of Rossendorf (SPONSOR) [59] where a variation of the positron energy E from 30 eV to 36 keV with a smallest step width

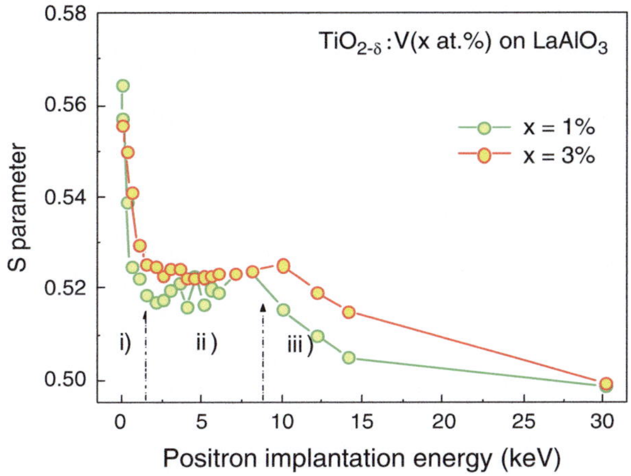

Fig. 5.11 Energy-resolved depth profiles of S parameter for TiO_2 thin films directly doped by vanadium atoms of 1÷3 at.% [51]

of 50 eV, if required, is possible. The energy resolution of the Ge detector at 511 keV was (1.09 ± 0.01) keV, resulting in a high sensitivity to changes in material properties from the surface to depths up to several micrometers. The S parameter was calculated within an energy range of (511 ± 0.928) keV.

Typical energy dependencies of S parameter for thin films recorded at room temperature for TiO_2:V(1÷3 at.%) on $LaAlO_3$ are shown in Fig. 5.11. Both curves clearly exhibit three characteristic regions related to (i) the surface region of the $TiO_{2\text{-}\delta}$:V film (below 2 keV), (ii) the bulk $TiO_{2\text{-}\delta}$:V film (2 ÷ 9 keV), and (iii) the interface region between the film and $LaAlO_3$ substrate (above 9 keV). The film with 3 at.% V has a smoother plateau pointing to a more homogeneous bulk layer, while for 1 at.% doping some dispersion in S parameter values has been obtained that could be related to the inhomogeneous film growth. Nevertheless, the presence of the plateau indicates about successful preparation of the TiO_2:V layer with a rather homogeneous distribution of the negatively charged defects inside.

TiO_2:V(1÷3 at.%) thin films were prepared on $LaAlO_3$ by RF magnetron sputtering technique in the argon–oxygen atmosphere and exhibit the ferromagnetism at room temperature according to SQUID measurements (see, e.g., [51] (for 1 at.%)). The presence of a noticeable amount of magnetic impurities like iron or cobalt has been ruled out by EDX analysis and MO spectroscopy. Since for such a system a hypothesis about d^0-ferromagnetism was suggested and since it is interesting to reveal a role of negatively charged defects like Ti (or V) vacancies separately from the positively charged oxygen ones (or their neutral complexes), the DBS has been exploited. The sensitivity of this spectroscopy method to the preparation conditions of TiO_2 thin film such as oxygen pressure and substrate temperature has been already confirmed within our prior work [60].

The prepared films have a mixed structure of rutile and anatase $TiO_{2\text{-}\delta}$ with grain sizes around several hundred nm, and it is expected that the majority of structural defects are concentrated at the grain boundaries. Using the software package VEPFit, it was possible to determine the positron diffusion length in the $TiO_{2\text{-}\delta}$

layer to (13 ± 3) nm. This value is much smaller than the obtained grain size. That means, positrons are mostly trapped by defects *inside* the grains; thus, the S parameter represents the measure of negatively charged defects in the $TiO_{2-\delta}$ matrix and not in the cross-border space. Furthermore, the fitted value is considerably lower than this one measured for metal oxides where the positron diffusion length of the undamaged material amounts to 100–200 nm. That allows the conclusion that the prepared $TiO_{2-\delta}$ layers contain a high defect concentration.

As follows from Fig. 5.11, DBS is sensitive to the doping level of studied $TiO_{2-\delta}$:V films. In the plateau region, the values of S parameter correlate well with the magnetization data: $TiO_{2-\delta}$:V(1 at.%) film has a magnetic moment of ~0.6 μ_B per impurity atom and smaller S parameter, while $TiO_{2-\delta}$:V(3 at.%) film possesses a magnetic moment of ~1.2 μ_B per impurity atom and a larger value for the S parameter.

The PAS technique has been also applied to studies of $TiO_{2-\delta}$:V(3 at.%) thin films with different conductivities. It was found that the energy dependence of the S parameter of the semiconductive film exhibits a visible notch for the surface region till the depth about 50 nm (see Fig. 5.12); that is related to a lower defect concentration in this region. For the conductive film, such a peculiarity is absent. Such a feature is most probably related to the preparation conditions: in the case of the semiconductive film, the oxygen flow was changed at the end of the film growth, while for the conductive one, it was kept constant for the whole preparation time. On depths larger than 50 nm, the values of the S parameter are comparable. Thus, the integral value of the S parameter and related amount of negatively charged defects is larger for the conductive film.

In both cases the higher saturation magnetization of the films is found to correlate well with a higher density of negatively charged structural defects as probed by PAS. Possible candidates for these defects are Ti vacancies or defect complexes involving Ti^{3+}. For TiO_2:V(3 at.%), the main contribution to the difference of the semiconductive and the conductive film magnetizations initiates by the surface region with fewer defects in the semiconductor film. Thus, it is proposed

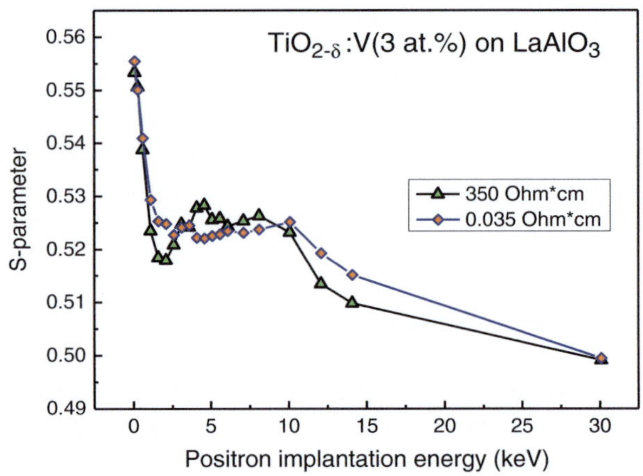

Fig. 5.12 Energy-resolved depth profiles of S parameter for semiconductive (350 Ω cm) and conductive (0.035 Ω cm) TiO_2:V(3 at.%) thin films

that these types of defects (and their agglomerations) should be taken into account when considering room temperature ferromagnetism in V-doped $TiO_{2-\delta}$ in addition to the influence of positively charged oxygen vacancies and their complexes. Nevertheless, further positron annihilation studies are necessary to clearly distinguish the role of positron-sensitive defects and conductive electrons in formation of room temperature ferromagnetic long-range order.

5.5 Comparison of Magnetic Properties of Co- and V-Doped $TiO_{2-\delta}$ Films

In this section we compare experimental results on the magnetic, magnetotransport and MO properties of Co- and V-doped $TiO_{2-\delta}$ magnetic oxides at the doping level around 1 at.%. The samples were prepared using RF magnetron sputtering in identical conditions that allows to compare the mechanisms of above room temperature ferromagnetism observed in both cases of doping. In spite of the comparable values of the magnetic moment around 1 $\div$ 2.5 μ_B per 3*d* impurity derived from macroscopic magnetic measurements for both systems, the MO response of $TiO_{2-\delta}$:V was at least 2 orders of magnitude weaker. The anomalous Hall effect was absent in V-doped $TiO_{2-\delta}$, and no appreciable magnetic moment on V impurities was found by X-ray magnetic circular dichroism technique in contrast to Co-doped $TiO_{2-\delta}$.

We consider properties of two thin film samples: $Ti_{0.987}Co_{0.013}O_{2-\delta}$ (pure anatase) and $Ti_{0.99}V_{0.01}O_{2-\delta}$ (mixture of anatase and rutile). The samples were obtained by RF magnetron sputtering as described above (see details in 5.2 and [52]). Accordingly to XRD patterns, there was no evidence of Co and V clusters present in studied films. To check the possible secondary parasitic magnetic phases in V-doped TiO_2 films, EDX analysis has been performed at Helmholtz-Zentrum Dresden - Rossendorf (HZDR) with high-resolution scanning electron microscope Hitachi S-4800 with microanalysis system (INCA, Oxford Instruments). The EDX spectrum of $Ti_{0.99}V_{0.01}O_{2-\delta}$ sample (Fig. 5.13) demonstrates the absence of $K_{\alpha,\beta}$ fluorescent lines of iron and cobalt (6–7 keV), and only Ti, O, and V lines are observed. Due to the low V concentration, only the K_α line of vanadium (4.95 keV) is well resolved, and K_β line (5.43 keV) has a nondetectable intensity. So, the samples under investigations do not contain any parasitic ferro- or paramagnetic inclusions and clusters.

The magnetic hysteresis of studied films measured at room temperature is presented in Fig. 5.14. The coercive field of about 200 Oe is rather typical for $TiO_{2-\delta}$:Co systems with different concentrations of impurities. The average magnetic moment per Co atom determined from the saturation magnetization at 300 K is about 2.5 μ_B, which is higher than the cobalt magnetic moment in bulk 1.7 μ_B/Co^{2+} obviously excluding possible segregation of Co metal, and slightly exceeds the most of available in literature data (see [61]). Superparamagnetic

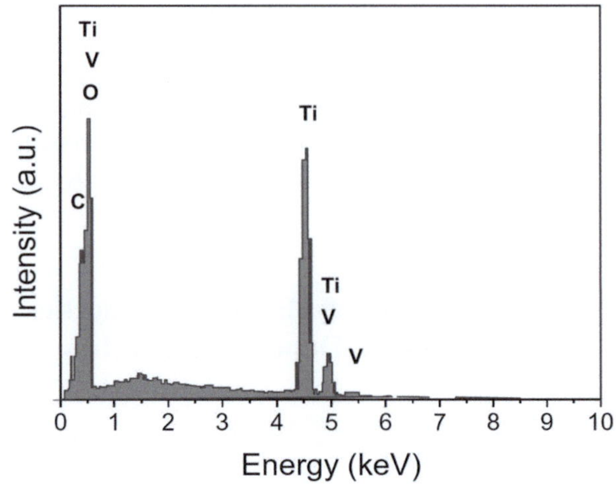

Fig. 5.13 EDX spectrum of $Ti_{0.99}V_{0.01}O_{2\text{-}\delta}$ thin film with mixed structure "anatase + rutile" [52]

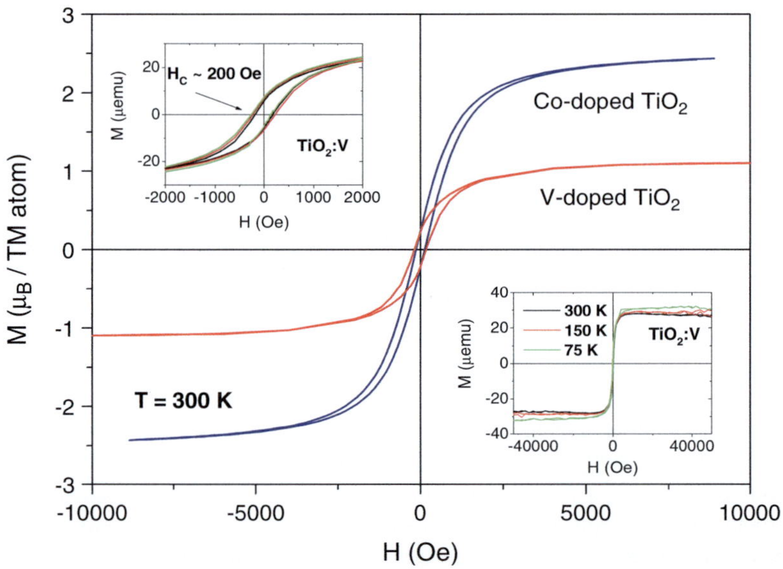

Fig. 5.14 Magnetic moment per TM atom vs. magnetic field curves at room temperature for Co- and V-doped TiO_2 thin films [52]

behavior was not observed in measured zero field cooled (ZFC) and field cooled (FC) curves. This supports the results of XRD, proving the absence of Co nanoparticles. It was shown in [41] (see also Sect. 5.2) that TKE spectrum of $Ti_{0.987}Co_{0.013}O_{2\text{-}\delta}$ produced by magnetron sputtering differs from the spectra of metallic Co clusters in nonmagnetic $TiO_{2\text{-}\delta}$ matrix and TKE signal is larger than for pure Co thin film. This confirms the results of XRD analysis about the absence of metallic Co clusters in $TiO_{2\text{-}\delta}$ at the concentrations around 1 at.%.

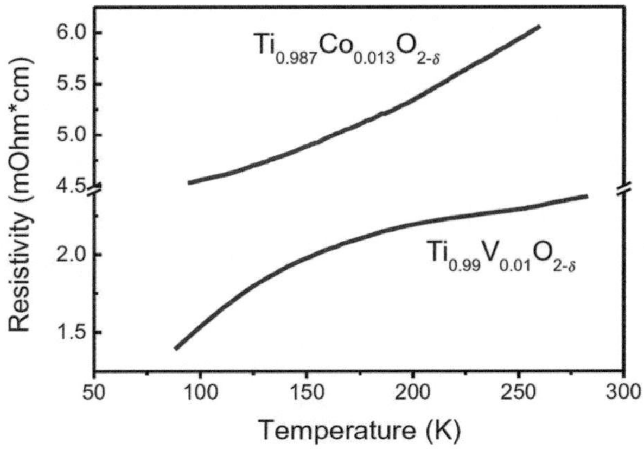

Fig. 5.15 The temperature dependence of electrical resistivity *ρ(T)* of $Ti_{0.987}Co_{0.013}O_{2-\delta}$ and $Ti_{0.99}V_{0.01}O_{2-\delta}$ [52]

The temperature dependence of electrical resistivity *ρ(T)* reveals a "metallic" character from 90 to 300 K (Fig. 5.15), which means all impurities are activated at $T > 90$ K, and therefore the carrier concentration *n* does not change too much in the studied temperature range. This conclusion is confirmed by Hall effect measurements.

The studied $Ti_{0.99}V_{0.01}O_{2-\delta}$ sample exhibits clear magnetic hysteresis (Fig. 5.14) at room temperature with a similar level of the coercivity (~200 Oe) as for $Ti_{0.987}Co_{0.013}O_{2-\delta}$. The saturation magnetization at room temperature corresponds to ~1.1 μ_B per vanadium atom. This value differs from the results obtained by Hong et al. [24] and from the maximum value of 4.8 μ_B per vanadium atom discussed above in Sect. 5.3. It is not of surprise because of different sputtering conditions and film thicknesses. Since neither bulk vanadium nor its oxides are ferromagnetic, it is hard to believe that possible vanadium clusters can show above room temperature ferromagnetism. XRD analysis and transmission electron microscopy did not indicate the presence of metallic vanadium clusters. Hysteresis loops measured at 150 and 75 K did not show the significant changes in the saturated magnetization. ZFC/FC curves recorded by SQUID (not shown) confirm the ferromagnetic ordering in the sample and do not reveal superparamagnetic behavior. We can also exclude that the ferromagnetic behavior of the studied $Ti_{0.99}V_{0.01}O_{2-\delta}$ sample is due to possible parasitic phases because of the results of EDX analysis and discussed underneath spectroscopy.

Unexpectedly, no MO signal has been detected for the $Ti_{0.99}V_{0.01}O_{2-\delta}$ sample with 300 nm thickness: the value of TKE signal was on the noise level (Fig. 5.16). So, in spite of a rather strong magnetic moment per impurity atom estimated from magnetometry data, more sensitive MO spectroscopy surprisingly could not confirm the magnetic behavior of this system. We also were unable to detect any nonlinearity in the field dependence of Hall resistivity which indicates a negligible AHE.

XANES and XMCD spectroscopy has been applied for $Ti_{0.97}V_{0.03}O_{2-\delta}$ DMS [46]. No XMCD signal was detected at the V K-edge even if the TKE was detected

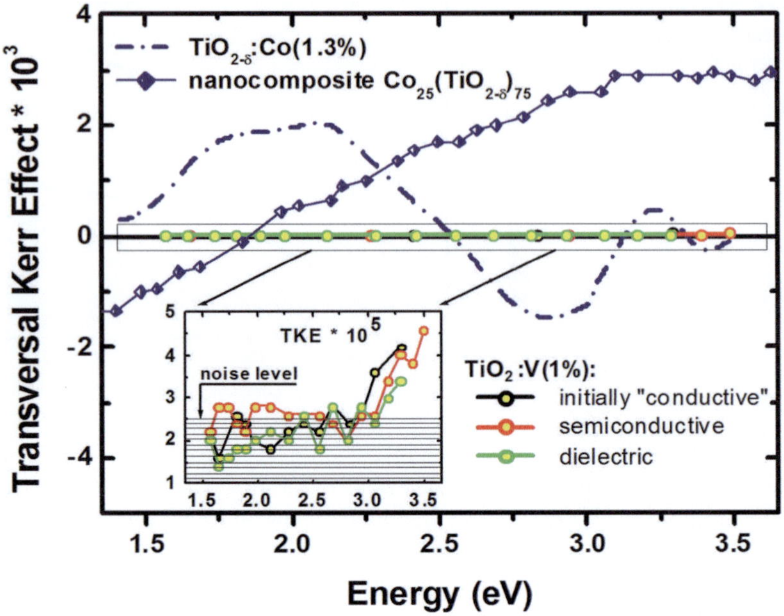

Fig. 5.16 TKE spectra of Co- and V-doped $TiO_{2-\delta}$ films. The presented spectra for nanocomposite $Co_{0.25}(TiO_{2-\delta})_{0.75}$ is significantly different, which is considered as indirect evidence of an absence of Co metallic clusters (from [51])

Fig. 5.17 TKE spectra for $Ti_{0.97}V_{0.03}O_{2-\delta}$ thin films of various resistivity

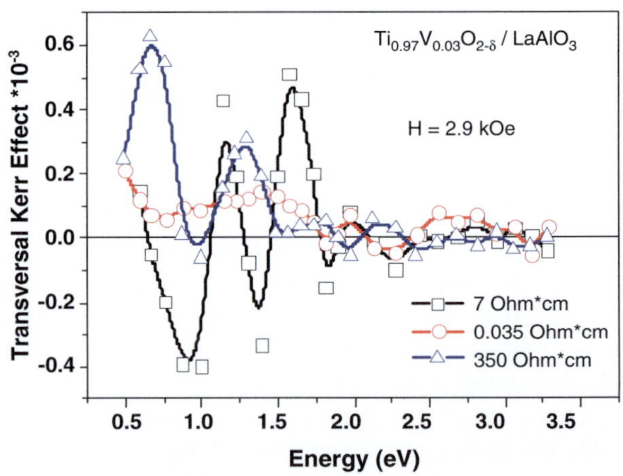

(Fig. 5.17). This XMCD result would indicate that the vanadium atoms have no local orbital magnetic moment associated with the hybridized *p* shell at least at the level of 2×10^{-5} μ_B. The same result was obtained for a $Ti_{0.99}V_{0.01}O_{2-\delta}$ thin film. The shape of XANES spectra at the V K-edge is significantly different and shifted along the energy scale as compared to the reference spectrum of vanadium foil which confirms the initial assumption of an ionic state of the V atoms. Thus, the absence of a magnetic signal in XMCD, TKE, and Hall effect measurements could be of principal nature.

Although we could not observe any TKE signal in $Ti_{0.99}V_{0.01}O_{2\text{-}\delta}$, it was found at 3 at.% V doping (Fig. 5.17), but the signal is quite weak, at least one order of magnitude smaller than in $TiO_{2\text{-}\delta}$:Co at low Co concentration (Fig. 5.4).

The value of magnetic moment per impurity atom derived from the macroscopic magnetization data can be determined by at least three different contributions (contamination is ruled out because undoped films prepared in the same chamber and handled in the same way as the doped films exhibit no ferromagnetism). The first is the local polarization of the 3*d* impurities (spin and orbital magnetic moments). The second one is magnetic polarization related to oxygen vacancies. Finally, the last contribution to magnetic polarization is determined by structural defects. As it was shown in Sect. 5.2 and this section in TiO_2 doped with Co, the first contribution is dominant even at 1 % of Co impurities. But in the case of TiO_2 doped with V, it does not follow from obtained experimental data, at least for $Ti_{0.99}V_{0.01}O_{2\text{-}\delta}$.

What are the possible reasons for room temperature ferromagnetism in $Ti_{0.99}V_{0.01}O_{2\text{-}\delta}$ (Fig. 5.14) which is not accompanied with ferromagnetic XMCD signal for V, TKE, and anomalous Hall effect? At the first glance, we obligate to conclude that V ions do not carry magnetic moments, so defects or oxygen vacancies are responsible for long-range ferromagnetic order. PAS experiments (Sect. 5.4) confirm the correlation between magnetization and concentration of negatively charged defects. Both MO and anomalous Hall effects are due to spin–orbit interaction. Taking into account that since both MO signal and anomalous Hall resistivity are negligible in $Ti_{0.99}V_{0.01}O_{2\text{-}\delta}$, one can suppose that spin–orbit interaction is very small. It confirms defect-driven magnetization and charge-transfer model [53] if the impurity band with Stoner-type ferromagnetism appears due to light impurities with a weak spin–orbit interaction. Since the formation of the impurity band, i.e., its position in the gap and density of states are very sensitive to structural parameters, it becomes clear why samples produced by different methods and/or under different conditions exhibited completely different properties. In this scenario V impurities do not bear magnetic moments themselves but induced defects with noticeable magnetic moments. This explanation looks consistent, but as a rule a defect-induced magnetization is small and decreases after long-term annealing at high temperatures [62] that is not the case for TiO_2 doped with V. Since XMCD, TKE, and anomalous Hall effects originate from spin–orbit interaction, perhaps, V ions when they replace Ti positions have a weak spin–orbit interaction. Besides, the TKE signal might be strong in another spectral range and anomalous Hall effect might be negligible if there is a weak spin polarization of current carriers. So we have no definite answer, and further experimental and theoretical work is needed to shed light on room temperature magnetism in DMO based on TiO_2.

Another important difference of magnetic properties of TiO_2 doped with Co and V should be emphasized. For room temperature ferromagnetism in TiO_2:Co, the carrier concentration plays a crucial role (see Sect. 5.2, [63] and references therein), but in the case of TiO_2:V, magnetization of samples with resistivity 350 Ω cm and 0.035 Ω cm (difference by 4 orders of magnitude) changes less than twice [46]. It clearly shows that the origin of ferromagnetism in TiO_2:Co and TiO_2:V is not the same.

5.6 Ferromagnetism in ZnO:TM

Zinc oxide is a well-known functional material with a wide range of applications [18, 64, 65]. When the Zener model has been applied to the description of ferromagnetism in Mn-doped semiconductors [12], ZnO and GaN became intensively studied semiconducting materials since above room temperature ferromagnetism was predicted for these two systems. The possibility to realize carrier-mediated ferromagnetism in TM-doped ZnO at room temperature was an effective motivation for keeping the experimental research of this material on the boil, which finally led to the series of the surprising and striking findings.

There is a continuing large interest in ferromagnetism in doped and undoped ZnO nanomaterials [11, 61, 66] because this semiconducting material is widely used in technological applications and relatively cheap. Desired combination of ferromagnetism at room temperature with well-known attractive optical and electronic properties of ZnO could lead to the realization of new devices of semiconductor spintronics. First examination of ferromagnetism in TM-doped ZnO thin films was reported in [67]. Later there were numerous published works about ferromagnetic zinc oxide doped with Co and Mn as well as other 3*d* metals [68, 69] and even with "nonmagnetic" ions [70]. The room temperature ferromagnetism was observed in Cu-doped ZnO nanoneedles, which was attributed to the distribution of Cu^{2+} ions and defect density in ZnO both affected by annealing. Mn-substituted ZnO thin films demonstrated ferromagnetism with Curie temperature above 375 K [71] and average magnetic moment below 0.1 μ_B per Mn atom; the columnar structure of the films was shown to determine the coercivity and uniaxial anisotropy.

However, already the very first observations of ferromagnetism in doped ZnO nanomaterials were followed by several reports where the possibility of intrinsic ferromagnetism in ZnO was denied. Alaria et al. [72] have shown that even 10 % doping of ZnO polycrystalline powder with Co is not sufficient for the formation of ferromagnetic order. Co-doping with aluminum was used with the idea to increase the free carrier concentration to stimulate the carrier-mediated ferromagnetism [12]; however, authors could observe only the mixture of paramagnetic and antiferromagnetic response. Similar findings have been reported in [73]—the study of the temperature dependence of magnetization revealed the antiferromagnetic coupling of Co and Mn magnetic moments in doped polycrystalline bulk ZnO. The bulk samples $Zn_{1-x}Co_xO$ prepared using the solid-state reaction method demonstrated paramagnetic behavior with antiferromagnetic coupling of Co^{2+} ions which was confirmed also with Monte Carlo simulation. In the detailed analysis of observed ferromagnetic hysteresis in TM-doped ZnO [74], the coexistence of several sources of ferromagnetic response was proven, mainly the formation of nanoscale magnetic clusters of metals and contamination of substrates used for the film growth. Secondary phase can be formed by metal clusters [75] and by chemical compound of doping element with atoms of hosting oxide. The presence of $(Mn,Zn)Mn_2O_4$ spinel was shown to be responsible for observed magnetic transition in polycrystalline ZnO samples [76]. In case of cobalt doping the secondary phase can be segregated ferromagnetic CoZn (Curie temperature of 400–450 K) [77], Zn-doped rocksalt CoO ($Co_{1-y}Zn_yO$) [78].

The non-monotonic dependence of the magnetization on the concentration as a common and well-known feature of doped zinc oxide was systematically studied by Straumal [79]. The developed approach to describe the ferromagnetic properties of ZnO is based on the concept of "ferromagnetic grain-boundary foam", quantified with specific grain boundaries (GB) area [10, 80]. Based on the idea of intrinsic ferromagnetism in oxides related to the oxygen defects [53, 81], the grain boundaries are considered as a source of oxygen vacancies, structural dislocations, and crystallographic imperfections carrying the magnetic moments. The role of the TM dopant ions in such systems can be limited to give only paramagnetic contribution which was demonstrated using sophisticated XMCD measurements [82]. If the amount of such localized magnetic moments is large enough for Stoner spin-splitting of an impurity band, a weak ferromagnetic response can be observed in ZnO even in absence of TM ions—for magnetic ordering, a specific GB area should exceed a certain threshold value [83]. In the case of Mn-doped ZnO, the intriguing non-monotonic behavior of the magnetic moment per Mn atom can be understood as a result of the change of the relative contributions of Mn^{2+}, Mn^{3+}, and Mn^{4+} ions and consequently the amount of unoccupied Mn $3d$ states [79]. Also the topology of the GB in polycrystalline and nanostructured ZnO was shown to play an important role for the specific GB area which confirms the strong dependence of the defect-related ferromagnetic response on the synthesis methods and growth conditions [79].

Ongoing intensive research work elucidated a wide variety of experimental and theoretical results of studying the ferromagnetism in ZnO. However the debates about the origin of the ferromagnetic order observed at room temperature are continued without cease. Special attention is being paid to d^0-ferromagnetism in undoped ZnO followed with the idea of obligatory existence of ferromagnetic response in nanostructured ZnO (in the form of polycrystalline, nanograined, and mosaic thin films as well as nanoparticles, rods, wires, etc.) in case of sufficient amount of defects on its surface or inside the crystal lattice [9, 84]. The influence of the morphology and nanostructure of Co-doped and undoped ZnO films grown using metal organic chemical vapor deposition (MOCVD) was carefully studied in the series of works [85–87]. Single-crystalline Al_2O_3 (r- and c-sapphire), $ZrO_2(Y_2O_3)$, and $MgAl_2O_4$ were used as substrates. No correlation between the magnetic properties and concentration of Co was observed, which reveals a minor role of TM ions for the ferromagnetic ordering of films. The influence of structural defects and morphology was carefully investigated. Conventional water-assisted MOCVD allowed the fabrication of films with various morphologies and structures—high-quality epitaxial or polycrystalline with a highly developed nanostructured surface, correspondingly [86]. In the latter case, the surface of the films shows a "crossed plates"-like or nanosized 200–300-nm-wide whisker structure, depending on the deposition temperature (see Fig. 5.18). The undoped ZnO films showed a different magnetic behavior—unlike the epitaxial films the nanostructured films (deposited using the water-assisted MOCVD) were found to be clearly ferromagnetic at room temperature. The more developed nanostructure of the film, the more pronounced was the observed ferromagnetic signal. The origin of such difference is believed to be due to the structural defects located at the surface of nanostructured films [84]. In order to prove the defect-induced nature of the room temperature ferromagnetism in undoped ZnO, the authors deposited films with in-plane variant structure. When the film is deposited on a cubic (111) $MgAl_2O_4$ substrate, nanosized structural epitaxial domains (variants) are formed [86]. Such variant structure leads to a large amount of structural imperfections in the crystal lattice at the boundaries of neighboring variants (Fig. 5.19). Thus, the

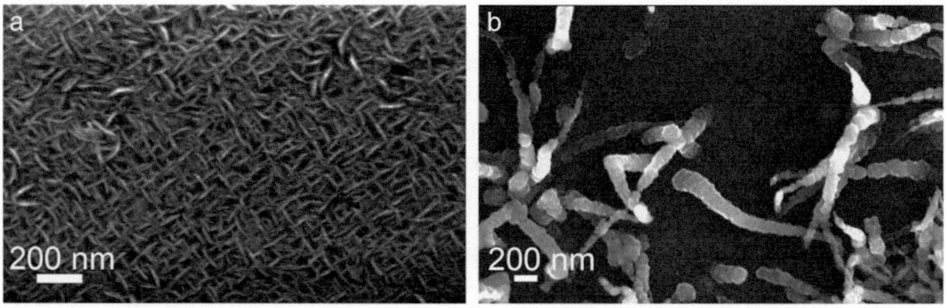

Fig. 5.18 SEM images of the ZnO films deposited by water-assisted CVD on r-sapphire: (**a**) at 300 °C (crossed plates structure); (**b**) at 500 °C (nanowhiskers). From [86]

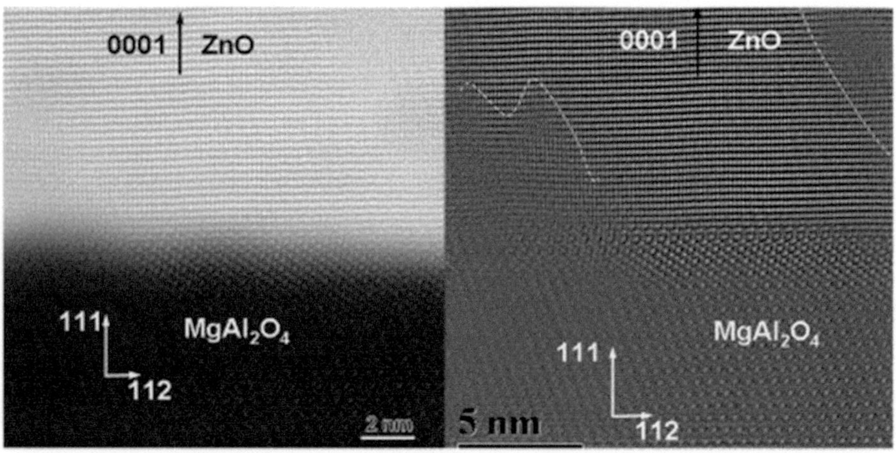

Fig. 5.19 HRTEM image (*left*) and its Fourier-transformation (*right*) for the ZnO film deposited on (111) $MgAl_2O_4$. Thin *white lines* show the imperfection areas between structural domains (variants) where defects are localized. From [86]

ferromagnetic response at room temperature is driven by defects localized at grain boundaries. The temperature dependence of the resistivity of the variant films reveals a hopping mechanism of electric conductivity. This points to the localization of the electronic states near the Fermi level which is typical for undoped semiconductors in case of disorder or point defects.

Both of these observations demonstrate the defect-induced nature of ferromagnetism at room temperature in undoped ZnO thin films and the strong correlation between nanostructure of films and their magnetic response. High-quality epitaxial films do not reveal any ferromagnetic response, whereas the highly developed structure of films (whiskers, nanoplates, or variants) leads to the stabilization of ferromagnetic ordering.

The correlation between magnetization and defects (oxygen vacancies) was shown also for TM-doped ZnO [88]. The authors have found that in the case of $Zn_{0.95}Co_{0.05}O$ nanoparticles, the ferromagnetism is induced by interfacial oxygen deficiency and the vacancies can initiate defect-related hybridization at the Fermi level and mediate the tunable long-range ferromagnetic order. Another interesting approach to tune the properties of ZnO was reported by Li et al. [89]. A variation of the morphology of oxidized ZnO films was achieved through the application of high magnetic field during the oxidation process. By variation of the concentration of oxygen vacancies and the ionic state of cobalt ions, the optical, electrical, and magnetic properties can be tuned.

Recently, the interest for ferromagnetism in ZnO is being shifted toward nanostructures, i.e., nanowires [90, 91], nanorods [92, 93], nanoparticles [94], etc. The synthesis of ZnO-based nanorods, wires, and particles with excellent magnetic properties and high crystallinity is a challenge for semiconductor spintronics devices. The fabrication methods for ZnO nanostructures have been comprehensively reviewed by Guo et al. [95]. Taking into account the correlation between structure and magnetic response of ZnO mentioned above, it is important to access the control of the morphology of the synthesized objects through the fabrication conditions. Nanostructures, such as nanowires or nanoparticles, allow to combine the effect of doping with TM ions and size effects—the influence of the surface in these systems plays an essential role for defect-driven magnetic response, and this can be successfully used for the realization of room temperature ferromagnetism in ZnO. However, the opinions about the magnetic nature of doped and undoped ZnO nanosized objects remain controversial. For example, zinc vacancies and corresponding dangling or unpaired O 2p states were shown to have a significant local magnetic moment and induce d^0-ferromagnetism in ZnO nanorods [92], while other group claimed the crucial role of Zn interstitials [96]. Finally, a ferromagnetic response with a Curie temperature near 400 K was observed only in case of Co doping, and no measurable signal was detected for undoped ZnO nanorods [93].

Based on the collective results of many research groups available in literature, no general idea on the source of ferromagnetic ordering in nanostructured ZnO can be inferred. There are always several contributions which should be taken into account for a correct description of the experimental observations. TM impurities (if present), oxygen vacancies (singly ionized oxygen vacancies, in particular), zinc vacancies as well as other types of defects and even structural peculiarities and the morphology of the ZnO-based nanomaterials play the crucial role in stabilization of ferromagnetism in these systems. Defect-induced ferromagnetism at room temperature is undoubtedly the fascinating finding in the story of ZnO, but one should understand that such type of ferromagnetic materials cannot be considered as really suitable to technological applications of semiconductor spintronics—firstly, the ferromagnetic response driven by defects cannot be enlarged significantly, and, secondly, this ordering is hard to control. Nevertheless, in spite of the recent apparent decrease of interest in ferromagnetic order in doped and undoped ZnO nanomaterials and magnetic semiconductors as a whole, it is, in fact, too early to talk about "the end of an era" with respect to these materials. New findings and insights into the multifaceted question concerning the nature of magnetic states of ZnO are expected to appear. In the same vein, belief in the realization of operating spintronic devices based on this oxide should be held fast by enthusiasts working in this puzzling and extremely challenging field of modern condensed matter physics.

5.7 Conclusions

As a result of 15-year investigations triggered by pioneering works of T. Dietl [12] and Matsumoto et al. [6], tremendous progress has been reached in the field of DMS and DMO. It has been demonstrated that it is possible to develop DMS and DMO with high Curie temperature and that these novel multifunctional materials exhibit a rich spectrum of outstanding properties which are attractive for practical applications in spintronics and magnetophotonics.

A number of findings reviewed in this Section documented that it is possible to obtain DMO semiconductors based on titanium dioxide doped with transition metal impurities and on zinc oxide, which reveal intrinsic ferromagnetism in the absence of magnetic aggregates or contaminations and have a high magnetization at temperatures exceeding the room temperature.

Titanium dioxide doped with vanadium impurities, in which the room temperature saturation magnetization exceeds 40 G, and the Curie temperature is 400 K and higher, shows the highest parameters in the oxide semiconductor group. However, the possible origin of high magnetic moments at low vanadium content, long-range ferromagnetic order, and high Curie temperature in this DMO is still under debates. Magnetic properties of titanium dioxide doped with Co and V are quite different, and the basic competitive approaches, namely, carrier-mediated ferromagnetism and defect-induced magnetism, fail to explain all available experimental data. Therefore, a further theoretical and experimental work should be done with a deeper insight in structural, magnetic, magnetotransport, and magneto-optical properties of DMO in a wide range of compositions.

It should be noticed that up to now there is no reliable experimental confirmation of high spin polarization of current carriers, which is of primary importance for spintronics. The low remanent magnetization of most developed DMO does not meet requirements for applications. Besides, reproducibility of magnetic properties, the level of their homogeneity along the surface, and thickness of DMO thin films are not satisfactory. Nevertheless, all these sufficiently complicated technical problems stated above do not seem to be insuperable.

Acknowledgments This work is partially supported by the Initiative and Networking Fund of the German Helmholtz Association, Helmholtz-Russia Joint Research Group HRJRG-314, and the Russian Foundation for Basic Research, RFBR #12-02-91321-SIG_a.

References

1. Dietl, T., Ohno, H.: Dilute ferromagnetic semiconductors: Physics and spintronic structures. Rev. Mod. Phys. **86**, 187–251 (2014)
2. Ohno, H.: A window on the future of spintronics. Nat. Mater. **9**, 952–954 (2010)
3. Dietl, T.: A ten-year perspective on dilute magnetic semiconductors and oxides. Nat. Mater. **9**, 965–974 (2010)
4. Olejnik, K., Owen, M.H.S., Novak, V., et al.: Enhanced annealing, high Curie temperature, and low-voltage gating in (Ga, Mn)As: A surface oxide control study. Phys. Rev. B **78**, 054403 (2008)
5. Wang, M., Campion, R.P., Rushforth, A.W., et al.: Achieving high Curie temperature in (Ga, Mn)As. Appl. Phys. Lett. **93**, 132103 (2008)
6. Matsumoto, Y., Murakami, M., Shono, T., et al.: Room-temperature ferromagnetism in transparent transition metal-doped titanium dioxide. Science **291**, 854–856 (2001)
7. Venkatesan, M., Fitzgerald, C.B., Coey, J.M.D.: Unexpected magnetism in a dielectric oxide. Nature **430**, 630 (2004)
8. Rylkov, V.V., Gan'shina, E.A., Novodvorskii, O.A., et al.: Defect-induced high-temperature ferromagnetism in $Si_{1-x}Mn_x$ ($x = 0.52–0.55$) alloys. EPL **130**, 57014 (2013)
9. Coey, J.M.D., Stamenov, P., Gunning, R.D., et al.: Ferromagnetism in defect-ridden oxides and related materials. New J. Phys. **12**, 053025 (2010)
10. Straumal, B.B., Mazilkin, A.A., Protasova, S.G., et al.: Magnetization study of nanograined pure and Mn-doped ZnO films: Formation of a ferromagnetic grain-boundary foam. Phys. Rev. B **79**, 205206 (2009)
11. Zhou, S.: Defect-induced ferromagnetism in semiconductors: A controllable approach by particle irradiation. Nucl. Instrum. Meth. Phys. Res. **326**, 55–60 (2014)
12. Dietl, T., Ohno, H., Matsukura, F., et al.: Zener model description of ferromagnetism in zinc-blende magnetic semiconductors. Science **287**, 1019–1022 (2000)
13. Janisch, R., Spaldin, N.A.: Understanding ferromagnetism in Co-doped TiO_2 anatase from first principles. Phys. Rev. B **73**, 035201 (2006)
14. Ivanov, V.A., Ugolkova, E.A., Pashkova, O.N., et al.: Ferromagnetism in dilute magnetic semiconductors and new materials for spintronics. J. Magn. Magn. Mater. **300**, e32–e36 (2006)
15. Calderon, M.J., Das Sarma, S.: Theory of carrier-mediated ferromagnetism in dilute magnetic oxides. Ann. Phys. **322**, 2618–2634 (2007)
16. Abraham, D.W., Frank, M.M., Guha, S.: Absence of magnetism in hafnium oxide films. Appl. Phys. Lett. **87**, 252502 (2005)
17. Fukumura, T., Toyosaki, H., Yamada, Y.: Magnetic oxide semiconductors. Semicond. Sci. Technol. **20**, S103–S111 (2005)
18. Pearton, S.J., Heo, W.H., Ivill, M., et al.: Dilute magnetic semiconducting oxides. Semicond. Sci. Technol. **19**, R59–R74 (2004)

19. Hong, N.H. (ed.): Magnetism in semiconducting oxides. Transworld Reseach Network, India (2007)
20. Griffin, K.A., Pakhomov, A.B., Wang, C.M., et al.: Intrinsic ferromagnetism in insulating cobalt doped anatase TiO_2. Phys. Rev. Lett. **94**, 157204 (2005)
21. Orlov, A.F., Perov, N.S., Balagurov, L.A., et al.: Giant magnetic moments in oxide ferromagnetic semiconductors. JETP Lett. **86**, 352–354 (2007)
22. Belghazi, Y., Schmerber, G., Colis, S., et al.: Extrinsic origin of ferromagnetism in ZnO and $Zn_{0.9}Co_{0.1}O$ magnetic semiconductor films prepared by sol–gel technique. Appl. Phys. Lett. **89**, 122504 (2006)
23. Ogale, S.B., Choudhary, R.J., Buban, J.P., et al.: High-temperature ferromagnetism with a giant magnetic moment in transparent Co doped $SnO_{2-\delta}$. Phys. Rev. Lett. **91**, 077205 (2003)
24. Hong, N.H., Sakai, J., Prellier, W., et al.: Ferromagnetism in transition-metal-doped TiO_2 thin films. Phys. Rev. B **70**, 195204 (2004)
25. Coey, J.M.D., Douvalis, A.P., Fitzgerald, C.B., Venkatesan, M.: Ferromagnetism in Fe-doped SnO_2 thin films. Appl. Phys. Lett. **84**, 1332–1334 (2004)
26. He, J., Xu, S., Yoo, Y.K., et al.: Room temperature ferromagnetic n-type semiconductor in $(In_{1-x}Fe_x)_2O_{3-\sigma}$. Appl. Phys. Lett. **86**, 052503 (2005)
27. Pemmaraju, C.D., Sanvito, S.: Ferromagnetism driven by intrinsic point defects in HfO_2. Phys. Rev. Lett. **94**, 217205 (2005)
28. Hong, N.H., Sakai, J., Huong, N.T., Brize, V.: Room temperature ferromagnetism in laser ablated Ni-doped In_2O_3 thin film. Appl. Phys. Lett. **87**, 102505 (2005)
29. Chambers, S.A., Droubay, T., Wang, C.M., et al.: Clusters and magnetism in epitaxial Co-doped TiO_2 anatase. Appl. Phys. Lett. **82**, 1257–1259 (2003)
30. Balagurov, L.A., Gan'shina, E.A., Klimonskii, S.O., et al.: Boundary conditions for the formation of a ferromagnetic phase during the deposition of $Ti_{1-x}Co_xO_{2-\delta}$ thin films. Crystallogr. Rep. **50**, 686–689 (2005)
31. Suzuki, T., Souda, R.: TiO epitaxial film growth on MgO(001) and its surface structural analysis. Surf. Sci. **445**, 506–511 (2000)
32. Kim, J.-Y., Park, J.-H., Park, B.-G., et al.: Ferromagnetism induced by clustered Co in Co-doped anatase TiO_2 thin films. Phys. Rev. Lett. **90**, 017401 (2003)
33. Seong, N.-J., Yoon, S.-G., Cho, C.-R.: Effects of Co-doping level on the microstructural and ferromagnetic properties of liquid-delivery metalorganic-chemical-vapor-deposited $Ti_{1-x}Co_xO_2$ thin films. Appl. Phys. Lett. **81**, 4209–4211 (2002)
34. Fukumura, T., Toyosaki, H., Ueno, K., et al.: Role of charge carriers for ferromagnetism in cobalt-doped rutile TiO_2. New J. Phys. **10**, 055018 (2008)
35. Tiwari, A., Bhosle, V.M., Ramachandran, S., et al.: Ferromagnetism in Co doped CeO_2: Observation of a giant magnetic moment with a high Curie temperature. Appl. Phys. Lett. **88**, 142511 (2006)
36. Orlov, A.F., Balagurov, L.A., Kulemanov, I.V., et al.: Intrinsic ferromagnetism created by vacancy injection in a semiconductor oxide $Ti_{1-x}Co_xO_{2-\delta}$. Phys. Sol. Stat. **53**, 482–484 (2011)
37. Singh, V.R., Ishigami, K., Verma, V.K., et al.: Ferromagnetism of cobalt-doped anatase TiO_2 studied by bulk- and surface-sensitive soft X-ray magnetic circular dichroism. Appl. Phys. Lett. **100**, 242404 (2012)
38. Fukumura, T., Yamada, Y., Tamura, K., et al.: Magneto-optical spectroscopy of anatase TiO_2 doped with Co. Jap. J. Appl. Phys. **42**, L105–L107 (2003)
39. Toyosaki, H., Fukumura, T., Yamada, Y., Kawasaki, M.: Evolution of ferromagnetic circular dichroism coincident with magnetization and anomalous Hall effect in Co-doped rutile TiO_2. Appl. Phys. Lett. **86**, 182503 (2005)
40. Weng, H., Dong, J., Fukumura, T., et al.: First principles investigation of the magnetic circular dichroism spectra of Co-doped anatase and rutile TiO_2. Phys. Rev. B **73**, 121201(R) (2006)
41. Gan'shina, E.A., Granovsky, A.B., Orlov, A.F., et al.: Magneto-optical spectroscopy of diluted magnetic oxides $TiO_{2-\delta}$: Co. JMMM **321**, 723–725 (2009)

42. Weakliem, H.A.: Optical spectra of Ni^{2+}, Co^{2+}, and Cu^{2+} in tetrahedral sites in crystals. J. Chem. Phys. **36**, 2117–2140 (1962)
43. Enomoto, M.: The O-Ti-V system (oxygen-titanium-vanadium). J. Phase. Equilibria **17**, 539–545 (1995)
44. Hai, N.N., Khoi, N.T., Vinh, P.V.: Preparation and magnetic properties of TiO_2 doped with V, Mn, Co, La. J. Phys. Conf. Ser. **187**, 012071 (2009)
45. Lü, X., Li, J., Mou, X., et al.: Room-temperature ferromagnetism in $Ti_{1-x}V_xO_2$ nanocrystals synthesized from an organic-free and water-soluble precursor. J. All. Comp. **499**, 160–165 (2010)
46. Orlov, A.F., Balagurov, L.A., Kulemanov, I.V., et al.: Magnetic and magneto-optical properties of $Ti_{1-x}V_xO_{2-\delta}$ semiconductor oxide films: Room temperature ferromagnetism versus resistivity. SPIN **02**, 1250011 (2012)
47. Le Roy, D., Valloppilly, S., Skomski, R., et al.: Magnetism and structure of anatase $(Ti_{1-x}V_x)$ O_2 films. J. Appl. Phys. **111**, 07C118 (2012)
48. Huang, D., Zhao, Y.-J., Chen, D.-H., Shao, Y.-Z.: Electronic structure and magnetic couplings in anatase TiO_2:V codoped with N, F, Cl. J. Phys. Condens. Matter **21**, 125502 (2009)
49. Osorio-Gillen, J., Lany, S., Zunger, A.: Atomic control of conductivity versus ferromagnetism in wide-gap oxides via selective doping: V, Nb, Ta in anatase TiO_2. Phys. Rev. Lett. **100**, 036601 (2008)
50. He, K.H., Zheng, G., Chen, G., et al.: Effects of single oxygen vacancy on electronic structure and ferromagnetism for V-doped TiO_2. Sol. Stat. Comm. **144**, 54–57 (2007)
51. Yildirim, O., Butterling, M., Cornelius, S., et al.: Ferromagnetism and structural defects in V-doped titanium dioxide. Phys. Status Solidi C **11**, 1106–1109 (2014)
52. Semisalova, A.S., Mikhailovsky, Y., Smekhova, A., et al.: Above room temperature ferromagnetism in Co- and V-doped TiO_2—revealing the different contributions of defects and impurities. J. Supercond. Nov. Magn. **28**, 805–811 (2015)
53. Coey, J.M.D., Wongsaprom, K., Alaria, J., Venkatesan, M.: Charge-transfer ferromagnetism in oxide nanoparticles. J. Phys. D. Appl. Phys. **41**, 134012 (2008)
54. Asoka-Kumar, P., Lynn, K.G.: Applications of positron annihilation spectroscopy. J. Phys. IV France **05**, C1-15–C1-25 (1995)
55. Gidley, D.W., Peng, H.-G., Vallery, R.S.: Positron annihilation as a method to characterize porous materials. Annu. Rev. Mater. Res. **36**, 49–79 (2006)
56. Puska, M.J., Corbel, C., Nieminen, R.M.: Positron trapping in semiconductors. Phys. Rev. B **41**, 9980 (1990)
57. Schultz, P.J., Lynn, K.G.: Interaction of positron beams with surfaces, thin films, and interfaces. Rev. Mod. Phys. **60**, 701–779 (1988)
58. Schultz, P.J., Massoumi, G.R., Simpson, P.J. (eds.): Positron Beams for Solids and Surfaces, Proceedings of the Fourth International Workshop on Slow-Positron Beam Techniques for Solids and Surfaces. American Institute of Physics, New York (1990)
59. Anwand, W., Brauer, G., Butterling, M., et al.: Design and construction of a slow positron beam for solid and surface investigations. Defect Diffus. Forum **331**, 25–40 (2012)
60. Butterling, M., Anwand, W., Cornelius, S., et al.: Optimization of growth parameters of TiO_2 thin films using a slow positron beam. J. Phys.: Conf. Ser. **443**, 012073 (2013)
61. Ogale, S.B.: Dilute doping, defects, and ferromagnetism in metal oxide systems. Adv. Mater. **22**, 3125–3155 (2010)
62. Orlov, A.F., Balagurov, L.A., Kulemanov, I.V., et al.: Structure, electrical and magnetic properties, and the origin of room temperature ferromagnetism in the Mn-implanted Si. JETP **136**, 703–711 (2009)
63. Fukumura, T., Yamada, Y., Ueno, K., et al.: Electron carrier-mediated room temperature ferromagnetism in anatase (Ti, Co)O_2. SPIN **2**, 123005 (2012)
64. Özgür, Ü., Alivov, Y.I., Liu, C., et al.: A comprehensive review of ZnO materials and devices. J. Appl. Phys. **98**, 041301 (2005)

65. Janotti, A., Van de Walle, C.G.: Fundamentals of zinc oxide as a semiconductor. Rep. Prog. Phys. **72**, 126501 (2009)
66. Pan, F., Song, C., Liu, X.J., et al.: Ferromagnetism and possible application in spintronics of transition-metal-doped ZnO films. Mater. Sci. Eng. R. Rep. **62**, 1–35 (2008)
67. Ueda, K., Tabata, H., Kawai, T.: Magnetic and electric properties of transition-metal-doped ZnO films. Appl. Phys. Lett. **79**, 988–990 (2001)
68. Sharma, P., Gupta, A., Rao, K.V., et al.: Ferromagnetism above room temperature in bulk and transparent thin films of Mn-doped ZnO. Nat. Mater. **2**, 673–677 (2003)
69. Venkatesan, M., Fitzgerald, C.B., Lunney, J.G., Coey, J.M.D.: Anisotropic ferromagnetism in substituted Zinc oxide. Phys. Rev. Lett. **93**, 177206 (2004)
70. Herng, T.S., Lau, S.P., Yu, S.F., et al.: Magnetic anisotropy in the ferromagnetic Cu-doped ZnO nanoneedles. Appl. Phys. Lett. **90**, 032509 (2007)
71. Diaconu, M., Schmidt, H., Hochmuth, H., et al.: Room-temperature ferromagnetic Mn-alloyed ZnO films obtained by pulsed laser deposition. JMMM **307**, 212–221 (2006)
72. Alaria, J., Bieber, H., Colis, S., et al.: Absence of ferromagnetism in Al-doped $Zn_{0.9}Co_{0.1}O$ diluted magnetic semiconductors. Appl. Phys. Lett. **88**, 112503 (2006)
73. Lawes, G., Risbud, A.S., Ramirez, A.P., Seshadri, R.: Absence of ferromagnetism in Co and Mn substituted polycrystalline ZnO. Phys. Rev. B **71**, 045201 (2005)
74. Potzger, K., Zhou, S.: Non-DMS related ferromagnetism in transition metal doped zinc oxide. Phys. Stat. Solidi **246**, 1147–1167 (2009)
75. Lotin, A.A., Novodvorsky, O.A., Rylkov, V.V., et al.: Properties of $Zn_{1-x}Co_xO$ films produced by pulsed laser deposition with fast particle separation. Semiconductors **48**, 538–544 (2014)
76. Han, S.-J., Jang, T.-H., Kim, Y.B., et al.: Magnetism in Mn-doped ZnO bulk samples prepared by solid state reaction. Appl. Phys. Lett. **83**, 920–922 (2003)
77. Kaspar, T.C., Droubay, T., Heald, S.M., et al.: Hidden ferromagnetic secondary phases in cobalt-doped ZnO epitaxial thin films. Phys. Rev. B **77**, 201303(R) (2008)
78. Mesquita, A., Rhodes, F.P., da Silva, R.T., et al.: Dynamics of the incorporation of Co into the wurtzite ZnO matrix and its magnetic properties. J. All. Comp. **637**, 407–417 (2015)
79. Straumal, B.B., Protasova, S.G., Mazilkin, A.A., et al.: Ferromagnetic properties of the Mn-doped nanograined ZnO films. J. Appl. Phys. **108**, 073923 (2010)
80. Straumal, B.B., Mazilkin, A.A., Protasova, S.G., et al.: Grain boundaries as the controlling factor for the ferromagnetic behaviour of Co-doped ZnO. Philos. Mag. **93**, 1371–1383 (2013)
81. Coey, J.M.D.: d^0 ferromagnetism. Solid State Sci. **7**, 660–667 (2005)
82. Tietze, T., Gacic, M., Schütz, G., et al.: XMCD studies on Co and Li doped ZnO magnetic semiconductors. New J. Phys. **10**, 055009 (2008)
83. Straumal, B.B., Protasova, S.G., Mazilkin, A.A., et al.: Ferromagnetic behaviour of Fe-doped ZnO nanograined films. Beilstein J. Nanotechnol. **4**, 361–369 (2013)
84. Sundaresan, A., Bhargavi, R., Rangarajan, N., et al.: Ferromagnetism as a universal feature of nanoparticles of the otherwise nonmagnetic oxides. Phys. Rev. B **74**, 161306(R) (2006)
85. Burova, L.I., Samoilenkov, S.V., Fonin, M., et al.: Room temperature ferromagnetic (Zn, Co)O epitaxial films obtained by low-temperature MOCVD process. Thin Solid Films **515**, 8490–8494 (2007)
86. Burova, L.I., Perov, N.S., Semisalova, A.S., et al.: Effect of the nanostructure on room temperature ferromagnetism and resistivity of undoped ZnO thin films grown by chemical vapor deposition. Thin Solid Films **520**, 4580–4585 (2012)
87. Kulbachinskii, V.A., Kytin, V.G., Reukova, O.V., et al.: Electron transport and low-temperature electrical and galvanomagnetic properties of zinc oxide and indium oxide films. Low. Temp. Phys. **41**, 116–124 (2015)
88. Gu, H., Zhang, W., Xu, Y., Yan, M.: Effect of oxygen deficiency on room temperature ferromagnetism in Co doped ZnO. Appl. Phys. Lett. **100**, 202401 (2012)
89. Li, G., Wang, H., Wang, Q., Zhao, Q., et al.: Structure and properties of Co-doped ZnO films prepared by thermal oxidization under a high magnetic field. Nanoscale Res. Lett. **10**, 112 (2015)

90. Wan, W., Huang, J., Zhu, L., et al.: Defects induced ferromagnetism in ZnO nanowire arrays doped with copper. CrystEngComm **15**, 7887–7894 (2013)
91. Chang, L.-T., Want, C.-Y., Tang, J., et al.: Electric-field control of ferromagnetism in Mn-doped ZnO nanowires. Nano Lett. **14**, 1823–1829 (2014)
92. Singh, S.B., Wang, Y.-F., Shao, Y.-C., et al.: Observation of the origin of d^0 magnetism in ZnO nanostructures using X-ray-based microscopic and spectroscopic techniques. Nanoscale **6**, 9166–9176 (2014)
93. Pal, B., Dhara, S., Giri, P.K., Sarkar, D.: Room temperature ferromagnetism with high magnetic moment and optical properties of Co doped ZnO nanorods synthesized by a solvothermal route. J. All. Comp. **615**, 378–385 (2014)
94. Luo, X., Lee, W.-T., Xing, G., et al.: Ferromagnetic ordering in Mn-doped ZnO nanoparticles. Nanoscale Res. Lett. **9**, 625 (2014)
95. Guo, T., Yao, M.-S., Lin, Y.-H., Nan, C.-W.: A comprehensive review on synthesis methods for transition-metal oxide nanostructures. CrystEngComm **17**, 3551–3585 (2015)
96. Wang, J., Hou, S., Chen, H., Xiang, L.: Defects-induced room temperature ferromagnetism in ZnO nanorods grown from ε-$Zn(OH)_2$. J. Phys. Chem. C **118**, 19469–19476 (2014)

Chapter 6
Soft Magnetic Wires for Sensor Applications

Valentina Zhukova

6.1 Introduction

First amorphous materials using rapid quenching from the liquid state were prepared nearly 50 years ago [1–4]. Development of the rapid-quenching technique allowed obtaining of new materials with metastable crystalline, amorphous, nanocrystalline, granular structures with a new combination of physical properties (mechanical, magnetic, electrochemical, etc.) and opening of new fields of research in material science, magnetism, and technology. During the next years, few rapid-quenching technologies allowing preparation of different types of rapidly quenched materials have been developed. At the beginning most attention has been paid to studies of planar rapidly quenched materials: rapidly quenched ribbons produced by quenching on the drum [4–6].

Excellent magnetic softness of amorphous materials obtained by the melt-spinning technique has attracted considerable attention, making them very attractive in potential applications in recording head and microtransformer industries. Such magnetic softness originates from the absence of magnetocrystalline anisotropy in these alloys [6].

Further development of the rapidly quenching fabrication techniques allowed preparation of rapidly materials with cylindrical symmetry: rapidly quenched wires [7–10]. Amorphous wires, typically around 125 μm in diameter, obtained by the so-called in-rotating-water quenching technique, have been firstly introduced in 1980.

The magnetostrictive compositions exhibit rectangular hysteresis loop, while the best magnetic softness is observed for the nearly zero magnetostriction composition.

V. Zhukova (✉)
Department of Materials Physics, Chemistry Faculty, Basque Country University, 20018 San Sebastián, Spain

Department of Applied Physics, EUPDS, UPV/EHU, 20018 San Sebastian, Spain
e-mail: valentina.zhukova@ehu.es

A. Zhukov (ed.), *Novel Functional Magnetic Materials*, Springer Series in Materials Science 231, DOI 10.1007/978-3-319-26106-5_6

Initially most attention of researchers has been paid to the spontaneous magnetic bistability (related to large and single Barkhausen jump exhibited by amorphous materials with rectangular hysteresis loops) of amorphous wires, although similar effect has been previously reported for crystalline wires (Wiegand wires, iron whiskers) [11, 12]. In fact magnetic bistability was previously reported also for annealed amorphous ribbons [13, 14], but this phenomena can be easily obtained in amorphous wires in as-prepared state without any special heat treatment [7–9].

A number of various magnetic sensors utilizing magnetically bistable behavior of amorphous wires have been proposed [9, 15].

Fast magnetization switching of amorphous wires through the large and single Barkhausen jump is in fact one of the most interesting properties of amorphous wires. The spontaneous magnetic bistability of amorphous wires related to perfectly rectangular hysteresis loop has been attributed to the nucleation or depinning of the reversed domains inside the inner single domain and the consequent propagation of head-to-head remagnetization front [7, 16].

Since 1994 the other feature of amorphous wires—giant magnetoimpedance (GMI) effect—renewed the interest in studies of amorphous wires [17, 18]. Studies of GMI effect attracted considerable attention during last two decades since. It is worth mentioning that in fact GMI has been first reported in 1936 by E. P. Harrison et al. [19]. But intensive research of GMI effect in various materials began only after the publication in 1994 of few papers on GMI effect in amorphous wires [17, 18, 20, 21].

Phenomenologically GMI effect consists of large change of the impedance of magnetically soft conductor on applied magnetic field [17–21].

GMI effect is particularly interesting for magnetic sensor applications basically due to extremely high magnetic field sensitivity suitable for low magnetic field detection [17, 18].

GMI effect has been successfully explained in the terms of classical electrodynamics through the influence of magnetic field on penetration depth of electrical current flowing through the magnetically soft conductor [17, 18]. Cylindrical shape and high circumferential permeability observed in Co-rich amorphous wires with vanishing magnetostriction constant are quite favorable for achievement of high GMI effect [17, 18].

Additionally special attention has been paid, in the last two decades, to the study of nanocrystalline phases obtained by suitable annealing of amorphous metallic ribbons owing to their attractive properties as soft magnetic materials [22–25]. Such soft magnetic character is thought to be originated because the magnetocrystalline anisotropy vanishes and the very small magnetostriction value when the grain size approaches 10 nm [22–24]. As was theoretically estimated by Herzer [24, 25], average anisotropy for randomly oriented α-Fe (Si) grains is negligibly small when grain diameter does not exceed about 10 nm. Thus, the resulting magnetic behavior can be well described with the random anisotropy model [24, 25].

Last advances in amorphous magnetic materials are based on the miniaturization of modern magnetic sensors. Consequently the alternative rapid-quenching technologies, like melt extraction [20, 26] or Taylor–Ulitovsky method [10, 21], for producing thinner metallic wires have been developed. Typical metallic wire diameters are of the order of 30–50 and 1–30 μm, respectively [10, 20, 21, 26].

In the case of glass-coated microwires, the glass coating introduces an additional magnetoelastic contribution to the magnetic anisotropy acting as a new parameter determining the magnetization process. Consequently additional efforts for minimization of the additional magnetoelastic anisotropy arising from the glass-coating are needed [21].

The aim of this chapter is to overview different families of soft magnetic wires exhibiting soft magnetic properties and GMI effect, paying attention on advantages and disadvantages of each family of magnetic wires.

6.2 Different Families of Magnetic Wires

As mentioned above starting from 1990, a novel family of rapidly quenched materials—rapidly quenched wires presenting amorphous or nanocrystalline structure—have been introduced [21–26]. First generation of wires deals with typical diameter typically around 125 μm in diameter, obtained by the so-called in-rotating-water quenching technique.

Demand in thinner rapidly quenched wires resulted in growing interest in alternative methods allowing preparation of thinner wires such as melt-extracted wires and glass-coated microwires.

All the fabrication processes of rapidly quenched wires have common features: all of them involve melting of the ingot of desirable chemical composition and consequent fast quenching from the melt. But each of the methods has some peculiarities that affect magnetic properties, morphology, and structure of each family of rapidly quenched wires. Below we'll briefly review the most known fabrication processes, allowing the preparation of rapidly quenched wires.

6.2.1 Magnetic Wires Produced by "In-Rotating-Water" Technique

The preparation of perfectly cylindrical rapidly quenched wires is reported at the beginning of 1980, although first studies were focused on mechanical properties and structure of the wires prepared by rapid quenching [27, 28]. Unusual magnetic properties like magnetic bistability and related fast domain-wall propagation and later GMI effect attracted great interest starting from the end of 1980 [7, 17, 18, 29]. Detailed description of the fabrication technique usually denominated as "in-rotating-water" can be find elsewhere [27–29]. Fist magnetic ($Fe_{100-x-y}Si_xB_y$) wire preparation is reported in 1981, although previously essentially the same method has been employed for the preparation of Pd–Cu–Si amorphous wires having a circular cross section (0.05–0.2 mm in diameter) [30].

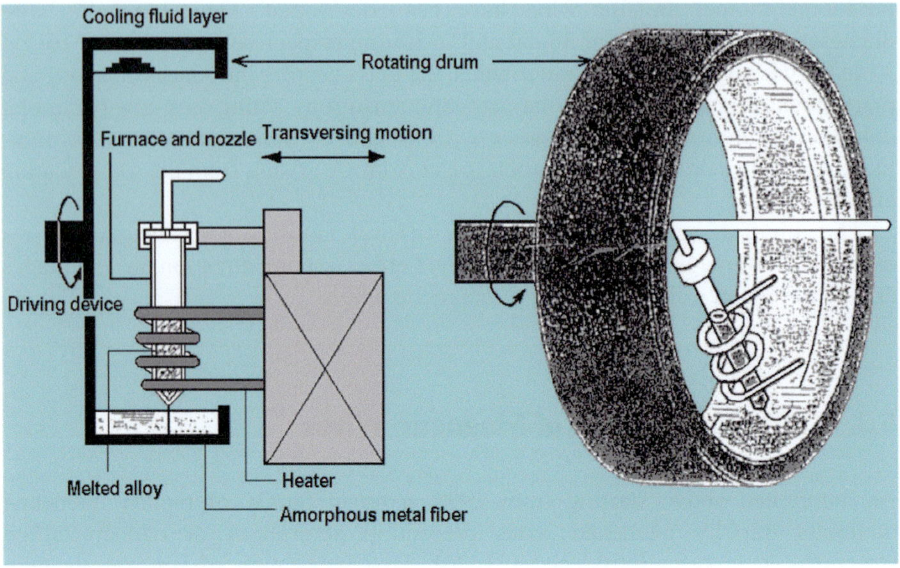

Fig. 6.1 Schematic picture of "in-rotating-water" rapid-quenching process

The fabrication process consists of the preparation of the master alloy and loading it into a quartz nozzle and subsequent induction melting. The temperature was usually monitored with an optical pyrometer. A jet of molten metal was ejected under pressure through the orifice of quartz nozzle into the rotating water layer (see scheme in Fig. 6.1). The orifice size of the quartz nozzles varied between 0.06 and 0.34 mm in diameter, allowing the preparation of amorphous wires with various diameters. The ejection pressure was in the range of 0.3–0.8 MPa. Adjusting the melt superheats, ejection angles, and drum rotation speeds allows production of continuous amorphous alloy wires with good shape and homogeneity. Continuous amorphous alloy wires over 100 m long with diameters from 60 up to 320 μm were prepared. One of the most important factors in this spinning process is to control the ratio of the jet velocity (V_j) to the water velocity (V_w).

Using this method, continuous amorphous wires with various diameters between 0.07 and 0.27 mm were prepared from the master alloys. Typically, the amount of alloy melted in one run was about 1 g, the rotation speed of the drum (50 cm in diameter).

6.2.2 *Melt-Extracted Microwires*

Melt-extracted wires can be obtained by melt-extraction technique initially developed by Maringer and Mobley [31] and later modified by Rudkowski et al. [32–34]. This technique typically takes the material in the shape of a rod a few mm in

Fig. 6.2 Experimental setup for fabrication of the wires using melt-extraction technique. Reprinted from [26] doi:10.1016/S0924-4247(03)00172-9. Copyright © 2003 Elsevier B.V. (Fig. 1)

diameter, melts the tip by a clean heat source such as RH induction or CW infrared, and extracts the wire from the melt by means of a rapidly rotating sharpened wheel (made from a refractory metal such as molybdenum) moving at tangential speeds between 10 and 50 m/s [31–34]. The process allows producing wires typically 10 m in length and 30–60 μm in diameters.

Later the process has been modified. The modified setup for the fabrication of wires by melt extraction was assembled with the quenching block positioning in the interior of a vacuum chamber. The main components of the quenching block are (1) an inductive coil with the quartz ampoule inside having a round orifice and served as the crucible for the alloy melting, (2) quenching disk made out of a refractory metal and positioned on the shaft of the micro-motor, and (3) a mechanism for the vertical displacement of the quartz ampoule. The linear velocity of the disk edge was about 30 m/s. An overview of the setup is shown in Fig. 6.2.

The ingot of about 5 g was placed in the quartz ampoule. After melting of the ingot and forming a liquid drop at the ampoule orifice, the ampoule was going down by means of the displacement mechanism up to contact with the rotating disk edge. As the contact was achieved, a thin continuous wire with the diameter of 30–60 μm has been extracted from the liquid drop [26].

6.2.3 Glass-Coated Microwires

The fabrication method denominated in most of modern publications as a modified Taylor–Ulitovsky and/or quenching-and-drawing method [21] is actually well known since 1960 [35–42]. Initially this method has been used mostly for fabrication of nonmagnetic microwires (e.g., Cu), but the casting method is principally the

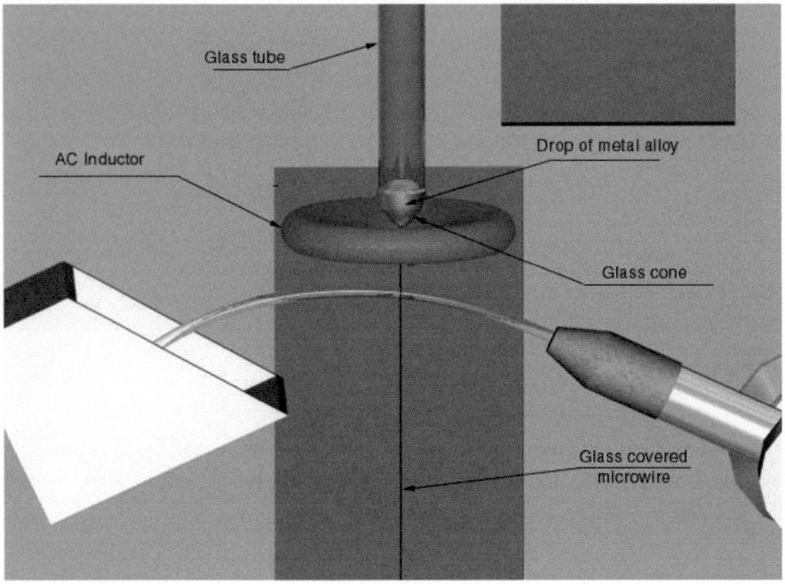

Fig. 6.3 Schematic drawing of microwire fabrication process by the Taylor–Ulitovsky method. Reprinted from [43] with permission of Elsevier (Fig. 1)

same. This method is based essentially on direct casting from the melt, as schematically depicted in Fig. 6.3.

The method basically consists of simultaneous drawing of the composite microwire (metallic nucleus inside the glass capillary) through the quenching liquid (water or oil) jet onto rotating bobbins [43].

In the laboratory process, an ingot containing few grams of the master alloy with the desired composition is put into a Pyrex-like glass tube and placed within a high-frequency inductor heater. The alloy is heated up to its melting point, forming a droplet. While the metal melts, the portion of the glass tube adjacent to the melting metal softens, enveloping the metal droplet. A glass capillary is then drawn from the softened glass portion and wound on a rotating coil. At suitable drawing conditions, the molten metal fills the glass capillary, and a microwire is thus formed where the metal core is completely coated by a glass shell. The amount of glass used in the process is balanced by the continuous feeding of the glass tube through the inductor zone, whereas the formation of the metallic core is restricted by the initial quantity of the master alloy droplet. The microstructure of a microwire (and hence, its properties) depends mainly on the cooling rate, which can be controlled by a cooling mechanism when the metal-filled capillary enters into a stream of cooling liquid (water or oil) on its way to the receiving coil.

The main advantages of this method of microwire fabrication are [21, 35, 36]:

1. Repeatability of microwire properties at mass production
2. Extended range of variation in parameters (geometrical and physical)

3. Fabrication of continuous long pieces of microwire up to 10,000 m
4. Control and adjustment of geometrical parameters (inner core diameter and glass thickness) during the fabrication process

In spite of many advantages, this method meets some complexities related with peculiarities of metallurgical processes in rapid quenching of the composite material and effect of electromagnetic field of inductor on alloy ingot and stability of the process. All these peculiarities affect magnetic and structural properties of cast microwires.

Some of these peculiarities listed below were detailed revised in recent book [36].

6.2.3.1 Chemical and Metallurgical Processes Related with Interaction of the Ingot Alloy and the Glass

The character of this interaction depends on chemical composition of the ingot as well as on the type of glass used for the casting [36]. Previously these processes have been studied mostly for Cu and Ag microwires. Only quite recently few reports on the interaction of the ferromagnetic alloy ingot and the glass have been reported [44–46].

In particularly, the evidence of the interfacial layer between the metallic nucleus and glass coating has been reported and studied for the case of Fe- and Co-based magnetic alloys.

Taking into account the aforementioned regarding the interfacial layer, glass-coated microwire consists of metallic nucleus, glass insulating shell, and interfacial layer which should be in an ideal case as more homogeneous as possible. It is worth mentioning that the character of the interfacial layer as well as the properties of microwires depends on properties and character of interaction between selected alloy and selected glass as well as on geometry (metallic nucleus diameter, *d*, and glass-coating thickness). The other source of instability of properties of cast microwire is related with gas content inside the microwire nucleus and glass coating. The sources of the gas are the atmosphere, the gas impurities in the alloy, and the glass.

6.2.3.2 Electromagnetic and Electro-Hydrodynamic Phenomena in the System of Inductor Ingot

The main peculiarity of the Taylor–Ulitovsky method is melting of the ingot placed inside the glass tube by the HF currents from the inductor with consequent formation of the capillary from the softened glass and filled by the molten alloy. This electrodynamic interaction determining the shape of the molten ingot and temperature regimes significantly affects the casting process and should be also taken into account. It is worth mentioning that the most critical point during the casting of microwires is the bottom part of the molten ingot. At special conditions determined by the shape of the HF inductor, electromagnetic field inside the molten ingot, weight of the ingot, viscosity, etc., the equilibrium shape of the ingot can be achieved. At these conditions the casting process is homogeneous.

Achievement of these conditions depends on many parameters, such as releasing power to the ingot, shape of the inductor, frequency of the electromagnetic irradiation created by the inductor, conductivity of the alloy, and the weight of the metal. Determination of these conditions allows establishing reliable casting process.

It should be mentioned again that these aspects have been well studied for the case of nonmagnetic alloys and metals, but for the novel family of magnetic materials based on Fe, Co, and Ni, these problems should be addressed.

6.2.3.3 Thermal Conditions of Formation of Cast Microwire

As the point of view of properties of cast microwire, the thermal conditions play an important role. It should be mentioned that there are two main problems related with thermal conditions: thermal phenomena inside the molten ingot under the effect of electromagnetic field of the inductor and thermal conditions related with solidification of cast microwire.

Main efforts related with the first point have been performed on the determination of the temperature distribution inside the ingot and the effect of the metallic bar dimensions on the temperature distribution inside the metallic ingot. It was found in the case of Cu that increasing the diameter of the bar from 2 till 6 mm, the temperature inside the ingot decreases up to 10–35 % [36].

The conditions of the cooling of cast microwire determine the crystalline structure of produced microwire.

6.2.3.4 Parameters of the Casting Process and Their Limits Are the Casting Rate, Diameter of a Microwire, Composition of the Metallic Nucleus, and Glass Coating

Typical limits for the metallic nucleus diameters are between 0.8 and 30 μm, while the thickness of the coating is in the range of 2 and 15 μm.

The geometrical characteristics of the microwire depend on the physical properties of the glass and metal, the diameter of the initial glass tube, and the parameters of the heating inductor. It has been proven that the strongest effect on the geometry of a microwire comes from the glass properties.

The present method allows pure metals such as Ag, Au, Cu, Ni, Sn, Pb, and Pt to be used. Semiconductors such as Ge, Si, or Bi can also be casted. The alloys of these metals can be also processed, but not over the whole range of compositions. For example, the content of Cr in Ni- and Co-based alloys can be up to 20 %; the content of Mo, W, and V in Ni-, Co-, and Fe-based alloys can be up to 8–10 %. Some limitations relate to the melting temperature, which must not be too high.

On the other hand, rare earth metals and metals which react with the glass and atmosphere (Ti, Cr, Mo, W, Nb, Al, Na, La, Nd) are not suitable for this technology.

Of principle interest is the fabrication of magnetic microwires with soft magnetic characteristics for which Fe, Ni, and Co are the alloy's main elements. In the case of amorphous microwire, elements such as Si, B, C, and Al are added to enable the amorphicity of the alloys.

Metastable supersaturated solid solutions can be obtained for immiscible metal systems in a solid and even liquid state, such as Cu–Co, Cu–Co–Ni, Ag–Fe, and Ag–Co. As-cast microwire can exhibit a structure of a supersaturated solid solution of transition metals in Cu or Ag. After suitable thermal treatments, small single-domain particles of Co or Fe can precipitate. Recently giant magnetoresistance (GMR) effect attributed to the scattering of the electrons on grain boundaries between ferromagnetic grains and paramagnetic matrix has been observed in such compositions [36, 47].

Mixed crystalline–amorphous structure can be also obtained by thermal annealing of initially amorphous microwires [48, 49] as well as in-prepared state using specially designed mixed compositions like (FeSiB)–Cu and (CoFeSiB)–Cu. These compositions exhibit unusual magnetic properties with irregular hysteresis loops. Additionally, recently Heusler-type compositions with near room Curie temperature have been successfully produced using this method [47, 50].

The properties of microwires are related to a great extent to the microstructure of the metal core. The analyzed method of casting offers the possibility of rapid cooling and solidification of liquid metal in a nonequilibrium process, allowing the formation and existence of metastable metallic phases. The microstructure of rapidly cooled metal depends on its chemical composition, the melting temperature of metal, and the cooling rate. Depending on the critical cooling rates, different metastable phases can be formed, i.e., supersaturated solid solution and microcrystalline and amorphous phases. In certain cases, a mixed microstructure consisting of nano- or microcrystals embedded in the amorphous matrix can be formed as well.

6.3 Magnetic Properties Relevant for Applications

As mentioned above, first generation of amorphous wires deals with typical diameter around 125 μm in diameter (the same technique allows fabrication of wires with diameters from 60 up to 200 μm), obtained by the so-called in-rotating-water quenching technique. Melt extraction technique allows the fabrication of thinner (roughly 30–60 μm in diameter) microwires. Additionally the diameter of wires produced by in-rotating-water technique can be reduced using either cold drawing or warm drawing [51–55]. The thinnest cast microwires can be produced using Taylor–Ulitovsky technique [21, 26, 56]. Thus, recently the preparation of microwires with submicrometric (less than 1 μm) diameters of metallic nucleus has been recently reported [56].

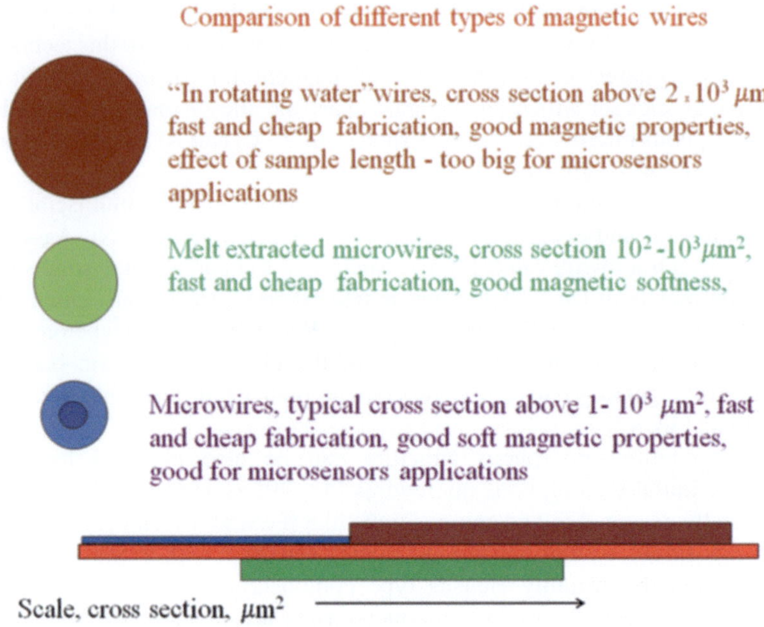

Fig. 6.4 Scheme illustrating comparison of different types of rapidly quenching wires

Comparison of the dimensions and properties of different families of rapidly quenched wires is presented in Fig. 6.4.

Rapidly quenched wires exhibit a number of unusual magnetic properties. Magnetostrictive wires prepared using "in-rotating-water" technique and Fe-rich glass-coated microwires exhibit rectangular hysteresis loop, while best magnetic softness is usually observed for the nearly zero magnetostriction compositions in all kinds of wires.

Two main features that attract main technological interest are related to the magnetic bistability observed in magnetostrictive compositions and magnetic softness and GMI effect in nearly zero magnetostriction compositions [36]. Below we'll overview magnetic properties promising from the viewpoint of applications.

6.3.1 Magnetic Bistability and Domain-Wall Propagation

One of the main technological interests for utilization of amorphous microwires is related with large and single Barkhausen jump (LBJ) observed above some value of applied magnetic field, called switching field between two stable remanent states. Therefore, this phenomenon is usually also referred to as magnetic bistability. Hysteresis loop of magnetically bistable samples presents perfectly rectangular shape (see Fig. 6.5). Such particular magnetization process has been observed

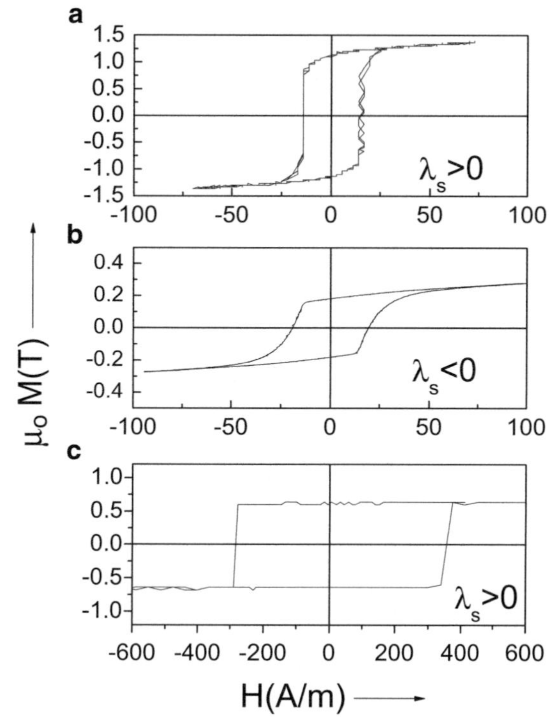

Fig. 6.5 Hysteresis loops of Fe-rich ($\lambda_s > 0$) (**a**) amorphous wires, Co-rich ($\lambda_s < 0$) (**b**) amorphous wires (diameters, $d = 125$ μm), and Fe-rich ($\lambda_s > 0$) $Fe_{70}B_{15}Si_{10}C_5$ (**c**) amorphous microwire $d = 6$ μm. Reprinted with permission from [57] Copyright (2012) SpringerOpen (Fig. 1c)

previously also in iron whiskers [12], amorphous Co-rich ribbons after special heat treatment [13, 14, 58], and different kinds of amorphous wires [7, 29]. Abrupt magnetization jump and sharp electrical pulses related with the magnetization switching during LBJ have been proposed in different kinds of magnetic sensors, such as magnetic markers and magnetoelastic sensors [7–9].

It is worth mentioning that the appearance of large and single Barkhausen jump takes place under magnetic field above some critical value (denominated as switching field) and also if the sample length is above some critical value denominated also as critical length. The switching field, H_s, that is, the magnetic field required to reverse magnetization is proportional to the energy required to form the domain wall γ involved in the bistable process. The wall energy is related to the magnetoelastic anisotropy and, therefore, to the applied tensile stress as given by [57, 59, 60]

$$H_s \approx \gamma \propto \left[3A\, \lambda_S(\sigma_a + \sigma_r)/2\right]^{1/2} / \cos\alpha \tag{6.1}$$

where α is the angle between magnetization and axial direction, A is the exchange energy constant, λ_S is the saturation magnetostriction constant, and σ_a and σ_r are the applied and residual tensile stress. Consequently, H_s must be proportional to $(\sigma_a + \sigma_r)^{1/2}$ for cos $\alpha \approx 1$.

Regarding the critical length, detailed studies of the ferromagnetic wire diameter on magnetization profile and size of the edge closure domains have been performed in [59]. Particularly, critical length, l_c, for magnetic bistability in conventional Fe-rich samples (120 μm in diameter) is about 7 cm. This critical length depends on saturation magnetization, magnetoelastic energy, domain structure, and magnetostatic energy [59, 61]. Thus, in Co-rich conventional amorphous wires (120 μm in diameter), such critical length is about 4 cm [61, 62]. Below such critical length hysteresis loop loses its squared shape.

The magnetostatic energy depends on the demagnetizing field, H_d, expressed as

$$H_d = NM_s \tag{6.2}$$

where N is the demagnetizing factor given for the case of long cylinder with length, l, and diameter, D, as

$$N = 4\pi[\ln(2l/D) - 1](D/l)^2. \tag{6.3}$$

On the other hand, the shape of the hysteresis loop of wires with a given length depends on the applied stress. As an example, Fig. 6.6 shows the effect of the applied stress on the hysteresis loop of the 8 cm long sample measured by the long secondary coil. It is worth to mention that the initially rectangular shape of the hysteresis loop is lost under the certain applied stress (around 70 MPa) but finally it recovers under stronger applied stress of around 320 MPa.

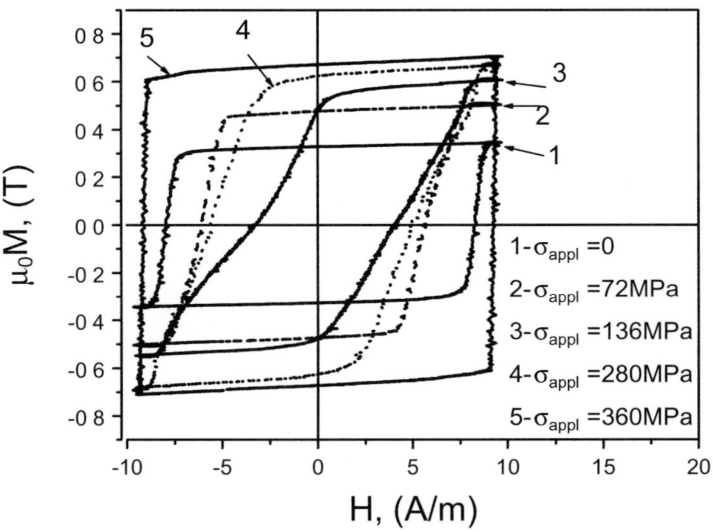

Fig. 6.6 Stress dependence of the hysteresis loops of 8 cm long $Fe_{77.5}B_{15}Si_{7.5}$ amorphous wire with 125 μm diameter, measured by a long pickup coil. Reprinted with permission from Ref. [63] Copyright (2003) AIP Publishing LLC (Fig. 1)

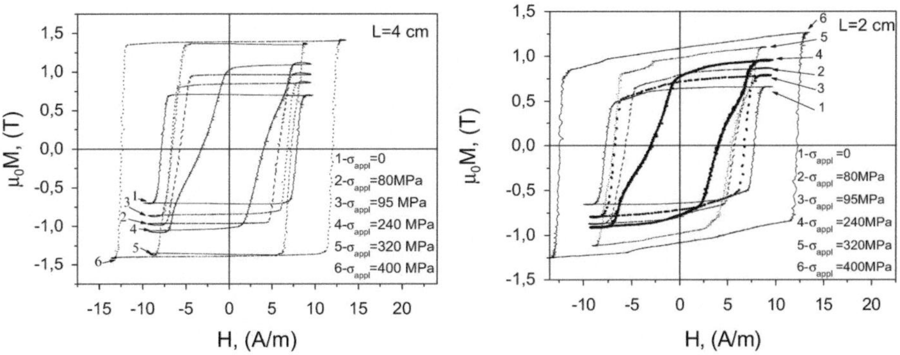

Fig. 6.7 Effect of stress applied on hysteresis loops of 8 cm long $Fe_{77.5}B_{15}Si_{7.5}$ amorphous wire, measured by a short movable coil placed at $L = 4$ cm (**a**) and 2 cm (**b**). Reprinted with permission from Ref. [63] Copyright (2003) AIP Publishing LLC (Fig. 2)

Similar experiments have been done by using short movable pickup coil which permitted to measure local hysteresis loops as well as to obtain the magnetization profile and its dependence on applied stress. Such dependence measured in the center of the 8 cm long wire, i.e., at the position of the movable pickup coil of $L = 4$ cm and at $L = 2$ cm, is presented in Fig. 6.7a, b, respectively. At $L = 4$ cm the hysteresis loop measured for $\sigma_{appl} < 60$ MPa and $\sigma_{appl} > 320$ MPa exhibits squared shape, while at $L = 2$ cm all hysteresis loops are not perfectly rectangular.

The remanence, $\mu_0 M_r$, profile has been acquired from the measurements of local hysteresis loops. Such profiles measured for the 8 cm long wire are shown in Fig. 6.8. The remanence profile has roughly symmetric character. The decrease of the remanent magnetization close to the wire ends (variation of $\mu_0 M_r$ with the position of the coil, L) as well as the variation of the penetration depth, l_d, of the closure domains with σ_{appl} is observed in Fig. 6.8. As can be assumed from Fig. 6.8, the penetration depth, l_d, first increases and then decreases with σ_{appl}. Below certain sample length, $L \approx l_c \approx 2\ l_d$, magnetic bistability disappears because the closure domains penetrate far enough inside the inner axially magnetized core and destroy it.

Consequently the critical length first increases with σ_{appl} but then again decreases, and finally magnetic bistability recovers under strong enough applied tensile stress.

General increase of $\mu_0 M_r$ with σ_{appl} should be attributed to the increasing volume of axially magnetized area under the effect of applied stresses due to the positive magnetostriction constant. On the other hand, in the range of $60\ \text{MPa} \leq \sigma_{appl} \leq 300$ MPa, magnetic bistability disappears, and the local hysteresis loops do not show rectangular character even in the central region of 8 cm long wire (see Fig. 6.6). Such behavior is related with the demagnetizing factor of the internal axially magnetized core: the diameter of such single-domain core increases which results in increasing of the demagnetizing factor given by Eq. (6.3). This affects the penetration depth of the closure domains, l_d, and in this way the magnetic bistability is destroyed even in the central zone of the wire.

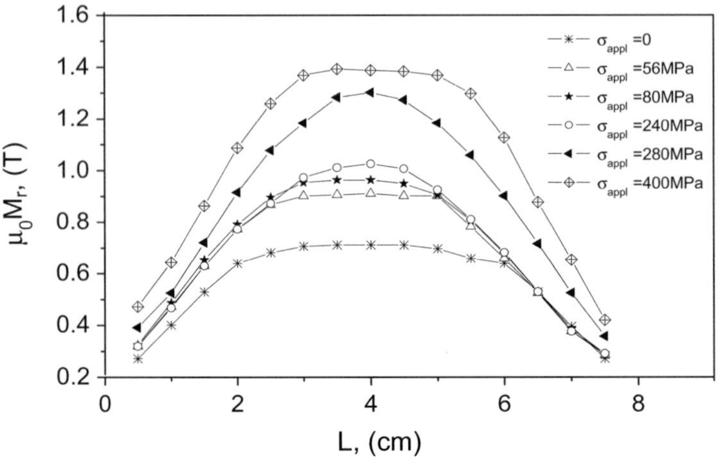

Fig. 6.8 Effect of stress on magnetization profiles measured in $Fe_{77.5}B_{15}Si_{7.5}$ amorphous wire measured for sample length $l = 8$ cm. Reprinted with permission from Ref. [63] Copyright (2003) AIP Publishing LLC (Fig. 5)

Further increase of applied stresses results in further increasing of the remanent magnetization and also in drastic increasing of the coercivity and recovering of magnetic bistability. We assume that the increase of the external tensile stress results in the increase of the domain-wall energy:

$$\gamma = 2(AK)^{1/2}, \tag{6.4}$$

where A is the exchange energy and K is the magnetic anisotropy constant, which in the case of amorphous state depends mainly on the magnetoelastic component, K_{me}, given by

$$K_{\text{me}} \approx 3/2\lambda_s\sigma, \tag{6.5}$$

where $\sigma = \sigma_i + \sigma_a$ is the total stress, σ_i is the internal stresses, σ_a is the applied stresses, and λ_s is the magnetostriction constant.

By applying external stress, the magnetoelastic energy and consequently the domain-wall energy, γ, increases. At such conditions there is a competition between magnetostatic and magnetoelastic energies. Such increasing of the magnetoelastic energy gives rise to the decreasing of the size (and consequently of the penetration depth) of closure domains and therefore results in new decreasing of the critical length and recovering of the magnetic bistability.

Phenomenon of magnetic bistability observed in different families of amorphous wires is certainly quite interesting for the magnetic sensor applications. But enhanced critical length, l_c (of the order of few cm), observed in conventional amorphous wires is limiting these possibilities. Consequently the wire diameter

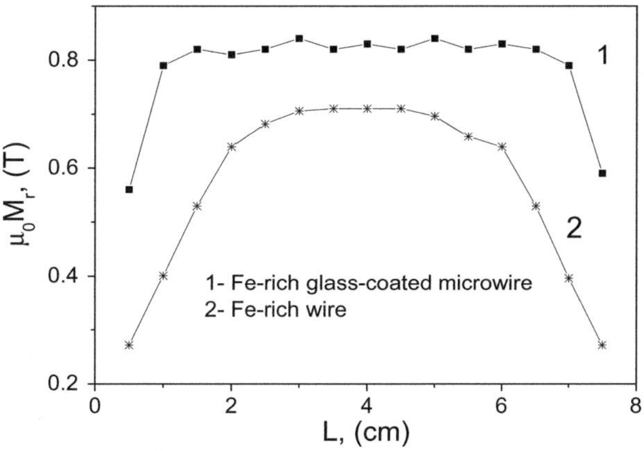

Fig. 6.9 Comparison of the magnetization profiles measured in Fe-rich wire with 125 μm diameter and Fe-rich microwire. Reprinted with permission from [59] Copyright (1995) Elsevier (Fig. 4)

reduction might be beneficial for the technological applications of spontaneous magnetic bistability typical for amorphous wires.

As mentioned above the most efficient reduction of the wire diameter is possible if we'll employ the Taylor–Ulitovsky technique for the wire preparation. This allows one order reduction of the wire diameter. Consequently in glass-coated microwires (with almost one order smaller diameters), drastic reduction of the critical length has been observed [59]. Thus, magnetic bistability for the sample length $L = 2$ mm has been observed for Fe-rich microwire with metallic nucleus diameter, d, about 10 μm [59]. A comparison of the remanent magnetization profile measured in Fe-rich microwires and conventional Fe-rich wire shows that Fe-rich mentioned profile exhibits deviations just near the sample ends (see comparison of the magnetization profiles in Fig. 6.9). In contrast the remanent magnetization profile measured in Fe-rich conventional wire exhibits considerable change with the distance up to 3 cm from each end. This difference in the remanent magnetization profiles is the reason of the different critical length of these two wires. As can be appreciated from Fig. 6.8, stresses can also significantly affect the remanent magnetization profile. Strong internal stresses (mostly of axial origin), appearing during the fabrication of glass-coated microwires, can also contribute to the reduction of the critical length for magnetic bistability.

One of the peculiarities of the Taylor–Ulitovsky technique for glass-coated microwires is that it involves the simultaneous solidification of composite microwire consisting of ferromagnetic nucleus surrounded by glass coating. Quite different thermal expansion coefficients of the glass and the metallic alloys introduce considerable internal stresses inside the ferromagnetic nucleus during simultaneous fast solidification of the composite microwire [64–67]. The strength of these internal stresses depends on ρ-ratio defined as the $\rho = d/D$, where d is the metallic nucleus diameter and D the total microwire diameter. The estimated values of the internal stresses in these glass-coated microwires arising from the difference in the thermal expansion coefficients of metallic nucleus and glass coating are of the order of 100–1000 MPa, depending strongly on the ρ-ratio [64–67], increasing with the glass-coating thickness.

Such large internal stresses give rise to a drastic change of the magnetoelastic energy, K_{me}, given by Eq. (6.5) even for small changes of the glass-coating thickness at fixed metallic core diameter. Additionally, such a change of the ρ-ratio should be related to the change of the magnetostriction constant with applied stress [57]:

$$\lambda_s = (\mu_o M_\sigma/3)(\mathrm{d}H_\kappa/\mathrm{d}\sigma), \tag{6.6}$$

where $\mu_o M_s$ is the saturation magnetization.

It is worth mentioning that residual stresses of glass-coated microwires arising during simultaneous solidification of metallic nucleus and glass coating mostly have been estimated from the simulations of the process of simultaneous solidification of metallic nucleus inside the glass tube [64–67], and experimental determination of such residual stresses is rather complex. One of the experimental evidence of the existence of such stresses is the dependence of hysteresis loops and particularly magnetic properties (coercivity, remanent magnetization) on ρ-ratio [57, 59] as well as the applied stress dependence of the hysteresis loops [60, 68] and effect of chemical etching of the glass on hysteresis loops [69, 70]. The effect of ρ-ratio on hysteresis loop of Fe-rich microwires is shown in Fig. 6.10. As can be appreciated, considerable increasing of coercivity with decreasing of ρ-ratio is observed.

Direct confirmation of the magnetoelastic origin of observed changes of the hysteresis loops is the dependence of the measured effect of the switching field, H_s, on applied stresses, σ_a. Switching field, H_s, is defined as the field at which the large Barkhausen jump starts. At low magnetic field amplitudes and frequencies, the H_s is almost as the same as coercivity, H_c. But the switching field, H_s, is usually almost independent on magnetic field amplitude, H_0 and frequency, f [71]. The difference between H_s and H_c was attributed to the DW dynamics, i.e., the time of domain-wall propagation through the entire wire [71]. Therefore, the use of H_s for studying stress dependence of hysteresis loops is more rigorous.

We observed a considerable and monotonic growth of H_s with σ_a (Fig. 6.11). Consequently from the comparison of Figs. 6.10 and 6.11, we can conclude that the internal stresses in glass-coated microwires have mostly tensile character (as also shown in previous papers [64–67]) and the strength of internal stresses increases with ρ-ratio decreasing.

Additionally, after chemical etching of the glass coating, we observed gradual changes of the hysteresis loop from inclined to almost rectangular in Co-rich microwires (Fig. 6.12). Both the last figures (Figs. 6.11 and 6.12) confirm the influence of nonmagnetic glass coating on hysteresis loops of ferromagnetic nucleus through the internal stresses induced by the glass coating inside the metallic nucleus.

Consequently, tailoring of the magnetoelastic energy, K_{me}, is essentially important for the optimization of magnetic properties of glass-coated microwires [59–68].

As pointed above, one of the characteristic features of the magnetic bistability is the appearance of rectangular hysteresis loop related with single and large Barkhausen jump appeared at low applied magnetic field. Such magnetic bistable

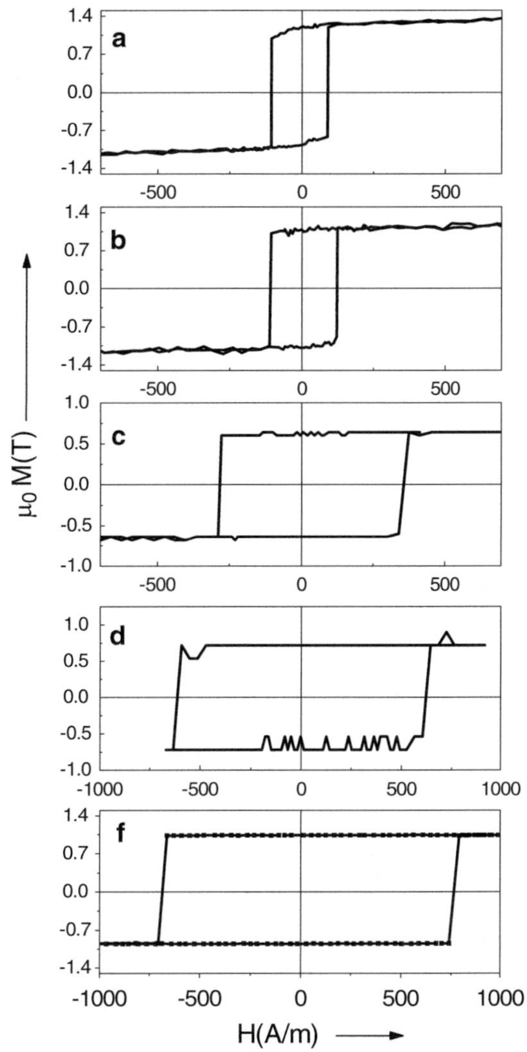

Fig. 6.10 Hysteresis loops of Fe-rich amorphous microwires with the same sample length and different metallic nucleus diameter d and total diameters D: $Fe_{70}B_{15}Si_{10}C_5$ microwires with $\rho = 0.63$; $d = 15$ μm (**a**); $\rho = 0.48$; $d = 10.8$ μm (**b**); $\rho = 0.26$; $d = 6$ μm (**c**); $\rho = 0.16$; $d = 3$ μm (**d**) and of $Fe_{72.75}Co_{2.25}B_{15}Si_{10}$ microwire with $\rho = 0.14$; $d \approx 1.4$ μm (**f**). Reprinted with permission from [57] Copyright (2012) SpringerOpen (Fig. 1)

behavior has been related to the presence of a single large Barkhausen jump, which was interpreted as the magnetization reversal in a large single-domain inner core of amorphous wire [16, 29, 57]. The rectangular hysteresis loop has been interpreted in terms of nucleation or depinning of the reversed domains inside the internal single domain and the consequent domain-wall propagation [16, 29]. Perfectly rectangular shape of the hysteresis loop has been related with a very high velocity of such domain-wall propagation. Along many years great attention has been paid to studies of domain-wall (DW) propagation in different wire families [16, 29, 57, 72–75]. Recent growing interest on DW propagation is related with proposals for prospective logic and memory devices [72, 73]. In these devices, information is encoded in

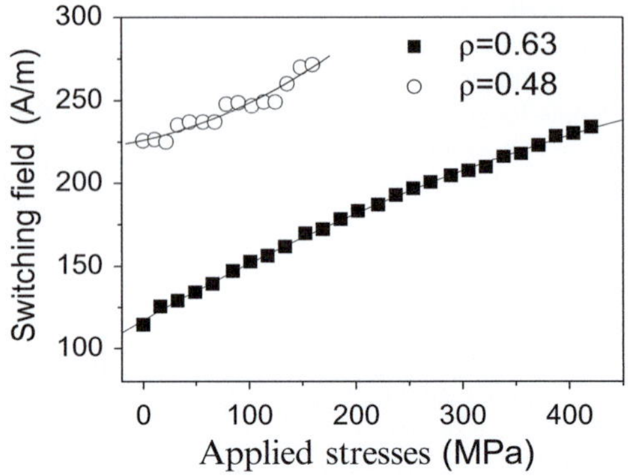

Fig. 6.11 Effect of applied stresses on switching field of $Fe_{70}B_{15}Si_{10}C_5$ amorphous microwires with the different ρ-ratio: $\rho = 0.63$, $d = 15$ μm and $\rho = 0.48$, $d = 10.8$ μm. Reprinted from [68]

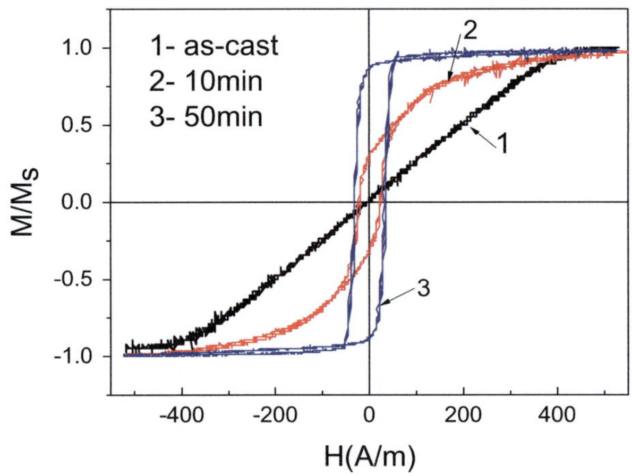

Fig. 6.12 Hysteresis loops of $Co_{70.5}Mn_{4.5}Si_{10}B_{15}$ microwire in the as-cast state and chemically treated during 10 and 50 min

the magnetic states of domains in lithographically patterned nanowires. DW motion along the wires allows for the access and manipulation of the stored information. The speed at which a DW can travel in a wire has an impact on the viability of many proposed technological applications in sensing, storage, and logic operation [75]. Special effort has been performed in nanowires to enhance the DW speed. Thus, application of transverse magnetic field proposed in [76] allowed to improve slightly the DW speed till about 600 m/s at about 120 A/m.

As mentioned above amorphous character of rapidly quenched wires produced at appropriate fabrication conditions allows improving considerably their magnetic softness. The DW velocities observed in different amorphous wires can easily achieve few hundred meters, as was first shown for conventional Fe-rich amorphous wires [29]. Moreover for conventional amorphous wires, the DW velocity increased with decreasing diameter. Consequently even higher DW velocity has been observed in Fe-rich thin glass-coated microwires [16, 57].

Thus, quite high DW velocities achieving up to 18 km/s have been reported in glass-coated microwires with few μm diameter [75]. Perfectly rectangular shape of

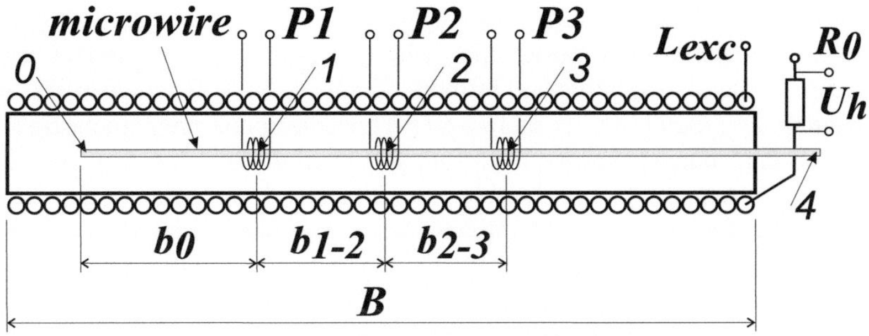

Fig. 6.13 Schematic picture of the experimental setup. Reprinted with permission from [81], Copyright [2009], AIP Publishing LLC (Fig. 1)

the hysteresis loop has been related with a very high speed of such domain-wall propagation. It is demonstrated by few methods that the remagnetization process of such magnetic microwire starts from the sample ends as a consequence of the depinning of the domain walls from closure domains and subsequent DW propagation from the closure domains [75, 77]. The magnetization process in axial direction runs through the propagation of the single head-to-head DW. It is worth mentioning that the micromagnetic origin of rapidly moving head-to-head DW in microwires is still unclear, although there are evidences that this DW is relatively thick and has complex structure [29, 78, 79].

In the case of conventional amorphous wires and microwires, the DW dynamics was measured using modified Sixtus–Tonks [80] experiments, as described recently elsewhere [16, 29, 78, 79, 81]. Usually the system consists of three coaxial pickup coils (Fig. 6.13). Additionally one end of the sample has been placed outside the solenoid in order to ensure domain-wall nucleation always near one of the microwire ends. In this way in contrary to the classical Sixtus–Tonks experiments [80], we do not need the nucleation coils to nucleate the DW, since the closure domain wall already exists. The small closure domains are created at the ends of the wire in order to decrease the stray fields [16].

Then, DW velocity in this case can be estimated as

$$v = \frac{l}{\Delta t} \tag{6.7}$$

where l is the distance between pickup coils and Δt is the time difference between the maximum in the induced emf.

In order to obtain the dependence of the DW velocity on magnetic field $v(H)$, it is necessary (1) to create a reverse domain in the certain, well-defined region of the sample and (2) apply a stable magnetic field, H, of the required value along the wire axis. The magnetic field is created by a long exciting coil L_{exc} (with length B of 140 mm, 10 mm in diameter) and tree pickup coils p_1, p_2, and p_3 (2 mm long and 1 mm inner diameter) with distances b_{1-2} and b_{2-3} between coils of 27 mm. Each pickup coil is connected to corresponding input of digital oscilloscope. More detailed description can be found in Ref. [81].

As described above, the preparation of glass-coated microwires involves simultaneous solidification of composite microwire consisting of ferromagnetic metallic nucleus inside the glass coating and introduces considerable residual stresses inside the ferromagnetic metallic nucleus [64–67]. This additional magnetoelastic anisotropy affects soft magnetic properties of glass-coated microwires (Figs. 6.10, 6.11, and 6.12). Consequently considerable attention has been paid to the effect of magnetoelastic anisotropy on DW dynamics in amorphous magnetically bistable microwires [57].

It is worth mentioning that the magnetostriction constant, λ_s, in system $(Co_xFe_{1-x})_{75}Si_{15}B_{10}$ changes with x from -5×10^{-6} at $x = 1$ to $\lambda_s \approx 35 \times 10^{-6}$ at $x \approx 0.2$ [82, 83]. Therefore, producing microwires with various Fe–Co-rich compositions, we were able to change the magnetostriction constant from $\lambda_s \approx 35 \times 10^{-6}$ for Fe-rich compositions ($Fe_{72.75}Co_{2.25}B_{15}Si_{10}$ and $Fe_{70}B_{15}Si_{10}C_5$) till $\lambda_s \approx 10^{-7}$ for $Co_{56}Fe_8Ni_{10}Si_{10}B_{16}$ microwire. Additionally within each composition of metallic nucleus, we also produced microwires with different ratio of metallic nucleus diameter and total diameter, D, i.e., with different ratios $\rho = d/D$. This allowed us to control residual stresses, since the strength of internal stresses is determined by ratio ρ [57, 64–67].

It is worth mentioning that the magnetoelastic energy, K_{me}, which is given by Eq. (6.5) depends on both total stress, σ (i.e., on applied, σ_a, and internal, σ_i, stresses), and magnetostriction constant λ_s.

In this way the effect of magnetoelastic contribution on DW dynamics has been experimentally studied, controlling the magnetostriction constant and applied and/or residual stresses.

As can be appreciated from Fig. 6.10, considerable increasing of switching filed (from about 80 till 700 A/m) is observed when ferromagnetic metallic nucleus diameter decreases from 15 till 1.4 μm (i.e., one order). At the same time, rectangular hysteresis loop shape is maintained even for smallest microwire diameters.

Usually it is assumed that domain wall (DW) propagates along the wire with a velocity:

$$v = S(H - H_0) \tag{6.8}$$

where S is the DW mobility, H is the axial magnetic field, and H_0 is the critical propagation field [16, 29].

Dependences of domain-wall velocity, v, on magnetic field, H, for $Fe_{16}Co_{60}Si_{13}B_{11}$ and $Co_{41.7}Fe_{36.4}Si_{10.1}B_{11.8}$ amorphous microwires with the same ρ-ratio are shown in Fig. 6.14. In this case, the effect of only magnetostriction constant is that higher magnetostriction constant (according to Refs. [82, 83]) for $Co_{41.7}Fe_{36.4}Si_{10.1}B_{11.8}$ microwire $\lambda_s \approx 25 \times 10^{-6}$ should be considered, while for $Fe_{16}Co_{60}Si_{13}B_{11}$ composition ($\lambda_s \approx 15 \times 10^{-6}$) results in smaller DW velocity at the same magnetic field and smaller DW mobility, S.

In order to evaluate the effect of ρ-ratio, i.e., effect of residual stresses on DW dynamics, we performed measurements of $v(H)$ dependences in the microwires with the same composition, but with different ρ-ratios. Dependences of DW

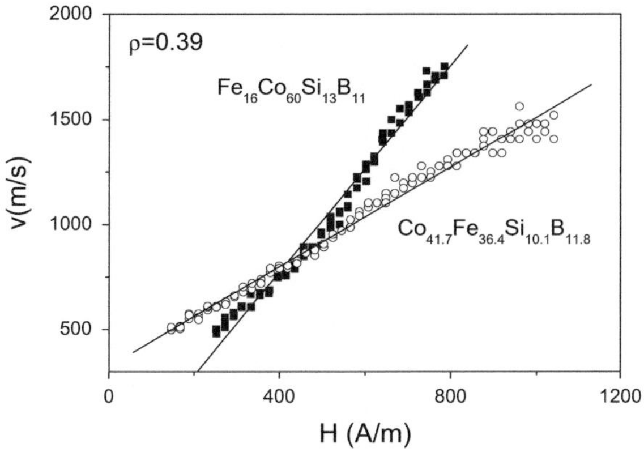

Fig. 6.14 $v(H)$ dependences for $Fe_{16}Co_{60}Si_{13}B_{11}$ and $Co_{41.7}Fe_{36.4}Si_{10.1}B_{11.8}$ microwires with $\rho = 0.39$. Reprinted with permission from [57] Copyright (2012) Springer Open (Fig. 2)

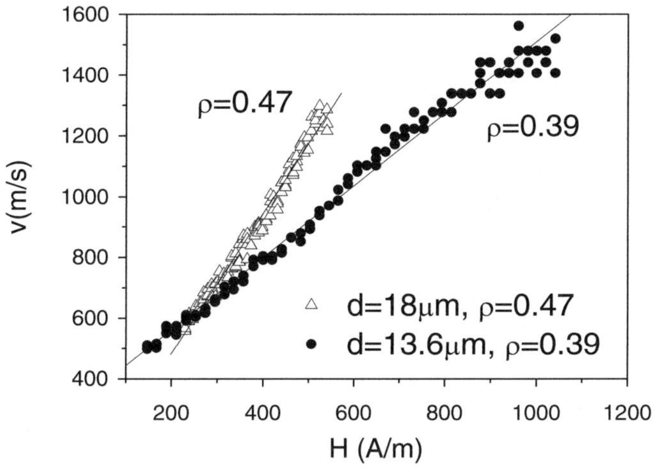

Fig. 6.15 $v(H)$ dependences for $Co_{41.7}Fe_{36.4}Si_{10.1}B_{11.8}$ microwires with different ratios ρ

velocity on applied field for $Co_{41.7}Fe_{36.4}Si_{10.1}B_{11.8}$ microwires with different ratios are shown in Fig. 6.15. Like in Fig. 6.14, at the same values of applied field, H, the domain-wall velocity is higher for microwires with higher ρ-ratio, i.e., when the internal stresses are lower [64–67].

The other way to manipulate the magnetoelastic energy is to apply stresses during measurements. Figure 6.16 shows $v(H)$ dependences for $Co_{41.7}Fe_{36.4}Si_{10.1}B_{11.8}$ microwire ($\rho \approx 0.55$) measured under applied tensile stresses. Considerable decreasing of domain-wall velocity, v, at the same magnetic field value, H, has been observed under the application of applied stress. Additionally, increasing of applied stress, σ_a, results in decreasing of DW velocity, v.

Consequently in low magnetostrictive $Co_{56}Fe_8Ni_{10}Si_{10}B_{16}$ microwire ($\lambda_s \approx 10^{-7}$), DW velocity values, achieved at the same values of applied field

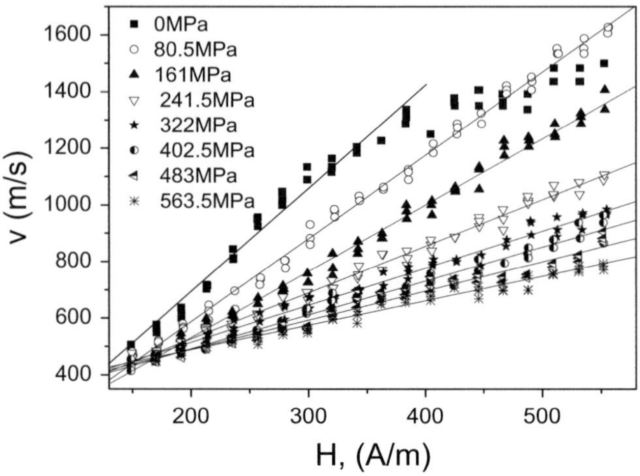

Fig. 6.16 $v(H)$ dependences for $Co_{41.7}Fe_{36.4}Si_{10.1}B_{11.8}$ microwires ($d \approx 13.6$ μm, $D \approx 24.6$ μm, $\rho \approx 0.55$) measured under application of applied stresses, σ_a. Reprinted with permission from [57] doi:10.1186/1556-276X-7-223 Copyright (2012) SpringerOpen (Fig. 4)

(see Fig. 6.17), are considerably higher (almost twice) than that observed for microwires with higher magnetostriction constant (compare with Figs. 6.14, 6.15, 6.16, and 6.17). Similar to the case of higher magnetostriction microwires, v-values drastically decrease with increasing applied stresses.

Regarding experimentally observed $v(H)$ dependences shown in Figs. 6.14, 6.15, 6.16, and 6.17, there are few typical features: linear extrapolation to zero domain-wall velocity gives negative values of the critical propagation field, H_o. Such a negative value, previously reported, for instance, in Ref. [75], has been explained in terms of the negative nucleation field of the reversed domain. In the case of amorphous microwires, the reversed domain already exists and does not need to be nucleated by the reversed applied magnetic field.

The domain-wall dynamics in viscous regime is determined by a mobility relation in Eq. (6.8), where S is the domain-wall mobility given by

$$S = 2\mu_0 M_\sigma / \beta \tag{6.9}$$

where β is the viscous damping coefficient and μ_0 is the magnetic permeability of vacuum. Damping is the most relevant parameter determining the domain-wall dynamics. Various contributions to viscous damping β have been considered, and two of them are generally accepted [57, 84]:

- Micro-eddy currents circulating nearby moving domain wall are the more obvious cause of damping in metals. However, the eddy current parameter β_ε is considered to be negligible in high-resistive materials, like thin amorphous microwires.

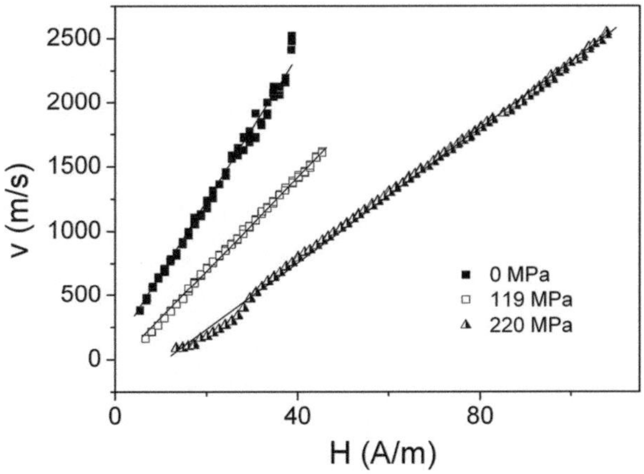

Fig. 6.17 $v(H)$ dependences for $Co_{56}Fe_8Ni_{10}Si_{10}B_{16}$ microwires measured under application of applied stresses, σ_a. Reprinted with permission from [57] doi:10.1186/1556-276X-7-223 Copyright (2012) SpringerOpen (Fig. 5)

– The second generally accepted contribution of energy dissipation is magnetic relaxation damping, β_ρ, related to a delayed rotation of electron spins. This damping is related to the Gilbert damping parameter and is inversely proportional to the domain-wall width δ_ω [57, 84]:

$$\beta_r \approx \alpha M_s/\gamma\Delta \approx M_s(K_{me}/A)^{1/2} \tag{6.10}$$

where γ is the gyromagnetic ratio, A is the exchange stiffness constant and K_{me} is the magnetoelastic anisotropy energy given by Eq. (6.5).

Consequently, the magnetoelastic energy can affect domain-wall mobility, S, as we experimentally observed in few Co–Fe-rich microwires.

Considering the aforementioned, we can suggest that DW velocity, v, should decrease with stress and magnetostriction constant increasing, and if only the magnetoelastic energy affects the DW dynamics, v should show an inverse square root dependence on stress or magnetostriction. Therefore, we tried to evaluate the $v(\sigma_{app})$ dependence. An example of $v(\sigma_{app})$ dependence, obtained for $Fe_{55}Co_{23}B_{11.8}Si_{10.2}$ microwires ($d = 13.2$ μm; $D = 29.6$ μm), is shown in Fig. 6.18a. Qualitatively, we observed decreasing of DW velocity, v, with applied stresses, σ_{app}. In order to evaluate, if obtained dependence $v(\sigma_{app})$ fits inverse square root dependence on applied stress, σ_{app}, we expressed obtained dependences as $\sigma_{app}(v^{-2})$. From Fig. 6.18b we can conclude that obtained $v(\sigma_{app})$ dependences cannot be described by single $v({\sigma_{app}}^{-1/2})$ dependence. On the other hand, at high enough σ_{app} observed $v(\sigma_{app})$ dependence probably can be fitted by 2 $v({\sigma_{app}}^{-1/2})$.

Previously a systematic analysis of mechanisms of DW dynamics in thicker (with diameters between 30 and 120 μm) magnetostrictive amorphous wires without glass has been performed in Ref. [85] on the basis of bubble domain dynamics. The systematic analysis method in this paper is also a strong basis for considering domain propagation dynamics in glass-covered thinner magnetostrictive

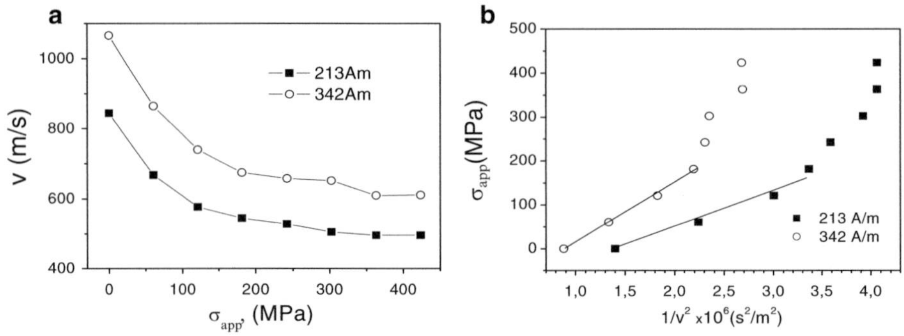

Fig. 6.18 $v(\sigma_{app})$ dependences of $Fe_{55}Co_{23}B_{11.8}Si_{10.2}$ microwires ($d = 13.2$ μm; $D = 29.6$ μm) (**a**) and σ_{app} $(1/v^2)$ dependence (**b**). Reprinted with permission from [57] doi:10.1186/1556-276X-7-223. Copyright (2012) SpringerOpen (Fig. 6)

amorphous wires. Main assumptions on domain-wall configuration in thicker wires have been performed considering that the DW length, l, is much more than its radius, r ($r/l \ll 10^{-3}$).

Recently the attempt to extend the analysis has been performed for thinner glass-coated microwires (typically with the diameter of the order of 10 μm) with strong internal stresses induced by the glass-coating [86]. Particularly analyzing the voltage peak forms and experimental data on DW dynamics, we demonstrated that a very high DW mobility observed in magnetically bistable amorphous microwires with a diameter of about 10 μm can be associated with elongated domain shape. The experimental results can be explained in terms of the normal mobility with respect to the domain surface, which is reduced by a factor representing the domain aspect ratio estimated to be in the range of 300 for considered wire samples. On the other hand, experimental data on DW dynamics in thin microwires and analysis of the voltages on pickup coils show that generally the structure of propagating DW is far from abrupt and is quite complex [78, 79]. Thus, the characteristic width of the head-to-head DW, δ, depends on many factors, such as the applied magnetic field, H: at $H = 60$ A/m, $\delta \approx 65\,d$, while at $H = 300$ A/m, $\delta \approx 40\,d$. Additionally, δ depends on magnetic anisotropy constant, K, being $\delta/d \approx 13.5$ for $K = 10^4$ erg/cm^3, $\delta/d \approx 20$ for $K = 5 \times 10^3$ erg/cm^3, $\delta/d = 30$–34 for $K = 2 \times 10^3$ erg/cm^3, and $\delta/d = 40$–50 for $K = 10^3$ erg/cm^3, respectively [78].

Regarding aforementioned, it is interesting to compare the velocity of DW propagation in thinnest microwire with the values observed in submicrometric planar nanowires reported elsewhere [87]. The DW velocity in thin microwire is ranging between 700 and 850 m/s (Fig. 6.19), which is still higher than for the same range of magnetic field as compared with submicrometric nanowires (maximum $v \approx 110$ m/s at 700 A/m) reported elsewhere [87, 88].

On the other hand, for such elevated magnetic fields (1000–1500 A/m), the domain-wall velocity, v, is significantly lower than for thicker wires of the same

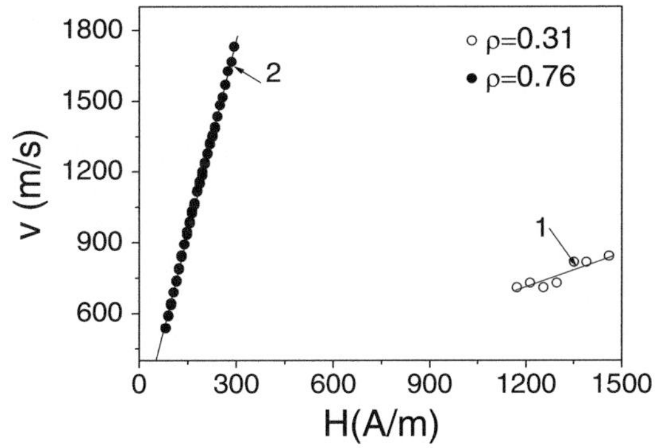

Fig. 6.19 $v(H)$ dependence for $Fe_{72.75}Co_{2.25}B_{15}Si_{10}$ amorphous microwires with metallic nucleus diameter, d, of 2.8 μm and total diameter $D \approx 9$ μm (*1*) and $d = 12.0/D = 15.8$ (*2*). Reprinted with permission from [57] doi:10.1186/1556-276X-7-223. Copyright (2012) SpringerOpen (Fig. 7b)

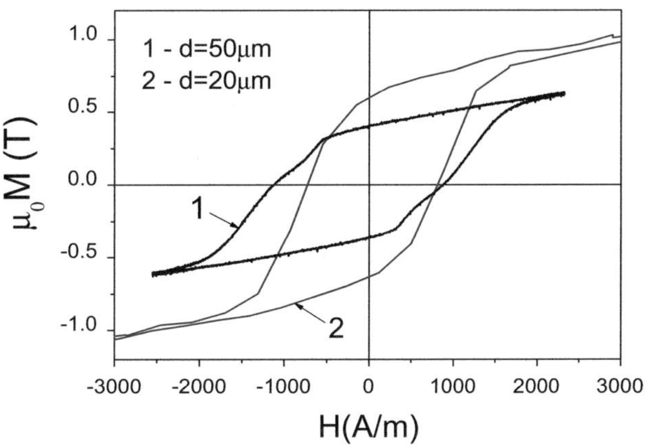

Fig. 6.20 Hysteresis loops of $Fe_{77.5}B_{15}Si_{7.5}$ microwires with diameter 50 μm (*1*) and 20 μm (*2*) obtained by cold-drawing technique

composition with lower ρ-ratio. The comparison $v(H)$ dependence for $Fe_{74}Si_{11}B_{13}C_2$ microwire with similar composition with metallic nucleus d and total D diameters 12.0 and 15.8 μm, respectively, is presented in Fig. 6.19. As can be deduced from the comparison of DW dynamics, thicker $Fe_{74}Si_{11}B_{13}C_2$ microwire at maximum achieved magnetic field (about 280 A/m) presented double higher velocity as compared with $Fe_{72.75}Co_{2.25}B_{15}Si_{10}$ amorphous microwire with metallic nucleus diameter, d, of 2.8 μm and total diameter $D \approx 9$ μm (Fig. 6.19).

It is worth mentioning that in the case of cold-drawn microwires Fe-rich compositions, the magnetization reversal process is more complex [54]: $Fe_{77.5}B_{15}Si_{7.5}$ cold-drawn wires with a diameter of 50 μm present roughly rectangular hysteresis loop (Fig. 6.20), while $Fe_{77.5}B_{15}Si_{7.5}$ cold-drawn microwires with a diameter of 20 μm do not present rectangular hysteresis loop (see example in Fig. 6.20). Both cold-drawn samples present rather high coercivities (above 1 kA/m). Decrease in the coercivity and increase in the remanent magnetization have been observed under the influence of applied stress in Fig. 6.21 for the case of $Fe_{77.5}B_{15}Si_{7.5}$ cold-drawn wires with the diameter of 50 μm.

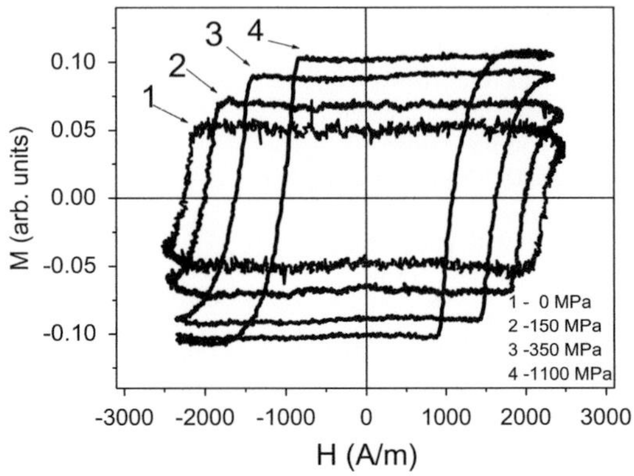

Fig. 6.21 Hysteresis loops of $Fe_{77.5}B_{15}Si_{7.5}$ microwires with diameter 50 μm measured under applied stress

This behavior must be associated with the complex stresses induced by cold-drawn technique. Moreover under the applied stress the hysteresis loops become similar to magnetically bistable wires and glass-coated microwires.

For the understanding of the remagnetization process of cold-drawn wire, the propagation of the remagnetization front has been studied using two pick-up coils (experiment similar to the Sixtus–Tonks described above).

Analyzing the propagation of the remagnetization front, it was assumed that in this case of cold-drawn wires, the magnetization reversal starts inside the wire (but not at the end of it) and exists as the motion of two domain walls is moving to the opposite sides.

Consequently similarly to the hysteretic properties, the magnetoelastic anisotropy considerably affects DW dynamics.

6.3.2 *Quasi Non-hysteretic Behavior*

In contrast to the bistable hysteresis loops, certain amorphous and nanocrystalline alloys (ribbons and wires) exhibit nearly non-hysteretic behavior [69, 70, 89, 90]. In particular, loops of nearly zero magnetostrictive compositions exhibit very low coercivities and quite large initial susceptibility. Figure 6.22a, b shows examples of such hysteresis loops with vanishing (Fig. 6.22a) and considerably negative (Fig. 6.22b) magnetic anisotropy field.

This quasi non-hysteresis behavior can be observed in wires and microwires with a well-defined transverse circular anisotropy. For the Co-rich microwires with low and negative magnetostriction coefficient, coercivities of the order of 4 A/m have been reported elsewhere [89]. For fixed metallic nucleus composition, the field of magnetic anisotropy is a function of the ratio, i.e., on the magnetoelastic anisotropy determined by the strength on internal stresses induced by the glass coating (see Fig. 6.23).

In this particular case, the magnetization process takes place by quasi-reversible magnetization rotation from the transverse (ribbons) or circumferential (wires and

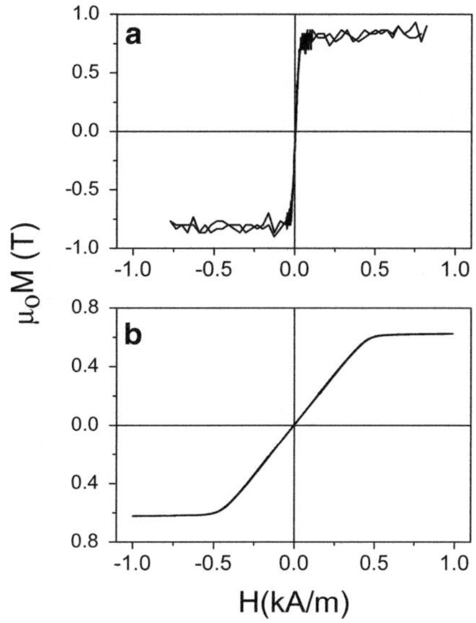

Fig. 6.22 Hysteresis loop of amorphous Co-rich microwires with nearly zero magnetostriction (**a**) and negative magnetostriction (**b**) coefficients

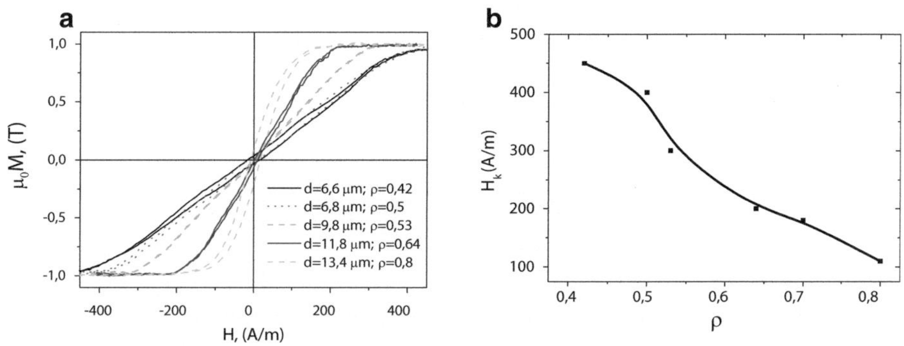

Fig. 6.23 Hysteresis loops of $Co_{67.1}Fe_{3.8}Ni_{1.4}Si_{14.5}B_{11.5}Mo_{1.7}$ microwires with different geometry (**a**) and dependence of H_k on ρ-ratio (**b**). Reprinted with permission from [90]. Copyright (2012) Springer Open (Fig. 5)

microwires) to the axial direction with increasing the axial applied magnetic field. It must be mentioned that in the case of glass-coated microwires, such circular anisotropy is directly induced by the internal stresses generated during fabrication.

6.3.3 Matteucci Effect

When a ferromagnetic rod is twisted, any change in the longitudinal magnetic field will induce a change of the circular magnetization, after Eq. (6.3), and hence a longitudinal electric field will appear in the rod. This is the effect discovered by

Wertheim and Matteucci, which provides a convenient experimental method for examining the initial magnetization process [91].

Due to the magnetoelastic energy, the easy magnetization direction is tilted by 45° from the rod axis, and neither longitudinal nor circular magnetization is observed in the demagnetized state. When a weak longitudinal magnetic field is applied, it induces both a longitudinal and a circular magnetization. This effect has been successfully used by Nielsen in determining the in-plane magnetic anisotropy in amorphous ribbons (by measuring the stress dependence of the inhomogeneous magnetization obtained by twisting the ribbon around its axis) [92].

The fact that the Matteucci effect in the case of magnetostrictive amorphous wires is prepared by water quenching is remarkable [9, 15, 29, 93]. Due to a peculiar domain configuration, these wires are known to present a bistable magnetic behavior as has been mentioned above. The switching field, H_s, results to be also very sensitive to twisting the wires. Hence, H_s is expected to vary as this was experimentally verified in an amorphous wire of FeSiB (Fe rich) of 128 μm [93]. A small torque exerted on such a wire conveniently biased by a static field might induce a very large electric pulse in the wire, by Matteucci effect.

Spontaneous Matteucci effect has been successfully observed also in glass-coated microwires [94], although the experiments on twisting of glass-coated microwires are difficult to realize. Spontaneous Matteucci effect observed in glass-coated microwires has been explained by the existence of helical magnetic anisotropy induced by stresses during the preparation of glass-coated microwires.

6.3.4 Giant Magnetoimpedance, Stress Impedance, and Torsion Impedance Effects

As already mentioned in the introduction, the GMI effect usually observed in soft magnetic materials phenomenologically consists of the change of the AC impedance, $Z = R + iX$ (where R is the real part, or resistance, and X is the imaginary part, or reactance), when submitted to an external magnetic field, H_0. The GMI effect was well interpreted in terms of the classical skin effect in a magnetic conductor, assuming the dependence of the penetration depth of the AC current flowing through the magnetically soft conductor on the DC applied magnetic field [17, 18]. Extremely high sensitivity of the GMI effect to even low magnetic field attracted great interest in the field of applied magnetism basically for applications for low magnetic field detection.

Generally, the GMI effect was interpreted assuming scalar character for the magnetic permeability, as a consequence of the change in the penetration depth of the AC current caused by the DC applied magnetic field. The electrical impedance, Z, of a magnetic conductor in this case is given by [17, 18]

$$Z = R_{dc} kr J_0(kr)/2J_1(kr) \tag{6.11}$$

with $k = (1 + j)/\delta$, where J_0 and J_1 are the Bessel functions, r is the wire's radius, and δ is the penetration depth given by

$$\delta = \sqrt{\pi \sigma \mu_\varphi f} \tag{6.12}$$

where σ is the electrical conductivity, f is the frequency of the current along the sample, and μ_Φ is the circular magnetic permeability assumed to be scalar. The DC applied magnetic field introduces significant changes in the circular permeability, μ_Φ. Therefore, the penetration depth also changes through and finally results in a change of Z [17, 18].

Usually for the quantification of the GMI effect, the magneto impedance ratio, $\Delta Z/Z$, is used. GMI ratio, $\Delta Z/Z$, is defined as

$$\Delta Z/Z = [Z(H) - Z(H_{max})]/Z(H_{max}), \tag{6.13}$$

where H_{max} is the axial DC field with maximum value up to few kA/m.

The main features of the GMI effect are the following:

1. Large change in the total impedance usually above 100 %. Usually for the case of amorphous wires with high circumferential permeability, the highest GMI effect is reported [17]. Thus, tailoring the magnetic anisotropy through the geometry of amorphous microwires, few researchers reported an achievement of about 600 % GMI ratio observation in Co-rich microwires with vanishing magnetostriction constant [95, 96]. In this case, it is quite promising for the application of magnetic sensors.
2. The GMI materials, whether wires, ribbons, or films, are usually extremely soft magnetic materials. It was pointed out [17, 18, 97] that the good magnetic softness is directly related to the GMI effect: the magnetic field dependence of the GMI spectra is mainly determined by the type of magnetic anisotropy. Thus, the circumferential anisotropy leads to the observation of the maximum of the real component of wire impedance (and consequently of the GMI ratio) as a function of the external magnetic field. On the other hand, in the case of axial magnetic anisotropy, the maximum value of the GMI ratio corresponds to zero magnetic fields [2–6], i.e., results in a monotonic decay of the GMI ratio with the axial magnetic field.
3. The alternating current plays an important part in the GMI effect. The main reason is that like magnetic permeability, GMI effect presents tensor character [56, 58]. Therefore, AC current flowing through the sample creates circumferential magnetic field. Additionally AC current produces the Joule heating [57]. There are many publications related with the origin of the GMI effect [12–16, 59–61]. It must be underlined that the GMI effect origin has been explained based on the theory of classical electrodynamics. The skin effect, which is responsible for GMI at medium and high frequencies, is a phenomenon

well described by the classical electrodynamics [62] many years ago. As a consequence of induced eddy currents, the high-frequency AC current is not uniformly distributed in the conductor volume but is confined to a shell close to the surface, with depth, δ, given in Eq. (6.12).

Cylindrical shape and high circumferential permeability observed in amorphous wires are quite favorable for achievement of high GMI effect [20, 21]. As a rule, better soft magnetic properties are observed for nearly zero magnetostrictive compositions. It is worth mentioning that the magnetostriction constant, λ_s, in system $(Co_xFe_{1-x})_{75}Si_{15}B_{10}$ changes with x from -5×10^{-6} at $x = 1$ to $\lambda_s \approx 35 \times 10^{-6}$ at $x \approx 0.2$, achieving nearly zero values at Co/Fe about 70/5 [17, 19, 63].

The main interest of the GMI effect is related with the high sensitivity of the impedance to an applied magnetic field, easy achieving up to 300 % relative change of impedance in conventional amorphous wires with vanishing magnetostriction (see Fig. 6.24 for the $(Co_{0.94}Fe_{0.06})_{72.5}B_{15}Si_{12.5}$ conventional amorphous wire) [17, 18, 98]. Circular domain structure with high circumferential permeability proved to be very favorable for highest GMI effect [99]. Such domain configuration is typical for the nearly zero magnetostrictive amorphous wires mainly produced by Unitika Ltd (Japan) [100].

As mentioned above, there are few parameters that affect the GMI ratio. Apart from the AC current amplitude, the frequency, f, of AC current is quite important. GMI effect measured at different frequencies is presented in Fig. 6.25a. As can be appreciated for $(Co_{0.94}Fe_{0.06})_{72.5}B_{15}Si_{12.5}$, conventional amorphous wire maximum GMI ratio, $\Delta Z/Z_m$, presents the highest value at relatively low frequencies: $\Delta Z/Z_m$ decreases when f increases (Fig. 6.25b). On the other hand, the field, H_m, at which $\Delta Z/Z_m$ is observed increases with f (Fig. 6.25c).

Fe-rich $Fe_{77.5}B_{15}Si_{7.5}$ conventional wires present rather lower GMI effect (Fig. 6.26a). Above 150 MHz $\Delta Z/Z(H)$ presents a maximum at certain magnetic fields. Optimum $\Delta Z/Z_m \approx 80$ % values can be achieved at relatively low frequencies (about 20 MHz). Similarly Co-rich wires H_m increases with f.

Fig. 6.24 $\Delta Z/Z$ dependences measured in $(Co_{0.94}Fe_{0.06})_{72.5}B_{15}Si_{12.5}$ conventional amorphous wire at different amplitudes of AC current

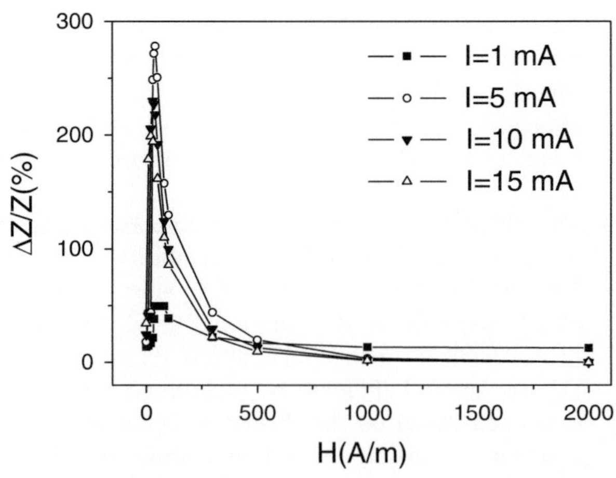

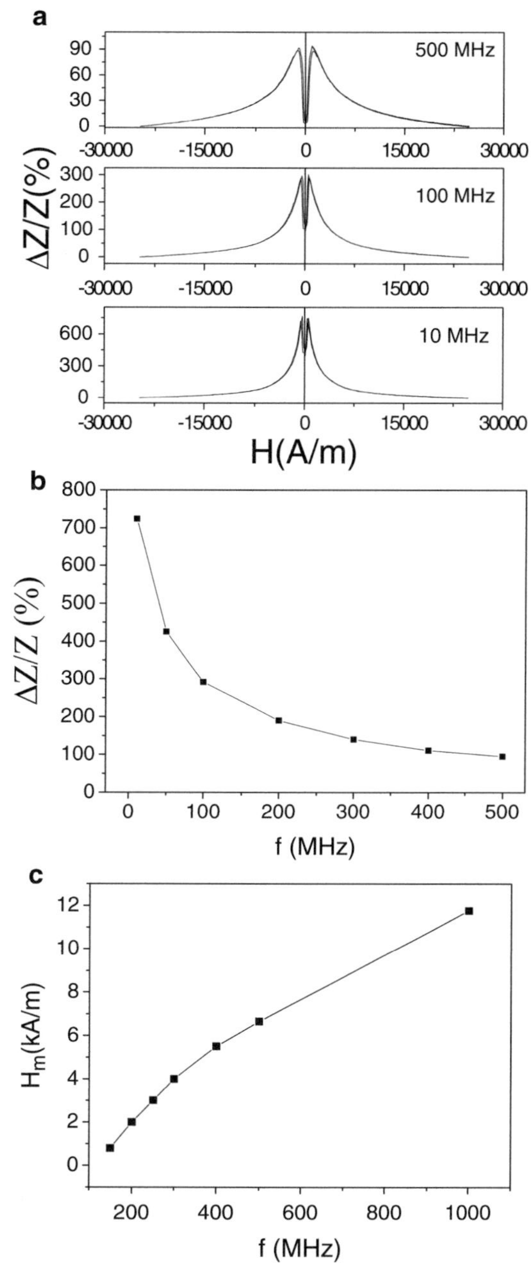

Fig. 6.25 $\Delta Z/Z$ dependences measured in $(Co_{0.94}Fe_{0.06})_{72.5}B_{15}Si_{12.5}$ conventional amorphous wire at different frequencies of AC current (**a**), dependence of $\Delta Z/Z_m$ (**b**), and H_m (**c**) on f. Reprinted with permission from [99] Copyright (2007) Bentham Science Publishers Ltd. (Fig. 2a, 5a, 6a)

At the meantime, generally an inferior GMI effect has been observed for the other amorphous magnetic materials without special treatment, such as amorphous melt-extracted cold-drawn and glass-coated microwires, where GMI ratio is

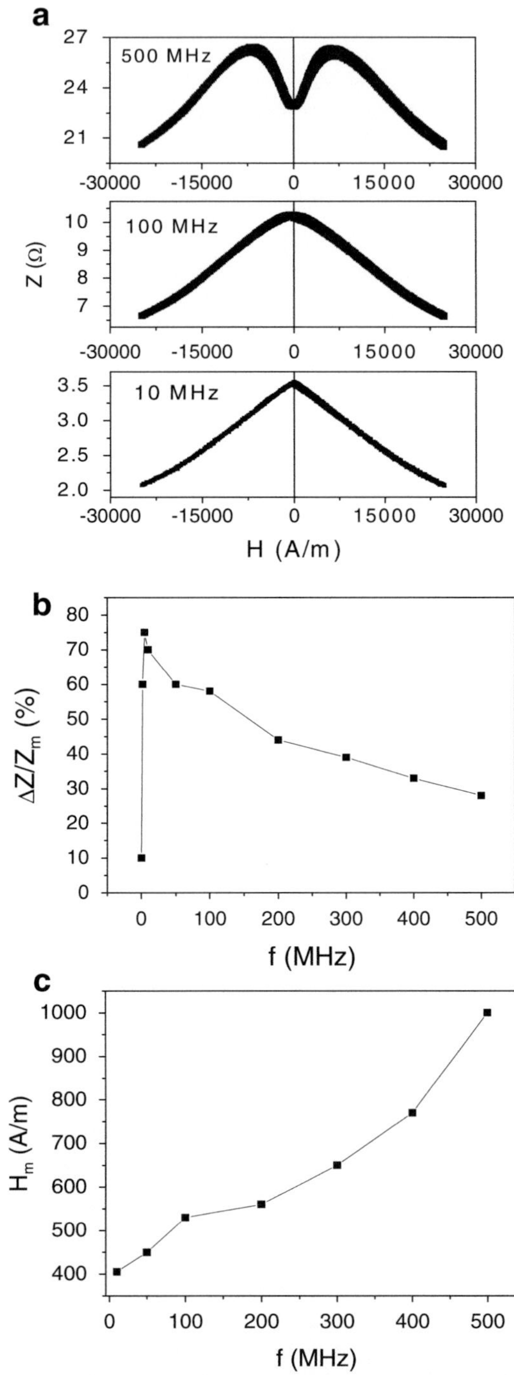

Fig. 6.26 GMI effect measured in $Fe_{77.5}B_{15}Si_{7.5}$ conventional amorphous wire at different frequencies of AC current (**a**), dependence of $\Delta Z/Z_m$ (**b**), and H_m (**c**) on *f*. Reprinted with permission from [99] Copyright (2007) Bentham Science Publishers Ltd. (Fig. 2b, 5b, 6b)

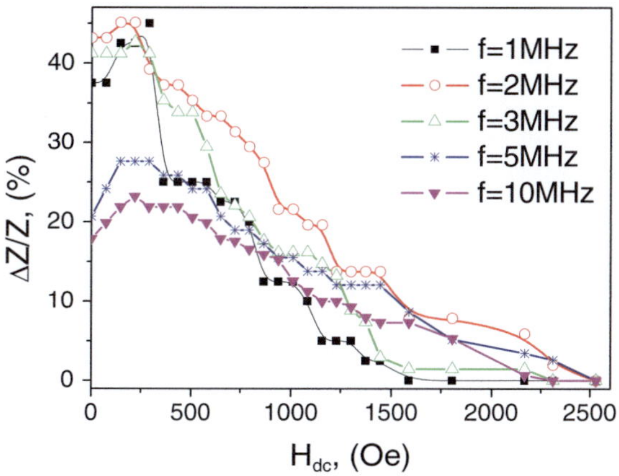

Fig. 6.27 $\Delta Z/Z(H)$ dependencies measured in as-extracted fiber for driving current amplitude $I = 1$ mA with the frequency, f, as a parameter. Reprinted from [26] doi:10.1016/S0924-4247(03)00172-9 Copyright © 2003 Elsevier B.V. (Fig. 5)

generally less than 60 % [29, 97, 98]. Thus, initially $\Delta Z/Z_m \approx 50$ % achieved at 2 MHz is reported for melt-extracted Co-rich microwires (see Fig. 6.27).

Similarly cold-drawn amorphous wire $Fe_{77.5}B_{15}Si_{7.5}$ with a diameter of 50 μm presents even lower GMI effect (Fig. 6.28). In this case considerable GMI hysteresis can be also observed (Fig. 6.28a). $\Delta Z/Z_m$ increase with f achieving about 40 % at $f = 500$ MHz and $H_m(f)$ presents non-monotonous dependence.

In the first report on GMI effect of glass-coated microwires, only $\Delta Z/Z_m \approx 16$ % at $f = 2$ MHz was reported [100].

Therefore, further processing or careful chemical composition selection of the other families of wires is needed [99]. Moreover, as shown elsewhere, AC current frequency for the achievement of optimum $\Delta Z/Z_m$ values must be higher.

Consequently extending the frequency range of measurements, higher $\Delta Z/Z_m$ values have been obtained for glass-coated microwires (see Fig. 6.29). Thus, for Fe-rich $Fe_{75.5}B_{13}Si_{11}Mo_{0.5}$ microwire with metallic nucleus diameter of about 18 μm (Fig. 6.29a, b), $\Delta Z/Z_m \approx 45$ % at $f = 500$ MHz has been reported [99]. Similarly to conventional "in-rotating-water" Fe-rich wire, a maximum on $\Delta Z/Z(H)$ dependence can be observed above $f \geq 250$ MHz, and increasing of H_m with f has been observed (Fig. 6.29c).

Co-rich microwires present even higher GMI effect (Fig. 6.30) that for $Fe_{3.7}Co_{69.8}Ni_1Si_{11}B_{13}Mo_{1.5}$ glass-coated microwire ($d \approx 19$ μm) achieves 140 % (Fig. 6.30a, b). Similarly to Co-rich conventional wires, this sample presents double-peak dependence on applied magnetic field (Fig. 6.30a). Optimum $\Delta Z/Z_m \approx 140$ % achieved at about 300 MHz (Fig. 6.30b), while H_m increases with f (Fig. 6.30c).

Afterwards GMI ratio of Co-rich microwires has been optimized through the minimization of the magnetoelastic energy either by variation of ρ-ratio of by annealing. The example is shown in Fig. 6.31, where $\Delta Z/Z(H)$ dependencies measured in $Co_{67.05}Fe_{3.85}Ni_{1.4}B_{11.55}Si_{14.5}Mo_{1.65}$ glass-coated microwires with the different sample geometry are shown.

Fig. 6.28 GMI effect measured in cold-drawn $Fe_{77.5}B_{15}Si_{7.5}$ microwire ($d \approx 50$ μm) at different frequencies of AC current (**a**), dependence of $\Delta Z/Z_m$ (**b**), and H_m (**c**) on f. Reprinted with permission from [99] Copyright (2007) Bentham Science Publishers Ltd. (Fig. 3a, 5c, 6c)

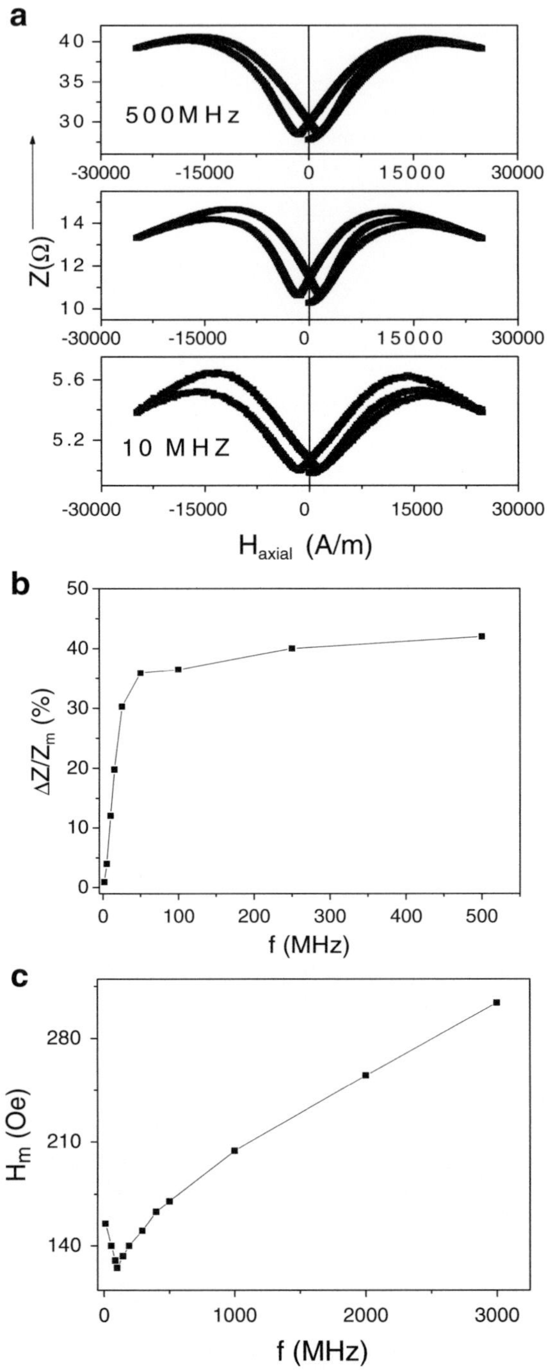

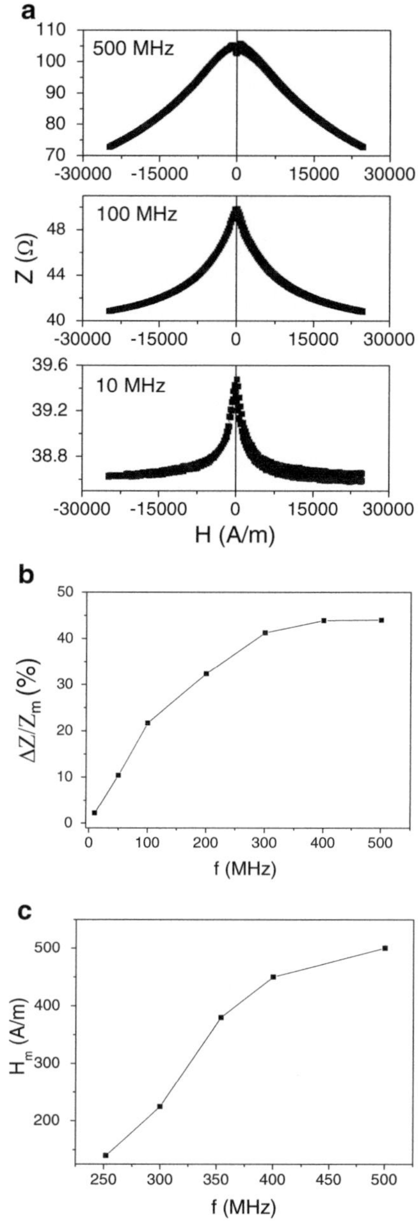

Fig. 6.29 GMI effect measured in $Fe_{75.5}B_{13}Si_{11}Mo_{0.5}$ glass-coated microwire ($d \approx 18$ μm) at different frequencies of AC current (**a**), dependence of $\Delta Z/Z_m$ (**b**), and H_m (**c**) on f. Reprinted with permission from [99] Copyright (2007) Bentham Science Publishers Ltd. (Fig. 4a, 5d, 6e)

Similarly considerable progress in the optimization of melt-extracted microwires is achieved [101–103]: maximum GMI ratio of Co-rich melt-extracted microwires has been enhanced up to 120 % [102] and even recently up to 250 % [103]. The problem with melt-extracted microwires is that due to the fabrication process, their cross section is not perfectly circular.

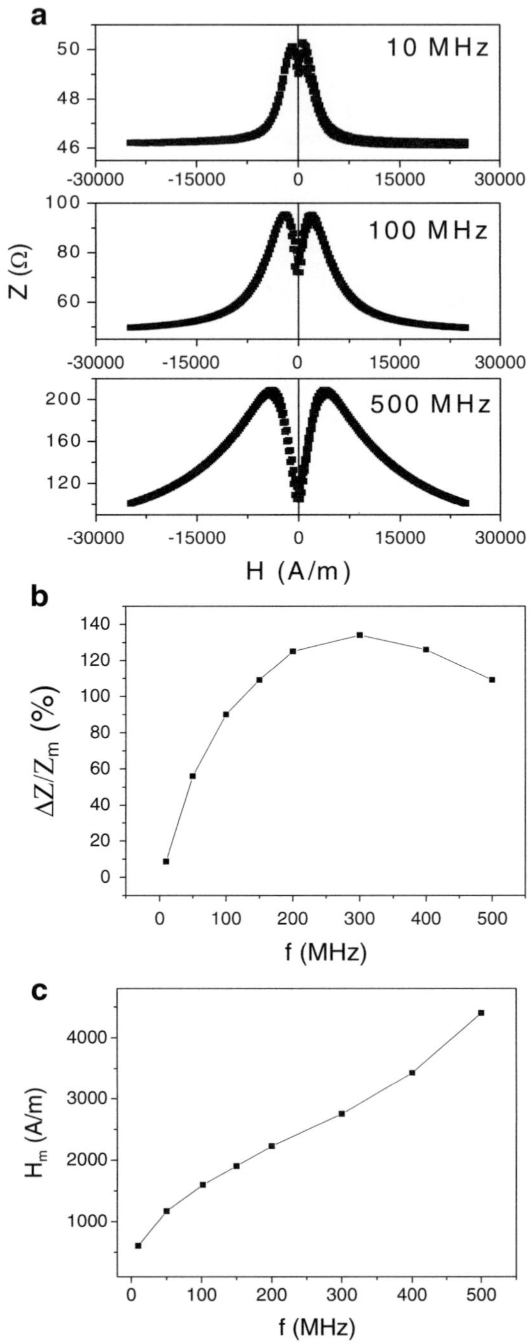

Fig. 6.30 GMI effect measured in $Fe_{3.7}Co_{69.8}Ni_1Si_{11}B_{13}Mo_{1.5}$ glass-coated microwire ($d \approx 19$ μm) at different frequencies of AC current (**a**), dependence of $\Delta Z/Z_m$ (**b**), and H_m (**c**) on f. Reprinted with permission from [99] Copyright (2007) Bentham Science Publishers Ltd. (Fig. 4b, 5f, 6g)

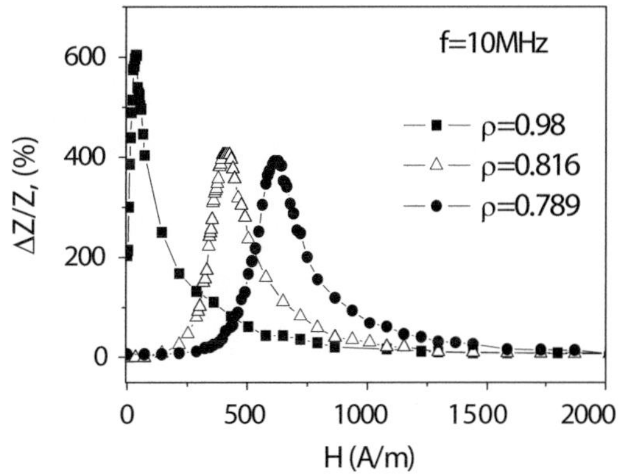

Fig. 6.31 GMI ratio measured in $Co_{67.05}Fe_{3.85}Ni_{1.4}$-$B_{11.55}Si_{14.5}Mo_{1.65}$ glass-coated microwires with the different sample geometry. Reprinted with permission from [99] Copyright (2007) Bentham Science Publishers Ltd. (Fig. 10)

The other possibility to diminish the magnetoelastic anisotropy and then increase the GMI effect of glass-coated microwires is the removal of the glass coating. Few attempts of glass removal consisting of preparation first relatively thick glass-coated microwires and then removing the glass coating either by cold-drawn or by warm-drawn techniques have been reported [104, 105]. As expected cold drawn considerably affected $\Delta Z/Z(H)$ dependencies: a transition from single-peak (SP) to double-peak (DP) GMI behavior was observed for the $Co_{69.5}Fe_{4.5}Cr_1Si_8B_{17}$ microwires after cold drawing. This has been explained by the induced circumferential stress during cold drawing, which changes the surface magnetic anisotropy of the wire from helical domains to circular domains [104]. On the other hand, considerable growth of magnetic anisotropy field is reported for $Co_{69}Fe_4Cr_4Si_{12}B_{11}$ amorphous glass-coated microwires after warm drawing [105]. Post processing annealing allowed the reduction of the anisotropy field (see Fig. 6.32) and the growing of remanent magnetization. Simultaneously improvement of the GMI ratio (as compared with as-prepared glass-coated microwire with $d \approx 90$ μm) up to about 240 % (at 10 MHz) has been observed. The effect of annealing temperature on $\Delta Z/Z(H)$ dependences of $Co_{69}Fe_4Cr_4Si_{12}B_{11}$ amorphous microwires subjected to different processing is presented in Fig. 6.33. Utilization of warm drawing allowed also improvement of the microwires ductility as pointed in [105].

As can be appreciated from aforementioned results on frequency dependence of GMI effect, optimal GMI ratio for different wire families can be achieved at rather different frequencies.

It is predicted that the diameter reduction must be associated with the increasing of resonance frequency and therefore the optimal GMI frequency range: a trade-off between dimension and frequency is required in order to obtain a maximum effect [106].

Fig. 6.32 Hysteresis loops of $Co_{69}Fe_4Cr_4Si_{12}B_{11}$ amorphous wires after glass removal, after drawing and series of heat treatment

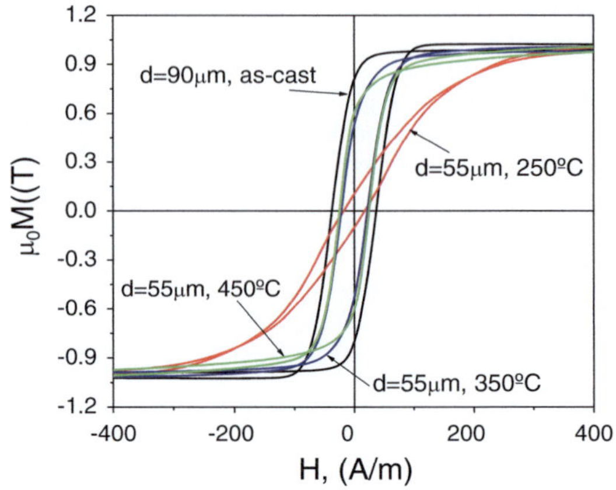

Fig. 6.33 $\Delta Z/Z\ (H)$ dependences of $Co_{69}Fe_4Cr_4Si_{12}B_{11}$ amorphous microwires subjected to different processing measured at 10 MHz

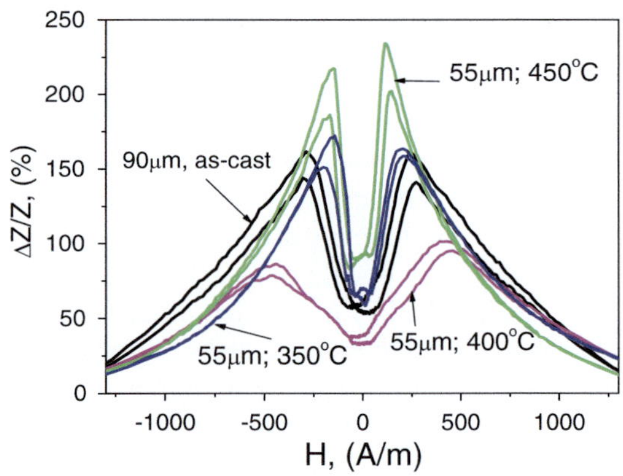

One of the last tendencies is the miniaturization of the GMI sensors. Consequently major attention is focused on high-frequency (GHz range) GMI applications owing to the development of thin magnetically soft materials and recent tendency in miniaturization of magnetic field sensors [21, 107].

For thin glass-coated microwires, considerable GMI effect at GHz frequencies has been recently reported elsewhere (see Fig. 6.34) [21, 108, 109].

It is worth mentioning that the magnetic fields corresponding to the impedance maximum, H_m, at GHz frequencies are much higher than the magnetic anisotropy field. At elevated frequencies, the magnetic field of Z maximum, H_m (1–10 kA/m,) while the magnetic anisotropy field, H_k, observed in quasi-static hysteresis loops of the same samples is about 0.1 kA/m. i.e., all the studied samples are magnetically saturated when exhibit an impedance maximum at GHz frequency range.

Usually the magnetic field of maximum is attributed to the magnetic anisotropy field [17, 18]. This considerable growth of H_m with increasing frequency can be explained considering the decrease of the skin depth with the increase of the

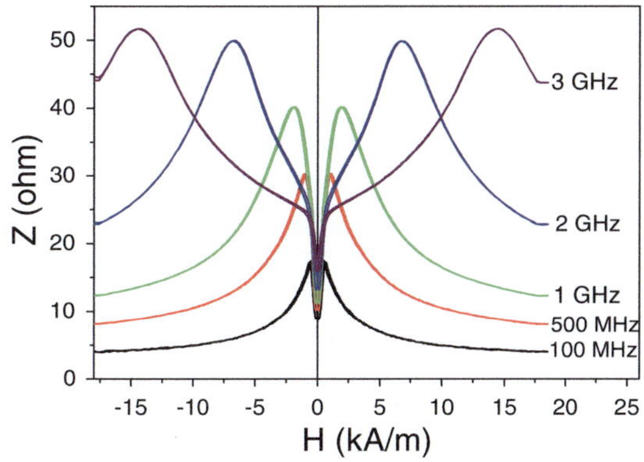

Fig. 6.34 *Z*(*H*) dependences of $Co_{66}Cr_{3.5}Fe_{3.5}B_{16}Si_{11}$

frequency. Consequently magnetic anisotropy of the surface area can be rather different from the bulk magnetic anisotropy. Indeed recently we reported on the formation of the interface layer between the metallic nucleus and glass coating [44]. This interfacial layer can present different magnetic anisotropy from the bulk.

On the other hand, previously several authors pointed out the fundamental link between ferromagnetic resonance, FMR, and GMI, both theoretically and experimentally [106, 108]. Indeed, for high-frequency impedance measurements of ferromagnetic wires in a longitudinal magnetic field, the electromagnetic field geometry is exactly what is required for FMR. Thus, the saturation magnetization can be estimated from this approach using the equation

$$M_s = 0.805 \times 10^{-9} \, \mathrm{d}f_0^2/\mathrm{d}H \tag{6.14}$$

where f_0 is the resonant frequency, H is the applied magnetic field, and M_s is the saturation magnetization.

Quite reasonable M_s values have been obtained for $Co_{67.05}Fe_{3.85}Ni_{1.4}B_{11.33}Si_{14.47}Mo_{1.69}$ microwires using Eq. (6.14) for MHz frequencies (see Fig. 6.35) [90].

Linear fit of the square of the resonance frequency on applied field is observed (Fig. 6.35b, d), although some deviation can be appreciated in low-field region. The origin of this deviation is unclear and can be related to low magnetic field GMI hysteresis recently discussed elsewhere [109]. Consequently some features of high-frequency GMI effect can be described using FMR-like approximation.

As mentioned above, magnetic field dependence of GMI ratio is intrinsically related with the magnetic anisotropy. Consequently application of the external stimuli that change the magnetic anisotropy induces changes in the GMI effect.

Thus, it was demonstrated [110–113] that the application of tensile and torsion stresses results in significant changes of the GMI effect through the modification of the domain structure under such stress (see Fig. 6.36).

Therefore, the stress impedance has been introduced in order to characterize the change of the electrical impedance under the stress [111, 112].

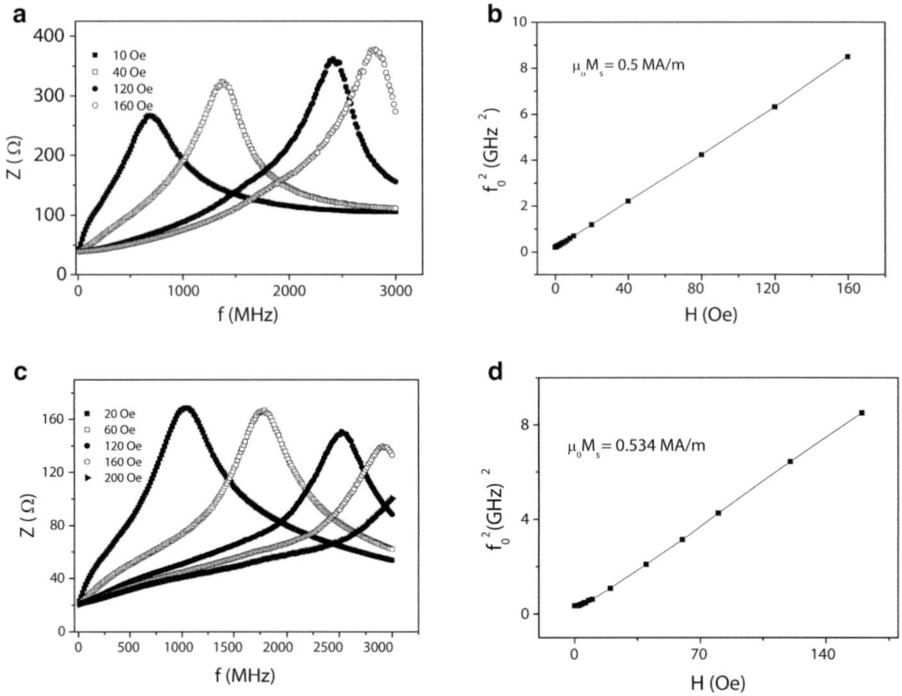

Fig. 6.35 · Frequency dependence of GMI effect (**a**, **c**) and $f_0^2(H)$ dependence (**b**, **d**) measured for $Co_{67.05}Fe_{3.85}Ni_{1.4}B_{11.33}Si_{14.47}Mo_{1.69}$ microwires with $d \approx 16.2$ μm $\rho \approx 0.7$ and $d = 21.4$ $\rho \approx 0.816$, respectively. Reprinted with permission from [90]. Copyright © Springer Science + Business Media New York 2012 (Fig. 4)

Particularly, it was found that the application of only stress (torsion or tensile) without any magnetic field gives rise to a significant change of the electrical impedance [111, 112].

When the torsion dependence of the impedance was investigated, the torsion giant impedance ratio (TGI), $(\Delta Z/Z)_\xi$, has been determined by the expression

$$(\Delta Z/Z)_\xi = [Z(\xi) - Z(\xi_{\max})]/Z(\xi_{\max}) \tag{6.15}$$

where $\xi_{\max}$ is the maximum torsion stress, ξ_{appl}, applied for the determination of this $(\Delta Z/Z)$.

Generally, torsion stress dependence of total impedance, Z, and $(\Delta Z/Z)_\xi$ have a shape of decay with applied torsion, ξ, beginning from some position corresponding to their maximum (Fig. 6.37). This decay has saturation, indicating that after application of large enough torsion, Z and $(\Delta Z/Z)_\xi$ are almost independent on ξ. The position of the maximum can change after different treatments. TGI has been measured in as-cast $(Co_{0.94}Fe_{0.06})_{72.5}B_{15}Si_{12.5}$ amorphous wire, after current annealing at 450 mA for up to 105 min under torsion (up to 33 πrad/min)

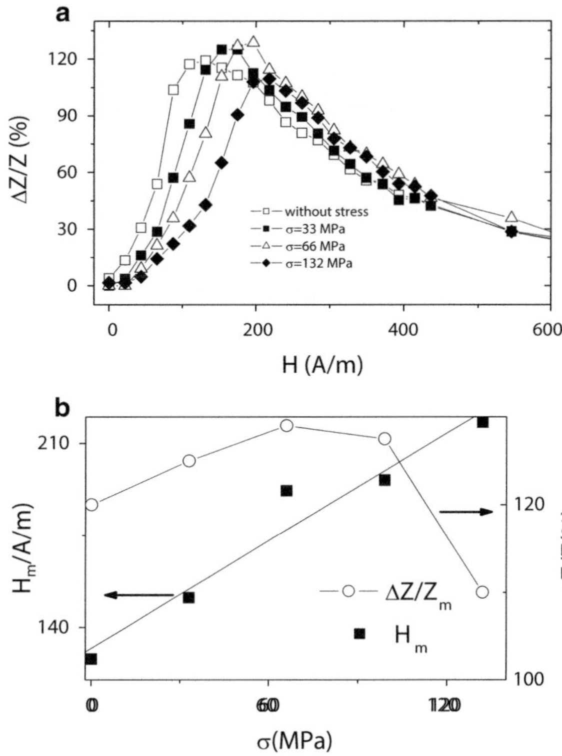

Fig. 6.36 $\Delta Z/Z(H)$ dependencies of $Co_{68.5}Mn_{6.5}Si_{10}B_{15}$ amorphous microwire measured at different σ (**a**) and H_m (σ) dependence (**b**). Reprinted from Ref. [114], Copyright (2005) with permission from Elsevier (Fig. 6)

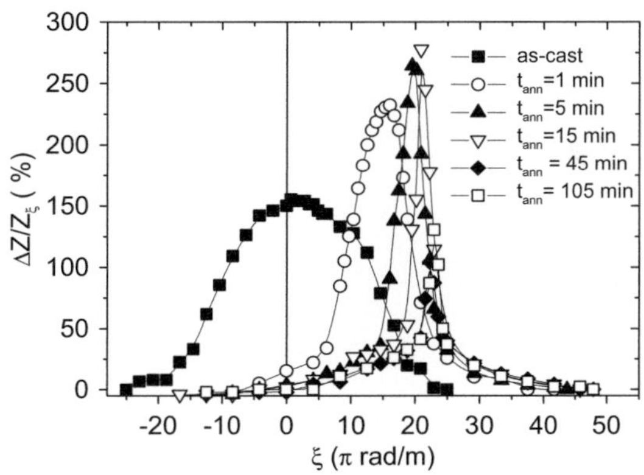

Fig. 6.37 Effect of current annealing time, T_{ann}, on $(\Delta Z/Z)_\xi$ (ξ) dependence

[111, 112]. Dependencies of $(\Delta Z/Z)_\xi$ on annealing time at fixed applied torsion during annealing and on applied torsion stress at fixed annealing time have been measured. Strong changes have been induced in the TGI ratio under Joule heating with these conditions.

Slightly asymmetric character of the $(\Delta Z/Z)_\xi$ dependence on applied torsion ξ with a broad maximum at around $\xi = 2.5$ πrad/m has been observed in as-cast state where maximum value of $(\Delta Z/Z)_\xi$ reaches about 155 % (see Fig. 6.37).

After Joule heating under applied torsion, $(\Delta Z/Z)_\xi$ has a tendency to achieve finally sharp and rather asymmetric shape with a sharp maximum at certain torsion, ξ_m. Evolution of $(\Delta Z/Z)_\xi$ dependencies with the time of current annealing, T_{ann}, as a parameter and at fixed applied torsion (25 πrad/min) during current annealing is presented in Fig. 6.37.

Consequently, the effect of the annealing time, T_{ann}, on ξ_m and $(\Delta Z/Z)_\xi$ is presented in Figs. 6.38 and 6.39. As can be observed, ξ_m increases and $(\Delta Z/Z)_\xi$ first increases and then decreases with T_{ann}.

Figure 6.40 presents an example of the evolution of $(\Delta Z/Z)_\xi$ (ξ) dependencies obtained after current annealing at 450 mA with fixed annealing time (10 min) but under different applied stress, ξ_α. Similarly in Fig. 6.37, this dependence shows an increase of the asymmetry with ξ. $(\Delta Z/Z)_\xi$ ratio shows significant improvement under certain conditions of current annealing. Pre-annealing by the Joule heating also introduces some changes in observed dependencies. Maximum $(\Delta Z/Z)_\xi$ ratio of 330 % is obtained after optimal conditions of current annealing under torsion (see Fig. 6.40).

Fig. 6.38 Dependence of ξ_m on T_{ann}

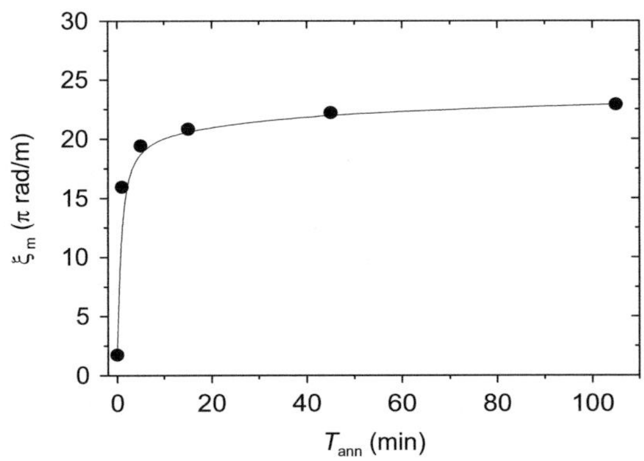

Fig. 6.39 Effect of T_{ann} on $(\Delta Z/Z)_\xi$

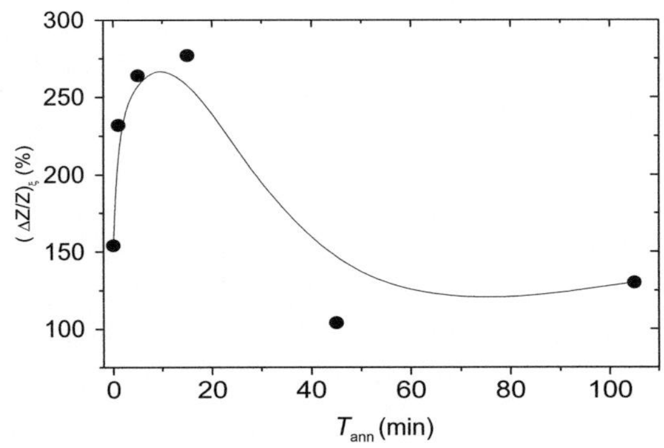

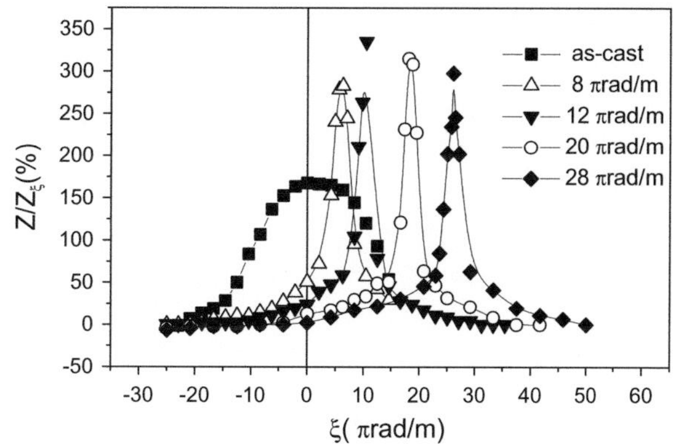

Fig. 6.40 Effect of applied torsion on $(\Delta Z/Z)_\xi(\xi)$ dependence

The asymmetry of the $(\Delta Z/Z)_\xi$ in as-cast state and after torsion annealing could be ascribed to the spontaneous or induced (after torsion annealing) helical anisotropy, which can be compensated by the application of certain torsion stress. Evolution of $(\Delta Z/Z)_\xi$ (ξ) dependencies with annealing under torsion reflects the kinetic of induced magnetic anisotropy in studied sample.

To explain the TGI effect, it was considered that the applied torsion strain induces a helical magnetic anisotropy of magnetoelastic character. Such helical anisotropy should be in competition with that ascribed to the complex internal stresses introduced during the fabrication process. Consequently, the circular magnetic permeability is enhanced by the torsion. That contribution should be connected with the tensor character of the magnetic permeability. Consequently, the contribution to the circular permeability due to the helical anisotropy should be negligible when the maximum $(\Delta Z/Z)_\xi$ is achieved.

As mentioned above Fe-rich microwires always present rectangular hysteresis loop associated with the strong axial magnetic anisotropy induced by residual stresses of mostly tensile origin [16]. Generally the conventional annealing of Fe-rich microwires does not affect the character of the hysteresis loops, changing slightly the coercive force (Fig. 6.41b). On the other hand, the stress annealing, SA, performed at the same annealing conditions ($T_{ann} = 400$ °C) but under applied stress, $\sigma = 458$ MPa, results in drastic changes of the hysteresis loop (see Fig. 6.41c): hysteresis loop becomes inclined with a magnetic anisotropy field about 1000 A/m.

A transverse magnetic anisotropy induced by the SA allows us to predict the existence of the magnetoimpedance effect in such samples. Indeed, if the samples with rectangular hysteresis loop do not present any significant GMI effect at 10 MHz, considerable GMI effect (about 13 %) has been observed in stress-annealed Fe-rich sample with inclined hysteresis loop (see Fig. 6.42).

A number of heat treatments with annealing temperatures between 100 and 400 °C have been performed (see Figs. 6.43 and 6.44) [110, 114]. Figure 6.43 shows the hysteresis loops of two samples annealed at the same temperature, T_{ann} (275 °C), with the same applied tensile stress but at different annealing time. As can

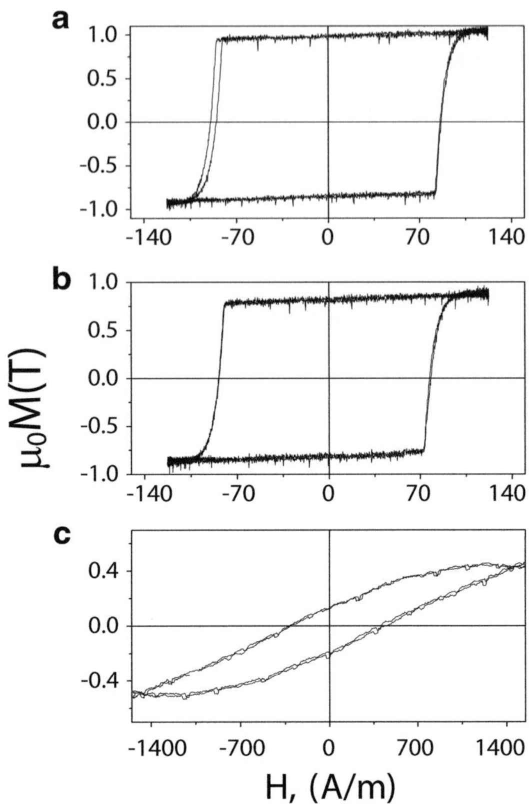

Fig. 6.41 Bulk hysteresis loops of $Fe_{74}B_{13}Si_{11}C_2$ microwire as prepared (**a**), furnace annealed at $T_{ann} = 400$ °C (**b**), and stress annealed at $T_{ann} = 400$ °C (**c**). Reprinted with permission from Ref. [114] Copyright Elsevier (2005) (Fig. 4)

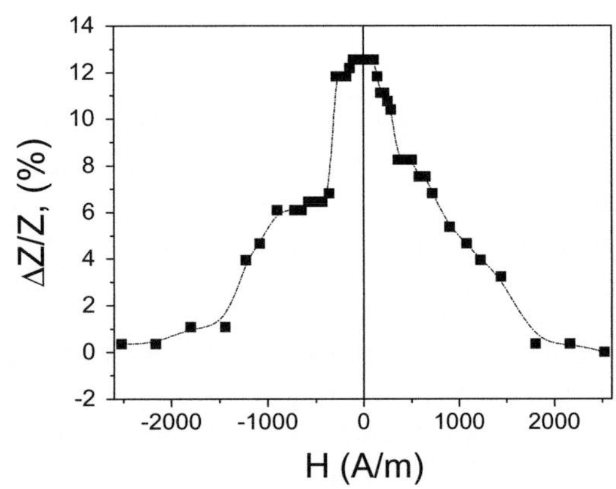

Fig. 6.42 GMI effect induced by stress annealing of Fe-rich microwire at $T_{ann} = 400$ °C. Reprinted with permission from Ref. [114] Copyright Elsevier (2005) (Fig. 12)

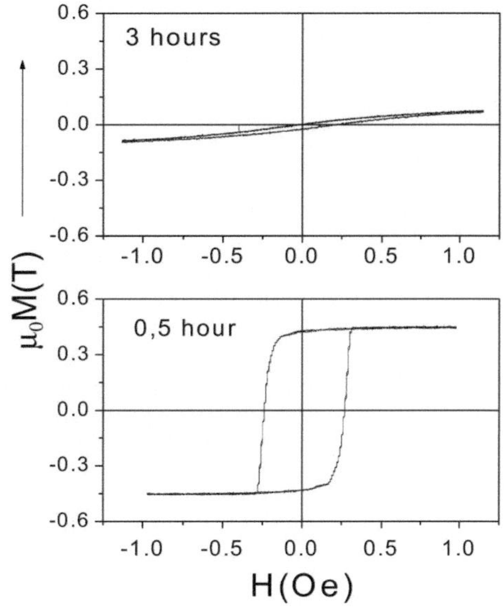

Fig. 6.43 Hysteresis loops of $Fe_{74}B_{13}Si_{11}C_2$ glass-coated microwire subjected to stress annealing at 275 °C for 0.5 and 3 h

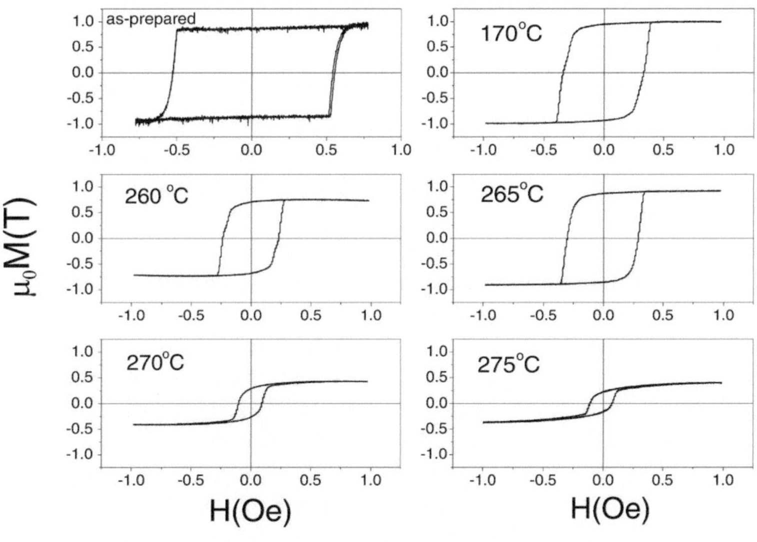

Fig. 6.44 Effect of annealing temperature on hysteresis loops of $Fe_{74}B_{13}Si_{11}C_2$ glass-coated microwire subjected to stress annealing for 0.5 h

be appreciated, the transverse magnetic anisotropy increases, increasing the SA time. Strong dependence of the transverse magnetic anisotropy reflected in the shape of axial hysteresis loop on annealing temperature has been observed. As it is shown in Figs. 6.43 and 6.44, the hysteresis loop becomes more and more rectangular, decreasing the annealing time (Fig. 6.44).

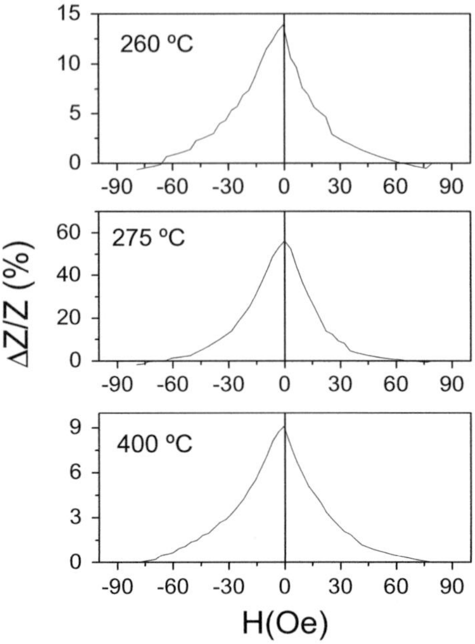

Fig. 6.45 GMI effect of stress annealed of $Fe_{74}B_{13}Si_{11}C_2$ glass-coated microwire subjected to stress annealing for 0.5 h at different temperatures

Observed changes of the shape hysteresis loops with transverse magnetic anisotropy changing from very slight to strong enough depending on the SA annealing temperature (see Fig. 6.45) allow us to predict the existence of the magnetoimpedance effect in such samples. Indeed, the samples subjected to SA exhibit significant magnetoimpedance ($\Delta Z/Z_{max} \approx 60$ %) (see Fig. 6.45). Another important result is that the application of tensile stresses to the samples subjected to the SA treatment results in the recovery of the rectangular hysteresis loop typical for the as-prepared state (see Fig. 6.46).

Consequently, disappearance of the inclined hysteresis loop under applied stress allows us to predict the stress-impedance effect, SI, in the sample subjected to SA treatment. Indeed considerable SI effect (about 60 %) has been observed under the application of tensile stress (see Fig. 6.46) in microwire previously stress annealed at 270 °C.

The origin of this creep-annealing-induced anisotropy has been attributed to both the (1) redistribution of the residual stresses during the stress annealing or (2) induced magnetic anisotropy related to the atomic ordering.

It has been assumed that after the thermal treatment under stress, the longitudinal stresses in stress-annealed samples become smaller than radial and tangential stresses, i.e., that the stress annealing results in redistribution of the internal stresses and/or local microstructure of the sample in order to minimize the magnetoelastic energy in the stressed state. As a result the easy magnetization axis in stress-annealed sample is aligned along the transverse direction.

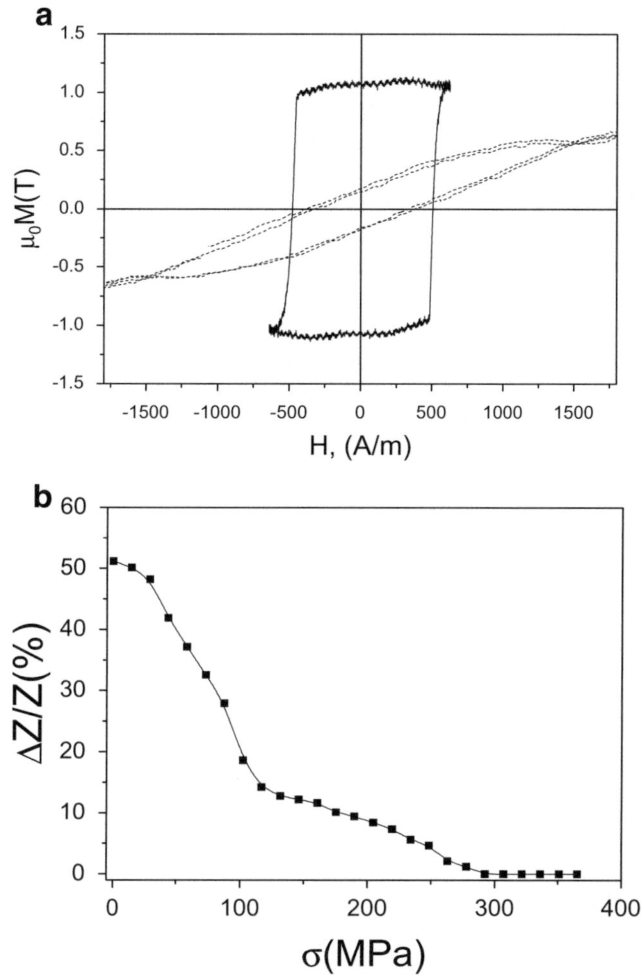

Fig. 6.46 Hysteresis loop of stress-annealed $Fe_{74}B_{13}Si_{11}C_2$ glass-coated microwire measured under applied stress of 500 MPa and hysteresis loop of $Fe_{74}B_{13}Si_{11}C_2$ subjected to SA without stress shown for the comparison by dashed line (**a**). Stress-impedance effect of stress-annealed $Fe_{74}B_{13}Si_{11}C_2$ glass-coated microwire under stress (468 MPa) at 275 °C for 0.5 h measured at 10 MHz for the driving current amplitude of 2 mA (**b**)

In this way, the removal of the mechanical load after stress annealing affects mostly the longitudinal stresses and results in drastic decrease of the longitudinal stress component and even in the appearance of the compressive longitudinal stresses (so-called back stresses).

Observed change of the hysteresis loop of the sample subjected to SA treatment when measured under applied tensile stress (Fig. 6.46a) confirms such assumption. This effect should be attributed to the increasing of the longitudinal stress component under the application of the tensile stress and consequently to the alignment of the easy magnetization axis along the highest stress component due to the positive magnetostriction constant of the studied sample. There is a competition between the magnetoelastic anisotropy induced by the applied stress (with a longitudinal easy axis) and the transverse anisotropy induced by the stress annealing. Upon overcoming the transverse anisotropy related to the creep-induced anisotropy, the longitudinal axis becomes an easy magnetization axis again and magnetic bistability recovers.

Although crystallization of amorphous materials usually results in degradation of their magnetic softness, in some cases crystallization can improve magnetically soft behavior. This is the case of so-called nanocrystalline alloys obtained by suitable annealing of amorphous metals. These materials have been introduced in 1988 by Yoshizawa et al. and later have been intensively studied by a number of research groups [22–25].

Recently considerable magnetic softening and great enhancement of GMI effect have been observed after devitrification of Finemet-type glass-coated microwires [115–119].

Magnetic softening has been achieved after annealing of the samples (see Figs. 6.47 and 6.48).

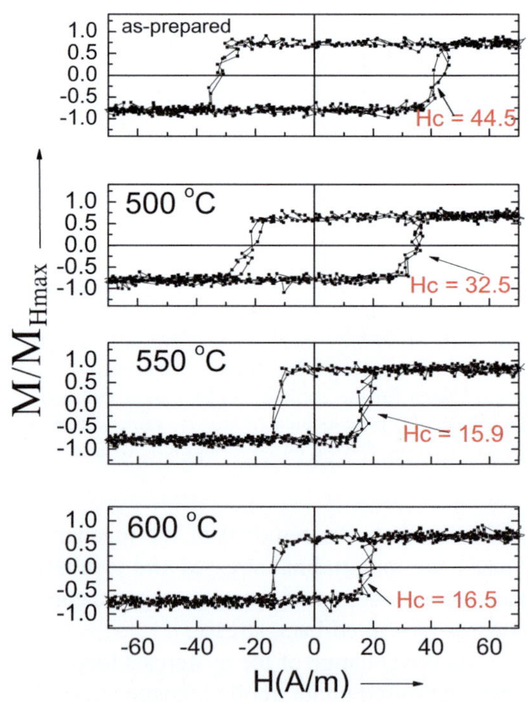

Fig. 6.47 Hysteresis loops of as-prepared and annealed $Fe_{70.8}Cu_1Nb_{3.1}Si_{14.5}B_{10.6}$ microwires, with $\rho = 0.81$, at different temperatures measured at fixed magnetic field amplitude of 225 A/m. Reprinted from [119] With kind permission from Springer Science and Business Media (2014), Fig. 6

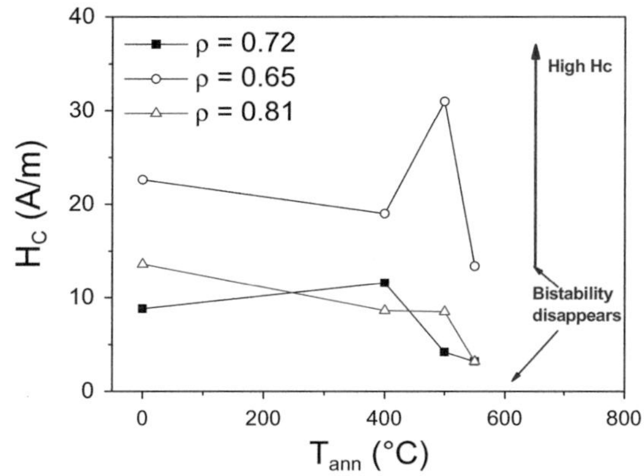

Fig. 6.48 Dependence of the coercivity, H_c, for the $Fe_{70.8}Cu_1Nb_{3.1}Si_{14.5}B_{10.6}$ glass-coated microwire with selected ρ-ratio on annealing temperature. Reprinted with permission from [120], Copyright (2013) Elsevier B.V. (Fig. 2)

In Fig. 6.48 we plotted coercivity dependence on annealing temperature for $Fe_{70.8}Cu_1Nb_{3.1}Si_{14.5}B_{10.6}$ glass-coated microwire with different ρ-ratio.

According to these data, a tendency towards decreasing H_c values is observed at the range of 500–550 °C. Above $T_{ann} \approx 600$ °C the magnetic bistability disappears (hysteresis loops cannot be considered perfectly rectangular).

Magnetic softening with the lowest value of coercivity and switching field is obtained in the samples treated at 500–600 °C which has been ascribed to the fact that the first crystallization process has been developed, leading to fine α-Fe (Si) nanocrystals with grain size around 10–20 nm, similarly to that one widely reported for *finemet* ribbons [22–25].

GMI ration has been measured in as-prepared and annealed samples. As it was expected, samples with amorphous structure in as-prepared samples exhibit a rather small GMI effect (below 5 %, see Fig. 6.49) similar to other Fe-based glass-coated microwires with positive magnetostriction. In contrast, nanocrystalline $Fe_{70.8}Cu_1Nb_{3.1}Si_{16}B_{9.1}$ microwires annealed at 550 °C exhibit higher GMI effect ($\Delta Z/Z \approx 90$ %, see Fig. 6.49).

As discussed elsewhere, after the nanocrystallization the average magnetostriction constant takes nearly zero values [22–25], thanks to the control of the crystalline volume fraction:

$$\lambda_{s,\text{eff}} = V_{cr}\lambda_{s,\text{cr}} + (1 - V_{cr})\lambda_{s,\text{am}} \tag{6.16}$$

where $\lambda_{s,\text{eff}}$ is the saturation magnetostriction coefficient and V_{cr} is the crystalline volume fraction.

As discussed elsewhere GMI effect is affected by the magnetic anisotropy [97]. Consequently, the shape of $\Delta Z/Z(H)$ dependence and maximum GMI ratio $\Delta Z/Z_{max}$ is considerably affected by annealing [97]. Additionally considerable enhancement of GMI ratio has been observed after devitrification of glass-coated microwires.

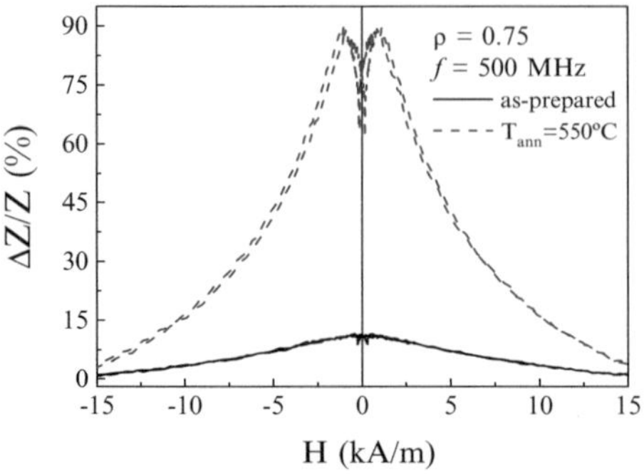

Fig. 6.49 Effect of annealing at 550 °C for 1 h on $\Delta Z/Z\,(H)$ dependence of $Fe_{70.8}Cu_1Nb_{3.1}Si_{16}B_{9.1}$ microwire with $\rho = 0.75$ measured at 500 MHz in comparison with the GMI response of the as-prepared sample. Reprinted from [119] with kind permission from Springer Science and Business Media (2014), Fig. 8

As has been shown, GMI, TGI, and SI effects have been observed in different kinds of magnetic wires. But generally cylindrical shape typical for magnetically soft wires is the most appropriate to achieve high GMI effect.

Consequently, GMI effect of different families of rapidly quenched wires can be tailored by the thermal treatment, allowing to change the magnetic field corresponding to the maximum of the GMI ratio as well as the maximum GMI ratio.

Existing and future applications of rapidly quenched wires are determined by magnetic and magnetotransport properties. Consequently, applications of magnetic microwires can be classified as follows:

1. Magnetic sensors and devices based on GMI effect
2. Magnetic sensors and devices based on magnetic bistability and domain-wall propagation
3. Magnetometers using almost zero magnetostrictive amorphous and nanocrystalline materials (flux sensors, current sensors, proximity sensors, and magnetometers)
4. Stress sensors using high magnetostriction constant ($\lambda_s = 2 - 4.5 \times 10^{-5}$). These include:
5. Thermal sensors utilizing the Curie temperature ($T = -50 \div 250$ °C)
6. Sensors based on magnetoelectric effects (inverse Wiedemann and Matteucci)
7. Stress-tunable and temperature-tunable composite materials based on thin ferromagnetic wires

At this moment the industrial application of magnetic microwires is related with high and linear response of the off-diagonal GMI effect achieved in cold-drawn and glass-coated microwires [21, 51, 52].

Additionally a number of magnetic sensors based on giant magnetoimpedance (GMI) effect and stress-impedance (SI) effect with the C-MOS IC circuitry and advantageous features compared with conventional magnetic sensors have been reported [51, 52]. The main possible applications are related with the detection of low magnetic fields, small weights, and vibrations in various branches of the industry as the car industry [51, 52, 121, 122].

On the other hand, magnetic domain-wall propagation becomes a hot topic of research because of the possibility of applications in magnetic devices, such as magnetic random access memory, integrated circuits, hard disks, domain-wall logics, etc. [72, 73]. High velocity of domain-wall propagation observed in glass-coated microwires can be attractive to transmit the information along the magnetic microwire, like it was observed recently in wires of submicrometer diameter [56]. The clear advantage of glass-coated microwires is that the domain-wall velocity is few times higher, achieving few km/s [75, 77].

There are many prototypes of magnetic sensors based on magnetoelastic properties, temperature dependence of magnetic properties, magnetic softness, and magnetic surveillance described in recent reviews [35, 36].

Among recent applications, new types of stress-tunable and temperature-tunable composite materials based on thin ferromagnetic wires with the effective microwave permittivity depending on an external DC magnetic field, applied stress, or temperature recently have been introduced [123, 124]. Such composites consist of short pieces of conductive ferromagnetic wires embedded into a dielectric matrix. The short wire inclusions play a role of "the elementary scatterers," when the electromagnetic wave irradiates the composite and induces a longitudinal current distribution and electrical dipole moment in each inclusion. A number of applications have been proposed, including the stress-sensitive media for remote nondestructive health monitoring of different structures, temperature-dependent media, and selective microwave coatings with field-dependent reflection/transmission coefficients [123, 124]. The important advantage of such applications is that the soldering problems are avoided because of the wireless detection of the signals. It is worth mentioning that thin wires with stress-sensitive magnetic anisotropy exhibiting stress-sensitive GMI effect and SI effect are quite necessary for designing such composites.

6.4 Conclusions

Rapidly quenched wires present a number of outstanding magnetic properties quite interesting from the viewpoint t of applications. Studies of magnetic properties and GMI effect of rapidly quenched wires and microwires reveal that choosing appropriate chemical composition and geometry, they can present either magnetic bistability or GMI effect. Hysteresis loops, domain-wall dynamics, and magnetic field dependences of GMI effect are affected by the magnetoelastic anisotropy. Annealing considerably affects the hysteresis loops and GMI effect of all families of wires and microwires.

The nanocrystallization of FeCuNbSiB microwires is key for the optimization of the GMI effect.

Acknowledgments Authors are grateful to Dr. A. Zhukov for the help with the chapter preparation and wish to acknowledge the contribution of Dr. J. M. Blanco, Dr. M. Ipatov, and Dr. A. Chizhik. This work was supported by the Spanish MINECO under MAT2013-47231-C2-1-P Project and by the Basque Government under Saiotek 13 PROMAGMI (S-PE13UN014) and DURADMAG (S-PE13UN007) projects.

References

1. Miroshnichenko, I.S., Salli, I.V.: A device for the crystallization of alloys at a high cooling rate. Ind. Lab. **25**, 1463 (1959) in English, Zav. Lab. **11**, 1398 (1959)
2. Klement, K., Wilens, R.H., Duwez, P.: Non-crystalline structure in solidified gold–silicon alloys. Nature **187**, 869–870 (1960)
3. Duwez, P., Williams, R.J., Klement, K.: Continuous series of metastable solid solutions in Ag-Cu alloys. Appl. Phys. **31**, 1136–1142 (1966)
4. Duwez, P.: Metastable phases obtained by rapid quenching from the liquid state. In: Reiss, H. (ed.) Progress in solid state chemistry of alloy phases, vol. 3, pp. 377–406. Pergamon Press, Oxford (1966)
5. Jones, H.: Splat cooling and metastable phases. Rep. Prog. Phys. **36**, 1425–1497 (1973)
6. Luborsky, F.E.: Amorphous metallic alloys. Butterworth & CoPublishers Ltd, London, UK (1983)
7. Humphrey, F.B., Mohri, K., Yamasaki, J., et al.: Re-entrant magnetic flux reversal in amorphous wires. In: Hernando, A., Madurga, V., Sánchez-Trujillo, M.C., Vázquez, M. (eds.) Magnetic properties of amorphous metals, pp. 110–116. Elsevier Science Publisher, Amsterdam, The Netherlands (1987)
8. Humphrey, F.B.: Article surveillance magnetic marker having a hysteresis loop with large Barkhausen discontinuities. US Patent 4,660,025, 21 Apr 1987
9. Mohri, K., Humphrey, F.B., Kawashima, K., et al.: Large barkhausen and matteucci effects in FeCoSiB, FeCrSiB, and FeNiSiB amorphous wires. IEEE Trans. Magn. **26**, 1789–1791 (1990)
10. Vázquez, M., Hernando, A.: A soft magnetic wire for sensor applications. J. Phys. D Appl. Phys. **29**, 939–949 (1996)
11. Wiegand, J.R.: Bistable magnetic device.US Patent 3,820,090, 1974
12. Heiden, C., Rogalla, H.: Barkhausen jump field distribution of iron whiskers. J. Magn. Magn. Mater. **26**, 275–277 (1982)
13. Ponomarev, B.K., Zhukov, A.: Fluctuations of start field exhibited by amorphous alloy. Sov. Phys. Solid State **26**, 2974–2979 (1984)
14. Ponomarev, B.K., Zhukov, A.P.: Effect of temperature on the start field distribution of amorphous $Co_{70}Fe_5Si_{10}B_{15}$ alloy. Sov. Phys. Solid state **27**, 444–448 (1985)
15. Mizutani, M., Katoh, H., Panina, L.V., et al.: Distance sensors utilizing a current-exciting large Barkhausen effect in twisted amorphous magnetostrictive wires. IEEE Trans. J. Magn. Japan **9**(2), 102–108 (1994)
16. Zhukov, A.: Domain wall propagation in a Fe-rich glass-coated amorphous microwire. Appl. Phys. Lett. **78**, 3106–3108 (2001)
17. Panina, L.V., Mohri, K.: Magneto-impedance effect in amorphous wires. Appl. Phys. Lett. **65**, 1189–1191 (1994)
18. Beach, R.S., Berkowitz, A.E.: Giant magnetic-field dependent impedance of amorphous FeCoSiB wire. Appl. Phys. Lett. **64**, 3652–3654 (1994)
19. Harrison, E.P., Turney, G.L., Rowe, H., et al.: The electrical properties of high permeability wires carrying alternating current. Proc. Roy. Soc. Math. Phys. Eng. Sci. **157**(891), 451–479 (1936)

20. Ciureanu, P., Rudkowska, G., Clime, L., et al.: Anisotropy optimization of giant magnetoimpedance sensors. J. Optoelectron. Adv. Mater. **6**, 905–910 (2004)
21. Zhukov, A., Ipatov, M., Churyukanova, M., et al.: Giant magnetoimpedance in thin amorphous wires: from manipulation of magnetic field dependence to industrial applications. J. Alloys Compd. **586**(Suppl. 1), S279–S286 (2014)
22. Yoshizawa, Y., Oguma, S., Yamauchi, K.: New Fe-based soft magnetic alloys composed of ultrafine grain structure. J. Appl. Phys. **64**, 6044–6046 (1988)
23. Yoshizawa, Y., Yamauchi, K.: Fe-based soft magnetic alloy composed of ultrafine grain structure. Mater. Trans. JIM **31**, 307–314 (1990)
24. Herzer, G.: Grain size dependence of coercivity and permeability in nanocrystalline ferromagnets. IEEE Trans. Magn. **26**, 1397–1402 (1990)
25. Herzer, G.: Anisotropies in soft magnetic nanocrystalline alloys. J. Magn. Magn. Mater. **294**, 99–106 (2005)
26. Zhukova, V., Zhukov, A., Kraposhin, V., et al.: Magnetic properties and GMI of soft magnetic amorphous fibers. Sens. Actuators (A) **106**, 225–229 (2003). doi:10.1016/S0924-4247(03)00172-9
27. Ohnaka, I., Fukusako, T., Ohmichi, T., et al.: Production of amorphous filament by in rotating- liquid spinning method. In: Masumoto T., Suzuki K. (eds.) Proc. 4th Inter. Conf. on rapidly quenched metals. pp. 31–34 (1982)
28. Hagiwara, M., Inoue, A., Masumoto, T.: Mechanical properties of Fe-Si-B amorphous wires produced by in-rotating water spinning method. Metall. Trans. **13A**, 373–382 (1982)
29. Ogasawara, I., Ueno, S.: Preparation and properties of amorphous wires. IEEE Trans. Magn. **31**(2), 1219–1223 (1995)
30. Masumoto, T., Ohnaka, I., Inoue, A., et al.: Production of Pd-Cu-Si amorphous wires by melt spinning method using rotating water. Scr. Metall. **15**, 293–296 (1981)
31. Maringer, E., Mobley, C.E. US Patent 3,871,439, Mar 1975
32. Rudkowski, P., Rudkowska, G., Strom-Olsen, J.O.: The fabrication of fine metallic fibers by continuous melt extraction and their magnetic and mechanical properties. Mater. Sci. Eng. A **133**, 158–161 (1991)
33. Rudkowski, P., Rudkowska, G., Strom-Olsen, J.O., et al.: The magnetic properties of sub 20 m metallic fibers formed by continuous melt extraction. J. Appl. Phys. **69**, 5017–5019 (1991)
34. Rudkowski, P., Strom-Olsen, J.O., Rudkowska, G., et al.: Ultra fine, ultra soft metallic fibers. IEEE Trans. Magn. **31**, 1224–1228 (1995)
35. Zhukov, A., Zhukova, V.: Magnetic properties and applications of ferromagnetic microwires with amorphous and nanocrystalline structure, p. 162. Nova, NewYork (2009). ISBN 978-1-60741-770-5
36. Zhukov, A., Zhukova, V.: Magnetic sensors based on thin magnetically soft wires with tunable magnetic properties and its applications. International Frequency Sensor Association (IFSA) Publishing, Spain (2014). ISBN 10: 84-617-1866-6
37. Taylor, G.F.: A method of drawing metallic filaments and a discussion of their properties and uses. Phys. Rev. **23**, 655–660 (1924)
38. Taylor, G.F.: Process and apparatus for making filaments. United States Patent Office, 1,793,529, 24 Feb 1931
39. Ulitovsky, A.V.: Micro-technology in design of electric devices, vol. 7. p. 6 Leningrad (1951)
40. Ulitovski, A.V., Avernin, N.M.: Method of fabrication of metallic microwire. Patent No161325 (USSR), 19.03.64. Bulletin No7, p. 14 (1964)
41. Ulitovsky, A.V., Maianski, I.M., Avramenco, A.I.: Method of continuous casting of glass coated microwire. Patent No128427 (USSR), 15.05.60. Bulletin. No10, p. 14 (1960)
42. Badinter, E.Ya., Berman, N.R., Drabenko, I.F. et al.: Cast microwires and its properties. Shtinica, Kishinev (1973)
43. Larin, V.S., Torcunov, A.V., Zhukov, A., et al.: Preparation and properties of glass-coated microwires. J. Magn. Magn. Mater. **249**(1–2), 39–45 (2002)

44. Zhukov, A., Kostitcyna, E., Shuvaeva, E., et al.: Effect of composite origin on magnetic properties of glass-coated microwires. Intermetallics **44**, 88–93 (2014)
45. Zhukov, A., Shuvaeva, E., Kaloshkin, S., et al.: Influence of the defects on magnetic properties of glass-coated microwires. J. Appl. Phys. **115**, 17A305 (2014)
46. Zhukov, A., Shuvaeva, E., Kaloshkin, S., et al.: Studies of the defects influence on magnetic properties of glass-coated microwires. IEEE Trans. Magn. **50**(11), 2006604 (2014)
47. Zhukov, A., Garcia, C., Ilyn, M., et al.: Magnetic and transport properties of granular and Heusler-type glass-coated microwires. J. Magn. Magn. Mater. **324**, 3558–3562 (2012)
48. Zhukov, A., García, C., Zhukova, V., et al.: Fabrication and magnetic properties of $Cu_{50}(Fe_{69}Si_{10}B_{16}C_{5})_{50}$ thin microwires. J. Non Cryst. Solids **353**, 922–924 (2007)
49. García, C., Zhukov, A., González, J., et al.: Fabrication, structural and magnetic characterization of thin microwires with novel composition $Cu_{70}(Co_{70}Fe_{5}Si_{10}B_{15})_{30}$. J. Alloys Compd. **483**, 566–569 (2009)
50. Varga, R., Ryba, T., Vargova, Z., et al.: Magnetic and structural properties of Ni-Mn-Ga Heusler-type microwires. Scr. Mater. **65**(8), 703–706 (2011)
51. Shen, L.P., Uchiyama, T., Mohri, K., et al.: Sensitive stress-impedance micro sensor using amorphous magnetostrictive wire. IEEE Trans. Magn. **33**, 3355–3359 (1997)
52. Mohri, K., Uchiyama, T., Shen, L.P., et al.: Amorphous wire and CMOS IC-based sensitive micro-magnetic sensors (MI sensor and SI sensor) for intelligent measurements and controls. J. Magn. Magn. Mater. **249**, 351–356 (2002)
53. García, C., Zhukova, V., Gonzazez, J., et al.: Magnetic and transport properties of Fe-rich thin cold-drawn amorphous wires. J. Alloys Compd. **488**, 5–8 (2010)
54. Garcia, C., Chizhik, A., del Val, J.J., et al.: Structural, magnetic and electrical transport properties in cold-drawn thin Fe-rich wires. J. Magn. Magn. Mater. **294**, 193–201 (2005)
55. Zhukova, V., Umnov, P., Molokanov, V., et al.: Magnetic properties and giant magnetoimpedance effect of ductile amorphous microwires without glass coating. Sens. Lett. **10**(3/4), 712–716 (2012)
56. Chiriac, H., Corodeanu, S., Lostun, M., et al.: Magnetic behavior of rapidly quenched submicron amorphous wires. J. Appl. Phys. **107**, 09A301 (2010)
57. Zhukov, A., Blanco, J.M., Ipatov, M., et al.: Manipulation of domain wall dynamics in amorphous microwires through the magnetoelastic anisotropy. Nanoscale Res. Lett. **7**, 223 (2012). doi:10.1186/1556-276X-7-223
58. Zhukov, A.P.: The remagnetization process of bistable amorphous alloys. Mater. Des. **5**, 299–305 (1993)
59. Zhukov, A.P., Vázquez, M., Velázquez, J., et al.: The remagnetization process of thin and ultrathin Fe-rich amorphous wires. J. Magn. Magn. Mater. **151**, 132–138 (1995)
60. Aragoneses, P., Blanco, J.M., Dominguez, L., et al.: The stress dependence of the switching field in glass-coated amorphous microwires. J. Phys. D Appl. Phys. **31**(1998), 3040–3045 (1998)
61. Zhukova, V., Zhukov, A., Blanco, J.M., Gonzalez, J., et al.: Effect of applied stress on remagnetization and magnetization profile of Co-Si-B amorphous wire. J. Magn. Magn. Mater. **242–245**, 1439–1442 (2002)
62. Zhukova, V., Zhukov, A., Blanco, J.M., et al.: Effect of applied stress on remagnetization and magnetization profile of Co-Si-B amorphous wire. J. Magn. Magn. Mater. **258–259**, 189–191 (2003)
63. Zhukova, V., Zhukov, A., Blanco, J.M., et al.: Effect of applied stress on magnetization profile of Fe-Si-B amorphous wire. J. Appl. Phys. **93**, 7208–7210 (2003)
64. Antonov, A.S., Borisov, V.T., Borisov, O.V., et al.: Residual quenching stresses in glass-coated amorphous ferromagnetic microwires. J. Phys. D Appl. Phys. **33**, 1161–1168 (2000)
65. Chiriac, H., Ovari, T.A., Zhukov, A.: Magnetoelastic anisotropy of amorphous microwires. J. Magn. Magn. Mater. **254–255**, 469–471 (2003)
66. Chiriac, H., Ovari, T.A., Pop, G., et al.: Internal stress distribution in glass-covered amorphous magnetic wires. Phys. Rev. B **42**, 10105–10113 (2005)

67. Velázquez, J., Vazquez, M., Zhukov, A.: Magnetoelastic anisotropy distribution in glass-coated microwires. J. Mater. Res. **11**(10), 2499–2505 (1996)
68. Zhukov, A., Ipatov, M., Blanco, J.M., et al.: Fast magnetization switching in amorphous microwires. Acta Phys. Pol. A **126**, 7–11 (2014)
69. Garcia Prieto, M.J., Pina, E., Zhukov, A.P., et al.: Glass coated Co-rich amorphous microwires with improved permeability. Sens. Actuators A **81**(1–3), 227–231 (2000)
70. Zhukov, A., Gonzalez, J., Blanco, J.M., et al.: Induced magnetic anisotropy in Co-Mn-Si-B amorphous microwires. J. Appl. Phys. **87**, 1402–1408 (2000)
71. Zhukov, A., Vázquez, M., Velázquez, J., et al.: Frequency dependence of coercivity in rapidly quenched amorphous materials. J. Mater. Sci. Eng. A **226–228**, 753–756 (1997)
72. Allwood, D.A., Xiong, G., Faulkner, C.C., et al.: Magnetic domain-wall logic. Science **309**, 1688–1692 (2005)
73. Hayashi, M., Thomas, L., Rettner, C., et al.: Dependence of current and field driven depinning of domain walls on their structure and chirality in permalloy nanowires. Phys. Rev. Lett. **97**, 207205(4) (2006)
74. Chen, D.-X., Dempsey, N.M., Vázquez, M., Hernando, A.: Propagating domain-wall shape and dynamics in iron-rich amorphous wires. IEEE Trans. Magn. **31**, 781–790 (1995)
75. Varga, R., Zhukov, A., Zhukova, V., et al.: Supersonic domain wall in magnetic microwires. Phys. Rev. B **76**, 32406 (2007)
76. Kunza, A., Reiff, S.C.: Enhancing domain wall speed in nanowires with transverse magnetic fields. J. Appl. Phys. **103**, 07D903 (2008)
77. Zhukov, A., Blanco, J.M., Chizhik, A., et al.: Manipulation of domain wall dynamics in amorphous microwires through domain wall collision. J. Appl. Phys. **114**, 043910 (2013)
78. Gudoshnikov, S.A., Grebenshchikov, Y.B., Ljubimov, B.Y., et al.: Ground state magnetization distribution and characteristic width of head to head domain wall in Fe-rich amorphous microwire. Phys. Stat. Sol. A **206**(4), 613–617 (2009)
79. Ekstrom, P.A., Zhukov, A.: Spatial structure of the head-to-head propagating domain wall in glass-covered FeSiB microwire. J. Phys. D Appl. Phys. **43**, 205001 (2010)
80. Sixtus, K.J., Tonks, L.: Propagation of large Barkhausen discontinuities. II. Phys. Rev. **42**, 419 (1932)
81. Ipatov, M., Zhukova, V., Zvezdin, A.K., Zhukov, A.: Mechanisms of the ultrafast magnetization switching in bistable amorphous microwires. J. Appl. Phys. **106**, 103902 (2009)
82. Konno, Y., Mohri, K.: Magnetostriction measurements for amorphous wires. IEEE Trans. Magn. **25**(5), 3623–3625 (1989)
83. Zhukov, A., Zhukova, V., Blanco, J.M., et al.: Magnetostriction in glass-coated magnetic microwires. J. Magn. Magn. Mater. **258**, 151–157 (2003)
84. Zhukov, A., Blanco, J.M., Ipatov, M., Zhukova, V.: Fast magnetization switching in thin wires: magnetoelastic and defects contributions. Sens. Lett. **11**(1), 170–176 (2013)
85. Panina, L.V., Mizutani, M., Mohri, K., et al.: Dynamics and relaxation of large Barkhausen discontinuity in amorphous wires. IEEE Trans. Magn. **27**(6), 5331–5333 (1991)
86. Panina, L.V., Ipatov, M., Zhukova, V., Zhukov, A.: Domain wall propagation in Fe-rich amorphous microwires. Physica B **407**, 1442–1445 (2012)
87. Beach, G.S.D., Tsoi, M., Erskine, J.L.: Current-induced domain wall motion. J. Magn. Magn. Mater. **320**, 1272–1281 (2008)
88. Blanco, J.M., Zhukova, V., Ipatov, M., Zhukov, A.: Magnetic properties and domain wall propagation in micrometric amorphous microwires. Sens. Lett. **11**(1), 187–190 (2013)
89. Vázquez, M., Zhukov, A.: Magnetic properties of glass-coated amorphous and nanocrystalline microwires. J. Magn. Magn. Mater. **160**, 223–228 (1996)
90. Zhukov, A., Ipatov, M., Garcia, C., et al.: From manipulation of giant magnetoimpedance in thin wires to industrial applications. J. Supercond. Nov. Magn. **26**(4), 1045–1054 (2013)
91. Hernando, A., Barandiarán, J.M.: The initial Matteucci effect. J. Phys. D Appl. Phys. **8**, 833–840 (1975)

92. Nielsen, O.V.: Magnetic anisotropy determined by differential magnetization measurements in twisted amorphous ribbons. J. Magn. Magn. Mater. **24**, 81–92 (1981)
93. Mohri, K., Humphrey, F.B., Yamasaki, J., Okamura, K.: Jitter-less pulse generator elements using amorphous bistable wires. IEEE Trans. Magn. **20**, 1409 (1984)
94. Cobeño, A.F., Blanco, J.M., Zhukov, A., et al.: Matteucci effect in glass coated microwires. IEEE Trans. Magn. **35**, 3382–3384 (1999)
95. Zhukova, V., Chizhik, A., Zhukov, A., et al.: Optimization of giant magneto-impedance in Co-rich amorphous microwires. IEEE Trans. Magn. **38**(5, part I), 3090–3092 (2002)
96. Pirota, K.R., Kraus, L., Chiriac, H., Knobel, M.: Magnetic properties and giant magnetoimpedance in a CoFeSiB glass-covered microwire. J. Magn. Magn. Mater. **221**, L243–L247 (2000). doi:10.1109/TMAG.2002.802397
97. Usov, N.A., Antonov, A.S., Lagar'kov, A.N.: Theory of giant magneto-impedance effect in amorphous wires with different types of magnetic anisotropy. J. Magn. Magn. Mater. **185**, 159–173 (1998)
98. Aragoneses, P., Zhukov, A., Gonzalez, J., et al.: Effect of AC driving current on magneto-impedance effect. Sens. Actuators A **81**(1–3), 86–90 (2000)
99. Zhukova, V., Ipatov, M., García, C., et al.: Development of ultra-thin glass-coated amorphous microwires for high frequency magnetic sensors applications. Open Mater. Sci. Rev. **1**, 1–12 (2007)
100. Vázquez, M., Zhukov, A., Aragoneses, P., et al.: Magneto-impedance of glass-coated amorphous CoMnSiB microwires. IEEE Trans. Magn. **34**(3), 724–728 (1998)
101. Phn, M.-H., Peng, H.-X.: Giant magnetoimpedance materials: fundamentals and applications. Prog. Mater. Sci. **53**, 323–420 (2008)
102. Sun, J.-F., Liu, J.-S., Xing, D.-W., Xue, X.: Experimental study on the effect of alternating-current amplitude on GMI output stability of Co-based amorphous wires. Phys. Status Solidi A **208**(4), 910–914 (2011)
103. Liu, J., Shen, H., Xing, D., Sun, J.: Optimization of GMI properties by AC Joule annealing in melt-extracted Co-rich amorphous wires for sensor applications. Phys. Status Solidi A **211** (7), 1577–1582 (2014)
104. Zhao, Y., Hao, H., Zhang, Y.: Preparation and giant magneto-impedance behavior of Co-based amorphous wires. Intermetallics **42**, 62–67 (2013)
105. Zhukova, V., Umnov, P., Molokanov, V., et al.: Magnetic properties and GMI effect of ductile amorphous microwires. IEEE Trans. Magn. **48**(11), 4034–4037 (2012)
106. Ménard, D., Britel, M., Ciureanu, P., Yelon, A.: Giant magnetoimpedance in a cylindrical conductor. J. Appl. Phys. **84**, 2805–2814 (1998)
107. Zhukov, A., Ipatov, M., Zhukova, V.: Giant magneto-impedance effect of thin magnetic wires at elevated frequencies. J. Appl. Phys. **111**, 07E512 (2012)
108. Zhukov, A., Talaat, A., Ipatov, M., Zhukova, V.: Tailoring of high frequency giant magnetoimpedance effect of amorphous Co-rich microwires. IEEE Magn. Lett. (2015)
109. Ipatov, M., Zhukova, V., Zhukov, A., et al.: Low-field hysteresis in the magnetoimpedance of amorphous microwires. Phys. Rev. B **81**, 134421 (2010)
110. Zhukov, A.: Design of magnetic properties of Fe-rich glass – coated magnetic microwires for technical applications. Adv. Funct. Mater. **16**(5), 675–680 (2006)
111. Blanco, J.M., Zhukov, A., Gonzalez, J.: Torsional stress impedance and magneto-impedance in $(Co_{0.95}Fe_{0.05})_{72.5}Si_{12.5}B_{15}$ amorphous wire with helical induced anisotropy. J. Phys. D Appl. Phys. **32**, 3140–3145 (1999)
112. Blanco, J.M., Zhukov, A., Gonzalez, J.: Asymmetric torsion stress giant magnetoimpedance in nearly-zero magnetostrictive amorphous wires. J. Appl. Phys. **87**(9), 4813–4815 (2000)
113. Zhukov, A.: Glass-coated magnetic microwires for technical applications. J. Magn. Magn. Mater. **242–245**, 216–223 (2002)
114. Zhukov, A., Zhukova, V., Blanco, J.M., Gonzalez, J.: Recent research on magnetic properties of glass-coated microwires. J. Magn. Magn. Mater. **294**, 182–192 (2005)

115. Talaat, A., Ipatov, M., Zhukova, V., et al.: Giant magneto-impedance effect in thin Finemet nanocrystalline microwires. Phys. Status Solidi C **11**(5–6), 1120–1124 (2014)
116. Zhukov, A.P., Talaat, A., Ipatov, M., et al.: Effect of nanocrystallization on magnetic properties and GMI effect of microwires. IEEE Trans. Magn. **50**(6), 2501905 (2014)
117. Talaat, A., Zhukova, V., Ipatov, M.: Effect of nanocrystallization on giant magnetoimpedance effect of Fe-based microwires. Intermetallics **51**, 59–63 (2014)
118. Talaat, A., Zhukova, V., Ipatov, M., et al.: Optimization of the giant magnetoimpedance effect of Finemet-type microwires through the nanocrystallization. J. Appl. Phys. **115**, 17A313 (2014)
119. Zhukova, V., Talaat, A., Ipatov, M., et al.: Effect of nanocrystallization on magnetic properties and GMI effect of Fe-rich microwires. J. Electron. Mater. **43**(12), 4540–4547 (2014). doi:10.1007/s11664-014-3370-4
120. Churyukanova, M., Zhukova, V., Talaat, A., et al.: Correlation between thermal and magnetic properties of glass coated microwires. J. Alloys Compd. **615**(SUPPL 1), S242–S246 (2014). doi:10.1016/j.jallcom.2013.11.191
121. Honkura, Y.: Development of amorphous wire type MI sensors for automobile use. Magn. Magn. Mater. **249**, 375–381 (2002)
122. Mohri, K., Honkura, Y.: Amorphous wire and CMOS IC based magneto-impedance sensors—Origin, topics, and future. Sens. Lett. **5**(2), 267–270 (2007)
123. Peng, H.X., Qin, F.X., Phan, M.H., et al.: Co-based magnetic microwire and field-tunable multifunctional macro-composites. J. Non Cryst. Solids **355**, 1380–1386 (2009)
124. Panina, L., Ipatov. M., Zhukova. V. et al.: Tuneable composites containing magnetic microwires, chapter 22: 431-460 DOI: 10.5772/21423. In: Cuppoletti, J. (ed) Metal, ceramic and polymeric composites for various uses, InTech - Open Access Publisher (www.intechweb.org), Janeza Trdine, 9, 51000 Rijeka, Croatia, DOI: 10.5772/1428 ISBN: 978-953-307-353-8 (ISBN 978-953-307-1098-3) (2011)

Chapter 7
Bimagnetic Microwires, Magnetic Properties, and High-Frequency Behavior

Manuel Vázquez, Rhimou ElKammouni, Galina V. Kurlyandskaya, Valeria Rodionova, and Ludek Kraus

7.1 Outlook Around Multilayer Microwires

Amorphous magnetic metals are being investigated because of their outstanding magnetic behavior that makes them especially suitable as sensing elements in various devices [1]. Their particular magnetic behavior is a consequence of the intrinsic atomic disordering that in addition results in very interesting fundamental phenomena. Glassy metals are prepared by rapid solidification techniques that enable their preparation in planar (ribbons or thin films [2]) or cylindrical (wires) shapes [3]. Alternatively, amorphous alloys with interesting magnetic behavior can be also obtained as bulk material or can give rise to ultrasoft magnetic alloys after

M. Vázquez (✉)
Instituto de Ciencia de Materiales de Madrid, ICMM/CSIC, 28049, Madrid, Spain
e-mail: mvazquez@icmm.csic.es

R. ElKammouni
Instituto de Ciencia de Materiales de Madrid, ICMM/CSIC, 28049, Madrid, Spain

Laboratory of Magnetic Sensors, UrFU, 620002 Ekaterinburg, Russia
e-mail: elkammounirhimou@gmail.com

G.V. Kurlyandskaya
Laboratory of Magnetic Sensors, UrFU, 620002 Ekaterinburg, Russia

Departamento de Electricidad y Electrónica, Universidad del País Vasco, UPV/EHU, 48080 Bilbao, Spain
e-mail: galina@we.lc.ehu.es

V. Rodionova
Immanuel Kant Baltic Federal University, 236041 Kaliningrad, Russia
e-mail: vrodionova@innopark.kantiana.ru

L. Kraus
Institute of Physics, Academy of Sciences of the Czech Republic, 182 21 Prague 8, Czech Republic
e-mail: kraus@fzu.cz

A. Zhukov (ed.), *Novel Functional Magnetic Materials*, Springer Series in Materials Science 231, DOI 10.1007/978-3-319-26106-5_7

suitable thermal treatments resulting in nanocrystallization [4, 5]. Amorphous magnetic wires show specific characteristics deriving from their cylindrical symmetry, and they are prepared by two different techniques: (1) by in-rotating-water quenching (100–150 μm diameter) and (2) by quenching and drawing (1–30 μm diameter). The microwires with larger diameter were systematically investigated along the 1980s, showing fascinating magnetic effects as magnetic bistability, characterized by the propagation of a single-domain wall resulting in a giant Barkhausen jump, and very high initial susceptibility that gives rise to giant magnetoimpedance effect [6]. Along the last 20 years, interest has been readdressed toward thinner glass-coated microwires [7], profiting of the similar effects and rather smaller size of interest for miniaturization. Particularly, they have been shown to be very relevant because of their dynamic properties around the ferromagnetic resonance and for a number of applications in electronic devices as sensing element [8].

Early attempts to fabricate multilayer microstructures with cylindrical geometry were performed on in-rotating-water-quenched microwires [9, 10] or by electrodeposition and sputtering techniques showing an interest for their magnetic bistability and microwave absorption [11–13]. Microwires with multilayer geometry structure containing two magnetic phases separated by an insulating layer have been introduced more recently [14, 15]. A schematic view of such multilayer microwire is depicted in Fig. 7.1. The fabrication process with micrometer size diameter requires an intermediate step where noble metal is sputtered onto the glass which serves as electrode in the subsequent electroplating as well as buffer layer insuring the substrate roughness reduction. Multilayer microwires present interesting properties in terms of magnetic couplings, while the electrical insulation between the two metallic layers plays an important role for certain technological applications. The fabrication process allows one the selection of a wide range of alloy compositions with tailored magnetic character for the magnetic nucleus and shell – soft/soft, soft/hard, or hard/hard – where the nucleus is typically amorphous or nanocrystalline [16] and the shell polycrystalline. The presence of the external shell induces significant mechanical stresses (i.e., magnetoelastic anisotropy) in the internal nucleus, and in addition, its stray field can bias the magnetic behavior of the system [17–19]. These magnetic interactions between magnetic layers give rise to tailored magnetic behavior as asymmetrical magnetoimpedance [20] and multiabsorption phenomena or enhanced stress sensitivity to mechanical stresses [21]. These properties make multilayer microwires very suitable for several applications as elements in field, stress or temperature sensor devices [22], orthogonal flux-gates [23, 24], biomedical applications [25], or microactuators [26]. For an overall information about the state of the art and applications, the reader is addressed elsewhere [27, 28].

This article reviews most relevant magnetic properties of bimagnetic microwires based on glass-coated microwires. The introductory sections focus on the preparation techniques and the phenomenology of magnetization process, aspects that have been thoroughly studied in recent manuscripts. Afterwards, we pay particular attention to the latest experimental results about the microwave properties and the

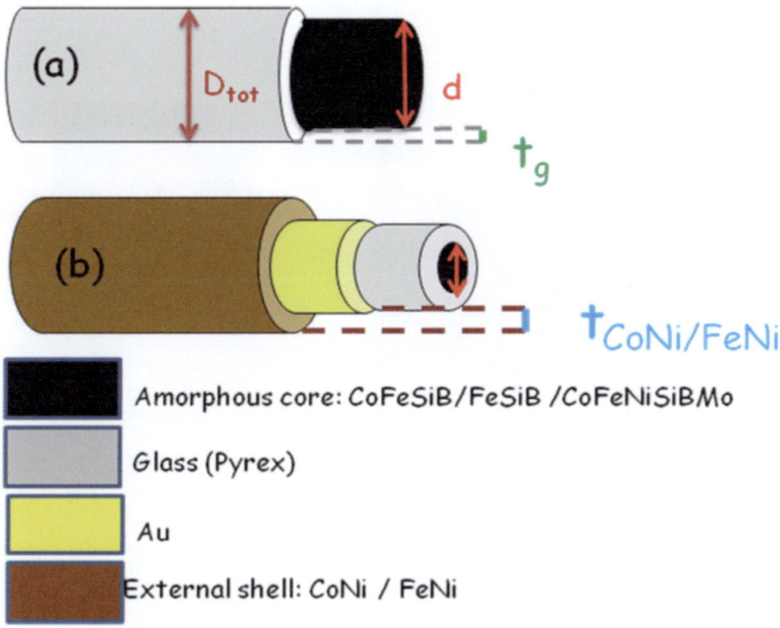

Fig. 7.1 Scheme of glass-coated (single-phase) microwire (**a**) and multilayer (biphase) microwire (**b**), with indication of thickness and composition of respective layers

influence of the measuring temperature. We conclude with a general overview about present trends around fundamental and applied research on this family of microwires.

7.2 Synthesis of Biphase Magnetic Microwires

Multilayer microwires are designed to consist of two electrically isolated magnetic phases, namely, a ferromagnetic nucleus and a ferromagnetic outer shell separated by an intermediate insulating dielectric layer. They are prepared by the combination of rapid quenching, sputtering and electrodeposition techniques. Firstly, the precursor magnetic nucleus covered by a Pyrex layer is directly obtained by quenching and drawing, a modified Taylor–Ulitovsky technique. Figure 7.2 shows images of the fabrication facility and a picture during the quenching process. This rapid solidification procedure to fabricate amorphous microwires has been described in detail in earlier works [3, 5]. The quenching rate is of the order of 10^5 K/s to enable the magnetic metallic nucleus to exhibit amorphous structure. The diameter of the metallic nucleus ranges typically between 1 and 20 μm while the thickness of the Pyrex-like glass cover can be tailored roughly between 2 and 10 μm. Three types of alloys with soft magnetic behavior have been considered in this chapter according to its saturation magnetostriction: Fe-based ($\lambda_s = 3 \times 10^5$), Co-based ($\lambda_s = -1 \times 10^6$), and CoFe-based alloys ($\lambda_s = -1 \times 10^7$). The amorphous microstructure of these alloys is confirmed by X-ray diffraction analysis.

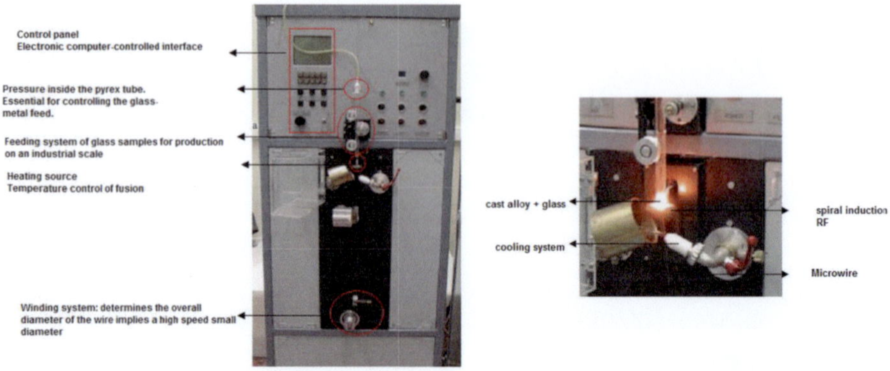

Fig. 7.2 Fabrication of glass-coated microwires. Images of the facility for quenching and drawing at the Instituto de Ciencia de Materiales de Madrid, ICMM/CSIC

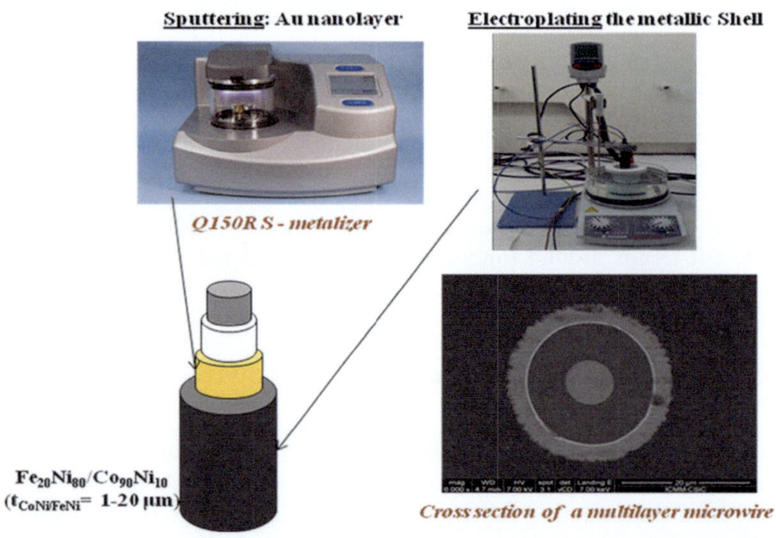

Fig. 7.3 Images illustrate the process of sputtering an intermediate Au nanolayer on top of the glass-coated microwire as well as the final electrochemical deposition of the external magnetic shell to obtain the multilayer microwire (its cross section is also visualized by optical microscopy after polishing)

Afterwards, an Au nanolayer (typically 20–30 nm thick) is grown on top of the glass surface using commercial sputtering (metallizer) system which is later used as an electrode for the final electroplating of the magnetic outer layer. Figure 7.3 illustrates the process of Au sputtering and electrochemical deposition of the external shell. Pictures of the simple equipments employed at the ICMM/CSIC, Madrid, are shown together with the cross-section image of a final multilayer bimagnetic-phase microwire. For the final galvanostatic electrodeposition, the sputtered Au is used as cathode and a Pt mesh with cylindrical geometry as anode. The control of the current density ($j = 12$ mA/cm^2 in the experiments) was carried out using a potentiostat/galvanostat power supply AMEL Instruments 2053, the temperature was adjusted to 40 °C and 55 °C, and the electrodeposited was performed under magnetic stirring for a maximum time between 60 and 90 min.

Regarding the composition of the external shell, we have selected FeNi and CoNi alloys with soft and relatively hard magnetic behavior, respectively. As for the electrolytes, in the case of FeNi plating, the bath was composed of $FeSO_4 \times 7H_2O$ (8 g/l), $NiSO_4 \times 6H_2O$ (125 g/l), $NiCl_2 \times 6H_2O$ (20 g/l), H_3BO_3 (40 g/l), and saccharin (6 g/l) in demineralized water [29]. Boric acid (H_3BO_3) is employed as an agent to stabilize the pH which in this case was adjusted to 2 and 2.80, by adding KOH to the solution. In the case of CoNi, the electroplating was performed in an aqueous solution of $NiSO_4 \times 6H_2O$ (150 g/l), $NiCl_2 \times 6H_2O$ (22.5 g/l), H_3BO_3 (45 g/l), $7H_2O$ $CoSO_4$ (150 g/l), and $CoCl_2 \times 6H_2O$ (22.5 g/l) [30]. CoNi solution has a pH of 4.4. The particular composition of the alloys is controlled tailoring the current density of the electroplating, and verified by EDS, microanalysis at surface and X-ray fluorescence analysis. The thickness, varied up to around 10 μm, is nearly proportional to both electrodeposition time and current density. Crystalline structure of FeNi and CoNi alloys shows *fcc* (face-centered cubic) and *hcp* (hexagonal close-packed) crystalline phases, respectively, as determined by XRD.

Several biphase magnetic configurations have been selected. The amorphous nucleus exhibits always soft magnetic behavior either with vanishing (CoFe-based) or large positive (Fe-based) magnetostriction, while the polycrystalline shell shows soft (FeNi) or hard (CoNi) behavior:

1. Soft/soft (*CoFe/FeNi*): $(Co_{0.94}Fe_{0.06})_{72.5}Si_{12.5}B_{15}/Fe_{20}Ni_{80}$ and $Co_{67.1}Fe_{3.8}Ni_{1.4}Si_{14.5}B_{11.5}Mo_{1,7}/Fe_{20}Ni_{80}$
2. Soft/soft (*Fe/FeNi*): $Fe_{76}Si_9B_{10}P_5/Fe_{20}Ni_{80}$ and $Fe_{77,5}Si_{7,5}B_{15}/Fe_{20}Ni_{80}$
3. Soft/hard (*CoFe/CoNi*): $(Co_{0.94}Fe_{0.06})_{72.5}Si_{12.5}B_{15}/Co_{90}Ni_{10}$ and $Co_{67.1}Fe_{3.8}Ni_{1.4}Si_{14.5}B_{11.5}Mo_{1,7}/Co_{90}Ni_{10}$
4. Soft/hard (*Fe/CoNi*): $Fe_{77,5}Si_{7,5}B_{15}/Co_{90}Ni_{10}$

For simplicity, along the chapter *nuclei* composition is labeled as CoFe and Fe, while the shell alloys are termed as CoNi or FeNi.

7.3 The Magnetization Reversal in Bimagnetic Microwires

7.3.1 Room Temperature Hysteresis Loops

A first magnetic characterization of the magnetic multilayer microwires is performed through the hysteresis loops taken in vibrating sample magnetometer (KLA Tencor EV7 VSM, LOT-Oriel) installed at the ICMM/CSIC, Madrid, under

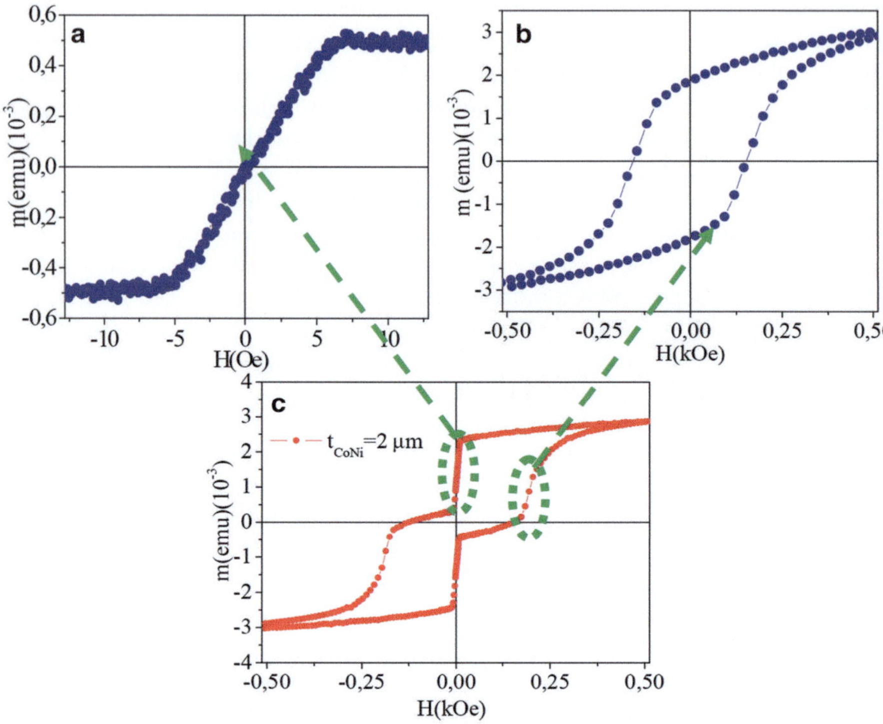

Fig. 7.4 Low-field (**a**) and high-field (**b**, **c**) hysteresis loops corresponding to single-phase CoFe soft amorphous nucleus, thickness $t_{CoNi} = 8.5$ μm (**a**), to single-phase CoNi external shell, $t_{CoNi} = 2$ μm (**b**), and CoFe/CoNi biphase microwire

applied magnetic field parallel to the microwire axis (temperature range 100–400 K). Biphase microwires present a double magnetization process whose individual contributions are better differentiated for soft/hard magnetic configuration. A typical example is shown in Fig. 7.4 for a soft/hard CoFe/CoNi microwire. Figure 7.4a shows the hysteresis loop of the precursor soft CoFe glass-coated amorphous microwire. This loop is typical of a microwire with relatively small and negative magnetostriction: this is nearly non-hysteretic and exhibits a well-defined transverse (circular) magnetic anisotropy with anisotropy field of around 7 Oe. Its domain structure is characterized by a main circumferential domain but containing an inner axial vortex structure [31]. Figure 7.4b depicts the loop for the CoNi hard magnetic shell (it was prepared for the experiment on a Pyrex capillary with similar diameter as in the sample of Fig. 7.4a). Figure 7.4c shows the hysteresis loop for the bimagnetic microwire where the magnetization processes of the two magnetic phases are clearly identified. Note that the fractional volume of each magnetic phase is given by the fractional magnetization jump (very similar in this particular example).

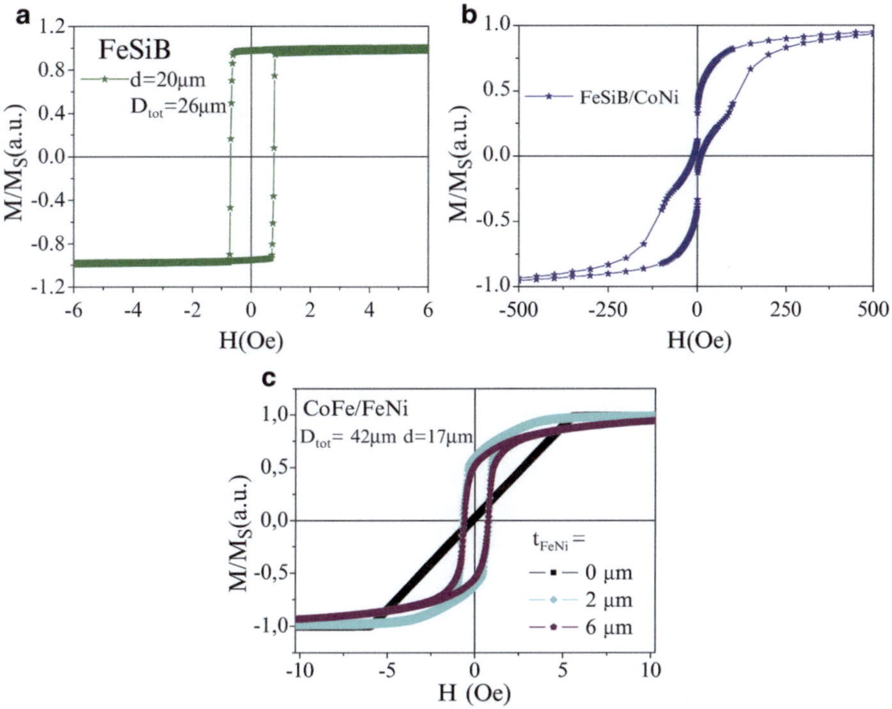

Fig. 7.5 Hysteresis loops corresponding to several microwires with indicated metallic, *d*, and total, D_{tot}, diameters: Fe single-phase (**a**), Fe/CoNi biphase, $t_{CoNi} = 3$ μm with magnetostrictive nucleus (**b**), and CoFe/FeNi biphase microwire with non-magnetostrictive nucleus (**c**)

The overall magnetic behavior of the bimagnetic microwires is determined by the nature of the phases as well as by the strength and nature of magnetic interactions, magnetoelastic and magnetostatic, between them. The magnetoelastic contribution appears in all the cases as a consequence of the stresses induced by the shell. The magnetostatic interactions are observed in the case of soft/hard microwires after submitting the sample to a saturating field so that, after its release, at remanence or under low field, the soft phase is magnetically biased by the stray field. In this overview we will consider the effect of magnetoelastic interactions in more detail. Figure 7.5a shows the hysteresis loop of FeSiB single-phase microwire with typical bistable magnetic behavior originating in its high and positive magnetostriction. The influence of the external shell is deduced in Fig. 7.5b, where the bistable behavior is deteriorated owing to the compressive stresses induced in the nucleus as has been reported elsewhere (see [27] and references inside).

In the case of soft/soft biphase microwires, the reversal process of each phase can overlap. As an example, Fig. 7.5c shows the hysteresis loops for CoFe/FeNi soft/soft microwire. This figure includes the nearly non-hysteretic loop for the small and negative magnetostriction CoFe nucleus with well-defined circular anisotropy. For the biphase microwires, we observe a large Barkhausen jump at around 1 Oe

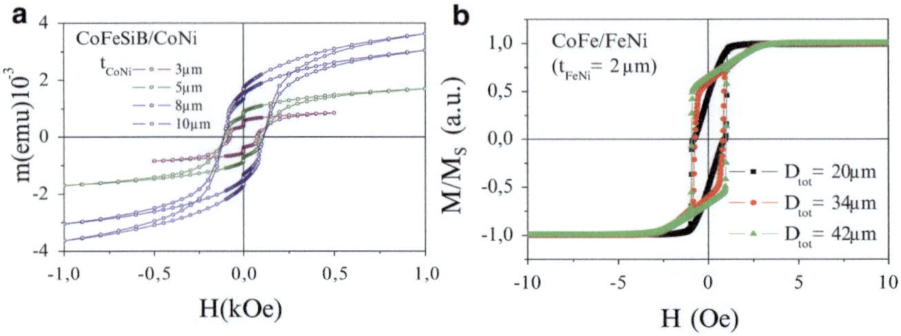

Fig. 7.6 Hysteresis loops of soft/hard CoFe/CoNi biphase microwire with increasing CoNi thickness (**a**) and of soft/soft CoFe/FeNi with increasing thickness of intermediate Pyrex layer, glass-coated CoNi diameter, D_{tot}, (**b**) [32]

applied field ascribed to the FeNi (Permalloy) external shell plus a nearly non-hysteretic region corresponding mainly to the CoFe nucleus. In this case, the determination of each magnetic fractional volume is not so straightforward.

7.3.2 *Influence of Layers Thickness*

While the general magnetic behavior of bimagnetic microwires is obviously determined by the magnetic nature of each magnetic phase, their particular properties can be tailored through the geometry (i.e., thickness) of each layer. The increase of the external layer gives rise to an enhanced fractional volume and to mechanical stresses in the nucleus that couple with the magnetostriction constant to result in the corresponding magnetoelastic anisotropy [32].

Figure 7.6 shows two examples about the influence of the thickness of external layers for the same CoFe soft amorphous nucleus. In Fig. 7.6a, we observe the variation of the high-field hysteresis loop of CoFe/CoNi with an increase of the thickness of the hard shell. Obviously, the total magnetic signal increases with the fractional volume of the CoNi shell which results in an enhanced remanence. Macroscopic coercivity also increases with that thickness. In fact, both remanence and coercivity correspond to those magnitudes of the shell for a sufficiently high thickness.

Apart from the external shell, the thickness of the intermediate insulating Pyrex layer also plays a significant role. Figure 7.6b shows the hysteresis loops of soft/soft CoFe/FeNi bimagnetic microwires for different values of the total diameter, D_{tot}, of the precursor glass-coated CoFe base microwire with constant metallic diameter ($d = 17$ μm) and increasing the thickness of Pyrex. Note that after the irreversibility determined by the FeNi shell, the nearly reversible region of the loop corresponding to the nucleus shows a susceptibility that decreases with the thickness of the intermediate Pyrex layer. Such hardening of the soft nucleus is actually determined by the mechanical stresses induced by the Pyrex.

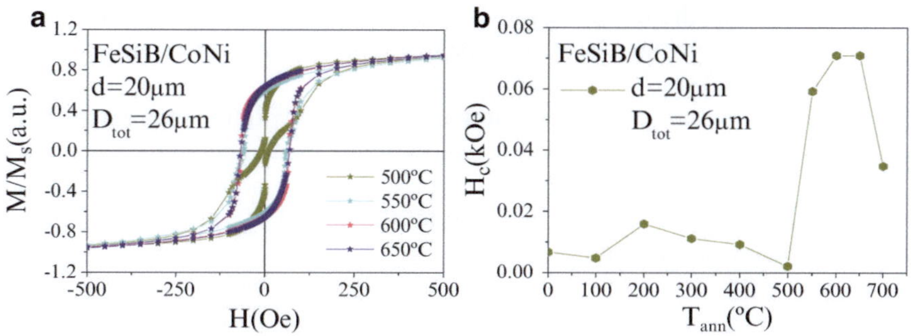

Fig. 7.7 Hysteresis loops of soft/hard FeSiB/CoNi (**a**) biphase microwires after annealing at indicated temperatures above the crystallization of the amorphous nucleus. Evolution of the coercive field (**b**) with annealing temperature (adapted from [33])

7.3.3 *Influence of Thermal Treatments*

Thermal treatments at high enough temperature and sufficiently long duration modify irreversibly the microstructure of the materials giving rise to significant changes in the magnetic properties. In the present case, we are dealing with biphase microwires whose internal nucleus is structurally amorphous while the shell shows polycrystalline structure. Their response to thermal annealing is thus somehow different. In this section we will describe some effects of low-temperature heating in the soft nucleus while in a subsequent section we will consider also some modifications in the external shell.

Thermal treatments were performed in the temperature range up to 700°C ranging from 100 to 700 °C in Argon atmosphere for 1 h. The cooling to room temperature took between 1 and 2 h, depending on the annealing temperature. The influence of thermal treatments has been determined from the hysteresis loops. We can distinguish two main ranges of annealing temperature, below and above the crystallization temperature of the soft amorphous nucleus. The crystallization temperature of Fe and CoFe-based amorphous alloys is in the range 500–600 °C for heating rates of around 10 °C/min. We consider here the case of soft/hard Fe/CoNi bimagnetic microwires whose Fe-base single-phase nucleus exhibits bistable behavior.

Figure 7.7a shows the hysteresis loops as a function of annealing temperature in the range from 500 to 700 °C. The two magnetization regions are still observed after annealing at 500 °C. After annealing below that temperature, the hysteresis loops retain very similar magnetic behavior, and only relatively small variations in the coercivity and remanence are observed which are ascribed to the partial relaxation of the amorphous structure and consequently of its internal mechanical stress. However, after annealing above 500 °C, a quite noticeable magnetic hardening is observed, and finally, magnetization takes place in apparently a single process.

This is ascribed to the crystallization of the amorphous nucleus. In the present example particularly, the result of crystallization is the growth of α-FeSi and Fe-borides grains [1].

Figure 7.7b summarizes the evolution of the coercivity of the biphase microwire as a function of annealing temperature. At low annealing temperatures, coercivity is determined by that of the soft nucleus, where even a kind of small softening of stresses is detected just before annealing at above 500 °C where the mentioned crystallization process takes place.

7.4 Temperature Dependence of Magnetic Properties

This family of multilayered biphase magnetic microwires has been recently introduced. The strongest attention has been paid in previous reports to the magnetization process and the magnetic interactions between phases, to some microwave properties, and to their technological applications in sensor devices. However, there are only very few publications around their temperature dependence below (in - single-phase microwires [34, 35]) and above room temperature [36, 37]. In this section we summarize most recent results obtained by the coauthors on the temperature dependence of static properties, while in the next section we will deal with the matter at the microwave frequency range.

7.4.1 *Low-Temperature Behavior*

The hysteresis loops at low frequency were measured in the temperature range 10–300 K for biphase microwires with different compositional configuration in the VSM facilities at the ICMM/CSIC, Madrid, and the University of the Basque Country, Bilbao. Figure 7.8 shows the data for selected alloy compositions. In Fig. 7.8a, the loops correspond to the soft/hard Fe/CoNi biphase system with magnetostrictive soft nucleus. The different contributions of the two phases to the magnetization process are clearly observed by the different values of the applied fields at which we observe irreversible jumps of the magnetization. Note that low-field jump ascribed to the soft Fe-based nucleus occurs at around 1 Oe applied field while the high-field irreversible jump ascribed to the CoNi shell is observed at around 100 Oe. Figure 7.8b shows the temperature dependence of the coercivity ascribed to the CoNi hard magnetic phase. The corresponding coercivity for the soft nucleus is barely observable at the same scale. The inset shows that variation for the soft phase in the biphase microwire (a significant error is found in its quantification) together with the result for the single-phase glass-coated microwire (which follows a standard monotonic evolution).

In the second example, given in Fig. 7.9, we present the data for the soft/soft CoFe/FeNi system. Here, the overall temperature dependence of the low-field loops

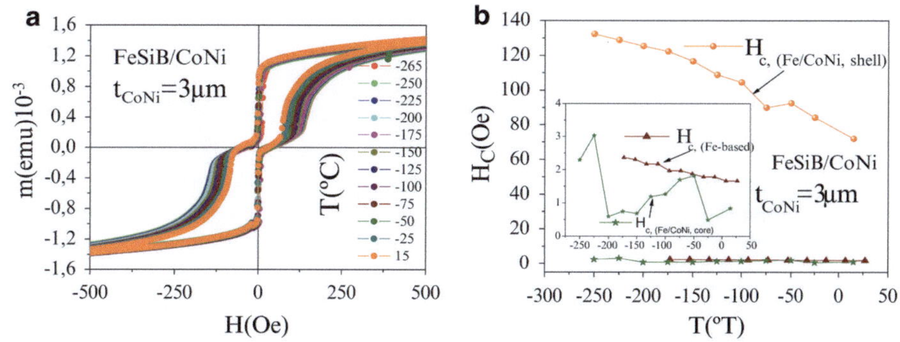

Fig. 7.8 Low-temperature dependence of hysteresis loops for the soft/hard Fe/CoNi biphase system (**a**) and quantitative values for the irreversibility field (coercivity) ascribed to the hard phase (**b**). *Inset* shows data at an enlarged scale for the soft nucleus which is compared to the single-phase microwire

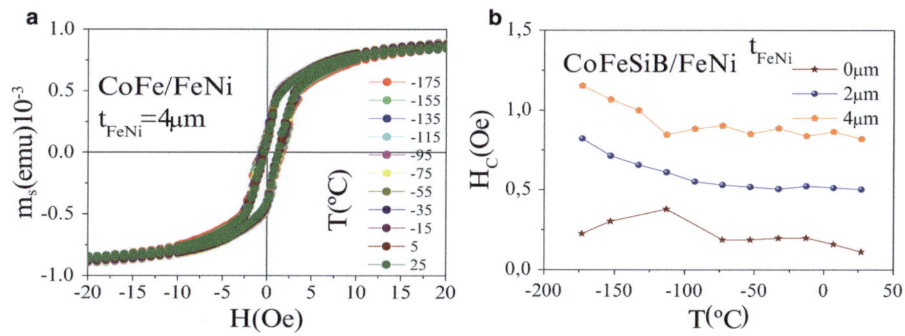

Fig. 7.9 Low-temperature dependence of hysteresis loop for the soft/soft CoFe/FeNi biphase system (**a**) and coercivity, ascribed to the FeNi soft phase (**b**) [39]

(see Fig. 7.9a) is much reduced. As indicated above, the irreversibility corresponds to the FeNi external shell while the higher-field nearly non-hysteretic is ascribed to the soft nucleus. The temperature dependence of coercivity is presented in Fig. 7.9b, corresponding to CoFe single-phase microwire and the biphase microwire with two thicknesses of the shell.

We should underline that the magnetization process in these biphase systems is actually determined in a significant manner by the differential temperature dependence of the metallic layers and that of the intermediate insulating layer. Mechanical thermal stresses are introduced as the measuring temperature is reduced as a consequence of the different thermal expansion coefficients of the layers. That contributes to the thermal dependence of coercivity and other magnetic magnitudes. A proper systematic study is thus required for a proper quantification of the presented behavior.

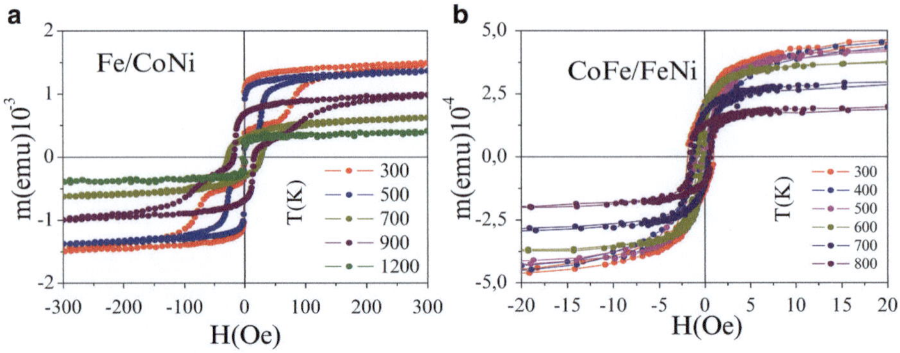

Fig. 7.10 High-temperature dependence of hysteresis loops for soft/hard Fe/CoNi (**a**) and soft/soft CoFe/FeNi (**b**) biphase microwire systems

7.4.2 High-Temperature Behavior

The high-temperature-dependent properties have been independently measured for the whole set of biphase microwires in the temperature range from 25 to 925 °C at the Lake Shore VSM (7400 series) magnetometer installed in Immanuel Kant Baltic Federal University, Kaliningrad. The commercial equipment was optimized for a high magnetic field resolution of 0.02 Oe to measure the magnetic moment at high temperature. Figure 7.10 shows the temperature dependence for two selected biphase systems, namely, soft/hard Fe/CoNi (Fig. 7.10a) and soft/soft CoFe/FeNi (Fig. 7.10b). Very complex dependence is observed in both families of microwires where various magnetic behaviors appear as a consequence of the structural changes occurring during the heating.

In order to analyze in more detail those changes, the temperature dependence of the received magnetic moment under 10 Oe applied magnetic field is shown for the CoFe-based and Fe-based single- and biphase systems in Fig. 7.11a, b, respectively. That enables us to follow the structural and magnetic-phase transformations sensitively detected by the thermomagnetic analysis. Figure 7.11a depicts the results for CoFeSiB single-phase and biphase microwires. We firstly identify the Curie point of the amorphous CoFe nucleus, $T_{C,am\text{-}CoFe} \approx 377$ °C, in the cases of CoFe single-phase and of the CoFe/FeNi biphase microwires. Data for similar CoFeSiB amorphous alloy ribbons give values in the same range [41, 42]. The crystallization into Co-rich phases is expected to occur at the temperature $T_{x,CoFe} \approx 567$ °C [43]; however, it seems that the applied field is likely not large enough to receive a significant magnetic response. From the data for the CoFe/FeNi biphase system, we can also identify the magnetic-phase transition of the FeNi shell, at $T_{c,FeNi} \approx 601$ °C, that agrees well with data reported in the literature [44, 45]. The reduction of the magnetic signal at the Curie point of the CoFe amorphous phase is less pronounced owing to its small fractional magnetic weight. We should mention that the Curie point, 1075 °C [37], of the crystalline CoNi external shell is not reached at the highest measuring temperature.

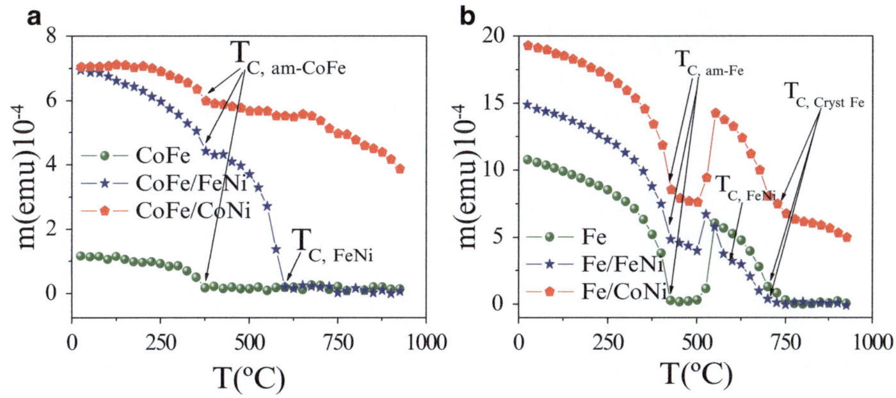

Fig. 7.11 Temperature dependence of the magnetic moment for single, CoFe, and CoFe/FeNi biphase or CoFe/CoNi biphase systems (**a**) and for single, Fe, and Fe/FeNi biphase or Fe/CoNi biphase systems (**b**). *Arrows* denote the Curie temperature, T_c, of different phases [40]

The results in Fig. 7.11.b correspond to Fe-based microwires. In the case of Fe/FeNi biphase microwire, we evaluate the magnetic-phase transitions for the amorphous nucleus at $T_{c,am\text{-}Fe} \approx 427$ °C (similar to that reported for ribbons in [41]) and $T_{C,FeNi} \approx 601$ °C for the crystalline FeNi shell. Also, we estimate the crystallization temperature of the amorphous core at around $T_{x,Fe} \approx 525$ °C and its Curie point at $T_{C,Fe\text{-}crys} \approx 645$ °C. Finally, for the Fe/CoNi biphase microwire, we evaluate $T_{c,Fe} \approx 427$ °C and $T_{C,Fe\text{-}cryst} \approx 675$ °C. Note that similar values of Curie and crystallization temperatures are experimentally measured for individual magnetic phases in the different single- and biphase microwires.

7.5 Network Analyzer-Ferromagnetic Resonance in Biphase Magnetic Microwires

The ferromagnetic resonance, FMR, or, to be more precise, the resonant absorption under external electromagnetic radiation is a technique that has been successfully employed for the investigation of magnetic substances, not only about their fundamental magnitudes but also for their technological applications. The microwave properties of amorphous magnetic alloys have been reported by several groups [12, 46–53]. The main features of FMR in amorphous microwires have been analyzed in several reports [54, 55]. The interpretation of rather complex experimental data for multilayer wires with and without intermediate glassy layer has been, however, sometimes contradictory. In the case of biphase wires with glassy interlayer, several difficulties have been pointed out to interpret their multiabsorption spectra [33, 56]. While single-phase microwires are characterized by single FMR absorption, biphase microwires show two or more different absorptions depending on the soft/hard nature of the two magnetic phases. The evolution of the resonance frequency with DC applied field has been fitted to Kittel's equation for thin films which is applicable also to metallic wires if the skin depth is small compared to the wire diameter [51]. Thus, the diversity of interpretation specifically occurs if the skin effect in ferromagnetic metal is not properly taken into account. This fact justifies the present updating of recent progress in understanding FMR aspects of ferromagnetic metallic wires.

FMR experiments collected in the present overview are basically divided into two categories: (1) under a fixed DC magnetic field varying the microwave frequency of the AC electromagnetic field by means of so-called network analyzer-FMR (NA-FMR) and coaxial or microstrip microwave circuits and (2) at constant frequency varying the amplitude of the DC magnetic field making use of classical FMR spectrometers and waveguide microwave techniques. The first type of measurements is presented in this section while the second ones are collected in the next one.

The microwave characterization was carried out at room temperature with a network analyzer (Agilent, model E8362B) and a transmission coaxial line in the frequency range between 10 MHz and nominally 20 GHz. DC magnetic field (up to 5 kOe) was applied parallel to wire axis by an electromagnet. SMA connectors and adapters are used suitable for measurements at a maximum frequency of 20 GHz. The adapted sample holder is based on a commercial SMA (SubMiniature version A) connector where the inner pin was removed to avoid radiation effects. The inner and outer conductors of the holder are shorted by means of the microwire nucleus: the Au and magnetic coatings are removed from the wire ends and the amorphous nucleus (around 50 Ω DC resistance) is welded using silver paint. The reflection coefficient S 11 is analyzed, from which real R and imaginary X components of impedance are determined.

Pieces of microwires 5 mm in length were taken for these measurements. The electric contacts between the inner metallic core and the measuring circuit were made by a silver paint. The microwave current passing through the wire induced a circumferential AC field in the core and the surrounding external shell (FeNi or CoNi) microtube. A schematic view of the whole measurement system is depicted in Fig. 7.12, while additional experimental details can be found elsewhere [33, 38].

7.5.1 Effect of Two Phases into the FMR Spectrum

Let us first summarize the main characteristics of FMR absorption spectra for single- and biphase magnetic systems. Figure 7.13a shows typical spectra for the real component of impedance corresponding to non-magnetostrictive CoFe-based single-phase glass-coated microwire for a range of indicated DC applied field. As observed, a clear resonance peak, FMR1, is observed which is naturally ascribed to the soft magnetic glass-coated microwire. Note that the amplitude of DC applied field is high enough so that the microwire is assumed to be magnetically saturated.

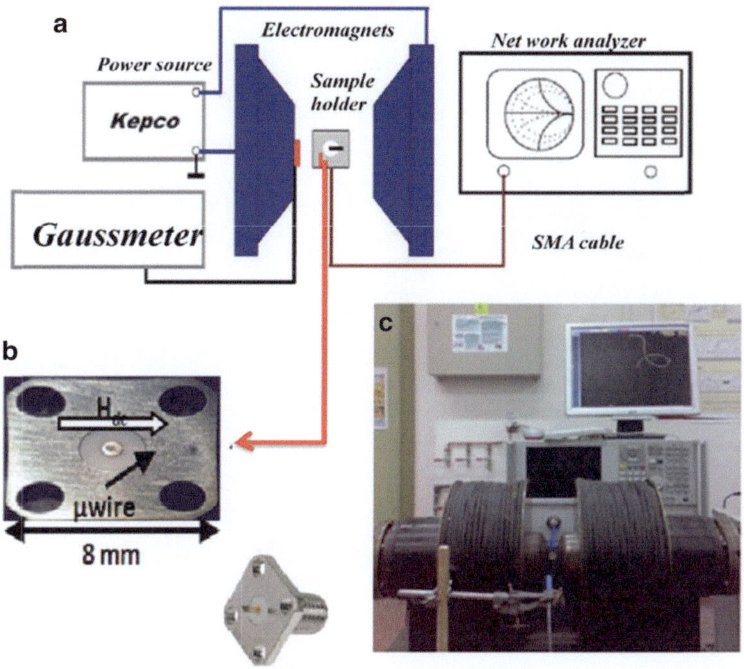

Fig. 7.12 Scheme of the experimental setup for FMR measurements in transmission coaxial line in the network analyzer: Diagram of measurement (**a**) with detail of the sample holder (**b**) and view of the whole system and electromagnet (**c**)

The evolution of the resonance frequency with applied field has been typically performed using the Kittel's condition of resonance for a tangentially magnetized planar film which corresponds to the skin-depth layer at the metallic microwire surface [39, 57–59]. The following equation holds:

$$\left(\frac{\omega}{\gamma}\right)^2 = (H_r + H_K)(H_r + H_K + 4\pi M_S) \tag{7.1}$$

where $\omega = 2\pi f_r$ is the angular frequency of the microwave field and γ is the gyromagnetic ratio ($\gamma/2\pi = 2.8 \times 10^6$ Hz/Oe). The evolution of f_r^2 is typically represented as a function of the DC applied field and fitted top, a linear behavior which allows one to determine a fitting value for the anisotropy field, H_k, and of the saturation magnetization. As deduced from the fitting in Fig. 7.14, the fitted value for the saturation magnetization is $4\pi M_s = 7.1$ kG which agrees well with the expectations for the CoFe-based alloy composition.

In the case of biphase magnetic microwires, we obtain in general multipeak spectra. For example, for the soft/hard CoFe/CoNi biphase microwire, we observe

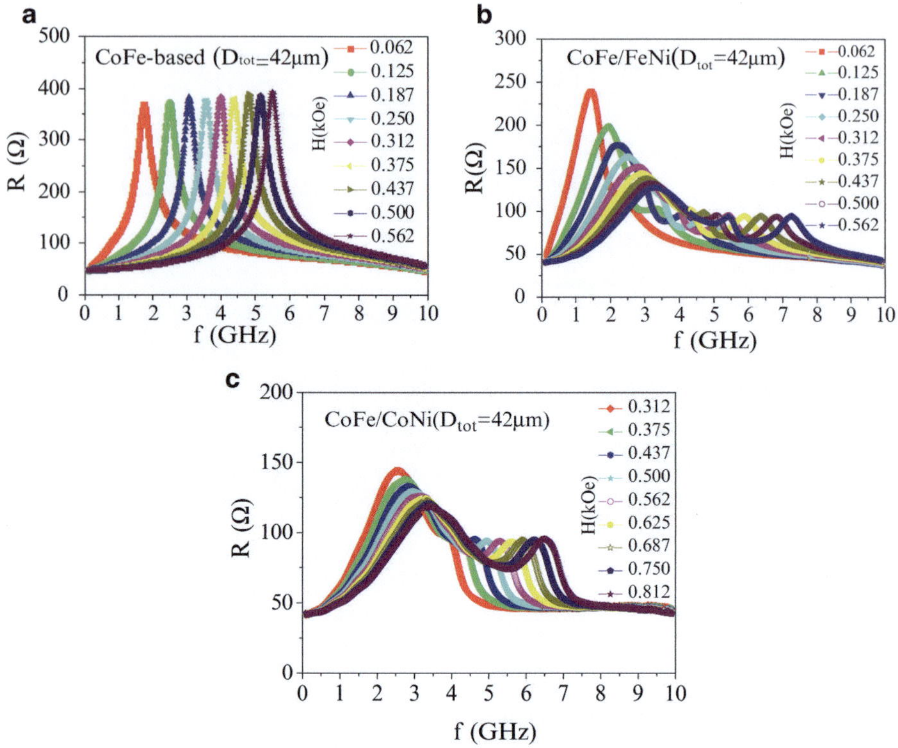

Fig. 7.13 Evolution of FMR spectra (real component of impedance) with static magnetic field for CoFe single-phase glass-coated microwire (17 and 42 μm metallic and total diameter) (**a**), CoFe/CoNi (2 μm CoNi thick) (**b**), and CoFe/FeNi (2 μm FeNi thick) (**c**). Adapted from [32]

in Fig. 7.13c the presence of two peaks that are ascribed to two absorption phenomena. That one at higher frequency can be ascribed to the soft CoFe nucleus as this is observed at similar frequency as FMR1, and it follows a similar trend with applied field as can be deduced in Fig. 7.14 as well as the value of the saturation magnetization fitted according to Eq. (7.1). However, the peak observed at the lower frequency range in Fig. 7.13c, that we will label FMR2, follows a different evolution (see Fig. 7.14) with applied field which cannot be ascribed to any magnetic phase as the fitted parameters would give us nonsense values for the external CoNi shell.

In soft/soft CoFe/FeNi microwire in Fig. 7.13b, we detect three peaks; FMR1 is ascribed to the CoFe nucleus, while the new peak, FMR3, should correspond to the soft FeNi shell as the fitted value (see Fig. 7.14) for the saturation magnetization is $4\pi M_s = 11.5$ kG, near to the expected value for Permalloy. Again, we observe a FMR2 peak which cannot be fitted to any of the two magnetic phases.

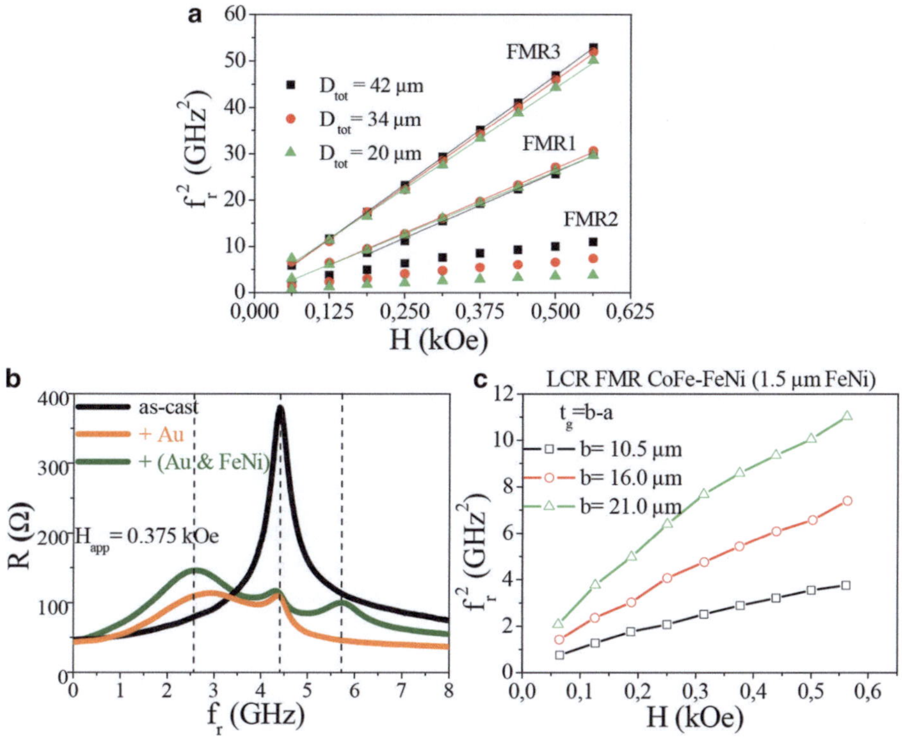

Fig. 7.14 Evolution of the square frequency at resonance as a function of applied magnetic field for CoFe/FeNi biphase microwires (CoFe glass-coated microwire with total diameter, $D_{tot} = 42$, 34, and 20, and 2 µm thick FeNi layer) where the linear fits correspond to Eq. (7.1) (**a**). Absorption spectra of CoFe, CoFe/Au, and CoFe/FeNi (2.5 µm thick) (**b**). Dependence of the absorption frequency for FMR2 with different dielectric thickness, t_g (**c**). Adapted from [32]

7.5.2 Influence of Layers Thickness: The Microwire as a Capacitor

The FMR study has been performed for biphase microwires with different thicknesses of the layers. First, we consider the influence of the thickness of the intermediate Pyrex layer. In fact, this thickness introduces mechanical stresses in the internal nucleus as has been commented in a previous section. Variations of that thickness also modify the FMR behavior of single- and biphase microwires. Measurements taken in CoFe-based single- and biphase non-magnetostrictive microwires show not very significant influence in the FMR peaks ascribed to each magnetic phase. Figure 7.14a depicts the fitting to Kittel's Eq. (7.1) for the data collected for soft/soft CoFe/FeNi biphase microwires with different total diameter, D_{tot}, of precursor glass-coated microwire (i.e., thickness of Pyrex layer).

However, that is relevant to unveil the nature of the FM2 absorption. Since it depends on the insulating intermediate layer, it seems reasonable to correlate the effect to a geometrical feature, particularly to the capacitance formed between the two magnetic metallic conductors and the insulating Pyrex layer [60, 61]. The multilayer microwire can be taken as a cylindrical capacitor of internal and external radii a and b, respectively, filled by a dielectric (Pyrex) with a given capacity, C:

$$C = \frac{2\pi\varepsilon_0\varepsilon_r l}{\ln(b|a)} \tag{7.2}$$

Therefore, the microwire and the measurement system form a LRC resonant circuit which resonance frequency is given by:

$$f_r = \frac{1}{2\pi\sqrt{L(H)C}} \tag{7.3}$$

where $L(H)$ is the inductance of the two magnetic-phase structure. This circuit reproduces qualitatively the FMR behavior observed for the biphase microwires CoFe/FeNi. As a result, the alternating current passes through both the inner metallic nucleus and the outer metallic shell of the microwires. Thus, the biphase microwire can be taken as two impedances in parallel (one much larger than the other since for a 5 mm long microwire, the resistance values are $R_{\text{nucleus}} \approx 50\ \Omega$, $R_{\text{shell}} = 1.5\ \Omega$). Thus, the observed shift of FMR2 resonance with the applied field can be understood as due to the field dependence given in Eq. (7.3). Figure 7.14b presents the absorption spectra at a constant applied field ($H_{\text{ap}} = 0.375$ kOe) for single-phase microwire CoFe (black), for the same microwire sputtered with a thin Au nanolayer (orange), and after electroplated by FeNi soft external shell (green). Indeed, the resistance of biphase microwires (electroplated by magnetic or not magnetic metal layer) is much smaller than the single-phase microwires. On the other hand, we observe that the FMR2 appears also in the case of a nonmagnetic coating of Au, which confirms the capacitance origin of its absorption peak. This phenomenon illustrates that the composite microwire can be taken by itself as a LRC circuit, and that its absorption should depend on the geometry of the associated capacitor as confirmed in Fig. 7.14c.

7.5.3 *Effect of Thermal Treatments: The Influence of the Not-Saturated Phase*

The objective of this section has been to investigate the effect of the thermal treatments on the FMR behavior of Fe/CoNi soft/hard bimagnetic microwires whose static magnetic properties were analyzed in a previous section. An additional objective has been trying to understand why no apparent absorption is received from the external hard shell.

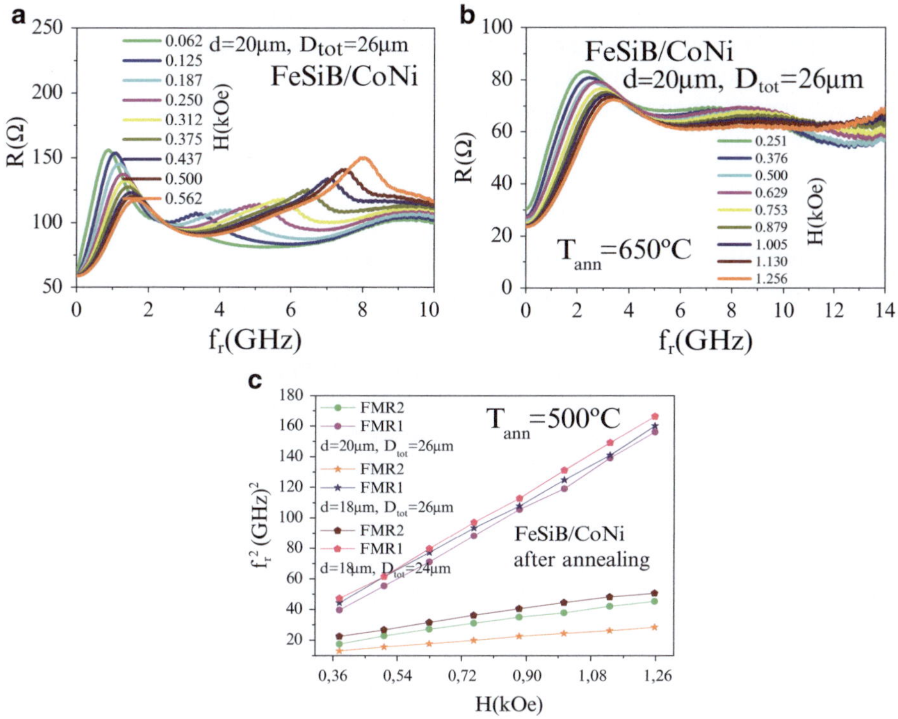

Fig. 7.15 Evolution of FMR spectra (real part of impedance, R) with applied field for FeSiB/CoNi biphase microwires in as-prepared state (**a**) and after annealing at 650 °C (**b**). Square of the resonance frequency as a function of applied static magnetic field for FeSiB/CoNi biphase microwires before and after annealing at 500 °C. Adapted from [33]

Measurements have been performed in Fe/CoNi microwires with different metallic, $d = 18$, 18, and 20 μm, and total diameters, $D_{tot} = 24$, 26, and 26 μm, ($\rho = d_m/D_{tot}$), respectively, for the glass-coated nucleus and a polycrystalline CoNi shell 3 μm thick. Annealing treatments were performed up to 700 °C and, as in the static case, we find two annealing temperature ranges for which the microwave response is clearly identified. After annealing up to around 500 °C, only relatively small changes are observed in the FMR absorption characteristics in comparison with as-prepared microwire (see Fig. 7.15a). Two absorption spectra, labeled above as FMR1 and FMR2, are observed. FMR1 is again ascribed to the Fe-based nucleus through the fitting to Kittel's Eq. (7.1) (see Fig. 7.15c).

However, after annealing at higher temperatures, the two absorption peaks can be still visible as observed in Fig. 7.15b for the sample after annealing at 650 °C. FMR1 corresponding to the soft nucleus now shows much less amplitude, and eventually, it should be ascribed to the partially crystallized FeSiB core. Fitting to Kittel's equation cannot be properly performed.

Regarding the data for FMR2, they do not follow the resonance condition and are ascribed to the mentioned capacitive effect. We should finally underline that no absorption can be ascribed to the CoNi external shell even after those thermal treatments. Thus, we understand that owing to its harder magnetic character, the CoNi shell is not sufficiently saturated magnetically. In this case, it does not show properly ferromagnetic absorption, and to observe its FMR a different experiment should be designed.

7.6 Ferromagnetic Resonance Through Cavity-Perturbation Measurements

In this section we summarize the experimental work and its analysis performed in biphase microwires with hard external shell obtained by means of cavity-perturbation technique at two different microwave frequencies, 9.5 GHz (X-band) and 69 GHz (K-band). Apart from the intrinsic interest of this technique to achieve broader and complementary information, an additional reason to perform this study was related to the fact that the maximum available magnetic field in the network analyzer equipment was not high enough to saturate magnetically neither the CoNi shell nor the crystallized FeSiB nucleus after the thermal annealing.

By this classical FMR experiment, we obtain the DC magnetic field dependence of the microwave absorption as a function of strong enough DC applied field (up to 30 kOe) which was parallel to the wire (with field modulation 1 Oe at 100 kHz) at given frequency. Measurements were taken at X-band at an extended temperature range (−269 to 25 °C) at the microwave cavity setup installed at the University of the Basque Country in Bilbao (a scheme of the experimental equipment is depicted in Fig. 7.16 together with a photograph of the whole commercial setup). Experiments at the Czech Academy of Sciences in Prague were performed at room temperature at X-band (rectangular TE10 waveguide) and K-band (on small pieces of microwires, around 2 mm long, cut from the selected wires, inserted into quartz

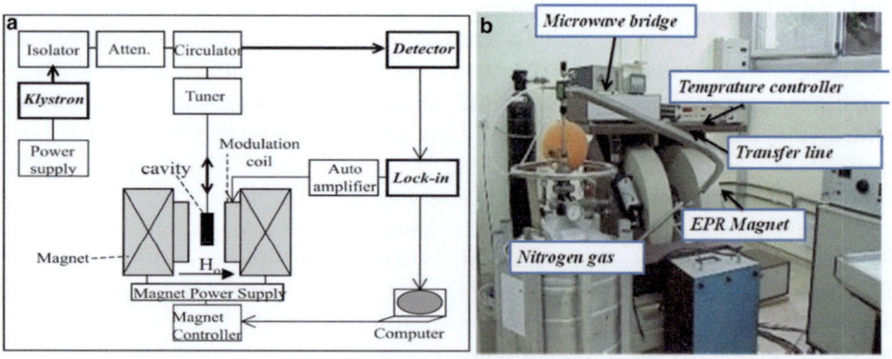

Fig. 7.16 Schematic view (**a**) and image (**b**) of the cavity-perturbation measurement setup installed at SGIker services of the University of the Basque Country

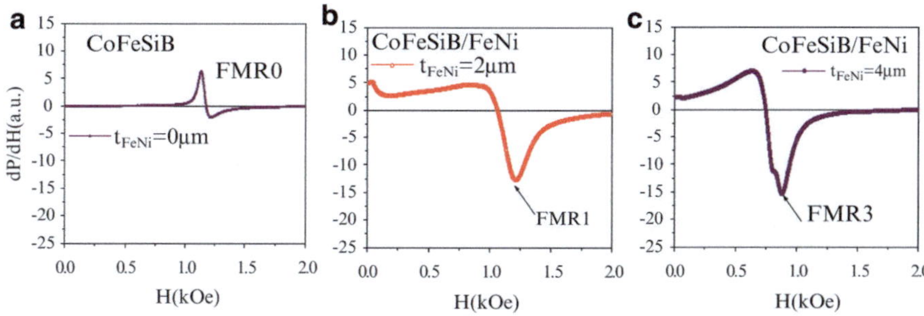

Fig. 7.17 Resonance spectra at 9.5 GHz of soft/soft CoFe/FeNi biphase microwires with different thickness (0, 2 and 4 μm) of the external FeNi shell [39]

capillary, and placed into the middle of circular waveguide with sample axis along the electric field vector).

Measurements were taken on two families of microwires based on single-phase glass-coated microwires of nominal composition FeSiB (positive and large magnetostriction) and CoFeSiB (vanishing magnetostriction), with metallic nucleus diameter $d = 12.5$ and 8 μm and total diameters $D_{tot} = 40$ and 24 μm, respectively. The corresponding biphase microwires contain a hard CoNi external shell with 3 μm thickness.

7.6.1 Temperature Dependence of Microwave Properties

The objective in this section has been to obtain a deeper knowledge on the microwave phenomena in single- and biphase microwires through the temperature dependence of their FMR absorptions peaks. To reach this goal, we have selected a soft/soft biphase microwire. It consists of an 8 μm diameter non-magnetostrictive CoFeSiB single-phase glass-coated microwire and a 2 or 4 μm thick FeNi external shell.

The results corresponding to the single- and biphase microwires are shown in Fig. 7.17, and they can be compared to the data obtained with the NA-FMR, at 9.5 GHz frequency. As discussed in [39], the spectrum displays one resonant peak for CoFe single-phase microwire at 1.08 kOe which compares with 1.17 kOe of FMR1 as deduced from the data obtained with the network analyzer. For $t_{NiFe} = 2$ μm biphase microwire, a main absorption is observed at 1.10 kOe which should correspond to that at 1.16 kOe from FMR1 in NA-FMR measurements which was correlated to the CoFe metallic core. However, for biphase microwires we should expect in principle that the main absorption should be ascribed to the external FeNi phase. This is the case of $t_{NiFe} = 4$ μm, where the main absorption at 0.77 kOe corresponds to $H_r = 0.87$ kOe for FMR3 in NA-FMR measurements. Small additional absorptions at applied magnetic field below the main peak can be hardly identified in Fig. 7.17, and its origin is not clear.

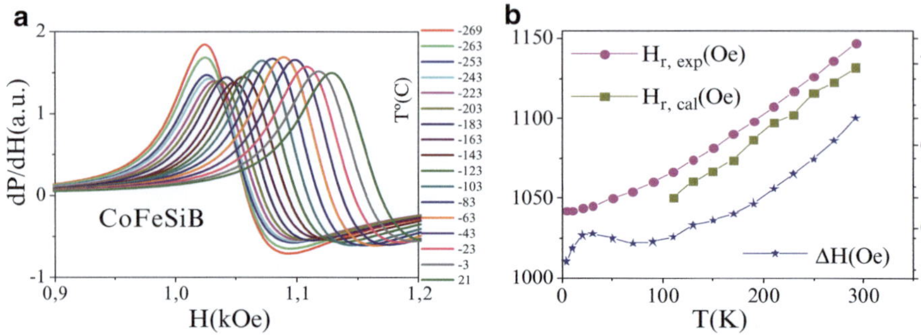

Fig. 7.18 Resonance spectra at 9.5 GHz of CoFe-based single-phase glass-coated microwire at selected temperatures (**a**) and temperature dependence of applied field at resonance, H_r (experimental and calculated), and total line width, ΔH (**b**) [39]

The derivative of microwave absorption versus external magnetic field resonance spectra of single-phase microwire measured at selected temperatures is shown in Fig. 7.18a. A single absorption peak is observed which resonance field, H_r, and the total line width, ΔH, are plotted in Fig. 7.18b as a function of temperature. As observed, both parameters increase monotonically with temperature, although ΔH shows a small maximum at low temperature.

Considering Eq. (7.1) and that $4\pi M_s \gg H_r \gg H_k$ in the first approximation, we derive the following expression for the temperature dependence of the resonance field, H_r:

$$H_r(T) \approx \frac{f_r^2}{\pi \gamma^2 M_S(T)} - H_K(T) \tag{7.4}$$

where the two contributions to H_r on the right-hand side increase with increasing temperature (note the negative-circular anisotropy field value). Figure 7.18b shows the experimental temperature dependence of $H_{r,exp}$ as deduced from data in Fig. 7.18a which is compared with calculated value, $H_{r,cal}$, obtained using Eq. (7.4) (taking saturation magnetization and anisotropy field values deduced from low-temperature hysteresis loops). The difference in both series of data can be justified after taking into consideration that hysteresis loops were obtained at very low frequency while those from FMR absorption were measured at the GHz frequency range.

The resonance spectra of biphase microwires with different thickness of external shell at selected temperatures are shown in Fig. 7.19. In the case of 2 μm thick FeNi shell, we observe a main absorption peak, FMR1, whose applied field at resonance, H_r, changes very noticeably at the low-temperature range and more moderately at higher temperatures. Also, a small pronounced peak can be observed at low field. For the 4 μm thick FeNi microtube, the variation of the corresponding H_r is much more reduced, while the low-field absorption is detected clearly at low temperatures. Inset in Fig. 7.19b depicts an enlarged view of the low-field peak.

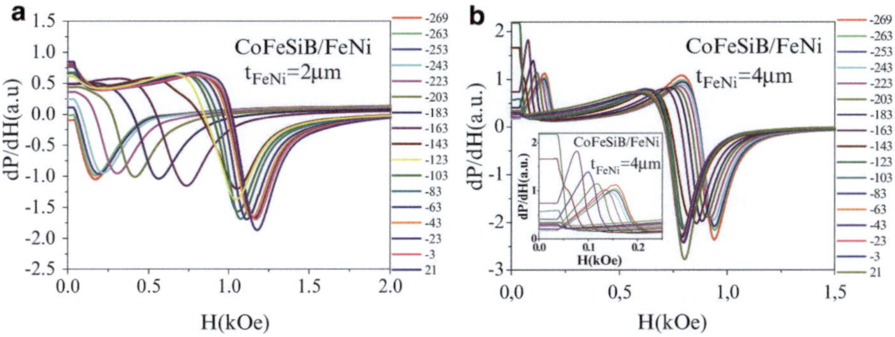

Fig. 7.19 Derivative resonance spectra of CoFe/FeNi biphase microwires with 2 μm (**a**) and 4 μm (**b**) thickness of the external FeNi shell. *Inset* in (**b**) depicts and enlarged view of the low-field peak [39]

Fig. 7.20 Temperature dependence of the applied field at resonance, H_r, for the different microwires [39]

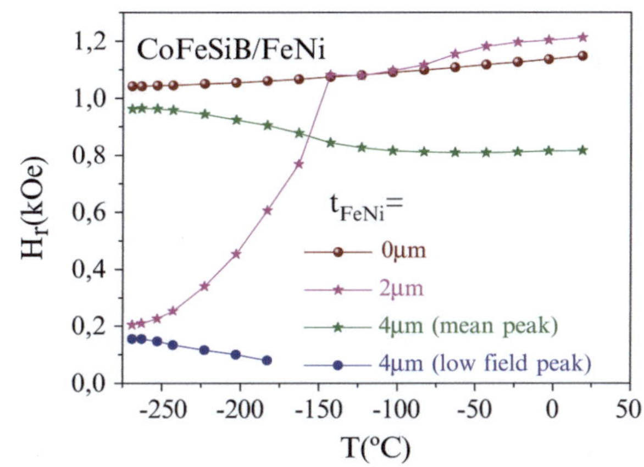

Figure 7.20 collects the evolution of the main resonance field, H_r, as a function of temperature. For the CoFe single-phase microwire, a moderate monotonic increase of H_r with temperature is observed. In the case of 2 μm thick NiFe shell microwire, at temperatures above −143 °C, we observe a similar behavior as that of CoFe single-phase microwire. This similarity, maybe accidental, could lead us to ascribe that absorption to the CoFe core, although we should consider as well that it corresponds to the external FeNi microtube. Note that a similar question was found in the interpretation of data obtained with the network analyzer at room temperature. Below −143 °C a pronounced reduction of H_r is observed that could be eventually interpreted if we would consider a change of sign of anisotropy field in Eq. (7.4). That is, assuming that the low-temperature axial anisotropy field ($H_k > 0$) evolves to circular anisotropy field ($H_k < 0$) at high temperatures.

An overall opposite evolution of H_r with temperature is observed in the case of the 4 μm thick FeNi microtube, including a change of trend at around −143 °C which again could be ascribed to the change of sign of the anisotropy field. Now, it should be circular at low temperature ($H_k < 0$) and become axial ($H_k > 0$) at temperatures higher than around −143 °C. The low-field absorption is observed at higher temperatures in the 2 μm thick FeNi biphase microwire, while it appears only at low temperatures in the 4 μm biphase microwire, where a circular magnetic anisotropy would be expected. The origin of this phenomenon could be connected with the absorption in non-saturated samples. Note that its correlation with ferromagnetic antiresonance, FMAR, is in principle discarded since the antiresonance could be observed only above some critical frequency ($\omega/\gamma > 4\pi M_s$) which for FeNi lies in the order of 30 GHz. The origin for the anomalous behavior observed in biphase microwires can be ascribed to the presence of strong induced anisotropy in FeNi layers at low temperatures, changing its sign at around −143 °C. The substantial difference in the behavior of the two biphase samples can be a consequence of the different magnetostriction constants, in both sign and magnitude, of the two FeNi layer (2 and 4 μm thick).

In order to properly determine the magnetostriction constant of the FeNi external shell, a final test of their compositions was performed by SEM-EDX (FEI Nova NanoSEM 230 high-resolution scanning electron microscope). That analysis confirmed in fact a different composition for the 2 and 4 μm thick external shell, namely, $Fe_{8.2}Ni_{91.8}$ and $Fe_{24.4}Ni_{75.6}$, respectively. If we consider that for bulk FeNi alloys [44], values of saturation magnetostriction at room temperature are 8×10^{-6} and -16×10^{-6} for $Fe_{8.2}Ni_{91.8}$ and $Fe_{24.4}Ni_{75.6}$, respectively, which supports our previous concern and assumption.

Summarizing all data, we can conclude that FMR1 definitely corresponds to CoFe single-phase wire. In the case of the biphase microwire with 2 μm thick FeNi (positive magnetostriction) external shell, FMR1 is hard to be ascribed, clearly owing to the similar data for CoFe nucleus and FeNi shell in both network analyzer and cavity measurements. In the case of the microwire with 4 μm thick FeNi (negative magnetostriction) shell, the main peak in cavity measurement is associated to FMR3 in the network analyzer measurements from the FeNi shell because of the similarity of H_r fields (0.77 and 0.87 kOe) at 9.5 GHz. We should finally comment that a straightforward comparison of network analyzer and classical FMR cavity-perturbation measurements on the biphase wires would require a fully rigorous theoretical analysis. In the cavity measurement, the microwave current passes through both the core and the FeNi external shell. In contrast, in the network analyzer measurement, the microwave current passes mostly through the CoFe core but also through the FeNi shell because it is transmitted via the capacitance bridge between core and shell. However, it seems that in classical FMR cavity experiment, the core is more effectively screened by the shell.

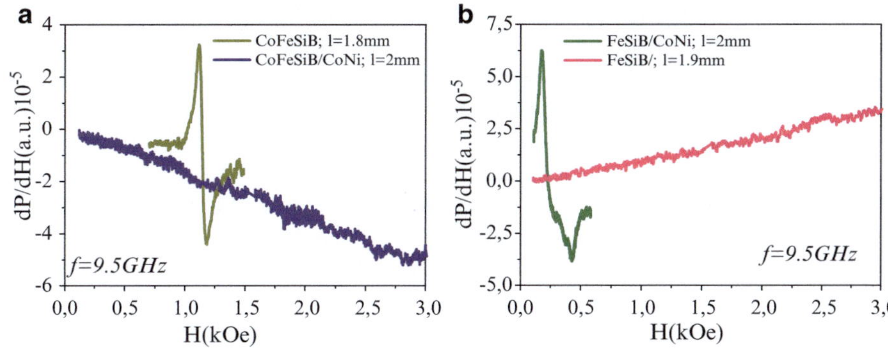

Fig. 7.21 Resonance absorption spectra at 9.5 GHz for single- and biphase microwires with CoFe-based (a) and Fe-based (b) nucleus

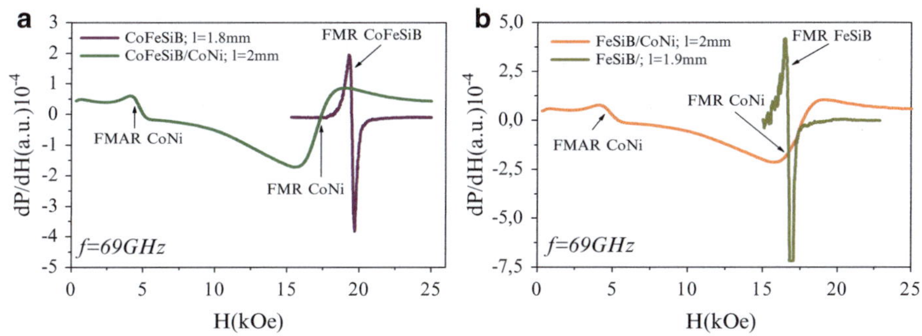

Fig. 7.22 Resonance absorption spectra at 69 GHz for single- and biphase microwires with CoFe-based (**a**) and Fe-based (**b**) nucleus

7.6.2 Room Temperature Analysis: The Role of the Hard-Phase Response

The FMR spectra for single- and biphase microwires measured at frequency 9.5 GHz (X-band) are shown in Fig. 7.21. Absorption is observed in single-phase wires at applied fields of around 1.2 and 0.3 kOe for CoFeSiB and FeSiB microwires, respectively. We should note that according to Eq. (7.1), the difference in applied field to reach resonance comes from the different saturation magnetization and magnetic anisotropy (arising from the magnetostriction constant) of the microwires. For biphase microwires, no absorption is observed up to the maximum applied of 3 kOe. That should be understood as a consequence of a double effect of the presence of CoNi shell: (1) the CoNi shell has not reached again its magnetic saturation and (2) it completely screens the internal core.

The results obtained from measurements in the microwave frequency of 69 GHz (K-band) are shown in Fig. 7.22. In the case of single-phase microwires, we observe clean FMR absorption at applied fields of 19 and 17 kOe for CoFeSiB and FeSiB single-phase microwires, respectively. Again, the different saturation magnetization and anisotropy fields of microwires account for the distinct applied fields to observe FMR. In the case of biphase microwires, two resonance peaks are observed at the same frequency (around 17.5 kOe) in both biphase microwires, which consequently lead us to ascribe them to the same magnetic phase, that is, the CoNi external shell, while no resonance can be correlated to the internal nucleus. That would confirm that in this experiment, the amorphous internal nucleus is screened by the CoNi external shell. Further analysis of the spectra indicates that in each sample, we are dealing with a symmetric antiresonance FMAR peak observed at the lower frequency at around 4 kOe together with the mentioned FMR peak. The experimental derivative FMR curves are distorted in both biphase microwires which occur most probably because the sample represents a large load to the microwave circuit [51].

The experiments performed with the cavity-perturbation method enable a broader overview to the analysis of experiments performed in the network analyzer-FMR in previous sections:

1. The lack of FMR absorption ascribed to CoNi hard shell in biphase wires is a consequence of the fact that it is not magnetically saturated under the maximum applied field.
2. The low-frequency FMR2 absorption arises from a capacitive effect.
3. The screening of the core is detected in a complementary way in the cavity experiments.

Further systematic experiments are in perspective.

7.6.3 Angular Dependence of Microwave Absorption

The interest of the determination of the angular dependence of the FMR characteristics is in many cases related to its capability to determine the contributions to the total anisotropy field, H_k. In this kind of experiments, the magnetic field is applied making a variable angle with a particular orientation of the investigated sample. Such angular dependence of microwave behavior has been successfully employed in samples with various magnetic and geometry characteristics as amorphous alloy ribbons, multilayer thin films, and ferrites [62, 63]. In the present study, we introduce preliminary results on the angular dependence of microwave absorption in CoFeSiB and FeSiB single-phase microwires. The microwave absorption measurements were carried out by the cavity technique at X-band spectrometer (9.5 GHz) at room temperature.

Figure 7.23 presents the derivative resonance spectrum of both samples for applied field parallel to microwire. In the case of the CoFeSiB glass-coated microwire, the spectrum displays just one peak at the $H_r = 1.3$ kOe which relates to the circular anisotropy field of the microwire. Note that such circular anisotropy

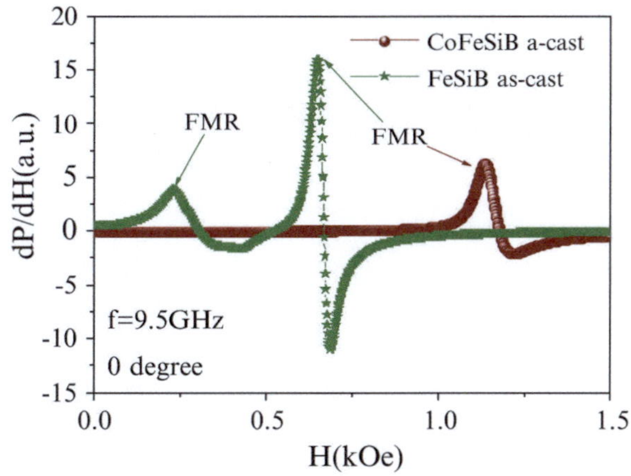

Fig. 7.23 Derivative resonance spectrum of CoFe- and Fe-based single-phase glass-coated microwires

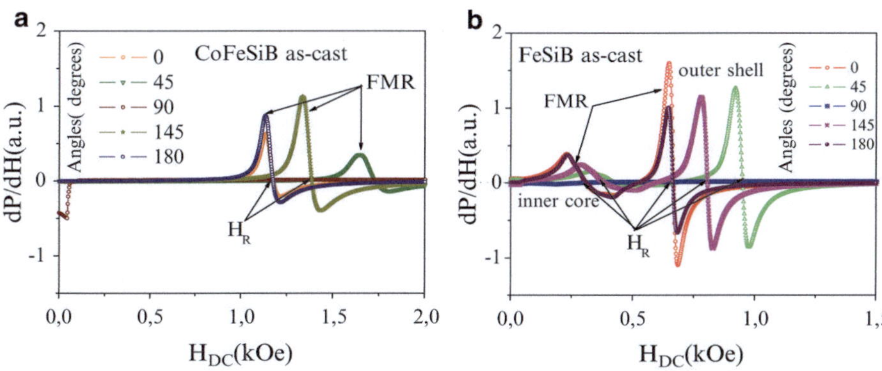

Fig. 7.24 Angular dependence (at selected angles) of the derivative resonance spectrum of CoFe- and Fe-based single-phase glass-coated microwires

arises from the coupling between internal stresses with the small but negative magnetostriction of that alloy, which takes a value of $\lambda_s \sim -1 \times 10^{-7}$. The spectrum in the case of FeSiB glass-coated microwire displays two peaks, both at lower field than in the CoFe-based microwire: a smaller but broader peak is observed at the lower field of 0.25 kOe, and the steep larger one is observed at higher field of 0.65 kOe. These peaks should be ascribed to two different regions inside the metallic Fe-based microwire, namely, the low-field resonance peak would correspond to the FMR response of an inner core which is known to exhibit with strong axial anisotropy, while the peak at high field is seemingly the response of an outer shell of the microwire with transverse anisotropy.

The angular dependence of FMR behavior is shown in Fig. 7.24 for both samples at selected angles between 0° and 180° (note that for 0°, the AC microwave field is

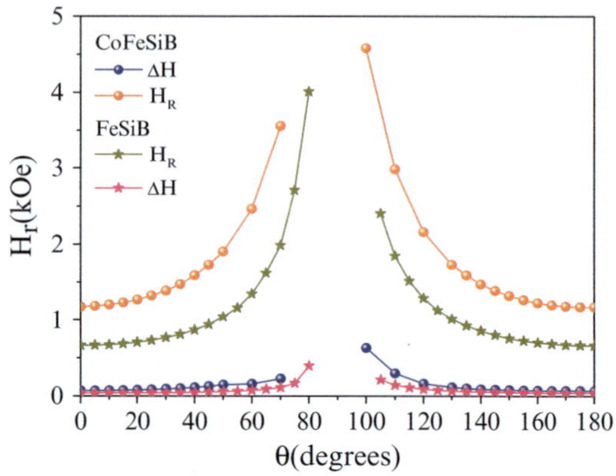

Fig. 7.25 Angular dependence of applied field at resonance, H_r, and of absorption width, ΔH, in CoFe- and Fe-based single-phase glass-coated microwires

in the plane of the wire and the DC field is parallel to the wire axis, while for 90°, the AC field remains parallel to the plane and the DC field is normal to the wire). Measurements in Fig. 7.24a for CoFeSiB ($\lambda \leq 0$) microwire show an increase in the resonant field and a reduction of absorption amplitude as the angle approaches 90°. Similarly, resonance linewidth, ΔH, sharply increases for angles close to 90°. For angles $90° < \theta < 180°$, a symmetric evolution is observed (see also Fig. 7.24). The angular dependence for the Fe-based microwire is presented in Fig. 7.24b with the two mentioned FMR spectra. A similar angular evolution as in the case of CoFe-based microwire is observed.

Figure 7.25 collects all the results for the angular dependence of the field at resonance, H_r, and the absorption width, ΔH, for both samples CoFeSiB and FeSiB (high-field resonance peak). In order to understand that angular broadening, we should consider that because of skin effect, the Kittel resonance conditions for an infinite cylinder cannot be used in general for metallic wires. For an oblique magnetization of the wire, the resonance curve is inhomogeneously broadened. It means that different parts of the wire exhibit different resonance fields, and the resulting resonance curve is given by the envelope of the local resonance curves. The inhomogeneous broadening increases with increasing angle θ and reaches maximum at $\theta = 90°$. However, for thick enough wires, where the skin depth is much smaller than the wire diameter, the skin layer can be approximated by a thin tube at the wire surface. Then, the local resonance field can be calculated from the Kittel resonance condition of an obliquely magnetized thin film [64]. For the transversally magnetized wire ($\theta = 90°$), the minimum and maximum local resonance fields are:

$$H_{\min} = \sqrt{(\omega/\gamma)^2 + (2\pi M_s)^2}, \quad H_{\max} = \omega/\gamma + 2\pi M_s \tag{7.5}$$

This gives very broad resonance curve at fields above the measured field range.

7.7 Final Remarks and Conclusions: Future Perspectives

We have summarized in the chapter some of the most relevant recent results on the magnetic properties of bimagnetic multilayer microwires synthesized over glass-coated amorphous microwires. After describing the synthesis procedure, we have characterized for some selected biphase microwires with soft/soft and soft/hard magnetic behaviors based on magnetostrictive and non-magnetostrictive amorphous nucleus and polycrystalline soft and hard shell.

We have confirmed the role played by the thickness of different layers, particularly that of the intermediate glass cover and of the external shell. Both contribute to induce magnetoelastic anisotropy resulting in a relative hardening of the nucleus. In addition, the fractional volume of the phases is determined by the thickness of the shell which finally determines the magnetic response of the microwire. Less attention has been paid until now to the temperature effects in these microwires. Annealing at temperatures around 500–600 °C results in the crystallization of the amorphous nucleus and its subsequent magnetic hardening. Measurements at high temperature enable the determination of the Curie and crystallization temperatures of the different magnetic phases.

The measurements at microwave frequencies were firstly performed in network analyzer for samples with selected thickness of external layers and after heating at different temperatures. Multipeak absorption has been observed with two and three absorptions for soft/hard and soft/soft microwires. FMR absorption is detected in all the cases corresponding to the soft phases, the nucleus, the shell, or both of them. However, an additional low-frequency absorption is always detected, which does not follow Kittel's resonance condition, and is ascribed to a resonance of the microwire as bimetallic cylindrical capacitor.

Measurements in perturbation cavity allow us to confirm that the FMR absorption corresponding to the hard phase is only observed when working at high enough frequency (K-band). In addition, the shell produces a screening effect of the soft nucleus. Preliminary data on the angular dependence of FMR for single-phase microwires are analyzed. Further, measurements as a function of the measuring temperature are interpreted in terms of the temperature dependence of the magnetic anisotropy of soft phases. That has allowed us to forecast the particular evolution of FeNi soft shell with different alloy composition and magnetostriction.

Regarding new perspectives of this family of bimagnetic microwires, there is no doubt of the potential opportunities offered by this family of bimagnetic microwires as sensing elements of various devices as has been shown in previous reports. Present data confirm the possibility to employ such materials in sensing devices at moderately high working temperatures. Their magnetic response can be tailored through particular design of alloy composition and geometry (i.e., diameter and thickness of layers or especial coating of external shell) and specific preheating treatments. On the other hand, data at microwave frequencies demonstrate the specific response of bimetallic magnetic systems with cylindrical geometry. New advanced measurements are still open for further deepening into fundamental knowledge. The possibility to deal with two phases of tailored magnetic character opens new opportunities for electromagnetic shielding effects at different working frequencies and for high-frequency sensor devices.

Acknowledgments Authors thank I. Orue and Luis Lezama (UPV-EHU) for special support. Selected microwave measurements were made at SGIker services UPV-EHU. Work in Madrid has been supported by the Government of Madrid under Project S2013/MIT195, 2850 NANOFRONTMAG-CM. The work done in Prague was partly supported by the Grant Agency of the Czech Republic under the project P102/12/2177.

References

1. Luborsky, F.E.: Amorphous Metallic Alloys. Butterworths, London (1983)
2. Svalov, A.V., Fernandez, A., Barandiaran, J.M., Vas'kovskiy, V.O., Orue, I., Tejedor, M., Kurlyandskaya, G.V.: J. Magn. Magn. Mater. **320**, 734 (2008)
3. Vazquez, M.: Advanced Magnetic Microwires" in Handbook of Magnetism and Advanced Magnetic Materials, p. 2193. Wiley, Chichester (2007)
4. McHenry, M.E., Willard, M.A., Laughlin, D.E.: Prog. Mater. Sci. **44**, 291 (1999)
5. Zhukov, A., Zhukova, V.: (2009) Magnetic Properties and Applications of Ferromagnetic Microwires with Amorphous and Nanocrystalline Structure, pp. **117–162**. Nova Science, Hauppauge, NY (2009)
6. Knobel, M., Kraus, L., Vazquez, M.: Giant magnetoimpedance. In: Buschow, K.H. (ed.) Handbook of Magnetic Materials, vol. 15, pp. 1–69. Elsevier Science, Amsterdam (2003). Chap. 5
7. Chiriac, H., Ovari, T.A.: Prog. Mater. Sci. **40**, 333 (1996)
8. Vazquez, M., Chiriac, H., Zhukov, A., Panina, L., Uchiyama, T.: Phys. Status Solidi A **238**, 493 (2011)
9. Sinnecker, J.P., Garcia, J.M., Asenjo, A., Vazquez, M., Garcia-Arribas, A.: J. Mater. Res. **15**, 751 (2000)
10. Garcia, J.M., Asenjo, A., Vazquez, M., Yakunin, A.M., Antonov, A.S., Sinnecker, J.P.: J. Appl. Phys. **89**, 3888 (2001)
11. Kurlyandskaya, G.V., Garcia-Miquel, H., Svalov, A.V., Vas'kovskiy, V.O., Vázquez, M.: Phys. Met. Metallogr. **91**, S125 (2001)
12. Garcia-Miquel, H., Bhagat, S.M., Lofland, S.E., Kurlyandskaya, G.V., Svalov, A.V.: J. Appl. Phys. **94**, 1868 (2003)
13. Muñoz, A.G., Schiefer, C., Kisker, E.: J. Appl. Phys. **103**, 073904 (2008)
14. Pirota, K., Hernandez-Velez, M., Navas, D., Zhukov, A., Vazquez, M.: Adv. Funct. Mater. **14**, 266 (2004)
15. Pirota, K., Provencio, M., Garcıa, K., Mendoza, P., Hernandez-Velez, M., Vazquez, M.: J. Magn. Magn. Mater. **290**, 68 (2005)
16. Borza, F., Corodeanu, S., Lupu, N., Chiriac, H.: J. Alloys Compd. **554**, 150 (2013)
17. Kraus, L., Pirota, K., Torrejon, J., Vazquez, M.: J. Appl. Phys. **101**(2007), 063910 (2007)
18. Torrejon, J., Badini, G., Pirota, K., Vazquez, M.: Acta Mater. **55**, 4271 (2007)
19. Badini-Confalonieri, G., Infante, G., Torrejon, J., Vazquez, M.: J. Magn. Magn. Mater. **320**, 2443 (2008)
20. Torrejon, J., Vazquez, M., Panina, L.V.: J. Appl. Phys. **105**, 033911 (2009)
21. Torrejón, J., Pirota, K.R., Badini-Confalonieri, G.A., Vázquez, M.: Sens. Lett. **7**, 236 (2009)
22. Vazquez, M., Pfützner, H., Pirota, K., Badini, G., Torrejon, J. Patent PCT/ES2005/070173 (2006)
23. Vazquez, M., Badini, G., Infante, F., Butta, M., Ripka, P.: Patent PCT/ES2009/070417 (2009)

24. Butta, M., Ripka, P., Vazquez, M., Infante, G., Kraus, L.: Sens Lett **11**, 50 (2013)
25. Pfutzner, H., Kaniusas, E., Kosel, J., Mehnen, L., Meydan, T., Vazquez, M., Rhon, M., Merlo, A. M., Marquardt, B. Sensor. Actuat. **129**,154 (2006)
26. Kolesar, V., Vazquez, M. Patent P201431530 (2014)
27. Torrejon, J., Infante, G., Badini-Confalonieri, G., Pirota, K., Vazquez, M.: J.O.M. (2013). doi:10.1007/s11837-013-0614-3
28. Badini-Confalonieri, G., Navas, D.: Bimagnetic microwires and nanowires: synthesis and characterization. In: Vazquez, M. (ed.) Magnetic Nano and Microwires, pp. 275–310. Woodhouse Elsevier, Cambridge UK (2015)
29. Li, X.P., Seet, J., Fan, J.: J. Magn. Magn. Mater. **304**, 111 (2006)
30. Schlesinger, M., Paunovic, M.: Modern Electroplating. Wiley, Chichester (2000)
31. Torrejon, J., Thiaville, A., Adenot-Engelvin, A.L., Vazquez, M.: J. Magn. Magn. Mater. **333**, 144 (2013)
32. El Kammouni, R., Infante, G., Torrejon, J., Britel, M.R., Brigui, J., Vazquez, M.: Phys. Status Solidi A **208**, 520 (2011)
33. El Kammouni, R., Vazquez, M.: IEEE Trans. Magn. **49**, 34 (2013)
34. Lumá, H., Vázquez, M., Hernandez, M., Ruiz, J.M., García-Beneytez, J.M., Zhukov, A., Castaño, F.J., Zhang, X.X., Tejada, J.: J. Magn. Magn. Mater. **196–197**, 821 (1999)
35. Vázquez, M., Zhukov, A., García, K.L., Pirota, K.R., Ruiz, A., Martinez, J.L., Knobel, M.: Mater. Sci. Eng. A. **375–377**, 1145 (2004)
36. Rodionova, V., Nikoshin, A., Torrejón, J., Badini-Confalonieri, G.A., Perov, N., Vazquez, M.: IEEE Trans. Magn. **47**, 3787 (2011)
37. Iglesias, I., El Kammouni, R., Chichay, K., Vazquez, M., Rodionova, V.: Solid State Phenom. **233–234**, 265 (2015)
38. El Kammouni, R.: Ph.D., Autonomous University of Madrid (2015)
39. El Kammouni, R., Vazquez, M., Lezama, L., Kurlyandskaya, G., Kraus, L.: J. Magn. Magn. Mater. **368**, 126 (2014)
40. El Kammouni, R., Iglesias, I., Chichay, K., Svec, P., Rodionova, V., Vazquez, M.: J. Appl. Phys. **116**, 093902 (2014)
41. Mc Henry, M.E., Willard, M.A., Laughlin, D.E.: Progr. Mater. Sci **44**, 291 (1999)
42. Celegato, F., Coisson, M., Olivetti, E., Tiberto, P., Vinai, F.: Phys. Stat. Sol. A. **205**, 1745 (2008)
43. P. Svec, P. Svec, Sr., I. Matko, I. Skorvanek, J. Kovac, D. Janickovic and G. Vlasak, Solid State Phenom. **172–174**, 953 (2011)
44. Bozorth, R.M.: Ferromagnetism, p. 276. Van Nostrand, London (1968)
45. Yu, P., Jin, X.F., Kudrnovský, J., Wang, D.S., Bruno, P.: Phys. Rev. B **77**, 4431 (2008)
46. Kraus, L.: Phys. Lett. **99A**, 189 (1983)
47. Menard, D., Britel, M., Ciureanu, P., Yelon, A., Paramonov, V.P., Antonov, A.S., Rudkowski, P., Stroem-Olsen, J.O.: J. Appl. Phys. **81**, 4032 (1997)
48. Ledieu, M., Schoenstein, F., Deprot, S., Adenot, A.-L., Bertin, F., Acher, O.: IEEE Trans. Magn. **39**, 3046 (2003)
49. Antonenko, A.N., Baranov, S.A., Larin, V.S., Torcunov, A.V.: J. Mater. Sci. Eng. A **247**, 248 (1997)
50. Makhnovskiy, D.P., Panina, L.V.: J. Appl. Phys. **93**, 4120 (2003)
51. Kraus, L., Infante, G., Frait, Z., Vazquez, M.: Phys. Rev. B **83**, 174438 (2011)
52. García-Miquel, H., Kurlyandskaya, G.V.: Chin. Phys. B **17**, 1430 (2008)
53. Montiel, H., Alvarez, G., Betancourt, I., Zamorano, R., Valenzuela, R.: Appl. Phys. Lett. **86**, 072503 (2005)
54. Garcia Miquel, H., Garcıa, J.M., Garcıa-Beneytez, J.M., Vazquez, M.: J. Magn. Magn. Mater. **231**, 38 (2001)
55. Ipatov, M., Zhukova, V., Zhukov, A., Gonzalez, J., Zvezdin, A.: J. Phys. Conf. Ser. **200**, 082009 (2010)
56. Torrejón, J., Badini-Confalonieri, G.A., Vázquez, M.: J. Appl. Phys. **106**, 023913 (2009)

57. Kittel, C.: Introduction to Solid State Physics, p. 503. Wiley, New York (1996). Ch. 16
58. Kraus, L.: Ferromagnetic resonance in individual wires: from micro to nanowires. In: Vazquez, M. (ed.) Magnetic Nano and Microwires, pp. 449–482. Woodhouse, Elsevier, Cambridge, UK (2015)
59. Chiriac, H., Coleniuc, C.N., Ovari, T.A.: IEEE Trans. Magn. **35**, 3841 (1999)
60. Raposo, V., Vázquez, M., Flores, A.G., Zazo, M., Íñiguez, J.I.: Sens. Actuators. A **106**, 329 (2003)
61. Wu, Z.M., Zhao, Z.J., Liu, L.P., Lin, H., Cheng, J.K., Yang, J.X., Yang, X.L.: IEEE Trans. Magn. **43**, 3146 (2007)
62. Alvarez, G., Montiel, H., Garcia-Arribas, A., Zamorano, R., Barandiaran, J.M., Valenzuela, R.: J. Non Cryst. Solids **354**, 5195–5197 (2008)
63. Chiriac, H., Colesniuc, C. N., Ovari, T. A.: IEEE Trans. Magn. **15** (1999)
64. Kraus, L., Frait, Z., Ababei, G., Chayka, O., Chiriac, H.: J. Appl. Phys. **111**, 053924 (2012)

Chapter 8
Tuneable Metacomposites Based on Functional Fillers

Yang Luo, Faxiang Qin, Fabrizio Scarpa, Mihail Ipatov, Arcady Zhukov, and Hua-Xin Peng

8.1 Introduction

We propose to start this chapter with a glimpse at the history of metamaterials. In 1968, Veselago speculated that it is theoretically possible to acquire a media with simultaneously negative permittivity (ε) and permeability (μ) and foresaw the negative refractive index. Nevertheless, his work has remained unrecognised for almost 30 years [1]. The obscurity is how to realise this double-negative (DNG) characteristic which is impossible to host for any natural materials from the electromagnetics viewpoint. Also, the scientific and engineering significance was not clear for such DNG medium which made metamaterials research remain in silence until the late 1990s when Sir Pendry appears to be the first to find a practical way to realise a negative permittivity through arranging a parallel metallic wire

---

The original version of the book was revised because Arcady Zhukov's name was misspelled. An erratum explaining this can be found at DOI 10.1007/978-3-319-26106-5_11

Y. Luo • F. Scarpa
Advanced Composites Centre for Innovation and Science, Department of Aerospace Engineering, University of Bristol, University Walk, Bristol BS8 1TR, UK
e-mail: yang.luo@bristol.ac.uk; f.scarpa@bristol.ac.uk

F. Qin (✉) • H.-X. Peng (✉)
Institute for Composites Science Innovation (InCSI), School of Materials Science and Engineering, Zhejiang University, Hangzhou 310027, China
e-mail: iangqin@zju.edu.cn; hxpengwork@zju.edu.cn

M. Ipatov
Dpto. de Fisica de Materiales, Fac. Quimicas, Universidad del Pais Vasco, San Sebastian 20009, Spain
e-mail: mihail.ipatov@ehu.es

A. Zhukov
UPV/EHU, Basque Country University, San Sebastian, Spain
e-mail: arkadi.joukov@ehu.es

A. Zhukov (ed.), *Novel Functional Magnetic Materials*, Springer Series in Materials Science 231, DOI 10.1007/978-3-319-26106-5_8

array [2]. In fact, negative permittivity is hardly a tough issue because some natural materials (such as ferroelectrics) display negative ε [3]. The real challenge is achieving artificial negative permeability. Later on, a more revolutionary structure was developed via a periodic lattice structure consisting of the split ring resonators (SRRs) [4]. As such, an artificial magnetism from non-magnetic materials is discovered and can be deployed to meet negative μ per se. The experimental verification of a structure comprising of parallel wires and SRRs, where a DNG feature was remarkably identified, was further carried out in the microwave regime by Smith et al. [5] and instantly attracted tremendous interests in the research and industry community worldwide. To date, a great deal of research grants and academic programmes has been launched in the field of metamaterials to pursue their novel properties and possible future application perspectives. Global efforts have fertilised the field of metamaterials and extended the palette to a large extent from EM to mechanical and acoustic metamaterials [6, 7]. The blossom of metamaterials also invokes new and strong interactions of interdisciplinary subjects, e.g. material science, electromagnetics, optics, etc. Moreover, it spawned some nascent scientific branches. For example, transformation optics was invented as a tool to design the EM fields near/inside metamaterials [8, 9]. Nowadays, any research connected to metamaterials is considered fascinating. So questions come naturally: what are metamaterials and why metamaterials?

Metamaterials, by the latest definition, are artificial materials possessing exotic properties that are not accessible for natural materials. 'Meta-' means higher or beyond in Greek and reflects the exact meaning of metamaterials being the role of groundbreaking in every way [10]. To understand metamaterials, the DNG property is the primary and most important feature. The ε and μ are two basic parameters to characterise the EM response of a homogeneous material; thus, the detailed information of the intrinsic 'atoms' can be ignored while interacting with EM waves. This allows us to design the required EM properties at will, e.g. DNG or single negative (SNG) features, taking into account that the ε and μ can be tailored by macro-/mesostructural parameters, such as geometrical dimensions and periodicity. The glamour of EM metamaterials, such as negative refractive index [11], reversed Cherenkov radiation [12] and reversed Doppler shift [13], has conveyed a number of applications: artificial magnetism, perfect absorbers, high-resolution imaging, EM cloaking, etc. [14]. Likewise, a similar sense could be drawn to other types of metamaterial. For instance, a simultaneous negative longitudinal compressibility and a transverse compressibility are prerequisites to construct mechanical metamaterials for 2D/3D unfeelability cloaks [15, 16]. This is meaningful for absorbing blasting energy and attenuating earthquake waves. Figure 8.1 summaries the research advancements and potential application areas of metamaterials. It is certainly impossible to detail each stand in a single book chapter; here, we will mainly deal with metamaterials in the realm of electromagnetics.

Despite the flourishing of metamaterials, problems and challenges do stand in the way. Conventional metamaterials derive peculiar physical properties merely from a certain periodical structure rather than the intrinsic materials' properties of building blocks. However, this 'periodical structure' feature is disadvantageous when it comes to offer functionalities to satisfy the growing demands for

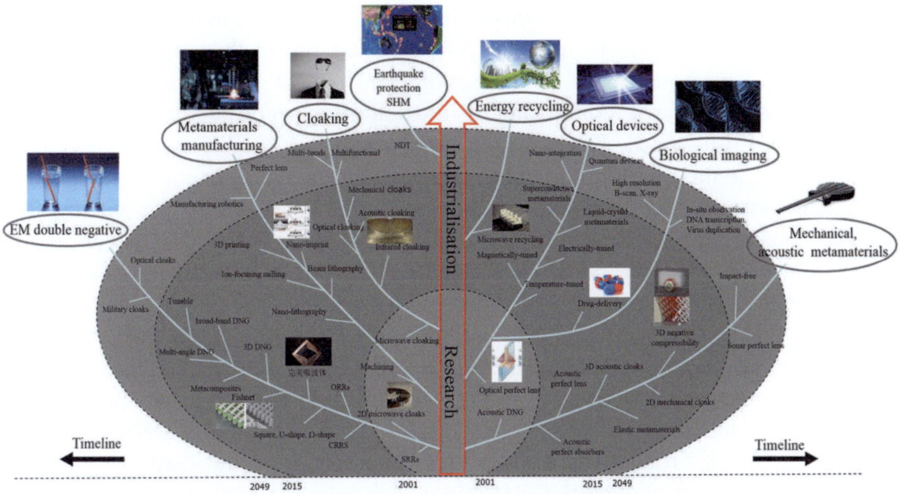

Fig. 8.1 A metamaterial tree covering current research activities and their future applications (Reprinted with permission from Ref. [17], copyright 2008 MRS. Reprinted with permission from Ref. [18], copyright 2008 Nature Publishing Group. Reprinted with permission from Ref. [19], copyright 2013 AIP. Reprinted with permission from Ref. [20], copyright 2013 Nature Publishing Group. Reprinted with permission from Ref. [7], copyright 2013 Nature Publishing Group. Reprinted with permission from Ref. [21], copyright 2008 Nature Publishing Group. Reprinted with permission from Ref. [22], copyright 2006 AAAS. Reprinted with permission from Ref. [23], copyright 2007 Nature Publishing Group)

metamaterials owing to the lack of flexibility and tunability with external stimuli applied (e.g. magnetic bias and mechanical stress). Another side effect for such structure-associated metamaterials is that generally a complicated internal topology is necessary to achieve DNG characteristics, which requires delicate processing techniques. In particular, in order to fabricate optical metamaterials, special nanolithography technology is obligatory to have precise control of geometrical dimensions, hence the metamaterial features [24]. Nevertheless, the trade-off between the precision and effectiveness (time, costs, etc.) confines the current metamaterials away from mass production. A solution is urgently needed.

Following the design principle for DNG properties, undoubtedly, the negative permittivity can still be realised from the plasmonic response of metamaterial system which is dominated by the geometrical parameters such as size and topology therein [2]. The negative permeability, nonetheless, can be excited via the ferrimagnetic or ferromagnetic resonance of metamaterial medium consisting of magnetic building blocks [25]. This is different from the conventional metamaterials where artificial magnetism is responded in non-magnetic materials. In this sense, the definition of metamaterials has evolved. By selecting proper materials as building blocks, the ultimate properties of metamaterials are dependent not only on their structure-affiliated parameters but more essentially on their intrinsic materials' properties of each component in the metamaterial context. The strong impetus of such new metamaterial category allows us to identify additional functionalities from the flexible response arising from the interplay between EM waves and internal constituents. This feature distinguishes from its conventional counterpart in that here a 'true' composite material is presented rather than a meta-structure. Therefore, we propose a term of 'metacomposite' where 'meta-' accounts for the metamaterial feature and 'composite' fits the definition according to the monograph authored by Professor R. Kelly, i.e. 'a combination of two or more materials in order to achieve the properties not obtainable with either component alone' [26]. Of particular note is that, among other requirements to realise an outstanding metacomposite, the magnetism from the functional fillers is the necessity to realise a negative μ. Among the ocean of magnetic materials, what categories could be potential candidates for metacomposites?

In this chapter, we will capitalise on the topic of metacomposite constructed by different magnetic materials together with their interesting EM behaviours and organise the remainder of the chapter as follows: In Sect. 8.2, we provide the fundamental aspects of metacomposite design in order to obtain a SNG or DNG feature. Section 8.3 reviews recent advances in metacomposites containing dielectric and/or magnetic nanoparticles featured with relevant metamaterial characteristics. Section 8.4 presents an improved strategy of an engineering metacomposite containing ferromagnetic microwires fabricated via an easy autoclave curing procedure. Section 8.5 details some essential efforts to optimise such microwire metacomposites with different geometrical configurations. The chapter is concluded in Sect. 8.6 with some future prospects of metacomposites.

8.2 Fundamentals to the Design and Fabrication of a Metacomposite

In the territory of electromagnetics, the EM behaviour of a material can be interpreted by ε and μ featured by their simultaneously negative values. From Maxwell equations of the differential form

$$\nabla \times \boldsymbol{E} = -j\omega\mu\boldsymbol{H}, \tag{8.1}$$

$$\nabla \times \boldsymbol{H} = j\omega\varepsilon\boldsymbol{E}, \tag{8.2}$$

it becomes rather clear that ε and μ are interconnected and are affected by each other through the entanglement between the electrical ($\boldsymbol{E}$) and magnetic component ($\boldsymbol{H}$) of incident EM waves [1]. This makes it extremely difficult to consider the quantity of ε and μ of every atom at microscopic or quantum level. However, it is of more practical significance to focus on the average performance of the EM medium, justified by the effective ε (ε_{eff}) and μ (μ_{eff}). Hence, Eqs. (8.1) and (8.2) can be transformed from the microscopic to macroscopic form and make the calculation much easier (for convenience, the ε and μ in this chapter denote ε_{eff} and μ_{eff}, respectively). The design of metamaterials is inspired by this idea based on the effective medium theory that a metamaterial should be recognised as a piece of homogeneous material electromagnetically so that its ε and μ can be deduced via the collective responses of all 'atoms' which give the macroscopic properties. However, as an artificial material, this requires that the size of the building blocks of metamaterials should be within the order of or much smaller than the wave length without which the EM homogeneity is simply collapsed [10]. Furthermore, at subwavelength region, the ε and μ can be independently treated, thanks to the decoupling of $\boldsymbol{E}$ and $\boldsymbol{H}$. It follows that respective design of the SNG ε and μ could be performed to further expect a DNG feature, which in turn is instrumental for metamaterial design and its engineering applications. In this section, we will discuss respectively the theoretical aspects of how to tackle negative ε and μ in the context of metacomposites in line with some successful paradigms. At last, a design example of the double-negative metacomposite containing ferromagnetic microwires is given.

8.2.1 Design a Negative Permittivity

Negative permittivity is not strange to us. In fact, continuous media with negative dielectric constant have long been known in the electromagnetic theory [27]. Although in some cases high losses usually prevent the onset of this property in common dielectrics, materials demonstrating negative ε can be found in nature. In fact, the Drude-Lorentz model, which is applicable for most materials, describes that the electromagnetic excitation of a condensed matter from elementary building blocks can be regarded as a response from a plasmon, i.e. a collective oscillation from electron density [28, 29]. A featured frequency is available representing the equilibrium where such excitation occurs:

$$\omega_{\mathrm{p}} = 2\pi f_{\mathrm{p}} = \sqrt{\frac{n_{\mathrm{eff}} e^2}{\varepsilon_0 m_{\mathrm{eff}}}}, \tag{8.3}$$

where $\omega_{\mathrm{p}} = 2\pi f_{\mathrm{p}}$ and f_{p} denotes the plasma frequency, n_{eff} is effective concentration of diluted electrons, m_{eff} is effective weight of electron and e is electron charge (1.6×10^{-19} C) [2]. The plasma frequency has profound influences on the dielectric profiles of metals and semiconductors, which are best-known examples for low-loss plasmas. Their effective permittivity function takes the following form on the interaction with electromagnetic radiation,

$$\varepsilon_{\mathrm{r}} = 1 - \frac{\omega_{\mathrm{p}}^2}{\omega(\omega + i\gamma)}, \tag{8.4}$$

where ω is angular frequency and the parameter γ represents the energy dissipation into the system. Of significant note is that the effective permittivity is essentially negative below the plasma frequency, provided the losses are small enough. So the design problem is reduced to how to design the plasma frequency in a metacomposite. Considering the f_{p} is determined by n_{eff} and m_{eff} which immediately link to the dimensions of the building blocks, we now discuss the metacomposites incorporated with functional micro- and mesoscaled fillers, respectively.

In the metacomposites containing nanoscaled fillers, at first sight, one can obtain f_p at various frequencies by altering the volume fraction of building blocks. For example, the higher the f_p, the larger the amount of fillers included in the metacomposites which dictates a higher electron concentration. Here matrix materials are assumed to exert negligible dielectric contribution. However, technically there exists a major issue. From a nanotechnology viewpoint, the possible aggregations of nano-fillers are detrimental to the plasmonic behaviour of metacomposites which makes the Drude model no longer applicable. Thus, by targeting a negative ε at optical or higher frequencies, nano-fillers must be distributed homogeneously, and certain techniques such as ultrasonic dispersion and surface chemical treatment are usually adopted [30]. Furthermore, s higher filler concentration causes a larger EM loss. As is known, a conductive network is formed in nanocomposites when fillers' amount is increased to a certain point, i.e. percolation threshold [31]. This will suppress the expected plasma frequency. This phenomenon is of paramount importance when one designs the configuration in the metacomposites to identify an f_p at higher frequencies, especially mid-infrared upwards.

Nonetheless, it is not an easy task to design a metacomposite system at lower frequencies from near infrared downwards, where Eqs. (8.1) and (8.2) can apply, in that EM energy dissipation asserts itself and the system cannot be recognised as a plasmonic one. The puzzle had confused scientists for long until Sir Pendry, the father of metamaterials, proposed a revolutionary strategy of an artificial medium made of very thin metallic wires (a few microns) in a periodical fashion [2]. Naturally, the n_{eff} can be expressed as

$$n_{\text{eff}} = n\frac{\pi a^2}{b}. \tag{8.5}$$

Here n, a and b denote the electron density of individual wire, wire radius and wire spacing, respectively. It can be assured that if $a \ll b$, $n_{\text{eff}} \ll n$. Thus, the f_p can be several orders of magnitude smaller than that of metacomposites containing nano-fillers. Substituting Eq. (8.5) into Eq. (8.3), the f_p and ε can be deduced as

$$f_p^2 = \frac{c^2}{2\pi b^2 \ln\left(\frac{b}{a}\right)}. \tag{8.6}$$

$$\varepsilon_r = 1 - \frac{\omega_p^2}{\omega\left(\omega + i\frac{\varepsilon_0 \gamma b^2 \omega_p^2}{\sigma \pi a^2}\right)}, \tag{8.7}$$

where σ is the conductivity of the wires. The wire medium has enormous significance in acquiring negative permittivity. First of all, a very simple configuration of a regular array of thin wires is requested. By adjusting the periodicity among wires

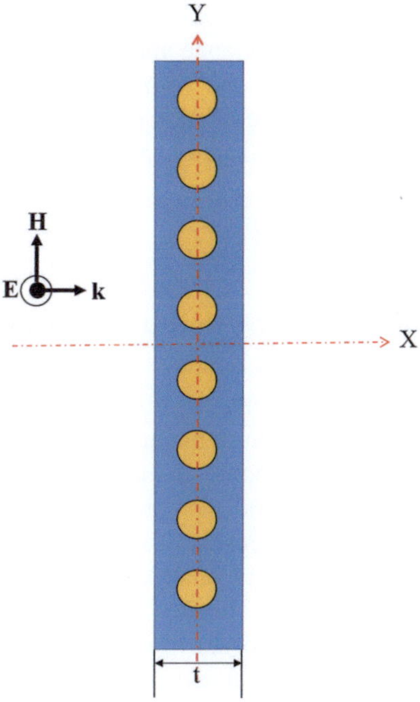

Fig. 8.2 Geometry of a slab material containing an array of wires

and selecting proper wire diameter, f_p is determined and tuned. Moreover, the dielectric response is now free of the filler aggregation issue often arising in the nano-metacomposites during the fabrication stage. In addition, the f_p is designed at low frequencies (normally microwave range). This drives us to think of using such wire medium as functional fillers to achieve negative ε in metacomposites.

Before we rush to the conclusion that metacomposites based on wire medium are able to generate negative permittivity, we must pause to consider the role of matrix materials. It has been introduced that a non-conductive material is preferential as matrix materials as it can be regarded as electromagnetically transparent medium at higher frequencies. At lower frequencies, on the contrary, the interaction between the matrix materials and EM waves, albeit very weak, would offer some contribution to the overall dielectric properties. Essentially, the above grid of wires is replaced by a slab of material containing such grid with constitutive parameters (ε_{eff}, μ_{eff}) and thickness of t, as illustrated in Fig. 8.2. Here it should be noted that ε_{eff} and μ_{eff} are average values considering wires and matrix materials from the effective medium theory. After homogenisation process of the composite medium, Eq. (8.7) is modified to

$$\varepsilon_{\text{eff}} = \varepsilon_0 - \frac{1}{\omega t b} \cdot \frac{(\omega L + X_{\text{w}}) + jR_{\text{w}}}{(\omega L + X_{\text{w}})^2 + R_{\text{w}}^2}, \tag{8.8}$$

where $L = -\mu_0/2\pi \cdot \ln(2\pi a/b)$ denotes the wire inductance in the array and X_w and R_w are the reactance and resistance along the wire, which determines the complex impedance as the form $Z_w = R_w + jX_w$ [32]. Nowadays, the wire medium is believed to be, if not the only, the easiest and most efficient way to obtain a negative permittivity, analogous to those exhibited by a solid metal at the UV frequencies.

8.2.2 To Design a Negative Permittivity

Compared to the negative permittivity, a negative permeability is more difficult to observe in natural materials. Only in some ferrimagnetic materials, where a natural magnetic response is identified, are negative μ possible yet in a weak excitation, rendering literally no practical value [33]. Generally, ferrites are magnetised to the saturation to present a tensor magnetic permeability with negative elements near the ferrimagnetic resonance. Although the negative permeability region is rather narrow for these ferrites, it enlightens that the ultimate purpose is to tailor the permeability tensor in the hope of obtaining some negative values.

Artificial magnetism becomes rather popular at the beginning of this century [11]. The huge significance is that a magnetic resonance can be derived from a non-magnetic material such as copper and silver. The effective μ of a metamaterial structure is defined as

$$\boldsymbol{B}_{\text{av}} = \mu_{\text{eff}}\mu_0\boldsymbol{H}_{\text{av}}, \tag{8.9}$$

where $\boldsymbol{B}_{\text{av}}$ and $\boldsymbol{H}_{\text{av}}$ represent the average **B**- and **H**-field along each of three axes in Cartesian coordinate. Again, here the scale of units must be much smaller than the wavelength to promise these average values are scientifically valid. If we consider the magnetic contribution along different axis, the μ_{eff} can be interpreted as [4]

$$(\mu_{\text{eff}})_x = (B_{\text{av}})_x/(\mu_0 H_{\text{av}})_x, \tag{8.10}$$

$$(\mu_{\text{eff}})_y = (B_{\text{av}})_y/(\mu_0 H_{\text{av}})_y, \tag{8.11}$$

$$(\mu_{\text{eff}})_z = (B_{\text{av}})_z/(\mu_0 H_{\text{av}})_z. \tag{8.12}$$

A negative effective permeability (MNG) is therefore justified and finds itself dependent on the homogeneity of the system. The next question is the configuration of the metamaterial system. Among numerous existing artificial structures, the split ring resonators (SRRs) attract considerable interests in research community and endorse themselves indispensable components in constituting conventional metamaterials [5]. Subsequently, some modified structures are invented such as open ring resonators (ORRs) [34], complementary split ring resonators (CSRRs) [35], double split ring resonators [36], etc. The electromagnetic dynamics of SRRs have been well described in many textbooks so here we will not harangue on this point. Interested reader could refer to Ref. [37]. However, the non-magnetic nature of the constituent materials of SRRs renders a double-edged sword in the realisation of metamaterials as they are electromagnetically lossy.

Now let us take a step back. What if those natural magnetic materials have much stronger magnetic activity? Most recently, attention has been paid to the use of ferromagnetic materials to generate negative permeability where the ferromagnetic resonance (FMR) takes place [38, 39]. Moreover, the FMR frequency can be easily and dynamically tuned by external magnetic fields. Naturally, ferromagnetic inclusions would be desirable candidates to fabricate the left-handed metacomposites.

The magnetic permeability of ferromagnetic materials is determined by the wave propagation mode. We assume that the excitation EM wave is the transversely propagating mode, where a dc magnetic bias $\boldsymbol{H}_{\mathrm{dc}}$ is applied perpendicularly to the wave propagation direction. To begin with, we consider a homogeneous slab material. In this scenario, the μ_{eff} can be defined as

$$\mu_{\mathrm{eff}} = \frac{\mu^2 - k^2}{\mu}, \tag{8.13}$$

where k is the wave number and

$$k = \frac{\omega_{\mathrm{m}}\omega}{(\omega_{\mathrm{FMR}} + j\alpha\omega)^2 - \omega^2}, \tag{8.14}$$

$$\mu = 1 + \frac{\omega_{\mathrm{m}}(\omega_{\mathrm{FMR}} + j\alpha\omega)}{(\omega_{\mathrm{FMR}} + j\alpha\omega)^2 - \omega^2}, \tag{8.15}$$

wherein $\omega_{\mathrm{FMR}} = \gamma\boldsymbol{H}_{\mathrm{dc}}$ is FMR frequency, α is damping factor, $\omega_{\mathrm{m}} = 4\pi\gamma M_{\mathrm{s}}$ is the characteristic frequency of the ferromagnets, γ is the gyromagnetic ratio and M_{s} is the saturation magnetisation [40]. Note that a slab material is not qualified to realise a negative permittivity, which makes the internal structure a burden for metamaterial design, not to mention the excessive reflection loss caused by the combination of these two materials, usually an array of metallic wires are still necessary. We switch to another solution to utilise conductive ferromagnetic microwires as building blocks. A simple configuration is illustrated in Fig. 8.2, where wires are parallelly arranged to the electric component (E_{k}) of incident waves. Still considering a TE mode of waves, Eq. (8.13) can be modified to

$$\mu_{\mathrm{eff}} = 1 + \frac{\mu_0\gamma M_{\mathrm{s}}(\mu_0\gamma H_0 - j\omega\alpha)}{-\omega^2 + \omega_{\mathrm{FMR}}^2 - j\omega\alpha\mu_0\gamma(2H_0 + M_{\mathrm{s}})}, \tag{8.16}$$

$$\omega_{\mathrm{FMR}} = \mu_0\gamma\sqrt{(H_0 + M_{\mathrm{s}})H_0}, \tag{8.17}$$

$$H_0 = H_{\mathrm{dc}} + H_{\mathrm{k}}, \tag{8.18}$$

where μ_0 is the vacuum and H_{k} is anisotropic magnetic fields of the microwires [41]. Of particular note is that, with such parallel wire arrangement, not only a negative permeability is enabled by selecting materials' properties such as H_{k} and

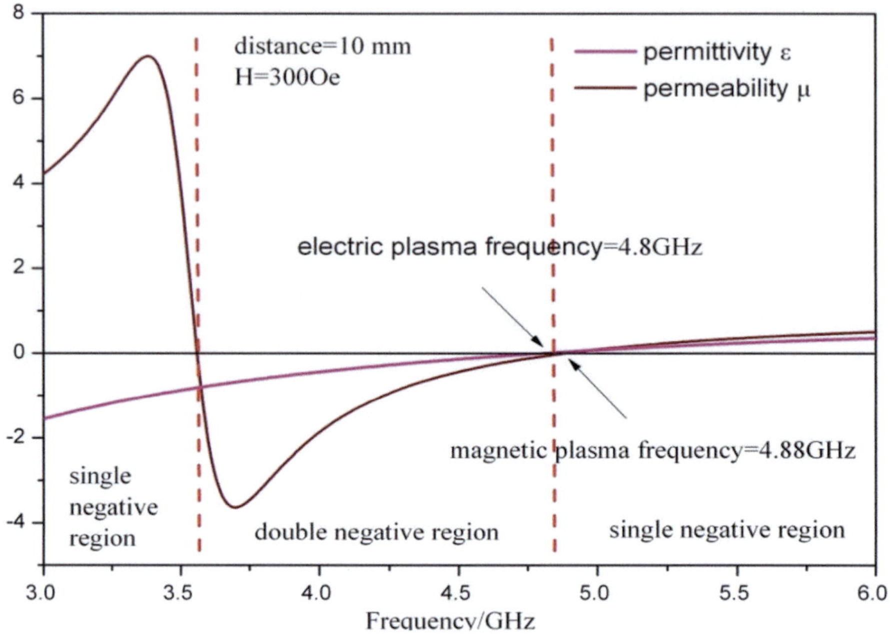

Fig. 8.3 Calculated results of permittivity and permeability of parallel $Fe_{77}Si_{10}B_{10}C_3$ wire array spaced by 10 mm in the presence of 300 Oe external magnetic field

M_s, which can be effectively controlled during the wire fabrication stage, but also a negative permittivity can be designed assuming the plasmonic behaviour is fulfilled in a wire medium. To this end, the double-negative metacomposites are now within our eyesight.

8.2.3 A Design Example of Double-Negative Metacomposites

Here we show a system consisting of an array of magnetic wires. Figure 8.3 displays the calculated real part of ε and μ in the frequency range of 3.0–6.0 GHz. To account for Eq. (8.7), the diameter of the metallic core is set as 16.6 μm and wire spacing is fixed as 10 mm, whereas $\gamma = 1.93 \times 10^{11}\ \mathrm{T}^{-1}$, $a = 0.02$, $H_{dc} = 300$ Oe and $M_s = 828\,\mathrm{Gs}$ determine the permeability in Eq. (8.16). It is clearly observed that in the frequency band of 3.6–4.85 GHz, the DNG features are identified. It is interesting to point out that the DNG region is 'moveable': One may apply moderate magnetic fields; thus, the frequency range of negative permeability can be tuned. On the other hand, an easy adjustment of changing wire spacing can be addressed to move plasma frequency, hence the frequency region of negative permittivity. The idea of metacomposites in turn provides an essential strategy that the DNG feature is now associated with the overlapping of topological metamaterial design at meso-/macro-scale and material properties design of building blocks at micro level. Being able to devise an SNG or DNG metacomposite, one needs to think what functional fillers and matrix materials are suitable and resolve details in the experimental stage.

8.3 Nano-metacomposites with SNG or DNG Characteristics

It is familiar to us that the application frequencies of metamaterials are dependent on the topological factors of the building blocks. For example, in a conventional metamaterial comprised of an array of SRRs and metallic rods, the size of the metamaterial units determines the artificial magnetic resonance in favour of a negative μ, whereas the periodicity among the rods dictates the dielectric behaviour where negative ε is realised [22]. In general, at higher frequencies (near infrared or higher), SNG or DNG features often demand metamaterials having nanoscaled units, whilst at microwave frequencies building blocks' dimensions are in millimetre range [23]. One can extend such requirement to metacomposites because a metacomposite is, first of all, a piece of metamaterial. Besides, although the actual magnetic properties of metacomposites' constituents are replacing the concept of non-magnetic magnetism in metamaterials, the size and shape of functional fillers are still at play to influence the negative permeability via the crucial magnetic parameters such as demagnetisation factors, anisotropic field, etc. From this section onwards, we will review the metacomposites featured with SNG or DNG properties containing different functional fillers following their nano- to mesoscales.

To discuss the nano-metacomposites, one needs to look back at the traditional nano-metamaterials. This century has witnessed the prosperity of photonic metamaterials since the DNG region was firstly reached above terahertz frequencies [42]. To date, nano-metamaterials with a great variety of structures are developed, to name a few, the fishnet, the antiparallel nanorods and the staple-shaped nanostructure [43–46]. These nanostructures require a precise control of the unit size which, later on, initiates the development of related fabrication nanotechnology [17]. Until now, fabrication techniques for photonic metamaterials have shifted from routine techniques such as nanolithography to more efficient and intelligent means, for example, nanoimprint, ion-focusing milling, 3D printing and so on [17, 47]. However, as the existing meta-structures are deployed to be tunable, broadband and multidimensional, these processing technologies can hardly prove to be cost-effective arising from the fact that normally a strictly clean environment is required to prevent possible oxidations and contaminations. In addition, either lithography or printing-associated techniques are time-assuming.

In a different area, nanocomposite materials have attracted substantial interests due to their outstanding mechanical and functional properties gained from the nanofillers/reinforcements in the context of a single matrix material [48]. It is natural to realise that a composite context would be a friendly medium to host SNG or DNG features via incorporating dielectric and/or magnetic nano-inclusions into base materials. In the following, metacomposites containing dielectric and magnetic fillers will be discussed, respectively.

8.3.1 Metacomposites Containing Dielectric Fillers

Conductive nano-fillers are favourable to generate the negative permittivity assuming they are uniformly dispersed in the insulating matrix materials as per the classic Lorentz-Drude model [49]. From the multifunctional standpoint, fillers with incredible electric conductivity, mechanical properties and chemical resistance are preferred owning to their ability to manipulate percolation threshold of the metacomposites, where the materials become fully conductive, and other dielectric parameters. From this perspective, by altering the volume fraction of fillers, the negative permittivity (ENG) properties of metacomposites can be effectively tuned. The matrix materials, however, should maintain electrically insulated while possessing excellent mechanical properties due to the necessity to provide essential constraint for nano-fillers. On the other hand, orderly arranged dielectric fillers are beneficial to induce a negative permeability as verified by numerous studies [50–52]. Therefore, through implementing nano-fillers into matrix materials, such negative permeability (MNG) features can be obtained and reasonably tuned by their content and shapes.

8.3.1.1 ENG Metacomposites

The ENG response has been a long-time research topic for natural materials and has become a nascent area for dielectric metamaterials via a well-managed structure, for instance, a wire medium. It stimulates a number of later researches on DNG metamaterials. In fact, the term ‘metacomposites’ also began as a piece of ENG metacomposites and continues to have an impact in the field. It was firstly realised that the combination of conductive nanoparticles and polymer composites would create amenable ENG characteristics. Zhu et al. achieved a metacomposite with ENG features via coating a homogeneous layer of tungsten oxide (WO_3) nanoparticles on the surface of conductive polypyrrole (PPy) polymer particulates [53]. The SEM morphology is presented in Fig. 8.4. The dipole resonance of the dispersed composite nanoparticles is responsible for the observed large amplitude of negative permittivity in the kHz frequency range. In specific, the negative permittivity can be tuned through changing the loading and morphology of fillers in addition to the chemical polymerisation process which can influence the electron transportation capability, hence the final dielectric properties. Furthermore, such tunable negative permittivity is validated in the WO_3/polyaniline (PANI) metacomposites [54]. It reveals that the switching frequency (frequency where

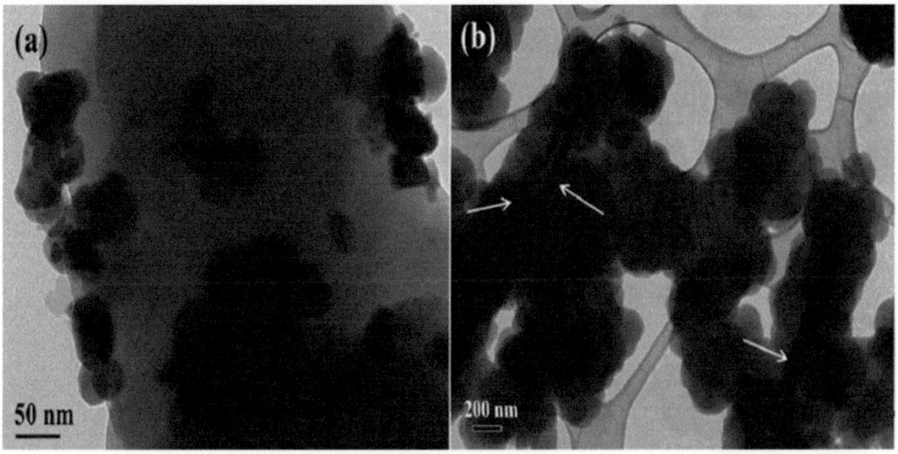

Fig. 8.4 TEM images of metacomposite particles of (**a**) 10 wt% WO_3 NPs/PPy and (**b**) 10 wt% WO_3 NRs/PPy (Reprinted with permission from Ref. [53], copyright 2010 ACS publication)

real part of effective permittivity switches from negative to positive) of metacomposites containing WO_3 nanorods is significantly higher than that containing WO_3 nanoparticles.

On the other hand, replacing WO_3 by magnetite (Fe_3O_4) nanoparticles, Guo et al. carried out a comprehensive study on PPy nanocomposites and showed that, apart from the similar negative permittivity feature to WO_3/PPy metacomposites, the negative magnetoresistance (MR) is also obtained and can be described by the orbital magnetoconductivity theory [55]. This contradicts to a parallel study which reported that a rather large positive MR response of 95 % is identified in the Fe_3O_4/PANI metacomposite [56]. This could be explained by the chemically induced dynamic spin polarisation due to different electrical involvement between fillers and polymer matrices. Among these studies, it should be stated that the negative permittivity would disappear with the increase of functional nanoparticles to a certain point that the percolation threshold is fulfilled and a conductive network is formed accompanied by a great enhancement of dielectric loss. These studies shed some light on chasing ENG features in the recipe of nanoparticles plus polymer system. Liu et al. explored the area from a different angle by using polymer/polymer composites. PANI nanoparticles are included into an epoxy prepolymer via an absorption-transferring process to yield a hybrid metacomposite [57]. The resultant composites reveal considerable magnitudes of the negative permittivity together with the electromagnetic interference (EMI) shield efficiency ca. 30–60 dB. Another unique advantage is that such hybrid composites are considered to be bulk materials which would be more useful from an engineering perspective, contrary to the above vulnerable metacomposites as they were synthesised to be discontinuous cluster of nanoparticles.

Carbon nanotubes (CNTs) are the most common fillers embedded in the nanocomposites to enhance their mechanical, electrochemical and other physical

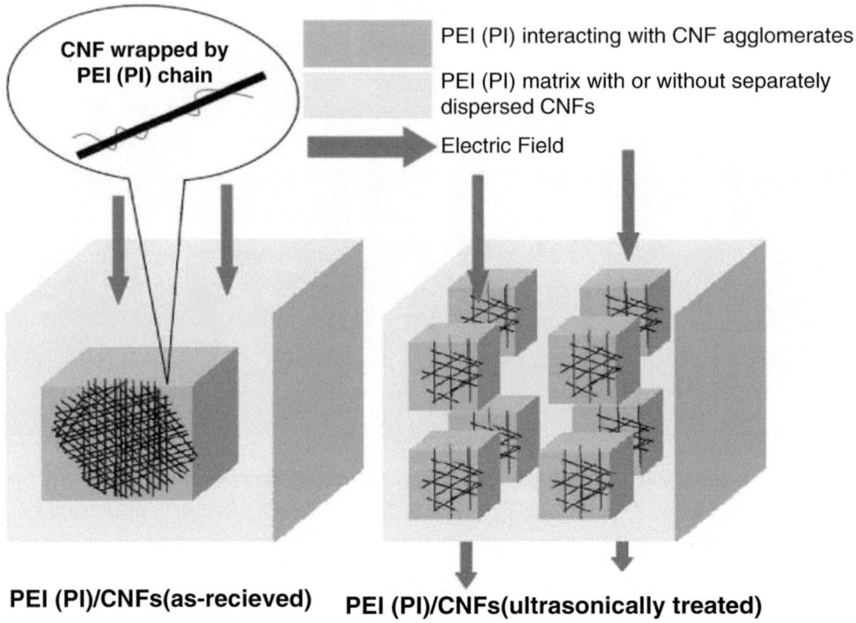

Fig. 8.5 Illustration of PEI-based metacomposites containing as-received and ultrasonically treated CNFs, respectively (Reprinted with permission from Ref. [60], copyright 2009 Wiley)

properties [31, 58]. These remarkable advantages push CNTs to the application frontiers, such as ultra-high mechanical composites, sensing devices, high-density data storage, etc. Rapid developments on nanocomposites based on CNTs have aroused great interests among researchers [59], and as a most recent one, CNTs are believed to be feasible fillers to construct negative permittivity. Alternatively, carbon nanofibres are another kind of 1D nano-fillers to possibly create a SNG medium due to comparable conductive properties but with much lower manufacturing costs compared with CNTs. A series of investigations were reported on the metacomposites containing CNTs or CNFs. As an innovative work, Li et al. selected polyetherimide (PEI) polymer as the mechanical host for CNFs [60]. The determinant structure of the metacomposites is the 3D network of CNFs wrapped with imide chains from polymer matrix (Fig. 8.5), which is similar to the Pendry's wire medium [2]. The displayed negative permittivity is strongly dependent on the compositional effects, including content, aspect ratio and distribution of CNFs. Moreover, it is revealed that the chemical structure of CNFs is of crucial importance on the overall dielectric performance of metacomposites via the chemical-interfacial bonding between CNFs and PEI. Generally, among other CNFs structures, the cup-stacked CNTs and a polymer matrix with large amount of amide groups are preferential to realise a metacomposite of larger negative permittivity response. Another study was carried out using CNFs/elastomer composites and reported a linear relation of electrical conductivity evolution and applied stresses as well as a negative permittivity dispersion [61]. Stress sensing and non-destructive testing (NDT) were suggested for potential applications. Compared with CNFs, CNTs have much larger surface energy due to stronger van der Waals bonding; hence, they are more difficult to homogeneously disperse in the composites. Studies on the CNTs reinforced metacomposites are listed here for interested readers [56, 62, 63].

2D nano-fillers can alleviate the aggregation issue of the metacomposites containing 0D or 1D fillers to a large extent. As one of the hottest materials at present, graphene has gained significant attention due to its exceptional properties which manifests itself suitable either as a filler to be included in the composite materials or a substrate to be decorated by other functional nanoparticles [64, 65]. An ENG metacomposite filled with graphene was reported by Zhu et al., and a higher electrical resistance of such metacomposites containing graphene is presented than those containing 1D CNFs or CNTs [66]. This reveals that a higher content of graphene is necessary to dispel the negative permittivity, which was observed in all the metacomposites containing carbonaceous nano-fillers.

In a brief summary, nano-fillers of different dimensions have been reported as constituents to realise metacomposites. As opposed to the conventional SNG metamaterial such as the wire medium, the negative permittivity can now be tunable via altering the filler content and managing a homogeneous microstructure. Moreover, the non-metallic nature of these fillers promises relevant applications advancing towards lightweight and miniaturisation directions. The problems of such metacomposite system, nonetheless, are also clear. Normally extra steps, e.g. ultrasonic dispersion or exfoliation, must be guaranteed to avoid filler clusters in addition to a fair interfacial bonding to matrices. This results in much manufacturing costs, although the SNG characteristics may be available eventually.

8.3.1.2 MNG Metacomposites Containing Dielectric Fillers

There have been a number of researches on achieving negative permeability through non-magnetic structures. Admittedly, MNG features realised from dielectric materials are scarcely reported. The Mie resonance, excited from dielectric materials, paves a path to create electric or magnetic resonance, thus providing a mechanism to produce MNG features. Previous efforts have testified that dielectric arrays like magnetodielectric spheres [67], ferroelectric spheres [68], polaritonic spheres [50] and ferroelectric rods [69] are favourable to obtain MNG response. But it was Zhou reporting the first piece of 3D MNG metacomposite consisting of an array of nanostructured dielectric cubes and Teflon polymers (Fig. 8.6) [70]. The size and the intrinsic permittivity of the used cubes are decisive to the Mie resonance and the pronounced MNG frequency band. Later, Zhou et al. revealed that such 3D metacomposites can also be thermally tuned in terms of their negative permeability [51]. However, the exhibited negative permeability is usually weak along a narrow frequency regime due to considerable reflection loss. A proper filler is still warranted to assure a better piece of MNG metacomposites.

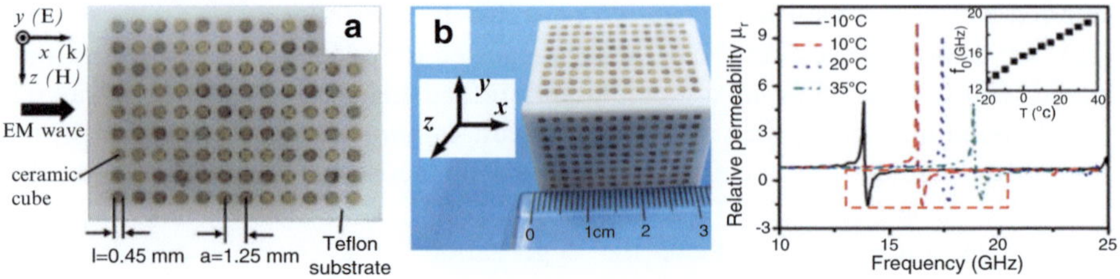

Fig. 8.6 Schematic views of periodical metacomposites containing nanostructured dielectric cubes. The thermal-tunable negative permeability is also obtained in the GHz range (Reprinted with permission from Ref. [70], copyright 2008 APS. Reprinted with permission from Ref. [51], copyright 2008 AIP)

8.3.2 Metacomposites Containing Magnetic Fillers

Unlike dielectric materials, magnetic fillers generate magnetic responses from their intrinsic properties. This defines a better option to realise MNG behaviour because it minimises the EM noises in particular the reflection losses. To think further, most of magnetic materials are metallic, which then makes them feasible to have ENG features; presumably, they are arranged in an ordered way to fit in plasmonic description.

8.3.2.1 MNG Metacomposites Containing Magnetic Fillers

Investigations on MNG metacomposites based on magnetic fillers originate from ferrimagnetic fillers in view of their unique magnetic resonances to EM waves. Among ferrimagnetic materials, the yttrium iron garnet (YIG) is a kind of synthetic garnet that has been widely used in microwave, acoustic, optical and magneto-optical devices [71, 72]. Their magnetic permeability has been extensively studied with or without external magnetic fields. As an important step to reach the MNG territory, He et al. reported that negative permeability can be built on the YIG-based system [73]. But strictly speaking, such MNG system is not a metacomposite because the ultimate structure is developed from the combination of bulk YIGs and a set of copper arrays (Fig. 8.7). Tsutaoka et al. improved this proposition by implanting the concept of composites. A metacomposite was established by fabricating the YIG-based granular composites and subsequently verifies a negative permeability resulted from the gyromagnetic spin resonance (Fig. 8.8) [74]. Compared with YIG, Permalloy is believed to be more competitive to act as magnetic fillers especially at high frequencies owing to its relatively higher conductive resistance. Kasagi et al. used $Ni_{45}Fe_{55}$ Permalloy particles thermally treated at different temperatures and hot-pressed into PPS resin powder matrix to yield a granular composite [75]. The results showed a remarkable negative permeability with 70 vol.% Permalloy particles included, indicating that high temperatures are in favour of suppressing detrimental eddy current effect at high frequencies.

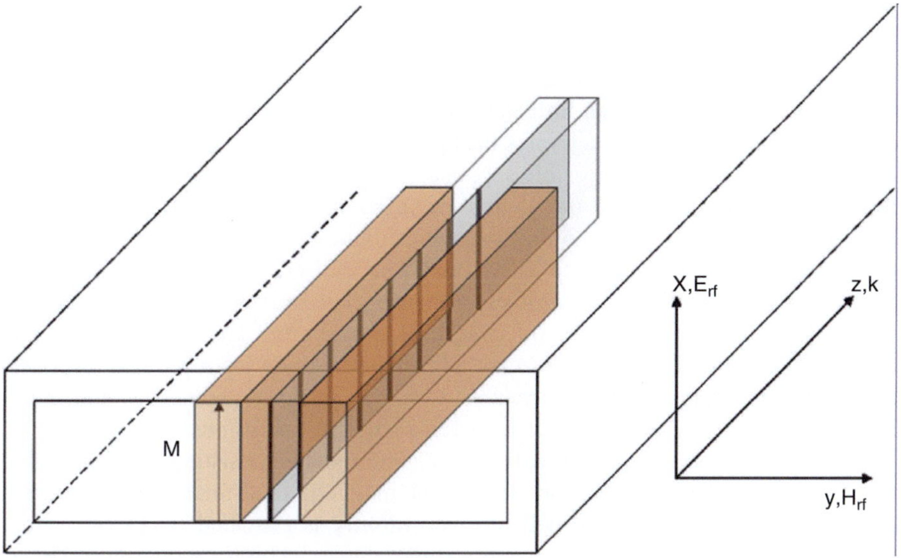

Fig. 8.7 Schematic view of the metacomposites consisting of multilayered YIG films and an array of eight copper wires spaced by 1 mm (Reprinted with permission from Ref. [73], copyright 2007 Elsevier)

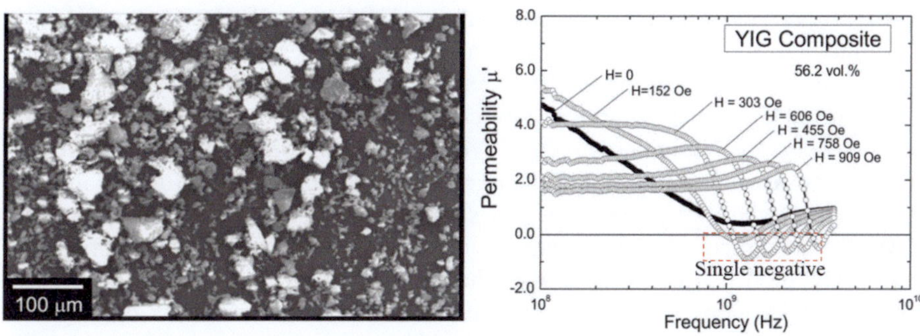

Fig. 8.8 SEM image of YIG-based composites and frequency plots of real part of permeability with presence of external magnetic bias (Reprinted with permission from Ref. [74], copyright 2011 AIP)

It is worth pointing out that the MNG response is generally weak aroused from metacomposites based on ferrimagnetic fillers due to their high magnetic remanence and coercivity. Thus, it is anticipated that soft magnetic materials, e.g. ferromagnetic materials, would be able to create more advantageous metacomposites in terms of more degrees of freedom to design friendly DNG features arising from their high magnetic sensitivity and broader working frequencies. This will be detailed in Sect. 8.4.

8.3.2.2 DNG Metacomposites

Most of the existing studies on metacomposites containing magnetic fillers reported MNG features have provided little details on their permittivity spectra thus far. For someone from the electromagnetics background, it is not difficult to imagine that the ENG features are also possible supposing magnetic fillers engaged in a periodical manner. In this way, the double-negative index can be obtained in the metacomposite medium containing magnetic fillers. In fact, Holloway et al. have theoretically presented that the DNG media is realised in a composite medium composed of insulating magnetodielectric particles embedded in a background matrix [67]. In this approach, no complicated scatterers are required, and this array-based composite also adds to the advantage of being isotropic EM response. Moreover, this approach can be readily extended to other geometries and, more importantly, other types of dielectric or ferromagnetic inclusions.

Regarding the experimental realisations, a composite consisting of a hexagonal Co_2Z slab and a YIG slab is fabricated, and a magnetic field-tunable DNG characteristic is observed [76]. The negative permittivity is provided by Co_2Z, which, nonetheless, prevents the negative permeability due to its high conductivity. This is why the YIG slab should be involved, i.e. offering magnetic resonance to further create MNG. It would be more attractive results if the microstructure of the composites is given and well expounded. In such metacomposites, the manufacturing technique is of critical importance to promise an optimal micro-/mesostructure and avert possible defects which otherwise would damage the DNG features.

In another perspective, some contributions are addressed from foam-ceramic-based composites containing ferromagnetic particles. Fan et al. delivered a double-negative Ni/Al_2O_3 metacomposite fabricated via a conventional wet-impregnation process where the final microstructure evidences that a discontinuous Ni network is formed and hosted in the Al_2O_3 ceramic foam (Fig. 8.9) [77]. It follows that Fe/Al_2O_3 composites were obtained in a similar way in line with DNG properties [78]. It is no surprise that the negative ε can be expected for the nickel network as per plasma model. It is further argued that the negative μ is excited by the current loop internally along the Ni or Fe particles realised from their weak diamagnetic properties in the presence of EM waves. However, it is more likely that the identified negative permeability is intrinsically realised from the ferromagnetic resonance of metallic network as Ni has pronounced soft magnetic properties. This understanding is confirmed in an earlier study: it was shown that only ENG is presented in the Ag/Al_2O_3 composites [79]. It is established that silver is a diamagnetic material, but no clear DNG profiles are received in the metacomposites. Before more materials are introduced into the metal/ceramic metacomposites, one should also admit some inherent problems. For example, the DNG band is still very narrow (within a few MHz), and an open porosity of the ceramic matrix predefines their metamaterial behaviour. This makes such DNG metacomposites inconvenient for practical applications. Besides, delicate processing techniques are necessary to acquire a homogeneous distribution of the

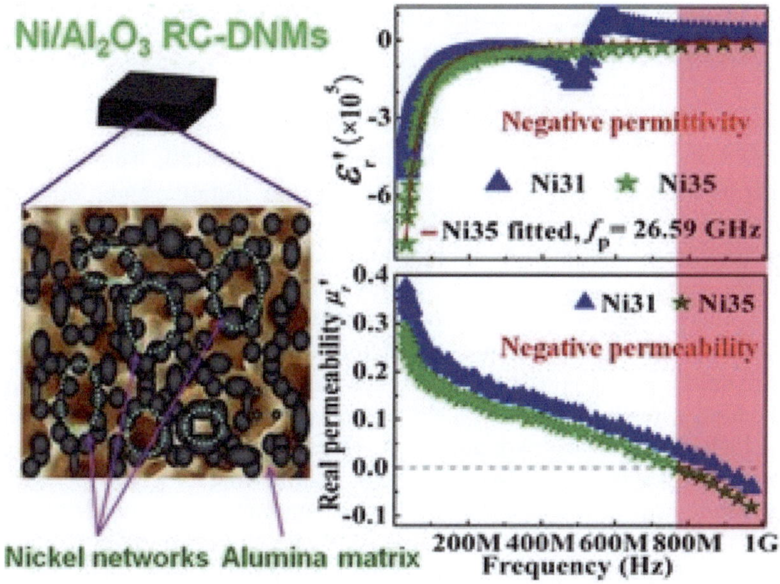

Fig. 8.9 Schematic image of the microstructure of the Ni/Al_2O_3 metacomposites and frequency plots of real part of effective ε and μ. The *pink area* characterises the double-negative region (Reprinted with permission from Ref. [77], copyright 2012 Wiley)

open pores of foam ceramics, which is of paramount significance to avoid percolation networks in the metacomposites.

Having appraised the SNG or DNG metacomposites containing functional fillers of different size and material properties, it is worth making some remarks with regard to the choice of fillers to fit in metacomposites. For the ENG properties, dielectric fillers are preferred in the context of periodical arrangement; otherwise, magnetic fillers would generate undesirable losses to overlap dielectric performance. For the MNG and DNG properties, on the other hand, one should select magnetic fillers as they have notable response to create a vital magnetic excitation for receiving the possible negative permeability. In addition, fillers at micro-scale are easy to handle during the manufacturing stage and can be feasibly applied for metacomposites at microwave frequencies through managing the composites' mesostructure. The EM properties can be conveniently tailored as opposed to the case of metacomposite containing nano-fillers that complex nanotechnologies are always a necessity. In conjunction with a highly efficient manufacturing strategy, as to be detailed in the following section, an engineered metacomposite has emerged and can potentially be developed for mass production.

8.4 Metacomposites Containing Ferromagnetic Microwires

Amorphous ferromagnetic microwires have aroused international interest in recent years due to their exceptional magnetic and EM properties [80]. These microwires, generally comprised of a layer of glass coat and metallic core, offer decent ferromagnetic properties because they are primarily of CoFeSiB or FeSiB in composition (doped with trace amount of other elements, e.g. Nb, Co, Cr). In particular, a remarkable magnetic phenomenon of giant magnetoimpedance (GMI) is discovered in some of the Co-based microwires, i.e. a large electric impedance change in a magnetic conductor with an ac current when a small magnetic field is applied [80]. A very large impedance change can also be attained in these microwires with the presence of external stress, i.e. giant stress-impedance (GSI) effect [81]. These two important effects provide unique signatures of microwires in response to EM waves and suggest that they can be built into microwave devices as highly sensitive components for sensing and NDT applications. In addition, microwires have fine geometrical size (ranging from a few microns to a few tens of microns in diameter) which is favourable for device miniaturisation. Additionally, the mechanical properties are fairly high owning to their fully amorphous microstructure, thanks to the metalloid elements and fast-cooling rate during fabrication process. All these outstanding characteristics persuade microwires as feasible fillers to constitute metamaterials. Recently, it has been reported that an array of Co-based microwires arranged in a waveguide can create a double-negative medium and the metamaterial behaviour can be tuned by external fields and stresses due to GMI and GSI [41, 82]. The advantage of this wire array is obvious: it is such a simple structure.

Most recently, a multifunctional polymer-based composite with embedded microwires is proposed [83]. The marriage between polymer composites and microwires showcases threefold merits: (1) a remarkable magnetic field and mechanical stress tunability in the microwave regime which offers excellent sensing properties, (2) a low filler content which guarantees minimum disturbance to the overall mechanical properties of the composites that is of primary concern for structural components such as aircraft fuselage, and (3) microwire composites which can be fabricated using a cost-effective technique. With all these benefits from wire composites, it is logical to ask the question: how about a microwire metacomposite? By incorporating microwires into the polymer-based composites, it is of great engineering interest to manufacture a metacomposite with possible DNG features, and this metacomposite is no doubt a true material rather than a structure.

In terms of microwire selection, there are two categories of ferromagnetic microwires in general: the Co based and the Fe based. Previous studies have elucidated the benefits of Co-based wires in realising a DNG feature arising from their superior GMI properties over the Fe-based counterparts and nearly zero magnetostriction constant attributed to their circumferential anisotropy

[80]. However, Fe-based wires might make more suitable building blocks for metacomposite with their competitive advantages that are in twofold: one is that, unlike Co-based microwires, Fe-based wires have natural FMR effect which forms negative μ without the aid of external fields and the other factor is that Fe-based wires are much cheaper, thereby more economic and desirable for large-scale productions. For the matrix materials, apart from the desirable mechanical properties to contain wires, they should be a commonly accepted polymer composite from an industrial viewpoint since the ultimate goal is to develop an engineering metacomposite [84, 85]. The carbon fibre-reinforced polymer composites are not compatible for such metacomposite system because their conductive nature would generate excessive reflection loss that prevents the further essential interaction between microwires and incident waves. The target is then switched to a familiar material of the glass fibre-reinforced composites in the form of thin-ply prepregs ($500 \times 500 \times 0.25$ mm [3]) as they are literally transparent to microwaves. For the manufacturing of the metacomposites, a conventional prepreg layup is employed with microwires incorporated between neighbouring prepregs, followed by a standard autoclave curing, which is justified as the most widely used manufacturing technique for commercial and industrial polymer composites.

Specifically, $Fe_{77}Si_{10}B_{10}C_3$ microwires with a total diameter of 20 μm and the glass layer thickness of 3.4 μm are fabricated via a modified Taylor-Ulitovskiy technique and provided by TAMAG, Spain. Figure 8.10 shows the mechanical properties and magnetisation curve of the used Fe-based wires in addition to a SEM image [86]. A rectangular magnetic hysteresis (M-H) loop is observed exhibiting a trivial coercivity of 0.31 Oe and a saturation magnetisation of 850 Gs. The inset stress-strain curve gives the tensile strength as high as 1297 MPa and the fracture strain of 2.88 %. These results validate the applicability of microwires as high-strength and soft magnetic functional fillers in the composites. The present microwires are embedded into two E-glass 950 aerospace-graded prepregs in a parallel geometry with 3, 7 and 10 mm wire-wire spacing, respectively, followed by adding the other two prepregs on top and bottom of this sandwiched wire-prepreg layer (Fig. 8.11). It should be noted that all the microwires are arranged along the direction of glass fibres to reduce hard contact between them if otherwise. The stacked wire prepregs are then cured in the autoclave to yield a resultant wire composite with dimensions of $500 \times 500 \times 1$ mm [3]. The curing conditions are detailed elsewhere [87]. The microwave characterisation is performed by a free-space measurement rig placed in an anechoic chamber in the frequency range of 0.9–17 GHz with or without a dc magnetic bias up to 3000 Oe. The electrical component (E_k) of incident waves is set along the microwires to maximise the EM response. The S-parameters representing the received signals of reflection and transmission are obtained and subsequently transformed into a built-in programme, i.e. Reflection/Transmission Epsilon Fast Model, to compute complex effective permittivity [88].

The transmission (S_{21}), reflection (S_{11}) and absorption (A) coefficients of composites containing wires spaced under different periodicity are presented in Fig. 8.12. A series of transmission windows are identified in the 1–7 GHz from

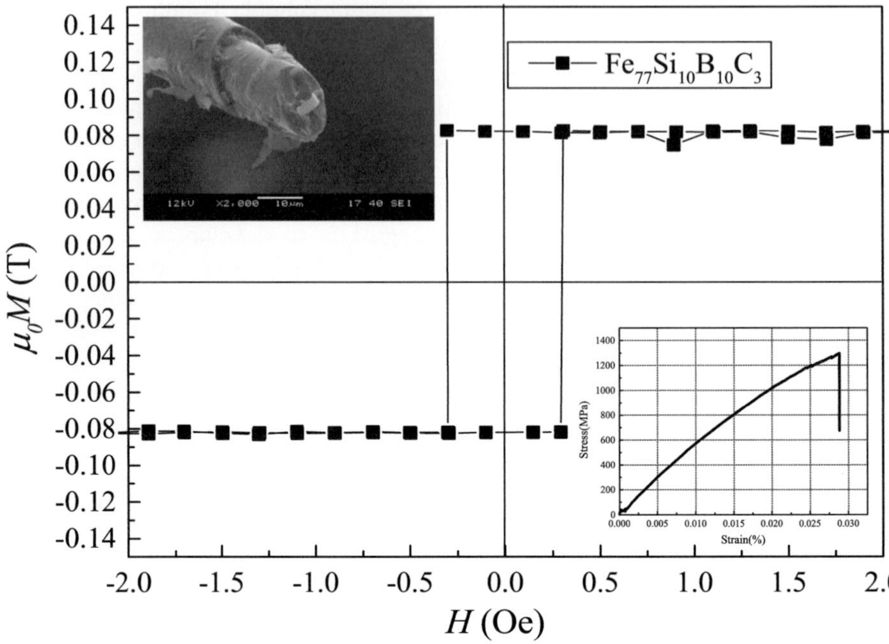

Fig. 8.10 Quasi-static magnetisation curve of $Fe_{77}Si_{10}B_{10}C_3$ microwires. The *insets* are cross-sectional SEM image and stress-strain curve of the microwire, respectively (Reprinted with permission from Ref. [86], copyright 2014 AIP)

the transmission spectra of metacomposites containing the 3 mm spaced microwire array (Fig. 8.12a) in addition to reflection dips and absorption peaks (Fig. 8.12b, c). The transmission windows suggest a negative ε and a negative μ are simultaneously obtained, i.e. wire arrays behave as transmitting structure in this condition. One should note that these windows are realised in the absence of magnetic fields, as such this is defined as the natural metacomposite feature. This eases the application of such materials in compact devices and lightweight structures where extra magnetic excitation components are considered as a burden. For the absorption spectra, at low frequencies, the resonance peaks are induced by FMR of microwires, in which Eq. (8.17) can be simplified to $f_{\text{FMR}} = \gamma(M_s + H_a/2\pi)$ [90, 91]. Substituting $M_s = 828$ Gs and $\gamma = 2.8\,\text{MHz/Oe}$ and assuming $H_a \ll 2\pi \cdot M_s$, f_{FMR} is calculated to be 2.3 GHz, which is quite close to the experimental peaks. The subtle difference is ascribed to the residual frozen-in stresses of the as-fabricated microwires. When the wire spacing is larger than 7 mm, the absorption peaks unanimously shift to higher frequencies according to Kittel's relations [92]. However, the transmission windows do not present any significant shift in addition to the identified absorption peaks. It has been reported that hard crystallites on the wire surface can dramatically degrade their magnetic properties [93, 94]. Considering the magnetic properties of microwires are temperature and stress-sensitive, the existing curing cycle (i.e. 125 °C heating for 2 h and 0.62 MPa for 4.5 h) [87]

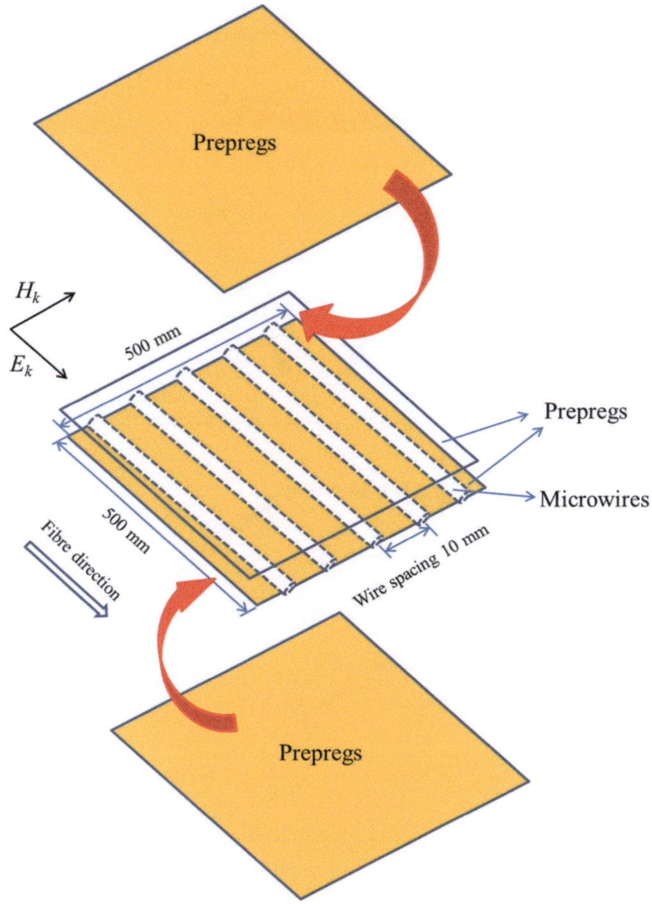

Fig. 8.11 Schematic view of manufacturing process of parallel wire array composite with wire spacing of 3, 7 and 10 mm, respectively (Reprinted with permission from Ref. [86], copyright 2014 AIP)

would have profoundly changed the stress distribution within the wires, hence conspiring to the formation of magnetically hard crystallites on the surface. This structural evolution would cancel out the expected window shifts in the transmission spectra of 3 mm array containing metacomposites where a relatively higher content of degraded microwires is included. Interestingly, such spacing-associated effect is also observed in determining the occurrence of the DNG features as evidenced by the disappearance of transmission windows if the wire spacing is larger than 3 mm, which links to the wires' unique dielectric performance.

The effective permittivity of metacomposites containing wire arrays of different spacing obtained from *S*-parameters is presented in Fig. 8.13. Surprisingly, one observes a notable discrepancy between the experimental results and theoretical values of 4.8, 6.6 and 16.6 GHz for composites with wire spacing of 10, 7 and

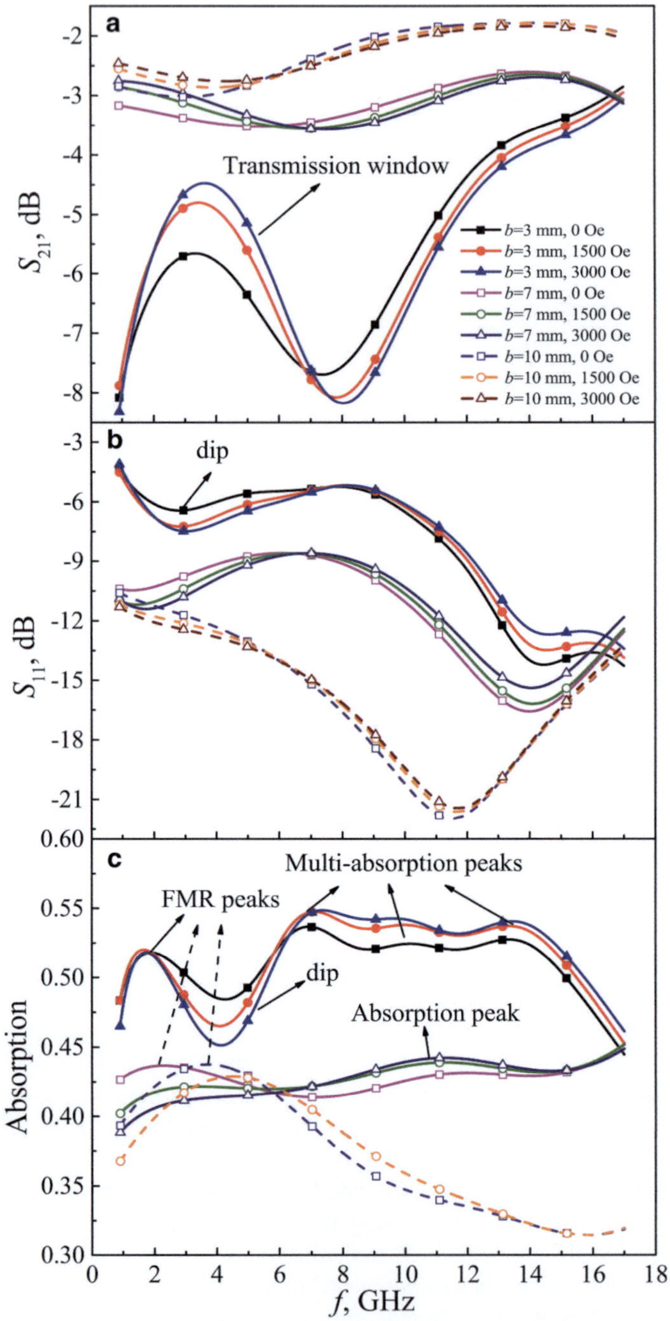

Fig. 8.12 Frequency plots of (**a**) transmission, (**b**) reflection and (**c**) absorption coefficients of metacomposites containing 3, 7 and 10 mm spaced parallel wire arrays (Reprinted with permission from Ref. [89], copyright 2013 AIP)

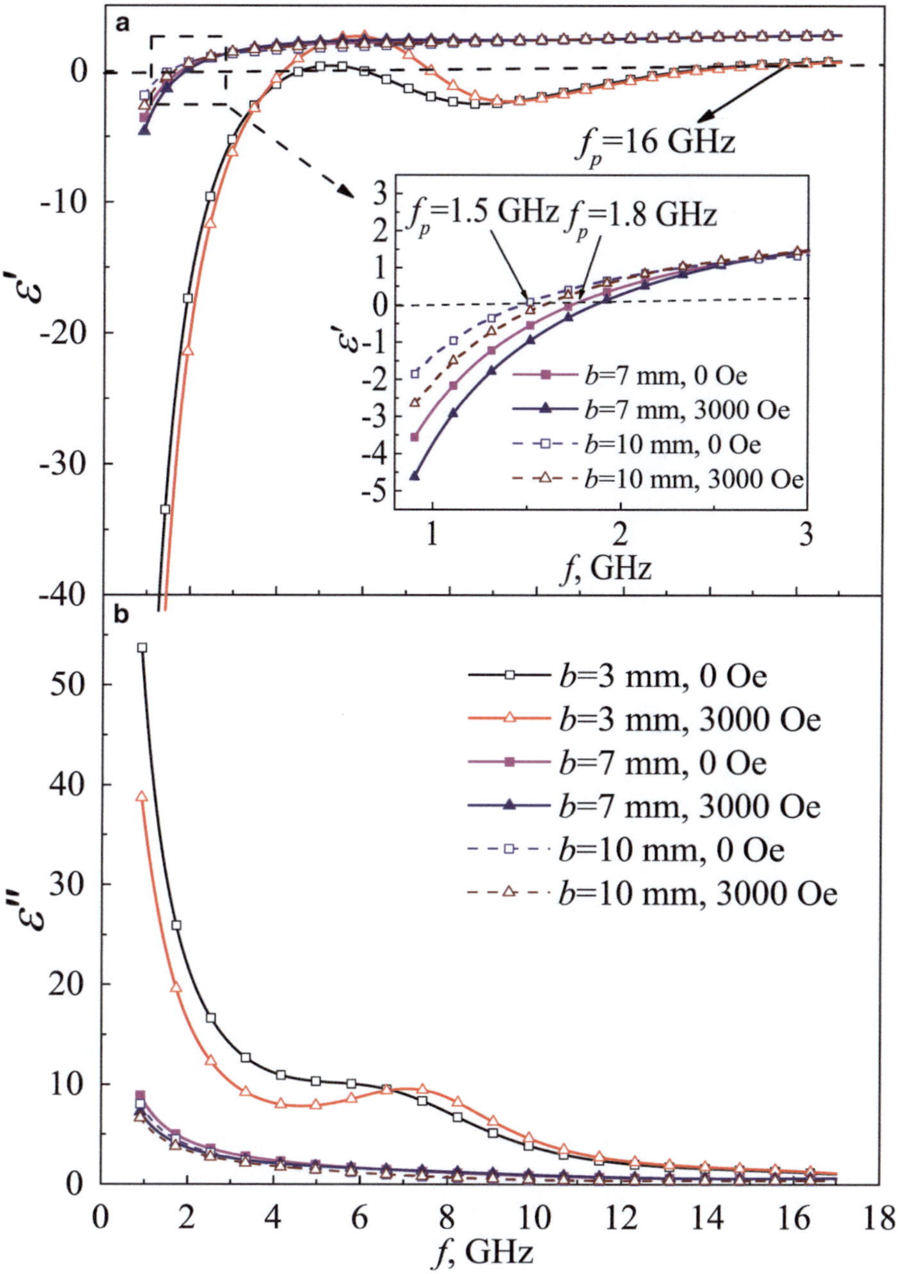

Fig. 8.13 Frequency dependence of (**a**) real part and (**b**) imaginary part of permittivity of parallel metacomposites with different spacing (Reprinted with permission from Ref. [89], copyright 2013 AIP)

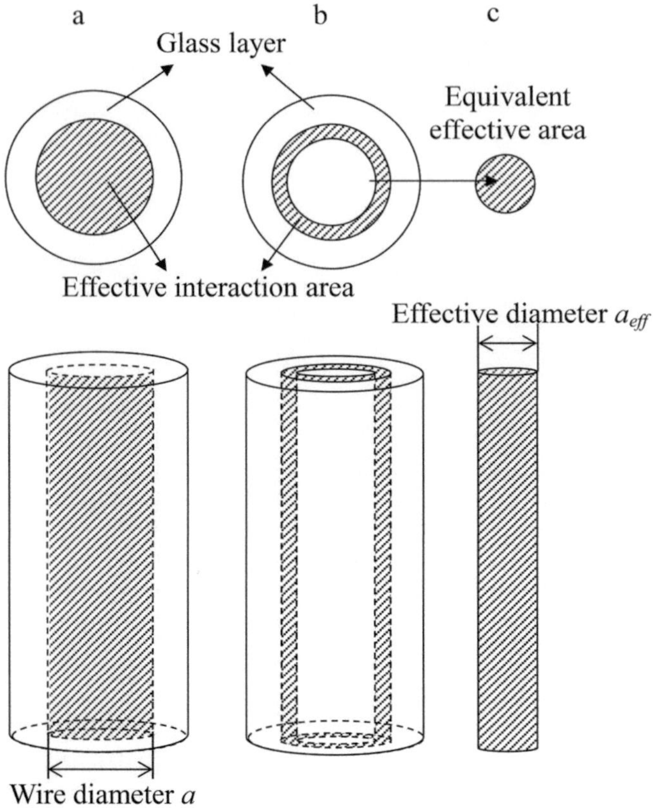

Fig. 8.14 Schematic view of microwave interaction area on (**a**) conductive wires, (**b**) Fe-based microwires and (**c**) the effective diameter accounting for the equivalent area in terms of Fe-based microwires (Reprinted with permission from Ref. [89], copyright 2013 AIP)

3 mm, respectively, as per Eq. (8.6). To explain such difference, it is acknowledged that the dielectric response of ferromagnetic microwires is controlled via managing their surface plasmons. This determines that the major permittivity contribution of a microwire comes from the outer shell of its domain structure [2]. However, Fe-based microwires only possess a trivial volume fraction of outer domain (Fig. 8.14a). This makes the dielectric wave-matter interaction area dramatically decreased; herein, a term of effective diameter a_{eff} is proposed in order to account for such equivalent area (Fig. 8.14b). In this sense, f_{p} should be modified to

$$f_{\mathrm{p}}^2 = \frac{c^2}{2\pi b^2 \ln\left(\frac{b}{a_{\mathrm{eff}}}\right)}. \tag{8.19}$$

Since $a_{\mathrm{eff}} \leq a$, the f_{p} is then significantly compromised (Fig. 8.14b), which makes no DNG feature for wire spacing exceeding 3 mm. A sudden increase of f_{p}

is drawn when the spacing is decreased to 3 mm. It can be explained that the dynamic interactions with microwaves occur in such closely packed microwires and have essentially compensated the excited circumferential magnetic fields, hence the a_{eff} [95], while 7 or 10 mm spacing proves too wide to induce either static or dynamic wire-wave interactions. This fundamentally determines why only in the case of 3 mm spacing the DNG features are available. Likewise, Eq. (8.19) can be deployed to resolve the small mismatch of f_{p} observed in the Co-based microwire composites. As is known, Co-based wires have large outer domain volume covering on a bamboo-like domain structure [80]. Thus, the a_{eff} of Co-based wires would be a touch smaller than the real a, rendering also a slightly smaller f_{p}. Further post-annealing treatment on wires are anticipated to make the f_{p} and DNG characteristics more predictable.

8.5 The Optimisation of Microwire Metacomposites

Although double-negative indices have been obtained in the parallel wire composites, one issue tarnishes their practical usage: as one can see from Fig. 8.14b that the reflection level is relatively high for 3 mm spaced wire array due to high wire content. A new geometry of wire arrangement with lower EM losses is therefore needed.

8.5.1 Metacomposites with Orthogonal Microwire Arrays

Liu et al. has theoretically predicted that an orthogonal array consisting of ferromagnetic wires and carbon fibres is possible to yield DNG features [96]. Bearing in mind that a simple structure is always of primary concern, owing to the excellent microwave properties of the Fe-based wires, using microwires alone for the orthogonal structure would be a much better way to construct a DNG system with other salient features.

A metacomposite containing orthogonal Fe-based wire arrays is designed and schematically shown in Fig. 8.15. Microwires with the same composition as in parallel metacomposites are used. Fabrication and characterisation details are given elsewhere [87, 88]. Compared with the metacomposite containing 3 mm spaced parallel wires, transmission windows with higher transmittance are obtained in the 1–6 GHz in the metacomposite containing orthogonal wire array with 10 mm spacing, which has a much lower wire loading (Fig. 8.16). The inserted horizontal microwires (Fig. 8.15) can be regarded as a typical wire medium for inducing negative permittivity. The vertical wires, on the other hand, can be considered as an integration of shortcut parallel wires between neighbour continuous wires resulting from the small axial amount of E_{k} in favour of creating circumferential magnetic fields that enhance the dynamic wire-wire interaction [97]. This explains why in the

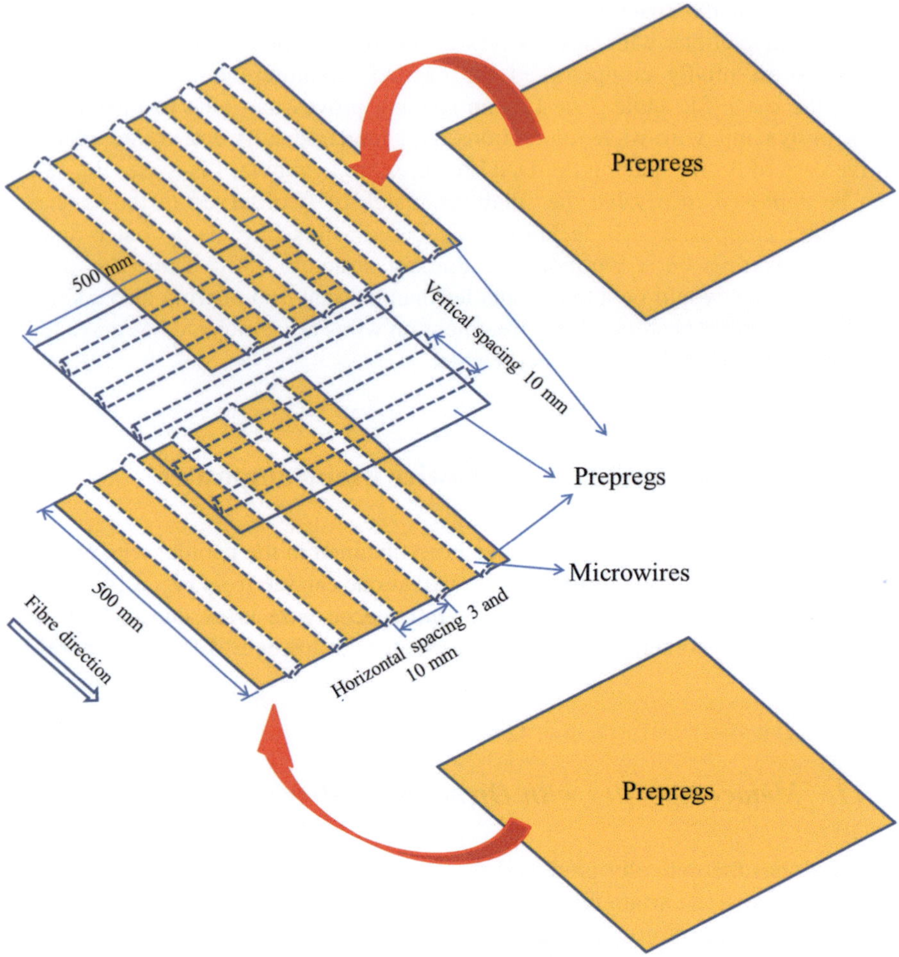

Fig. 8.15 Schematic view of manufacturing process of orthogonal wire array composite with fixed wire spacing 10 mm perpendicular to glass fibre direction and different horizontal wire spacing of 3 and 10 mm, respectively (Reprinted with permission from Ref. [86], copyright 2014 AIP)

orthogonal system the critical spacing (10 mm) is lower than that of parallel metacomposites (3 mm) (note that for orthogonal metacomposites with spacing larger than 10 mm, no transmission window is observed). Overall, the orthogonal metacomposites have the optimised configuration with improved transmission performance but less filler inclusion.

Interestingly, a slightly higher transmission of orthogonal metacomposite with 3 mm spaced array is noted compared with 3 mm spaced parallel metacomposites (Fig. 8.16). This transmission improvement is attributed to the additional wave-wire interaction from the vertical wire array. For such orthogonal structure, both

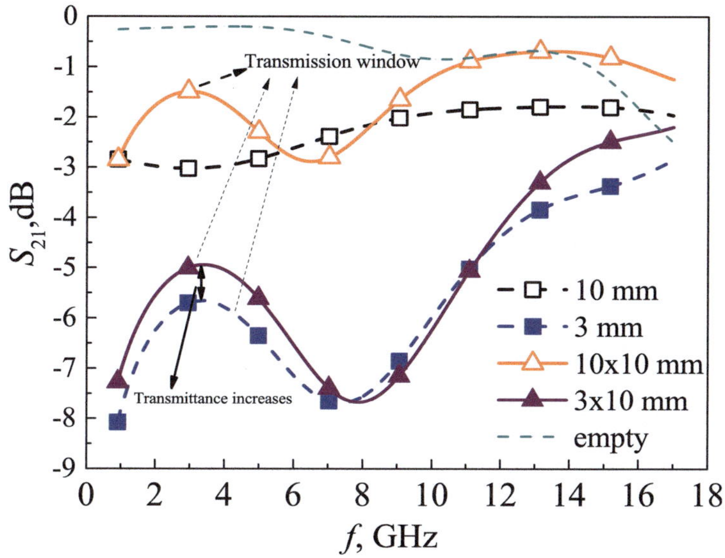

Fig. 8.16 Transmission spectra of metacomposites containing orthogonal and parallel wire arrays in the frequency band of 0.9–17 GHz. E_k is arranged along the glass fibres (Reprinted with permission from Ref. [86], copyright 2014 AIP)

permittivity and permeability are enhanced [97]. The improved permittivity comes from those 'imaginary shortcut wires' arising from the dielectric contribution of vertical wires, which enhances the permeability to a similar extent. An extra amount of permeability enhancement is excited along the vertical wires from the magnetic component (H_k) of incident waves [98]. Therefore, the transmission is eased by the improved impedance match in the orthogonal metacomposites taking into account

$$Z = \sqrt{\left(\frac{\mu}{\varepsilon}\right)}. \tag{8.20}$$

This makes the quantitative control of transmission possible through the understanding of the relation between transmission enhancement and the amount of inserted vertical wire arrays.

To this point, the role of glass fibres in the polymer matrix needs to be clarified. As the glass fibres in the matrix material are aligned perpendicularly to the vertical wires (Fig. 8.15), they would be influential to the overall microwave properties of the wire composites. Figure 8.17a presents the transmission coefficients of the orthogonal metacomposites measured with E_k perpendicular and along glass fibres, respectively. Note that the results from Fig. 8.16 regarding the DNG features are measured with E_k along the glass fibre direction. It is assumed that the horizontal wires lodge themselves into the in-prepreg and/or inter-prepreg gaps among glass fibres (Fig. 8.17b). In this scenario, the continuous wire configuration is preserved,

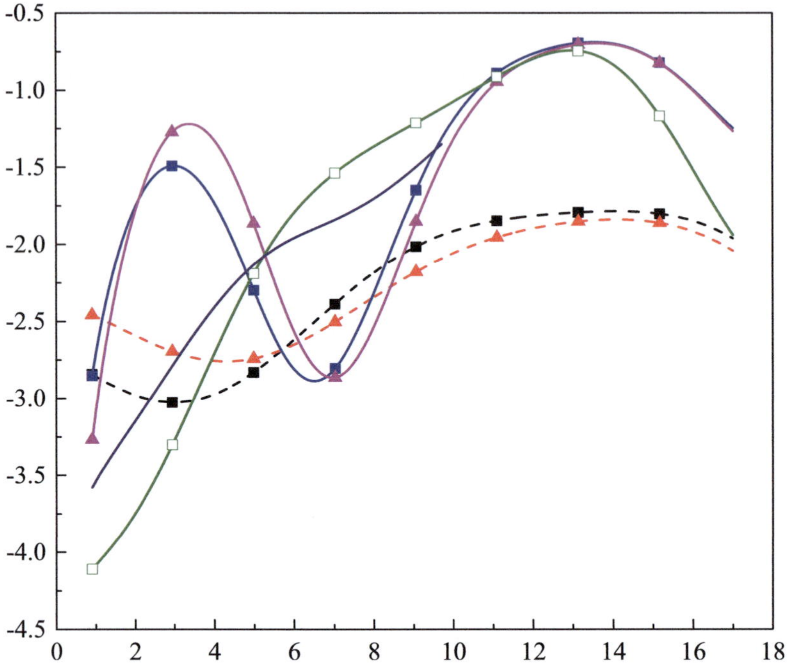

Fig. 8.17 (**a**) Frequency plots of transmission dispersion of orthogonal microwire array composite when E_k is perpendicular to glass fibres and (**b**) schematic view of vertical and horizontal microwires in the polymer matrices (Reprinted with permission from Ref. [86], copyright 2014 AIP)

which makes the effective medium theory and the assumptions of perfect ferromagnetic components still valid. On the one hand, the glass fibres in parallel to microwires exert synergistic effects in dictating the observed DNG features. On the other hand, the vertical microwires are likely in physical contact with glass fibres leading to considerable stress concentration on the wire surface (Fig. 8.17b). When E_k is parallel to glass fibres, the major dielectric contribution arises from the vertical microwires. However, the significant fibre/wire contact stress not only modifies the domain structure of wires, considering wires are extremely stress-sensitive [99–101], but also would damage the plasmonic behaviour of continuous wires, rendering the negative permittivity impossible. Hence, the wire orientation has profound influence on the total metacomposites behaviour, i.e. the DNG features can be turned on/off by simply rotating the composite panel by 90°. Such an angle-dependent DNG characteristic can be extended to future cloaking devices where mechanically tuning the EM behaviour of the fabricated metacomposites is possible. Nevertheless, further research is necessary to enunciate the quantitative relations between wire alignment angle and S_{21}.

In summary, an improved orthogonal configuration is introduced into the microwire metacomposites with a better transmission performance compared with parallel wire configuration. The vertical wires are beneficial to the enhancement of transmission coefficients with improved impedance match. Such orthogonal metacomposites also reveal strong competency in next-generation cloaking devices with additional functions such as field/mechanical tunability.

8.5.2 Hybrid Metacomposites

One challenging issue remains in the parallel and orthogonal Fe-based microwire metacomposites: Observation on the *S*-parameters gives transmission windows, but their position shift with respect to external fields is minor. This is a disadvantage when it comes to the development of multiband and/or tunable metamaterials which is of significant engineering interest. Fortunately, compared to Fe-based wires, Co-based microwires have distinguished GMI effect and highly sensitive field or stress tunable microwave response [80]. Therefore, in order to realise tunable windows by generating the external stimuli-controlled behaviour, the hypothesis of introducing Co-based wires to the existing Fe-based microwire metacomposites would be a very compelling viewpoint. It would be also of much scientific interest to investigate the interplay between Co- and Fe-based wires taking into account their different wire alignment, intrinsic microwave properties and geometrical dimensions. Additional magnetic resonances therein would be obtained by means of wire-wire dynamic interactions, which is beneficial to expand DNG operating frequencies and enhance microwave absorptions.

Technically, Co-based ($Fe_4Co_{68.7}Ni_1B_{13}Si_{11}Mo_{2.3}$) microwires with a total diameter of 15 μm and a glass coat thickness of 7 μm are added into the Fe-based wires containing prepregs in two topological arrangements, i.e. parallel Co-based and parallel Fe-based wire array (Fig. 8.18a) and shortcut Co-based and continuous orthogonal Fe-based wire array (Fig. 8.18b) [102]. To avoid excessive reflection loss from direct physical contact, Fe-based and Co-based wires should be integrated in separate prepreg layers. Also as a precaution to prevent microwave noise, the Co-based wire array should be intentionally mismatched to Fe-based wires by approximately 1 mm offset. A routine curing protocol is followed afterwards to fabricate composite samples having an in-plane size of 500×500 mm [2] and thickness of 1 mm.

The panel (a) to (i) in Fig. 8.19 describe the transmission, reflection and absorption coefficients of the parallel 10 mm Fe-based, 3 mm Co-based containing composites and their hybrid composite, in the frequency band of 1–6 GHz. First of all, it is familiar that the blueshifted absorption peaks in the composites containing parallel Fe-based wires are due to their FMR. We have expounded this phenomenon in Sect. 8.4. Furthermore, the field-tunable EM parameters of composite specimen containing Co-based wires are easily saturated with fields increased up to 600 Oe due to their soft magnetic properties. Specifically, the absorption coefficient of Co-based wires containing composites increases more significantly at low fields than that of Fe-based composites. This is due to the long-range dipole resonance between Co-based wires since the 3 mm spacing is below the critical spacing as

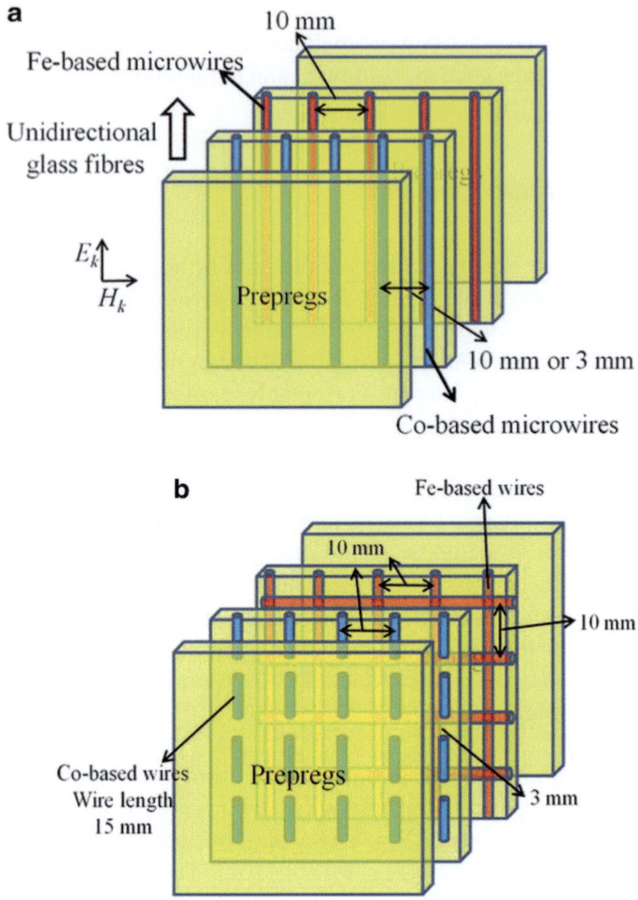

Fig. 8.18 Schematic illustration of the hybridisation of (**a**) continuous parallel Fe-based microwire array plus continuous Co-based microwire array and (**b**) orthogonal Fe-based microwire array plus shortcut Co-based microwire array [102]

discussed in parallel metacomposites [86, 89]. Interestingly, this identified resonance is seen to redshift with increasing frequency as can be seen from the absorption spectra in Fig. 8.19i. It can be explained as follows. As magnetic fields are increased, the skin depth of Co-based wires is drastically reduced due to the increased permeability as per the below equation, which is consistent with the enhanced absorption,

$$\delta_{\mathrm{res}} = \sqrt{\frac{\rho}{\pi f_{\mathrm{int}} \mu}}, \tag{8.21}$$

where ρ and f_{int} denote electrical resistivity and interaction resonance frequency, respectively [103]. Thin skin depth accompanies by undesirable eddy current loss.

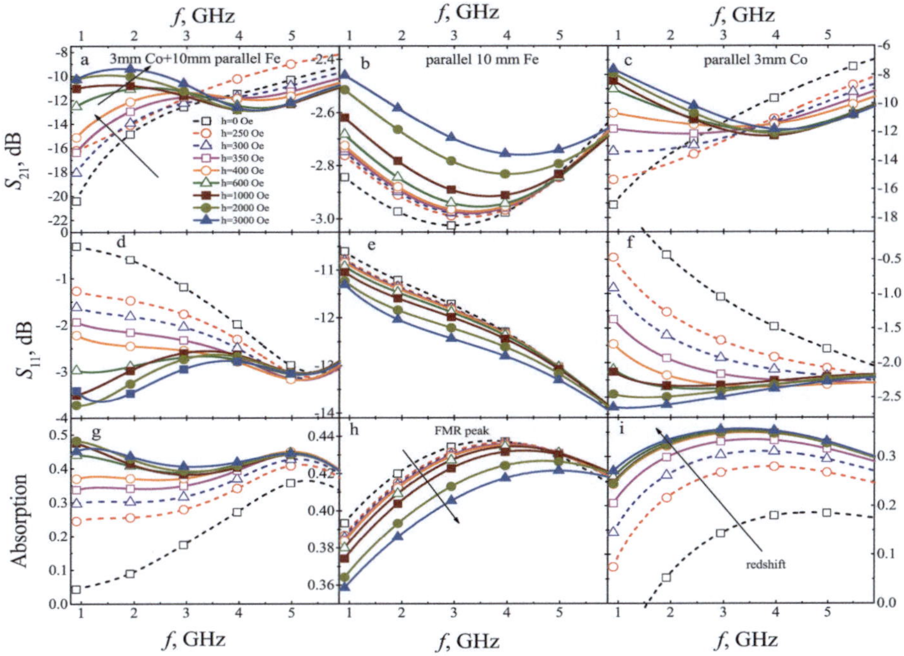

Fig. 8.19 Frequency plots of the transmission coefficients, S_{21}, of composite samples containing (**a**) hybrid wire arrays with 3 mm spaced Co-based wires, (**b**) pure Fe-based wires and (**c**) pure Co-based wires; the reflection coefficients, S_{11}, of composite samples containing (**d**) hybrid wire arrays with 3 mm spaced Co-based wires, (**e**) pure Fe-based wires and (**f**) pure Co-based wires; the absorption coefficients of composites containing (**g**) hybrid wire arrays with 3 mm spaced Co-based wires, (**h**) pure Fe-based wires and (**i**) pure Co-based wires [102]

The wire interaction resonance is thus decreased to compensate this loss. However, composites containing single Co-based wire array suppress the DNG properties because of overall high reflection, whereas 10 mm spaced Fe-based wires are too wide to induce such interaction. A logical idea would be the combination of Co-based and Fe-based wires. A transmission window is indeed displayed in the frequency band of 1–3.5 GHz for the hybridised wire array composite. This DNG feature is different from the natural metacomposite behaviour in parallel [89] and orthogonal [86] metacomposites because a small magnetic field of 300 Oe is required. Remarkably, the observed transmission windows display a redshift-blueshift evolution in terms of their peak positions. From the above analysis, it is implied that such effect is arising from the competition between the wire-wire dipole interaction at low fields and the FMR of Fe-based wires which prevail at high fields. Together, this magnetic field-tunable metacomposites enabled by a hybrid Co-based and Fe-based wire arrays satisfy such working environment that the DNG characteristics can be conveniently activated or deactivated by a dc magnetic field.

It should be mentioned that the DNG features are prohibited at high frequencies for above hybrid parallel metacomposites due to the narrow spacing between

Co-based wires conspiring to strong skin effect. It is inferred that a high-frequency transmission window would be preserved by diluting the concentration of Co-based wires in the hybrid metacomposites. The *S*-parameters of composites containing 10 mm spaced Co-based wires, 10 mm spaced Fe-based wires and their hybridised array are exhibited in Fig. 8.20 in 0.9–7 GHz frequency range. As expected, a transmission window is identified in the 1.5–5.5 GHz in the hybrid metacomposite. This is reminiscent of the parallel metacomposites enabled by single Fe-based wires; yet a slightly different window position is observed. This small deviation resulted from the involvement of Co-based wires and the modification of residual stresses on the as-cast wire surface. Further investigations on the hybridisation of Fe-based wire arrays with spacing of 7 or 3 mm plus 10 mm Co-based wire array reveal that the transmission window is cancelled out (*not shown here*). This is because both heavily loaded Co- and Fe-based wires show such high reflection level that prohibits the transmission window.

Figure 8.21 displays the high-frequency dependence of *S*-parameters of the above three composite specimens where a transmission enhancement is obtained in the frequency range of 9–17 GHz in the metacomposites containing diluted hybrid wire array in addition to some reflection and absorption dips. An absorption peak is also identified at 8.6 GHz, indicating that magnetic resonance occurred to account for the enhanced high-frequency transparency. We have pointed out that the inter-wire spacing is critical to create additional dynamic resonances [89]. It is believed that the 1 mm spacing between the Fe- and Co-based wire arrays would also excite such effect and significantly arouse microwave absorption. Moreover, the longitudinal anisotropy of Fe-based wires in response to E_k would generate a circumferential field that overlaps with the circumferential anisotropy of Co-based wires, subsequently enhancing the magnetic excitation and transmission. Overall, misalignment of microwires in the hybrid metacomposite system and wire-wire magnetic interaction greatly broadens the DNG frequency band, and these offer essential guidance for the design of microwire metacomposites.

Orthogonal configuration has proven to be an improved structure, thanks to the high transmission level but less wire loading compared with the parallel structure in the context of metacomposites. Could it become even better? The answer is yes. A shortcut array of Co-based microwires (Fig. 8.20b) is introduced into the orthogonal Fe-based wire array to realise the hybrid metacomposite. Fe-based and Co-based wires are arranged in separate prepreg layers to avoid formulation of continuous conductive network, hence high microwave loss. Figure 8.22 shows the transmission and absorption coefficients of composites containing single orthogonal Fe-based wires, shortcut Co-based wires and their hybridised wire arrays, respectively. A typical bandstop feature is identified from the composite containing shortcut Co-based wire array evidenced by a transmission valley and an absorption peak. According to Lorentz model, it is known that discontinuous wires behave as electrical dipoles while interacting with incident waves. Their dipole resonance f_{dr} can be interpreted as

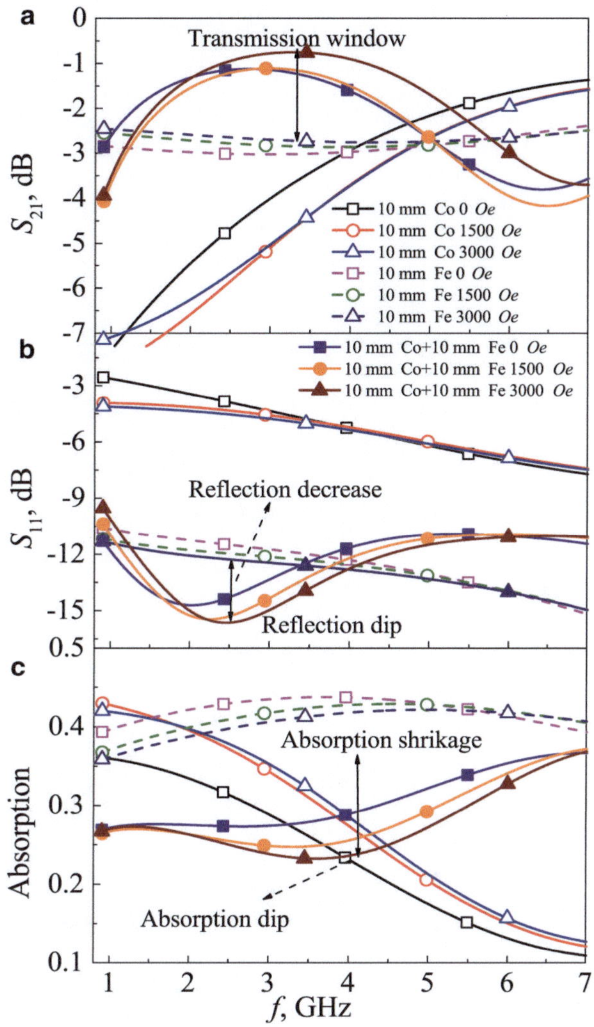

Fig. 8.20 Frequency plots in the 0.9–7 GHz of (**a**) transmission, (**b**) reflection and (**c**) absorption coefficients of composites containing 10 mm spaced Fe-based wire array, the 10 mm spaced Co-based wire array and their hybridised wire array [102]

$$f_{\mathrm{dr}} = \frac{c}{2l\sqrt{\varepsilon_{\mathrm{m}}}}, \tag{8.22}$$

where ε_{m} and l denote the permittivity of matrix materials and wire length, respectively [104]. Substituting l of 15 mm and ε_{m} of 3, f_{dr} is obtained as 5.8 GHz which is consistent with the absorption spectrum. Interestingly, this bandstop feature is maintained in the shortcut hybrid metacomposite along with a transmittance enhancement in the 1–6 GHz as compared with 10 mm spaced orthogonal metacomposite. Hence, it is fair to say that the introduction of shortcut Co-based wires can tune the transmission windows and DNG characteristics through a synergistic influence; this is due to the average low absorption of

Fig. 8.21 (**a**) Transmission, (**b**) reflection and (**c**) absorption coefficients of the same composites as shown in Fig. 8.4 measured in the frequency band of 7–17 GHz

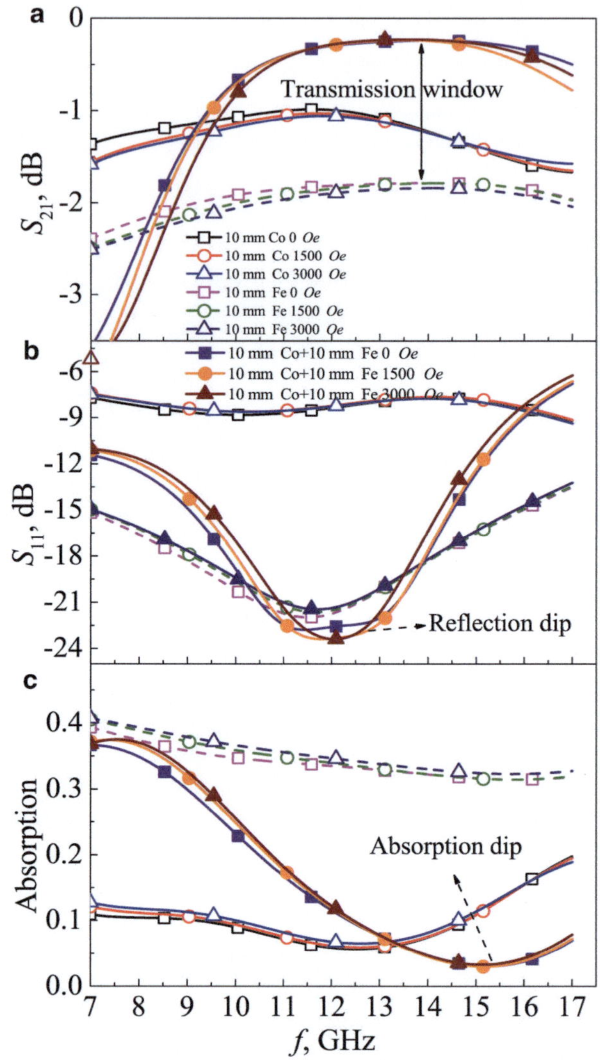

Co-based wires in the measuring frequency regime as confirmed in Fig. 8.23b. Besides, additional absorption peaks are induced by the magnetic interactions between adjacent Co- and Fe-wire couples.

This configuration-induced opaqueness in addition to the DNG features provide a clever route to realise tunable metacomposites from engineering perspective. The involvement of Co-based wires offers more degrees of freedom for composite design and the associated property control, which, at the mesostructural scale, dictates the DNG properties via a topological configuration of wires. At a smaller scale, four types of magnetic effects which are closely related to the intrinsic material properties of metacomposite components, i.e. FMR in Fe-based wires

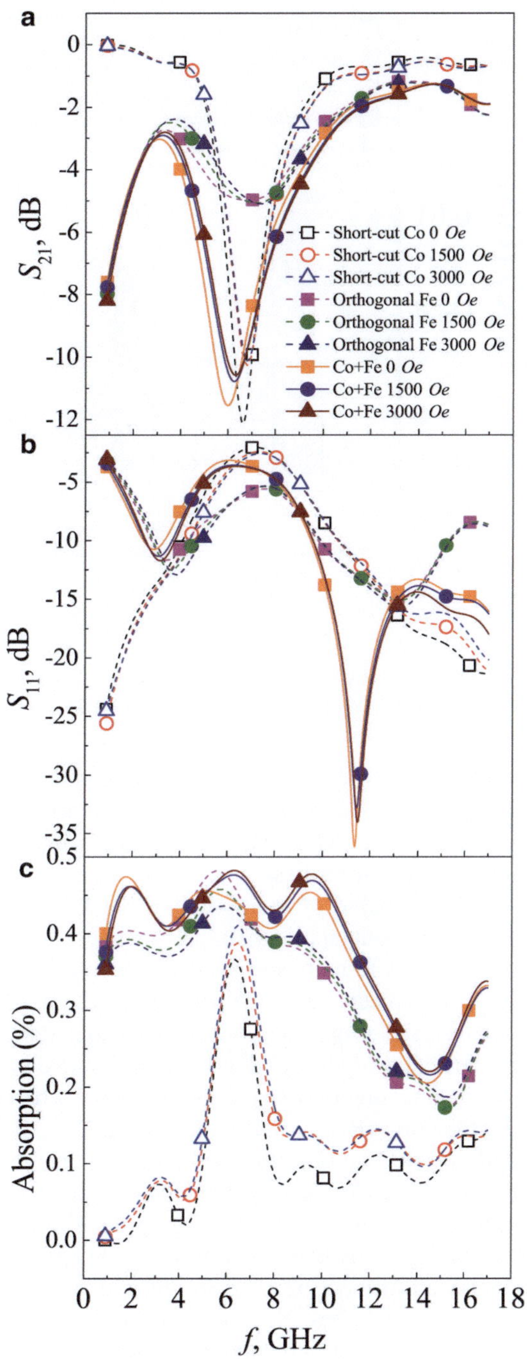

Fig. 8.22 Frequency plots of (**a**) transmission and (**b**) absorption coefficients of composites containing orthogonal Fe-based wire array, shortcut Co-based wire array and their hybridised wire arrays in 0.9–17 GHz [102]

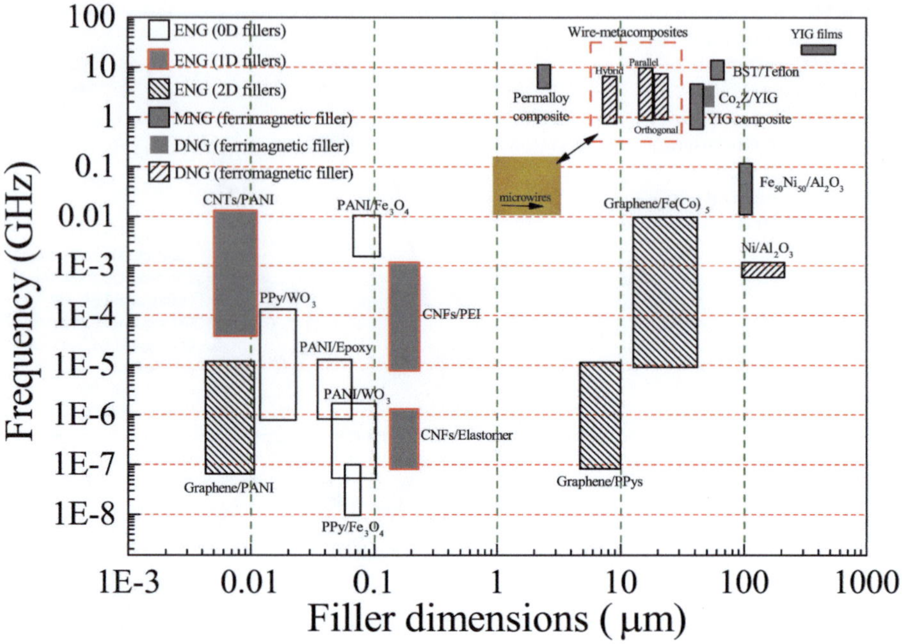

Fig. 8.23 Operating frequencies of metacomposites from kHz to GHz. The fillers' dimensions denote the unit or cell size in the metacomposites

and Fe-Fe, Fe-Co and Co-Co interactions, are revealed to dominate the absorption spectra, hence can effectively dictate the tunable transmission behaviour. In a brief sense, this is a course and fine control of metamaterial properties in the realm of hybrid metacomposites.

Being capable of manipulating the DNG features, now one can move on to speculate about application perspectives. The above results demonstrate strong impetus of wire metacomposites in developing cloaking devices due to the specifically enhanced transmission spectra. Stepping back from the frontier of metamaterial properties, one notices that the *S*-parameter spectra correspond to the particular topology of wires, therein the composites. Starting from the final EM coefficients, one may find application for wire metacomposites in radio-frequency identification (RFID). RFID is a contactless data capturing technique to automatically identify an object using radio-frequency waves. In principle, the object to be detected is coded with a tag that can reflect a unique pattern in terms of EM parameters. The ever-increasing applications of RFID for commercial inventory control in warehouses, supermarkets, hospitals as well as military friend-and-foe identification have resulted in considerable research interest on low-costs, long-range sensor design. Nevertheless, conventional RF tags consist of spacing-filling curves or capacitively tuned dipoles [105, 106], involving complicated shapes hence large manufacturing costs. Furthermore, the undesirable parasitic coupling effect of these structures when interacting with EM waves makes the precise analysis of their EM performance rather difficult. As a remedy for this, the objects can be embedded with a piece of wire metacomposites with a known wire configuration inside and thus have a unique identifiable 'barcode'. This is attractive to aerospace industry because the low wire inclusion, high microwave sensitivity and high mechanical performance of matrix context will maximise the RF detecting ability whereas mitigating any structural defects to the airplanes in service. The recent unfortunate 'MH370' disaster has called for urgent RFID applications for civil airplanes, and it is believed that the wire metacomposites are very promising. Such potential application appears to be more practical than cloaking invisibility at this stage and, importantly, of more civil significance for distinguishing civil from military planes or even identifying every airborne vehicle.

As a closing remark for the microwire metacomposites, we would like to emphasise their merits. First of all, a wire metacomposite is a piece of material; this is fundamentally different from the conventional metamaterial which normally gains purely structure-associated properties. From materials science point of view, properties are determined by microstructural factors. The SNG or DNG characteristics are readily available for wire metacomposites through the design of several multi-scale parameters. On the one hand, the resulting metamaterial properties are correlated to the magnetic and/or dielectric properties of microwires. Hence, one should carefully select the quality of wires regarding their roundness, surface smoothness, diameter and chemical composition. As a step further, a proper modulation, such as stress annealing and thermal annealing, is advantageous for microwires in favour of a uniform residual stress distribution and an improved domain structure. Thus, metacomposite response can be significantly enhanced via tuning the local properties of individual microwires in the metacomposites. Also, high mechanical properties, fine size and low content of wires rule out the possibility that the overall aerospace-graded performance of the polymer-based composites will be compromised. On the other hand, the meso-/macro-structural parameters are also at play in the wire-metacomposite system. As is shown in the above three kinds of wire metacomposites, i.e. parallel metacomposites, orthogonal metacomposites and hybrid metacomposites, the EM response can be effectively tuned by, e.g. wire spacing and alignment. From the viewpoint of composite design, it is of utmost significance to manage the material structure in order to further control its properties. This is why in popular areas of composite research, such as nanocomposites and heterogeneous composites, it is an eternal issue to discuss the reinforcement distribution [107]. The simple wire configuration is obviously easier to deal with compared to the great structure complexity in traditional metamaterials. One can now apprehend that a wire-metacomposite's EM properties are governed by microstructural parameters linking to intrinsic properties of wires and several conveniently adjustable mesostructural factors.

It has been mentioned at the beginning of this chapter that the expensive fabrication of nano-metacomposites confines their future usage. In fact, this is a universal technical issue facing all other fabrication techniques of metamaterials, no matter the top-down techniques, such as nanoimprint lithography [17, 108] and solid state superionic stamping [109, 110], or other bottom-up techniques, e.g. self-assembly [111]. But this is no longer an issue for wire metacomposites. The prepreg layup and autoclave curing processes have proven efficiency and cost-effectiveness for the composite manufacturing oriented towards aerospace and automobile industries while restraining modest defects.

8.6 Summary and Outlooks

Grand research programmes and the ever-quickly updated scientific findings have made metamaterials a truly interdisciplinary subject. Metamaterials research has therefore progressed exponentially, yet still in its infancy in some dimensions such as the gap between the beautiful theory and tangible applications, the delicate design and cost-effective industrial production, etc. The inherent problems of metamaterials arises as significantly as their merits, including, but not limited to, high manufacturing costs, complicated structures and poor tunability. The concept of metacomposites is proposed to address these issues. Let us recall its implications: a metacomposite is a true piece of material rather than a meta-structure, with its final properties engaged with the intrinsic properties of its constituents and mesoscopic parameters in terms of topological configuration, and that can be manufactured using conventional engineering techniques.

First of all, this chapter delivers a rigorous theoretical treatment on how to achieve a negative dielectric permittivity and a negative magnetic permeability. As such, the design principle for double-negative metacomposites is clear. Subsequently, we reviewed some of the previous studies on the metacomposites containing functional fillers of nano-/micro-scale. Metacomposites arise from efforts to realise ENG features. It was firstly proposed that 0D dielectric nanoparticles coated with a layer of polymer demonstrate a negative permittivity at kHz frequencies [53]. The idea is further extended to metacomposites containing carbonaceous fillers with 1D (CNTs, CNFs) or 2D (graphenes) dimensions. MNG metacomposites also attracted some attention [62, 66]. As a bold yet essential move, a negative permeability is achieved in the GHz range using the Mie resonance of dielectric fillers embedded regularly in the matrix [70]. Besides, ferrimagnetic materials, such as YIG, are feasible to attain a MNG response; nonetheless, their weak magnetic resonance tarnishes the operating frequency range [75]. In terms of DNG nano-metacomposites, sporadic studies showed that the DNG features could be obtained in a ceramic-based composite with Ni network architecture [77] as well as in a YIG-based composite [76]. These metacomposites are considered to be inspirational but unlikely to be adopted for engineering purposes due to the narrow SNG or DNG bandwidth and the delicate processing techniques.

Ferromagnetic microwires are perfect candidates for construction of metamaterials because of their excellent magnetic properties. Several metacomposites based on microwires are introduced to the family, i.e. parallel metacomposites [89], orthogonal metacomposites [86] and hybrid metacomposites [102]. In the parallel metacomposites, a natural DNG characteristic is evidenced by the transmission windows and a negative permittivity dispersion in the 1–7 GHz. As a measure to optimise the microwave properties, an orthogonal array of Fe-based wires is introduced into the metacomposites to enhance the transmittance while maintaining low wire donation. Furthermore, a metacomposite containing hybrid Co- and Fe-based wires has been designed and fabricated. Three major tunable DNG behaviours relating to the topological configuration are exhibited, namely, field-tunable transmission windows, dual DNG bandwidth and DNG/bandstop features. These microwire metacomposites are promising for microwave cloaking and radio-frequency barcoding applications.

To date, the reported SNG or DNG metacomposites have covered a wide range from kHz to GHz frequencies. Figure 8.23 summarises the operating frequencies of metacomposites with respect to the dimensions of fillers. As we will find out below, serious challenges remain to be addressed.

For the metacomposites containing nano-/micro-fillers, it is obvious that the general operating frequencies are lumped in kHz to MHz range. This seems to contradict with metamaterial theories which define a SNG or DNG feature for nanostructures at infrared or much higher frequencies. However, one should note that these traditional meta-structures have building blockings closely packed, making the unit-unit periodicity still within the nanometres range. f_p is dramatically increased with the decrease of spacing b (Eq. (8.6)); therefore, it is possible to have high-frequency negative permittivity and hence the DNG characteristics. Nevertheless, the microstructure of the metacomposites containing nano-fillers indicates that the inter-filler spacing is significantly higher than that of those meta-structures (often in the micron or millimetres), hence dwarfing the f_p to a tremendous extent. To anticipate higher operating frequencies, a higher loading of fillers is therefore necessary to decrease the spacing among fillers.

But the more the nano-fillers, the more likely they will aggregate. In fact, the dispersion of nano-fillers is always a critical issue in nanocomposites fabrication and so is in metacomposite fabrication [30, 107]. A good filler distribution is a prerequisite for metacomposites to attain controllable EM properties. Filler clusters are catastrophic as they will form continuous conductive network to damage the plasmonic behaviour in favour of ENG features. Strong reflections will also devastate the identifiable magnetic resonance for a negative permeability. In Zhu's work, the sonication technique is adopted to exfoliate surfactant-treated CNTs/CNFs/graphenes, but from the SEM images, a large amount of CNTs and CNFs are still entangled or crossed-linked like spaghettis because of van der Waals force [61, 66]. Other filler dispersion strategies such as surface functionalisation are welcomed to further improve the filler distribution, hence the EM response of metacomposites. In the DNG Ni/Al_2O_3 metacomposites [77], the question on how to precisely control the Ni discontinuous networks still remains. The emerging 3D printing technique might well be the answer. In recent years, 3D-printed porous ceramics have afforded illuminating perspectives for biological tissue engineering [112, 113]. It is our belief that a programmable DNG characteristic can be obtained via 3D-printed Al_2O_3 foam, controlled by, e.g. porosity density and pore size, subsequently impregnated by Ni injection.

For metacomposites containing ferromagnetic microwires, one critical issue is the significantly decreased f_p (cf. Figs. 8.13 and 8.14), which is linked to the distinct domain structure of wires and subsequent curing cycle. Although a modified Lorentz-Drude model is proposed, it would be even better to modulate the wire quality prior to incorporating them into polymer matrix. A rational annealing technique is encouraged, for example, current or stress annealing. Recently, we developed a combined current modulation annealing (CCMA) which demonstrates a remarkable influence to improve the circumferential anisotropy of wires and promise excellent soft magnetic properties [114]. This is particularly useful for increasing the volume of circular domain, which eventually increases a_{eff} and f_p. For 3D metacomposites, although some interesting results have been reported from 3D nano-metamaterials [21, 115, 116] they are merely structures with metamaterial features. Current research is basically focused on the thin-ply metacomposites containing one or two layers of wire arrays. The through-thickness EM response is neglected assuming dynamic weak coupling of microwires in different prepreg layers. It is envisaged that with increasing wire containing prepregs, through-thickness response will be gradually enhanced and likely to interact with in-plane metamaterial performances. However, as a side effect, the internal structural complexity would be increased, rendering the prepreg layup rather time-consuming. Alternatively, one can develop a 3D structure comprised of printed wires followed by impregnating into a polymer matrix. A roadmap may be needed to define the geometrical factors of this new wires and their composition.

Meanwhile, microwire metacomposites are defined as a category of engineering materials. It is essential to improve the manufacturing technique of wire metacomposites. At the current stage, an additional step of insertion of wires between prepreg layers is needed. To enable mass production, automation of fabricating microwire prepregs is advised, i.e. these microwires can be conveniently and automatically incorporated during the prepreg fabrication stage to realise a microwire prepreg. Depending on application needs, spacing and pattern of microwires can be tailored. As such, a standard production line can be commissioned for the mass production of microwire metacomposites.

References

1. Veselago, V.G.: The electrodynamics of substances with simultaneously negative values of ε and μ. Sov. Phys. Usp. **10**, 509 (1968)
2. Pendry, J.B., Holden, A.J., Stewart, W.J., Youngs, I.: Extremely low frequency plasmons in metallic mesostructures. Phys. Rev. Lett. **76**, 4773 (1996)
3. Raether, H.: Excitations of Plasmons and Interband Transitions by Electrons. Springer, Berlin (1980)
4. Pendry, J.B., Holden, A.J., Robbins, D.J., Stewart, W.J.: Magnetism from conductors and enhanced nonlinear. IEEE Trans. Microwave Theory Tech. **47**, 2075 (1999)
5. Smith, D.R., Padilla, W.J., Vier, D.C., Nemat-Nasser, S.C., Schultz, S.: Composite medium with simultaneously negative permeability and permittivity. Phys. Rev. Lett. **84**, 4184 (2000)

6. Chen, H., Chan, C.T.: Acoustic cloaking in three dimensions using acoustic metamaterials. Appl. Phys. Lett. **91**, 183518 (2007)
7. Bückmann, T., Thiel, M., Kadic, M., Schittny, R., Wegener, M.: An elasto-mechanical unfeelability cloak made of pentamode metamaterials. Nat. Commun. **5**, 4130 (2014)
8. Pendry, J.B., Aubry, A., Smith, D.R., Maier, S.A.: Transformation optics and subwavelength control of light. Science **337**, 549 (2012)
9. Vakil, A., Engheta, N.: Transformation optics using graphene. Science **332**, 1291 (2011)
10. Sihvola, A.: Metamaterials in electromagnetics. Metamaterials **1**, 2 (2007)
11. Smith, D.R., Pendry, J.B., Wiltshire, M.K.: Metamaterials and negative refractive index. Science **305**, 788 (2004)
12. Duan, Z., Wu, B.I., Xi, S., Chen, H., Chen, M.: Research progress in reversed Cherenkov radiation in double-negative metamaterials. Prog. Electromagn. Res. **90**, 75 (2009)
13. Lee, S.H., Park, C.M., Seo, Y.M., Kim, C.K.: Reversed Doppler effect in double negative metamaterials. Phys. Rev. B **81**, 241102 (2010)
14. Cui, T.J., Smith, D.R., Liu, R.P.: Metamaterials. Springer, Berlin (2014)
15. Florijn, B., Coulais, C., Hecke, M.: Programmable mechanical metamaterials. Phys. Rev. Lett. **113**, 175503 (2014)
16. Stenger, N., Wilhelm, M., Wegener, M.: Experiments on elastic cloaking in thin plates. Phys. Rev. Lett. **108**, 014301 (2012)
17. Chaturvedi, P., Hsu, K., Zhang, S., Fang, N.: New frontiers of metamaterials: design and fabrication. MRS Bull. **33**, 915 (2008)
18. Li, J., Fok, L., Yin, X., Bartal, G., Zhang, X.: Experimental demonstration of an acoustic magnifying hyperlens. Nat. Mater. **8**, 931 (2009)
19. Hawkes, A.M., Katko, A.R., Cummer, S.A.: A microwave metamaterial with integrated power harvesting functionality. Appl. Phys. Lett. **103**, 163901 (2013)
20. Landy, N., Smith, D.R.: A full-parameter unidirectional metamaterial cloak for microwaves. Nat. Mater. **12**, 25 (2013)
21. Valentine, J., Zhang, S., Zentgraf, T., Ulin-Avila, E., Genov, D.A., Bartal, G., Zhang, X.: Three-dimensional optical metamaterial with a negative refractive index. Nature **455**, 376 (2008)
22. Schurig, D., Mock, J.J., Justice, B.J., Cummer, S.A., Pendry, J.B., Starr, A.F., Smith, D.R.: Metamaterial electromagnetic cloak at microwave frequencies. Science **314**, 977 (2006)
23. Shalaev, V.M.: Optical negative-index metamaterials. Nat. Photonics **1**, 41 (2007)
24. Boltasseva, A., Shalaev, V.M.: Fabrication of optical negative-index metamaterials: Recent advances and outlook. Metamaterials **2**, 1 (2008)
25. Sodha, M.S., Srivastava, N.C.: Microwave propagation in ferrimagnetics. Springer, Berlin (1981)
26. Kelly, A.: Composites in context. Compos. Sci. Technol. **23**, 171 (1985)
27. Jackson, J.D., Jackson, J.D.: Classical electrodynamics. Wiley, New York (1962)
28. Pines, D., Bohm, D.: A collective description of electron interactions: II. Collective vs individual particle aspects of the interactions. Phys. Rev. **85**, 338 (1952)
29. Bohm, D., Pines, D.: A collective description of electron interactions: III. Coulomb interactions in a degenerate electron gas. Phys. Rev. **92**, 609 (1953)
30. Ma, P.C., Siddiqui, N.A., Marom, G., Kim, J.K.: Dispersion and functionalization of carbon nanotubes for polymer-based nanocomposites: a review. Compos. Part A **41**, 1345 (2010)
31. Bauhofer, W., Kovacs, J.Z.: A review and analysis of electrical percolation in carbon nanotube polymer composites. Compos. Sci. Technol. **69**, 1486 (2009)
32. Liberal, I., Nefedov, I.S., Ederra, I., Gonzalo, R., Tretyakov, S.A.: Electromagnetic response and homogenization of grids of ferromagnetic microwires. J. Appl. Phys. **110**, 064909 (2011)
33. Landau, L.D., Bell, J.S., Kearsley, M.J., Pitaevskii, L.P., Lifshitz, E.M., Sykes, J.B.: Electrodynamics of Continuous Media, 2nd edn. Pergamon Press, New York (1984)
34. Velez, A., Aznar, F., Bonache, J., Velazquez-Ahumada, M.C., Martel, J., Martin, F.: Open complementary split ring resonators (OCSRRs) and their application to wideband CPW band pass filters. IEEE Microwave Wireless Compon. Lett. **19**, 197 (2009)

35. Bonache, J., Gil, I., Garcia-Garcia, J., Martin, F.: Novel microstrip bandpass filters based on complementary split-ring resonators. IEEE Trans. Microwave Theory Tech. **54**, 265 (2006)
36. Azad, A.K., Dai, J., Zhang, W.: Transmission properties of terahertz pulses through subwavelength double split-ring resonators. Opt. Lett. **31**, 634 (2006)
37. Marqués, R., Martín, F., Sorolla, M.: Metamaterials with Negative Parameters: Theory, Design and Microwave Applications. Wiley-Interscience, Hoboken, NJ (2007)
38. Chui, S.T., Hu, L.: Theoretical investigation on the possibility of preparing left-handed materials in metallic magnetic granular composites. Phys. Rev. B **65**, 144407 (2002)
39. Pimenov, A., Loidl, A., Przyslupski, P., Dabrowski, B.: Negative refraction in ferromagnet-superconductor superlattices. Phys. Rev. Lett. **95**, 247009 (2005)
40. Liberal, I., Nefedov, I.S., Ederra, I., Gonzalo, R., Tretyakov, S.A.: On the effective permittivity of arrays of ferromagnetic wires. J. Appl. Phys. **110**, 104902 (2011)
41. Carbonell, J., García-Miquel, H., Sánchez-Dehesa, J.: Double negative metamaterials based on ferromagnetic microwires. Phys. Rev. B **81**, 024401 (2010)
42. Chen, H.T., Padilla, W.J., Averitt, R.D., Gossard, A.C., Highstrete, C., Lee, M., O'Hara, J.F., Taylor, A.J.: Electromagnetic metamaterials for terahertz applications. Terahertz Sci. Technol. **1**, 42 (2008)
43. Kafesaki, M., Tsiapa, I., Katsarakis, N., Koschny, T., Soukoulis, C.M., Economou, E.N.: Left-handed metamaterials: The fishnet structure and its variations. Phys. Rev. B **75**, 235114 (2007)
44. Shalaev, V.M., Cai, W., Chettiar, U.K., Yuan, H.K., Sarychev, A.K., Drachev, V.P., Kildishev, A.V.: Negative index of refraction in optical metamaterials. Opt. Lett. **30**, 3356 (2005)
45. Linden, S., Enkrich, C., Wegener, M., Zhou, J., Koschny, T., Soukoulis, C.M.: Magnetic response of metamaterials at 100 terahertz. Science **306**, 1351 (2004)
46. Ju, L., Geng, B., Horng, J., Girit, C., Martin, M., Hao, Z., Bechtel, H.A., Liang, X., Zettl, A., Shen, Y.R.: Graphene plasmonics for tunable terahertz metamaterials. Nat. Nanotechnol. **6**, 630 (2011)
47. Chanda, D., Shigeta, K., Gupta, S., Cain, T., Carlson, A., Mihi, A., Baca, A.J., Bogart, G.R., Braun, P., Rogers, J.A.: Large-area flexible 3D optical negative index metamaterial formed by nanotransfer printing. Nat. Nanotechnol. **6**, 402 (2011)
48. Thostenson, E.T., Li, C., Chou, T.W.: Nanocomposites in context. Compos. Sci. Technol. **65**, 491 (2005)
49. Zhang, D., Wang, P., Murakami, R., Song, X.: Effect of an interface charge density wave on surface plasmon resonance in ZnO/Ag/ZnO thin films. Appl. Phys. Lett. **96**, 233114 (2010)
50. Wheeler, M.S., Aitchison, J.S., Mojahedi, M.: Coated nonmagnetic spheres with a negative index of refraction at infrared frequencies. Phys. Rev. B **72**, 193103 (2005)
51. Zhao, Q., Du, B., Kang, L., Zhao, H., Xie, Q., Li, B., Zhang, X., Zhou, J., Li, L., Meng, Y.: Tunable negative permeability in an isotropic dielectric composite. Appl. Phys. Lett. **92**, 051106 (2008)
52. O'Brien, S., Pendry, J.B.: Photonic band-gap effects and magnetic activity in dielectric composites. J. Phys. Condensed Matter **14**, 4035 (2002)
53. Zhu, J., Wei, S., Zhang, L., Mao, Y., Ryu, J., Mavinakuli, P., Karki, A.B., Young, D.P., Guo, Z.: Conductive polypyrrole/tungsten oxide metacomposites with negative permittivity. J. Phys. Chem. C **114**, 16335 (2010)
54. Zhu, J., Wei, S., Zhang, L., Mao, Y., Ryu, J., Karki, A.B., Young, D.P., Guo, Z.: Polyaniline-tungsten oxide metacomposites with tunable electronic properties. J. Mater. Chem. **21**, 342 (2011)
55. Gu, H., Guo, J., Wei, S., Guo, Z.: Polyaniline nanocomposites with negative permittivity. J. Appl. Polym. Sci. **130**, 2238 (2013)
56. Gu, H., Guo, J., He, Q., Jiang, Y., Huang, Y., Haldolaarachige, N., Luo, Z., Young, D.P., Wei, S., Guo, Z.: Magnetoresistive polyaniline/multi-walled carbon nanotube nanocomposites with negative permittivity. Nanoscale **6**, 181 (2014)

57. Liu, C.D., Lee, S.N., Ho, C.H., Han, J.L., Hsieh, K.H.: Electrical properties of well-dispersed nanopolyaniline/epoxy hybrids prepared using an absorption-transferring process. J. Phys. Chem. C **112**, 15956 (2008)
58. Thostenson, E.T., Ren, Z., Chou, T.W.: Advances in the science and technology of carbon nanotubes and their composites: a review. Compos. Sci. Technol. **61**, 1899 (2001)
59. Breuer, O., Sundararaj, U.: Big returns from small fibers: a review of polymer/carbon nanotube composites. Polym. Compos. **25**, 630 (2004)
60. Li, B., Sui, G., Zhong, W.H.: Single negative metamaterials in unstructured polymer nanocomposites toward selectable and controllable negative permittivity. Adv. Mater. **21**, 4176 (2009)
61. Zhu, J., Wei, S., Ryu, J., Guo, Z.: Strain-sensing elastomer/carbon nanofiber "metacomposites". J. Phys. Chem. C **115**, 13215 (2011)
62. Zhu, J., Gu, H., Luo, Z., Haldolaarachige, N., Young, D.P., Wei, S., Guo, Z.: Carbon nanostructure-derived polyaniline metacomposites: electrical, dielectric, and giant magnetoresistive properties. Langmuir **28**, 10246 (2012)
63. Zhu, J., Zhang, X., Haldolaarachchige, N., Wang, Q., Luo, Z., Ryu, J., Young, D.P., Wei, S., Guo, Z.: Polypyrrole metacomposites with different carbon nanostructures. J. Mater. Chem. **22**, 4996 (2012)
64. Novoselov, K.S., Jiang, Z., Zhang, Y., Morozov, S.V., Stormer, H.L., Zeitler, U., Maan, J.C., Boebinger, G.S., Kim, P., Geim, A.K.: Room-temperature quantum Hall effect in graphene. Science **315**, 1379 (2007)
65. Nair, R.R., Blake, P., Grigorenko, A.N., Novoselov, K.S., Booth, T.J., Stauber, T., Peres, N., Geim, A.K.: Fine structure constant defines visual transparency of graphene. Science **320**, 1308 (2008)
66. Zhu, J., Luo, Z., Wu, S., Haldolaarachchige, N., Young, D.P., Wei, S., Guo, Z.: Magnetic graphene nanocomposites: electron conduction, giant magnetoresistance and tunable negative permittivity. J. Mater. Chem. **22**, 835 (2012)
67. Holloway, C.L., Kuester, E.F., Baker-Jarvis, J., Kabos, P.: A double negative (DNG) composite medium composed of magnetodielectric spherical particles embedded in a matrix. IEEE Trans. Antennas Propag. **51**, 2596 (2003)
68. Vendik, O.G., Gashinova, M.S.: Artificial double negative (DNG) media composed by two different dielectric sphere lattices embedded in a dielectric matrix. In Proceedings of the 34th European Microwave Conference, Amsterdam, vol. 3, pp. 1209–1212. IEEE Press, Piscataway, NJ (2004). http://ieeexplore.ieee.org/xpls/abs_all.jsp?arnumber=1411225&tag=1
69. Huang, K.C., Povinelli, M.L., Joannopoulos, J.D.: Negative effective permeability in polaritonic photonic crystals. Appl. Phys. Lett. **85**, 543 (2004)
70. Zhao, Q., Kang, L., Du, B., Zhao, H., Xie, Q., Huang, X., Li, B., Zhou, J., Li, L.: Experimental demonstration of isotropic negative permeability in a three-dimensional dielectric composite. Phys. Rev. Lett. **101**, 027402 (2008)
71. Bush, G.G.: The complex permeability of a high purity yttrium iron garnet (YIG) sputtered thin film. J. Appl. Phys. **73**, 6310 (1993)
72. Bush, G.G.: Functional dependence of the real part of the initial permeability of the YIG family of ferrites. J. Appl. Phys. **67**, 5515 (1990)
73. He, Y., He, P., Yoon, S.D., Parimi, P.V., Rachford, F.J., Harris, V.G., Vittoria, C.: Tunable negative index metamaterial using yttrium iron garnet. J. Magn. Magn. Mater. **313**, 187 (2007)
74. Tsutaoka, T., Kasagi, T., Hatakeyama, K.: Permeability spectra of yttrium iron garnet and its granular composite materials under dc magnetic field. J. Appl. Phys. **110**, 053909 (2011)
75. Kasagi, T., Tsutaoka, T., Hatakeyama, K.: Negative permeability spectra in Permalloy granular composite materials. Appl. Phys. Lett. **88**, 172502 (2006)
76. Xu, F., Bai, Y., Qiao, L., Zhao, H., Zhou, J.: Realization of negative permittivity of Co2Z hexagonal ferrite and left-handed property of ferrite composite material. J. Phys. D **42**, 025403 (2009)

77. Shi, Z.C., Fan, R.H., Zhang, Z.D., Qian, L., Gao, M., Zhang, M., Zheng, L.T., Zhang, X.H., Yin, L.W.: Random composites of nickel networks supported by porous alumina toward double negative materials. Adv. Mater. **24**, 2349 (2012)
78. Gao, M., Shi, Z.C., Fan, R.H., Qian, L., Zhang, Z.D., Guo, J.Y.: High-frequency negative permittivity from Fe/Al2O3 composites with high metal contents. J. Am. Ceram. Soc. **95**, 67 (2012)
79. Shi, Z.C., Fan, R.H., Zhang, Z.D., Gong, H.Y., Ouyang, J., Bai, Y.J., Zhang, X.H., Yin, L.W.: Experimental and theoretical investigation on the high frequency dielectric properties of Ag/Al2O3 composites. Appl. Phys. Lett. **99**, 032903 (2011)
80. Phan, M.H., Peng, H.X.: Giant magnetoimpedance materials: fundamentals and applications. Prog. Mater. Sci. **53**, 323 (2008)
81. Qin, F.X., Bingham, N.S., Wang, H., Peng, H.X., Sun, J.F., Franco, V., Yu, S.C., Srikanth, H., Phan, M.H.: Mechanical and magnetocaloric properties of Gd-based amorphous microwires fabricated by melt-extraction. Acta Mater. **61**, 1284 (2013)
82. García-Miquel, H., Carbonell, J., Boria, V.E., Sánchez-Dehesa, J.: Experimental evidence of left handed transmission through arrays of ferromagnetic microwires. Appl. Phys. Lett. **94**, 054103 (2009)
83. Qin, F., Peng, H.X.: Ferromagnetic microwires enabled multifunctional composite materials. Prog. Mater. Sci. **58**, 183 (2013)
84. Qin, F.X., Pankratov, N., Peng, H.X., Phan, M.H., Panina, L.V., Ipatov, M., Zhukova, V., Zhukov, A., Gonzalez, J.: Exceptional electromagnetic interference shielding properties of ferromagnetic microwires enabled polymer composites. J. Appl. Phys. **107**, 09A314 (2010)
85. Qin, F.X., Peng, H.X., Chen, Z., Wang, H., Zhang, J.W., Hilton, G.: Optimization of microwire/glass-fibre reinforced polymer composites for wind turbine application. Appl. Phys. A **113**, 537 (2013)
86. Luo, Y., Peng, H.X., Qin, F.X., Ipatov, M., Zhukova, V., Zhukov, A., Gonzalez, J.: Metacomposite characteristics and their influential factors of polymer composites containing orthogonal ferromagnetic microwire arrays. J. Appl. Phys. **115**, 173909 (2014)
87. Peng, H.X., Qin, F.X., Phan, M.H., Tang, J., Panina, L.V., Ipatov, M., Zhukova, V., Zhukov, A., Gonzalez, J.: Co-based magnetic microwire and field-tunable multifunctional macro-composites. J. Non-Crystal. Solids **355**, 1380 (2009)
88. Makhnovskiy, D., Zhukov, A., Zhukova, V., Gonzalez, J.: Tunable and self-sensing microwave composite materials incorporating ferromagnetic microwires. Adv. Sci. Technol. **54**, 201 (2009)
89. Luo, Y., Peng, H.X., Qin, F.X., Ipatov, M., Zhukova, V., Zhukov, A., Gonzalez, J.: Fe-based ferromagnetic microwires enabled meta-composites. Appl. Phys. Lett. **103**, 251902 (2013)
90. Baranov, S., Yamaguchi, M., Garcia, K., Vazquez, M.: Application of amorphous microwires for electromagnetic shielding. Moldavian J. Phys. Sci. **9**, 76 (2010)
91. Qin, F.X., Quéré, Y., Brosseau, C., Wang, H., Liu, J.S., Sun, J.F., Peng, H.X.: Two-peak feature of the permittivity spectra of ferromagnetic microwire/rubber composites. Appl. Phys. Lett. **102**, 122903 (2013)
92. Kittel, C.: On the theory of ferromagnetic resonance. Phys. Rev. **73**, 155 (1948)
93. Wang, H., Xing, D., Wang, X., Sun, J.: Fabrication and characterization of melt-extracted co-based amorphous wires. Metall. Mater. Trans. A **42**, 1103 (2011)
94. Wang, H., Qin, F.X., Xing, D.W., Cao, F.Y., Peng, H.X., Sun, J.F.: Fabrication and characterization of nano/amorphous dual-phase FINEMET microwires. Mater. Sci. Eng. B **178**, 1483 (2013)
95. Sukstanskii, A.L., Korenivski, V.: Impedance and surface impedance of ferromagnetic multilayers: the role of exchange interaction. J. Phys. D **34**, 3337 (2001)
96. Liu, L., Kong, L.B., Lin, G.Q., Matitsine, S., Deng, C.R.: Microwave permeability of ferromagnetic microwires composites/metamaterials and potential applications. IEEE Trans. Magn. **44**, 3119 (2008)

97. Kraus, L., Frait, Z., Ababei, G., Chiriac, H.: Ferromagnetic resonance of transversally magnetized amorphous microwires and nanowires. J. Appl. Phys. **113**, 183907 (2013)
98. Vázquez, M., Adenot-Engelvin, A.L.: Glass-coated amorphous ferromagnetic microwires at microwave frequencies. J. Magn. Magn. Mater. **321**, 2066 (2009)
99. Wang, H., Qin, F.X., Xing, D.W., Cao, F.Y., Wang, X.D., Peng, H.X., Sun, J.F.: Relating residual stress and microstructure to mechanical and giant magneto-impedance properties in cold-drawn Co-based amorphous microwires. Acta Mater. **60**, 5425 (2012)
100. Qin, F.X., Peng, H.X., Fuller, J., Brosseau, C.: Magnetic field-dependent effective microwave properties of microwire-epoxy composites. Appl. Phys. Lett. **101**, 152905 (2012)
101. Luo, Y., Peng, H.X., Qin, F.X., Adohi, B.J.P., Brosseau, C.: Magnetic field and mechanical stress tunable microwave properties of composites containing Fe-based microwires. Appl. Phys. Lett. **104**, 121912 (2014)
102. Luo, Y., Qin, F.X., Scarpa, F., Carbonel J., Ipatov, M., Zhukova, V., Zhukov, A., Gonzalez, J., Peng, H.X.: Hybridized magnetic microwire metacomposites towards microwave cloaking and barcoding applications. arXiv preprint arXiv:1506.07745 (2015).
103. Qin, F.X., Peng, H.X., Pankratov, N., Phan, M.H., Panina, L.V., Ipatov, M., Zhukova, V., Zhukov, A., Gonzalez, J.: Exceptional electromagnetic interference shielding properties of ferromagnetic microwires enabled polymer composites. J. Appl. Phys. **108**, 044510 (2010)
104. Makhnovskiy, D.P., Panina, L.V., Garcia, C., Zhukov, A.P., Gonzalez, J.: Experimental demonstration of tunable scattering spectra at microwave frequencies in composite media containing CoFeCrSiB glass-coated amorphous ferromagnetic wires and comparison with theory. Phys. Rev. B **74**, 064205 (2006)
105. McVay, J., Hoorfar, A., Engheta, N.: Space-filling curve RFID tags. In Proceedings of IEEE Radio Wireless Symposium, pp. 199–202. IEEE, San Diego, CA (2006). http://ieeexplore.ieee.org/xpls/abs_all.jsp?arnumber=1615129.
106. Jalaly, I., Robertson, D.: Capacitively-tuned split microstrip resonators for RFID barcodes. In Proceedings of European Microwave Conference, vol. 2, pp. 4–7. IEEE, Paris, France (2005). http://ieeexplore.ieee.org/xpls/abs_all.jsp?arnumber=1610138.
107. Thostenson, E.T., Chou, T.W.: Processing-structure-multi-functional property relationship in carbon nanotube/epoxy composites. Carbon **44**, 3022 (2006)
108. Jung, G.Y., Johnston-Halperin, E., Wu, W., Yu, Z., Wang, S.Y., Tong, W.M., Li, Z., Green, J. E., Sheriff, B.A., Boukai, A.: Circuit fabrication at 17 nm half-pitch by nanoimprint lithography. Nano Lett. **6**, 351 (2006)
109. Hsu, K.H., Schultz, P.L., Ferreira, P.M., Fang, N.X.: Electrochemical nanoimprinting with solid-state superionic stamps. Nano Lett. **7**, 446 (2007)
110. Nagpal, P., Lindquist, N.C., Oh, S.H., Norris, D.J.: Ultrasmooth patterned metals for plasmonics and metamaterials. Science **325**, 594 (2009)
111. Vignolini, S., Yufa, N.A., Cunha, P.S., Guldin, S., Rushkin, I., Stefik, M., Hur, K., Wiesner, U., Baumberg, J.J., Steiner, U.: A 3D optical metamaterial made by self-assembly. Adv. Mater. **24**, OP23 (2012)
112. Taboas, J.M., Maddox, R.D., Krebsbach, P.H., Hollister, S.J.: Indirect solid free form fabrication of local and global porous, biomimetic and composite 3D polymer-ceramic scaffolds. Biomaterials **24**, 181 (2003)
113. Hollister, S.J.: Porous scaffold design for tissue engineering. Nat. Mater. **4**, 518 (2005)
114. Liu, J., Qin, F., Chen, D., Shen, H., Wang, H., Xing, D., Phan, M.H., Sun, J.: Combined current-modulation annealing induced enhancement of giant magnetoimpedance effect of Co-rich amorphous microwires. J. Appl. Phys. **115**, 17A326 (2014)
115. Ma, H.F., Cui, T.J.: Three-dimensional broadband ground-plane cloak made of metamaterials. Nat. Commun. **1**, 21 (2010)
116. Soukoulis, C.M., Wegener, M.: Past achievements and future challenges in the development of three-dimensional photonic metamaterials. Nat. Photon. **5**, 523 (2011)

Chapter 9
Permanent Magnets: History, Current Research, and Outlook

R. Skomski

9.1 Introduction

Permanent magnets are used in an impressive range of applications, ranging from computer hard-disk drives, wind generators, and hybrid-car motors to everyday applications such as loudspeakers, windscreen wipers, locks, microphones, and toy magnets [1–3]. Their main task is to create a magnetic field outside the magnet. The performance of permanent magnets is epitomized by the maximum energy product $(BH)_{max}$, which is equal to twice the magnetostatic energy stored in free space, divided by the magnet volume. The division by the total volume of the magnet is important, as exemplified by exchange-bias magnets, which need a bulky antiferromagnetic phase and are therefore not suitable for permanent magnet applications. Energy products are usually measured kJ/m^3, but a more elegant unit is kPa = kJ/m^3. Gaussian units, MGOe for the energy product, remain frequently used in some countries (see Appendix).

Energy product never exceeds $\mu_o M_s^2/4$, a limit realized in hysteresis loops with perfect rectangular shape and coercivity $H_c \geq M_s/2$ (Curve A in Fig. 9.1). Here, M_s is the spontaneous or "saturation" magnetization, defined as magnetic moment m per unit volume. Poor hysteresis loops, having a nearly straight line in the second quadrant and $H_c \ll M_r$ (Curve B in Fig. 9.1), exhibit $(BH)_{max} = ¼\mu_o H_c M_s$. In practice, energy product development requires a high zero temperature magnetization, a high Curie temperature T_c, and a high magnetic anisotropy K_1. A high Curie temperature is required, because almost all permanent magnets are used at or above room temperature. A high anisotropy is necessary, because the coercivity scales as the anisotropy field $H_A = 2K_1/\mu_o M_s$, where K_1 is the first anisotropy constant (Sect. 2.2).

R. Skomski (✉)
Department of Physics and Astronomy and Nebraska Center for Materials and Nanoscience, University of Nebraska, Lincoln, NE 68508, USA
e-mail: rskomski@neb.rr.com

A. Zhukov (ed.), *Novel Functional Magnetic Materials*, Springer Series in Materials Science 231, DOI 10.1007/978-3-319-26106-5_9

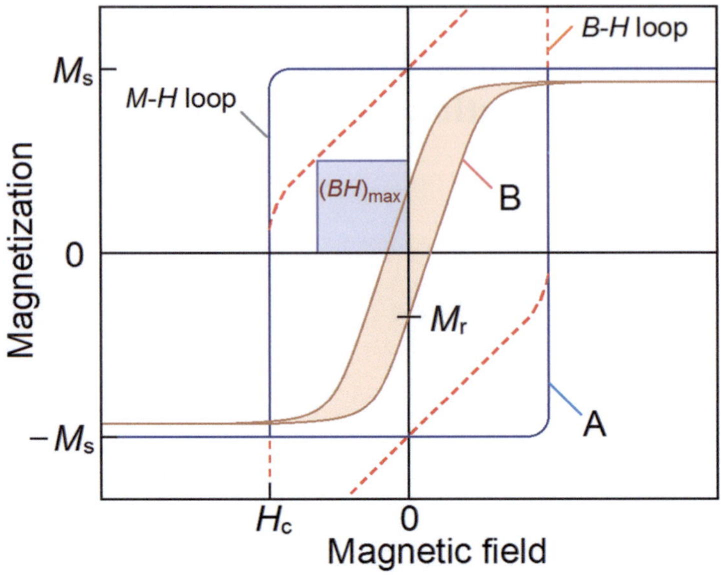

Fig. 9.1 Hysteresis loops: (*A*) nearly rectangular *M–H* loop and (*B*) S-shaped *M–H* loop. Indicated are coercive force or coercivity H_c (A), saturation magnetization M_s (*A*), and remanent magnetization or remanence M_r (*B*). The *B–H* loop and the corresponding energy product $(BH)_{max}$ refer to the nearly rectangular loop (*A*)

In the early stages of permanent magnetism, the anisotropy was the main limitation. Iron has a magnetization of $\mu_o M_s = 2.2$ T, corresponding to $\mu_o M_s^2/4 = 960$ kPa, but the coercivity of iron and steel magnets is very low, about 5 mT, and carbon steels have energy products of only about 1 kPa. These low energy products have resulted in cumbersome horseshoe-shaped magnets, a design no longer used except for alnico-type magnets (Sect. 2.3). Pure iron is very soft, whereas the mechanical and magnetic hardness of steel are caused by the tetragonal lattice distortion due to interstitial carbon. Moderate improvements were made around 1920 with the development of cobalt steels [5, 6], and these developments have attracted renewed attention under the keyword of interstitially modified tetragonal Fe–Co magnets (Sect. 3.1). Note that $Fe_{65}Co_{35}$, discovered by Preuss in 1912 [1], has a room temperature magnetization of 2.43 T—a record even today.

The first "true" or "compact-shaped" permanent magnets, namely, the $L1_0$ compound CoPt [7–10] and the hexagonal ferrite $BaFe_{12}O_{19}$ [11], date back to the 1930s–1950s, followed by the high-performance rare-earth transition-metal permanent magnets Sm–Co [12, 13] and $Nd_2Fe_{14}B$ [14–16] in the 1970s–1980s. This development has made it possible to enhance the energy product by two orders of magnitude to 460 kJ/m^3. The rare-earth compounds have a uniaxial crystal structure, where the crystallographic *c*-axis is the easy magnetization axis. Figure 9.2 summarizes the energy product development after the cobalt–steel era.

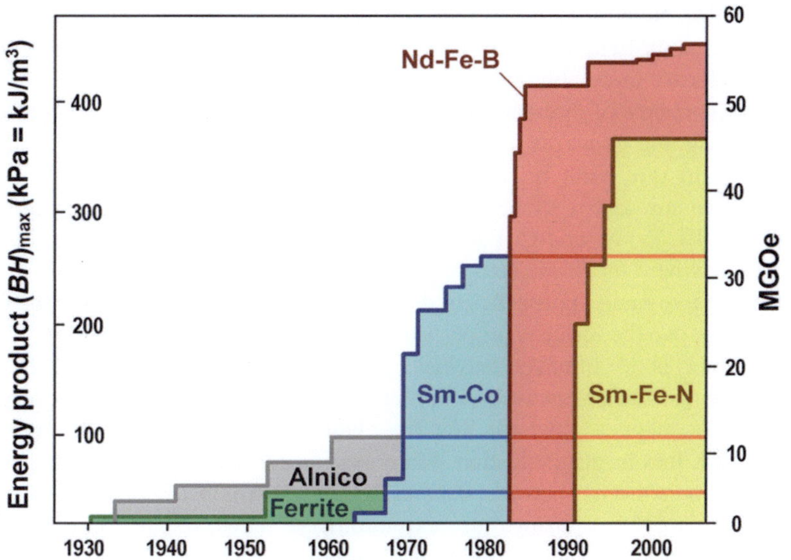

Fig. 9.2 Energy product development in the last 100 years

Today's permanent magnet market is dominated by (Ba, Sr)$Fe_{12}O_{19}$ at the low energy product end (ca. 35 % market share) and by Nd–Fe–B in the realm of high-performance materials (more than 50 % market share). The magnetization and Curie temperature of both materials are largely provided by the Fe, which is a very cheap element. Neodymium (Nd) is much less cheap, but its supply is not threatened at present. The remainder of the market is shared by Sm_2Co_{17}/$SmCo_5$ [13], alnicos, and a few other materials, such as $Sm_2Fe_{17}N_3$ [17] and vicalloy.

$L1_0$-alloys such as CoPt and FePt are very expensive due to the high Pt content (about 77 wt%). Alloys related to FePt are now used in ultrahigh-density magnetic recording, where the mass of required magnetic material is very low. CoPt was used in military and medical applications, but the former became rapidly replaced by Nd–Fe–B and Sm–Co. Permanent magnets for applications in dentistry, a very small market, did not undergo such a rapid shift to rare-earth alloys, because the mechanical strength and corrosion resistance of CoPt is superior to that of rare-earth alloys.

A similar problem is faced by top-end Nd–Fe–B magnets, which are used in motors for automobile applications. These magnets require dysprosium (Dy) to improve the anisotropy at operating temperatures up to about 200 °C. Dysprosium is much more expensive than Nd, and its supply is threatened by the scarcity of the element and by competing applications, for example, in high-performance light sources. Consequently, efforts are being made to save Dy by confining it to the surfaces of the 2:14:1 grains, where its coercivity-enhancing effect is largest. Grain boundary engineering to further perfect Nd–Fe–B magnets has therefore become an important research area [18–23], and similar approaches exist for melt-spun

Nd–Fe–B [24]. Note the rare-earth are not "rare" but merely difficult to mine and to extract. The natural abundance of rare earths is comparable to Co, Sn, and W, and all rare earths are more abundant than Hg.

The performance of rare-earth transition-metal intermetallics is difficult to beat. However, ongoing concerns about tight rare-earth supplies from China and huge fluctuations in rare-earth prices in recent years have reignited interest in new materials that are alloys of less critical and cheaper elements, such as Fe, Co, Mn, W, Al, Bi, Zr, N, and C. Furthermore, $Nd_2Fe_{14}B$ is presently used for many applications where moderate energy products are sufficient.

There are two main approaches to developing new permanent magnet materials. One is to improve the intrinsic properties (M_s, T_c, K_1) by atomic structuring [25–28], and the other is to improve extrinsic properties such as H_c and $(BH)_{max}$ by nanostructuring. The magnetization of metals such as Fe and $Fe_{35}Co_{65}$ is very high by permanent magnet standards, but the challenge is to enhance the anisotropy without much loss in magnetization. Many new materials considered in the present context have anisotropies between 0.5 and 2.0 MJ/m^3, which are sometimes qualified as "high," "giant," "surprising," or "extraordinary." However, this is true only by the standards of soft-magnetic Fe (0.05 MJ/m^3) and Ni (–0.005 MJ/m^3)—highly anisotropic materials have room temperature anisotropies ranging from 5 to 17.0 MJ/m^3 (Table 9.1 in Sect. 3).

K_1 is a difficult-to-improve atomic quantity. Since H_c is proportional to $2K_1/\mu_o M_s$, it is sometimes suggested to improve the coercivity by reducing the magnetization. A recent example is Rh-substituted ε-Fe_2O_3 [29]. This approach is counterproductive for permanent magnets, because it goes at the expense of the energy product, which scales as M_s^2 and thereby overcompensates the positive coercivity effect. Similar arguments apply to the dimensionless magnetic hardness parameter [3]

$$\kappa = \sqrt{K_1/\mu_o M_s^2} \tag{9.1}$$

A rule of thumb is that κ should be greater than one for an optimized permanent magnet of good loop shape. This is a necessary not sufficient condition, because the magnetization also needs to be large. Interestingly, a positive coercivity can also be achieved in some materials where K_1 is negative and κ is imaginary (Sect. 4.2). Typical values of κ are between about 0.45 (alnico, Co) and 4.4 ($SmCo_5$); $Nd_2Fe_{14}B$ and $BaFe_{12}O_{19}$ have $\kappa = 1.54$ and $\kappa = 1.37$, respectively.

The second approach is to work with existing compounds and to exploit nanostructuring. In particular, improving the energy product to about 1000 kJ/m^3 could be possible by combining the surplus anisotropy of rare-earth alloys with the high magnetization of Fe-based soft materials. In 1991, Kneller and Hawig [30] used layered structures to outline the positive effect of aligned hard–soft nanostructuring and to establish that the soft-phase regions cannot be much larger than twice the Bloch wall width of the hard phase, $\delta_h = \pi\ (A/K_h)^{1/2}$ [30–32]. The basic micromagnetics of layered two-phase nanostructures and the fundamental

Table 9.1 Extrinsic properties of some past or present commercial permanent magnets

Material	Composition (appr.) at%	$\mu_o M_r$ T	M_r kA/m	$\mu_o H_c$ T	H_c kA/m	$(BH)_{max}$ kJ/m^3	$(BH)_{max}$ MGOe
Magnetite	Fe_3O_4	0.15	120	0.025	20	0.75	0.1
Carbon steel	$Fe_{95}MnC_4$	1.0	800	0.005	4	1.6	0.2
Vicalloy	$Co_{50}Fe_{39}V_{11}$	0.75	600	0.025	20	7	0.9
Cunico	$Cu_{48}Ni_{22}Co_{30}$	0.34	270	0.066	53	7.2	0.9
Honda steel	$Fe_{58}Co_{34}Cr_3W_1C_4$	0.9	720	0.025	20	8.0	1.0
Co ferrite	$CoFe_2O_3$	0.25	200	0.065	52	9.6	1.2
Cunife	$Cu_{57}Ni_{21}Fe_{22}$	0.54	430	0.055	44	12	1.5
Lodex	$Fe_{65}Co_{35}$ in Pb	0.74	590	0.094	75	28	3.5
Mn–Bi	MnBi	0.42	330	0.33	260	33	4.1
Ba ferrite	$BaFe_{12}O_{19}$	0.43	340	0.21	170	36	4.5
Alnico 5	$Fe_{50}Co_{24}Ni_{14}Al_8Cu_2$	1.25	1000	0.064	50	44	5.5
Sr ferrite	$SrFe_{12}O_{19}$	0.42	330	0.35	275	43	5.4
Fe–Cr–Co	$Fe_{56}Cr_{25}Co_{14}V_3Ti_2$	1.35	107	0.055	44	44	5.5
Mn–Al	$Mn_{53}Al_{45}C_2$	0.56	450	0.23	180	44	5.5
Co–Pt	CoPt	0.65	520	0.45	360	73	8.8
Alnico 9	$Fe_{36}Co_{35}Ni_{13}Al_7Cu_3Ti_5$	1.12	890	0.14	110	84	10.6
Sm–Co (1:5)	$SmCo_5$	0.88	700	2.1	1700	150	19
Sm–Co (2:17)	Sm_2Co_{17}	1.08	860	1.4	1100	220	28
Nd–Fe–B	$Nd_2Fe_{14}B$	1.28	1020	1.3	1000	350	44

The first permanent magnet, lodestone or magnetite (Fe_3O_4), was described around 600 BC by Thales of Miletus, although no written accounts have survived before 400 BC

role of δ_h were well understood at that time, due to earlier papers by Goto [33], Kronmüller [34], and Nieber and Kronmüller [35]. Three-dimensional calculations, first performed by Skomski in 1992 [31, 32], confirmed the involvement of δ_h. However, it has become clear more recently that geometry effects are quantitatively important, and there exist nucleation field variations by factors of up to 16 for different geometries.

Aligned hard–soft nanocomposites continue to be an active research area [27, 36–44], with experimental proofs of principle [45–48], although processing requirements are demanding [30, 32]. In particular, it is a struggle to align the *c*-axes of the nanoregions of the hard phase and because real-structure imperfections reduce the coercivity [32, 49]. Isotropic two-phase nanostructures such as $Nd_2Fe_{14}B$:Fe [50–52] are much easier to produce but exhibit relatively low energy products, because the magnetization of the hard phase is reduced by a factor two (Sect. 4.3). Isotropic permanent magnetism is also exploited in polymer-bonded magnets, which have relatively low energy products but exhibit favorable mechanical and processing properties. Bonded magnets use powders made from a variety of materials, such as $Nd_2Fe_{14}B$, $Sm_2Fe_{17}N_3$, and $SrFe_{12}O_{19}$.

9.2 Permanent Magnet Physics

This section summarizes some basic principles of permanent magnetism, starting with magnetostatic energy considerations, followed by outlines of the physical principles of the intrinsic (atomic-scale) properties' magnetization, Curie temperature, and anisotropy. The final section deals with extrinsic (real-structure) phenomena, especially with the coercivity.

9.2.1 Energy Considerations

The primary purpose of permanent magnets is to store magnetostatic energy in free space, for example, in an air gap. Figure 9.2 illustrates the corresponding stray fields created by magnets of fixed magnetization **M** but different shapes. (In this subsection, the applied is assumed to zero.) To evaluate the energy contributions, it is convenient to start from the identity $\frac{1}{2}\int \mathbf{B}\cdot\mathbf{H}\,\mathrm{d}V = 0$ [3], where the integration extends over the whole space [4]. Splitting this integral into parts inside the magnet (index *in*) and outside the magnet (index *out*) where $\mathbf{B} = \mu_o\mathbf{H}$ yields the sought-for magnetostatic energy E_{out} outside the magnet

$$E_{\text{out}} = \frac{1}{2}\int_{\text{a}} \mu_o H^2\,\mathrm{d}V = -\frac{1}{2}\int_{\text{i}} \mathbf{B}\cdot\mathbf{H}\,\mathrm{d}V \tag{9.2}$$

This equation is normally interpreted as $(BH) = 2E_{\text{out}}/V$, where $(BH) = V^{-1}\int_{\text{in}} \mathbf{B}\cdot\mathbf{H}\,\mathrm{d}V$ is the energy product—twice the magnetostatic energy associated with a magnet of unit volume. The relationship between **H** and $\mathbf{B} = \mu_o(\mathbf{M}+\mathbf{H})$ is not straightforward, but a useful approximation is to assume that $\mathbf{M}_s$ is uniform and $\mathbf{H} = -D\mathbf{M}$, where D is a demagnetizing factor that depends on the magnet shape only [54]. This expression is exact for ellipsoids of revolution whose magnetization $M = M_s$ is homogeneous and parallel to the axis of revolution, but it is also a good approximation for some other shapes. These shapes include thin films (Fig. 9.3a, b), horseshoe-type flux closure configurations (Fig. 9.3c), and compact blocks or cylinders (Fig. 9.3d) [55]. Equation (9.2) then yields

$$(BH) = \mu_o D(1 - D)M_s^2 \tag{9.3}$$

This equation emphasizes that energy product is a shape-dependent "global" property of a magnet, rather than a materials' parameter.

Thin films with perpendicular and in-plane magnetization have $D \approx 1$ and $D \approx 0$, respectively, both corresponding to very low energy products. Figure 9.3a, b shows why a uniformly magnetized thin film produces no stray field, except around the edges, and most of the material is wasted. Similar considerations apply to

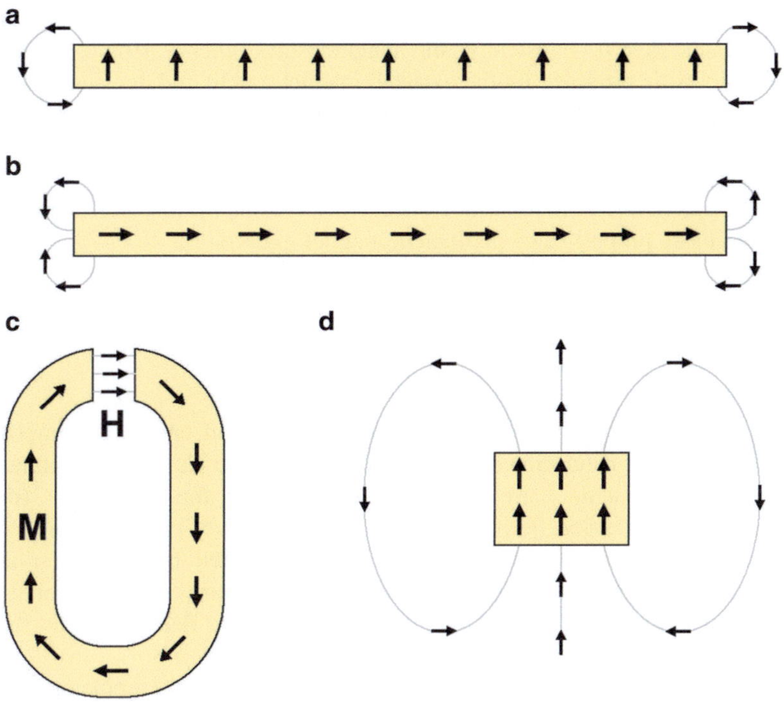

Fig. 9.3 Permanent magnetism and magnet shape: (**a**) thin film with perpendicular magnetization, (**b**) thin film with in-plane magnetization, (**c**) horseshoe-like flux closure geometry, and (**d**) compact magnet. The energy product increases as one goes from thin films (**a**, **b**) to horseshoe magnets (**c**) and to compact bulk magnets (**d**)

"horseshoe" shapes like that in Fig. 9.3c, characterized by $D \approx g/L$, where g is the air gap width and L is the contour lengths of the flux lines through the entire magnet.

Maximizing the energy product of Eq. 9.3 with respect to the shape of the magnet, $\partial(BH)/\partial D = 0$, yields $D = 1/2$ and $(BH)_{\max} = \frac{1}{4}\mu_o M^2$. A cylinder with its radius equal to its height has $D \approx \frac{1}{2}$ in good approximation, although a cylinder is not an ellipsoid of revolution and the magnetization is slightly inhomogeneous near the cylinder edges. This compact shape is typical for rare-earth permanent magnets, where $\kappa > 1$. When the hardness parameter κ is small, then the anisotropy is unable to stabilize the magnetization direction with respect to the magnets own demagnetizing field $-DM$, and shapes with $D < 1/2$ must be used. This is the reason for the iconic horseshoe shapes of steel magnets, similar to Fig. 9.3c, and leads to specific magnet designs for alnico magnets. However, the energy product $(BH)_{\max}$ remains the overriding figure of merit, and there is no point in double-counting magnet bulkiness in form of reduced $(BH)_{\max}$ and, separately, by emphasizing the magnet's elongated shape.

As indicated above, thin films cannot be used as permanent magnets, even if the "nominal" energy product extracted from the M–H hysteresis loop is very high.

In terms of Fig 9.3a, b, B and H are zero inside the magnet, respectively, so that Eq. 9.2 yields $E_a = 0$. Physically, the magnetostatic energy stored in thin-film hysteresis loops is provided by the external magnetic field and does not contribute to the energy product. To turn a thin-film magnet into a permanent magnet, thin-film layers must be mounted onto each other until a compact shape is achieved. This is a practical challenge and may also alter the hysteresis loop, thereby reducing the coercivity. However, thin films are widely used in laboratory-scale research to investigate basic properties of permanent magnet materials).

9.2.2 Intrinsic Properties

Intrinsic magnetic properties reflect crystal structure and chemical composition. The spontaneous or saturation magnetization M_s and the Curie temperature T_c of most permanent magnets are largely determined by the iron-series (or 3*d*) transition-metal sublattice. Some heavy atoms (4*d*, 5*d*, 4*f*, 5*f*) also carry a magnetic moment m, but their magnetization $M_s = m/V$ is diluted by the larger atomic volume V. For example, the atomic volume of rare-earth atoms is about three times that of Fe and Co. The 3*d* magnetization is largely determined by the spin (S), which is known as the orbital-moment quenching.

Atomic magnetic moments are caused by the Coulomb repulsion between electrons, in combination with Pauli's exclusion principle. Since a low-lying one-electron level can accommodate a ↑↓ electron pair but not a ↑↑ electron pair, one of the two ↑↑ electrons must occupy an excited one-electron level [56]. This promotion costs one-electron energy but reduces the Coulomb energy, which is largest for two electrons (↑↓) in a single orbital. In rare-earth atoms and in transition-metal oxides, the Coulomb interaction usually wins, and the atoms carry a magnetic moment. In transition-metal (TM) alloys, the one-electron or promotion energy is proportional to the bandwidth W, so that the ferromagnetism, referred to as itinerant ferromagnetism, occurs in narrow bands (Stoner criterion). The bandwidth depends on atomic composition and crystal structure, so that the itinerant TM moments are nontrivial to predict.

Similar considerations apply to interatomic exchange J and to the Curie temperature $T_c \sim J$. As a rule of thumb, elements in the middle of the iron TM series (Cr, Mn, sometimes Fe) exhibit a trend toward antiferromagnetic (AFM) order ($J < 0$), whereas those at the end (Co, Ni, and often Fe) prefer ferromagnetic (FM) order ($J > 0$). The preference of AFM order in half-filled bands reflects the interplay between bonding and antibonding states. FM order means that all bonding and antibonding states contain one ↑ electron, corresponding to a state that is neither bonding nor antibonding. AFM order means that all bonding states are occupied by ↑↓ spin pairs. AFM bonding is usually weaker than FM bonding, but if the net FM bonding is zero by occupancy, then the AFM bonding wins [3, 56].

Anisotropy means that the magnetic energy depends on the magnetization directions relative to the crystal axes. In the simplest case, the anisotropy energy density $E_a/V = K_1 \sin^2\theta$. Magnetocrystalline single-ion anisotropy, usually the most important contribution, is a relativistic effect involving spin–orbit coupling and therefore highest for heavy elements such as rare earths. Simplifying somewhat, the anisotropic electrostatic crystal field modifies the orbital motion of the electrons and affects, via the relativistic spin–orbit coupling $\lambda\, \mathbf{L} \cdot \mathbf{S}$ [57], the spin system. Most permanent magnet alloys have uniaxial (hexagonal, tetrahedral, or rhombohedral) crystal structures and an easy magnetization axis (*c*-axis) perpendicular to the basal plane (*a–b* plane) [3, 13, 16]. The magnetic anisotropy of rare-earth transition-metal alloys is largely provided by the rare-earth sublattice. The rare-earth anisotropy energy is equal to the electrostatic interaction energy between the 4*f* ions and the local crystal field. The 4f ions obey Hund's rules, described by Stevens coefficients, and are easily predicted across the rare-earth series.

The magnitude of anisotropy of itinerant magnets scales as λ^2/W, where λ is the spin–orbit coupling constant. An upper limit to 3*d* anisotropy is of order of a few MJ/m^3 [3]. In contrast to rare-earth anisotropy, there are no well-defined rules for the 3*d*, 4*d*, and 5*d* anisotropies as a function of the atomic number (or *d*-band filling) *n*. The anisotropy tends to strongly oscillate as a function of *n*, and these oscillations depend on Fermi-level-dependent *k*-space summations [58] which can only be treated numerically. However, there are some crude rules for nearly filled 3*d* band. For example, isomorphic compounds of Fe and Co often exhibit anisotropies of opposite sign.

Two-ion anisotropies are often equated with magnetostatic dipole interactions, although there also exist small two-ion anisotropy contributions of electronic origin. There are two types of magnetostatic two-ion anisotropies: a "magnetocrystalline" contribution caused by near atomic neighbors and shape anisotropy. The former is important, for example, in some Gd compounds, such as $GdCo_5$. The latter is actually a micromagnetic phenomenon, limited to nanoscale grains, or particles [59].

9.2.3 *Extrinsic Properties and Alnicos*

Extrinsic properties, such as the coercivity H_c, are usually realized on a length scale of several nanometers and on macroscopic time scales, as epitomized by the nonequilibrium character of magnetic hysteresis [59]. It is always a challenge to turn a material with favorable intrinsic properties M_s, T_c, and K_1, into a usable permanent magnet with high coercivity and high energy product. The fully magnetized state is unstable, because the stray field can be greatly reduced by adopting a multidomain state with zero net magnetization. Complicated metallurgical treatments are usually necessary to create a microstructure that avoids the nucleation of reverse domains and impede the propagation of domain walls. Any given material usually requires several years of optimization. Table 9.1 shows extrinsic properties of some industrial permanent magnets.

A semiempirical expression for the coercivity is the Kronmüller equation [34]

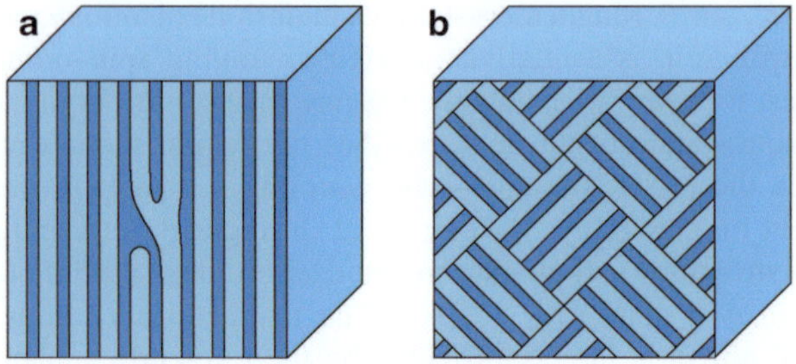

Fig. 9.4 Typical alnico nanostructures. FeCo-type magnetic regions (bright) are embedded in a AlNi-type nonmagnetic matrix (dark). The differences between (**a**) and (**b**) are caused by different field-annealing conditions. Compared to (**a**), the structure (**b**) has a higher coercivity and a lower remanence

$$H_c = a2K_1/\mu_o M_s - D_{eff} M_s \tag{9.4}$$

where α is dimensionless and D_{eff} is an effective *local* demagnetizing factor. The Kronmüller factor α parameterizes the magnet's real or "microstructure," which essentially means the magnet's nanostructure. In perfect ellipsoids of revolution, $\alpha = 1$, but in reality often $\alpha \ll 1$. Even in optimally processed magnets, it is difficult to achieve values in excess of 0.1–0.3. This finding, known as Brown's paradox [60, 61], is explained by real-structure imperfections. Coercivity mechanisms can be roughly divided into nucleation and pinning, but there are many material-specific variants [59, 62, 63].

The effective demagnetizing factor D_{eff} is unrelated to the macroscopic demagnetizing factor D and may have either sign. It describes how local dipole–dipole interactions affect the coercivity, for example, by creating stray fields near grain boundaries, and is often of the order of 1/3. A notable exception is alnico [2, 64–66], which basically consists of elongated soft regions of FeCo embedded in a nonmagnetic NiAl matrix (Fig. 9.4) and where D_{eff} is *negative* on account of the columnar microstructure. In idealized alnico [65], an optimum FeCo volume fraction of 2/3 maximizes the energy product, yielding $(BH)_{max} = \mu_o M_s^2/12$ and $D_{eff} = -1/6$.

Alnico anisotropy is a type of shape anisotropy, $K_{sh} = ¼(1 - 3D)\mu_o M_s^2$. Unlike single-ion anisotropy, shape anisotropy is a nanoscale phenomenon, fully developed only if the radius R of the particle or nanowire is smaller than about $l_{coh} = 5\sqrt{A/\mu_o M_s^2}$ or about 10 nm for a broad range of ferromagnets [59]. For large radii, shape anisotropy decreases as l_{coh}^2/R^2. In alnico, the elongated structures are created by spinodal decomposition, and the underlying thermodynamics makes it difficult to reduce R below about 20 nm. This explains the relatively low coercivity of alnicos, about 100 mT. However, depending on details of the FeCo–NiAl interface, including chemical variations in the various alnico grades, it may be possible to exploit a positive interface K_1 contribution to the anisotropy and coercivity [27]. This is important, because the small coercivity is the bottleneck for the use of this otherwise good permanent magnet material. Some other metallic magnets, such Fe–Co–Cr magnets and variants (chromindur) are structurally and magnetically similar to alnicos [8, 9].

As an explicit example, let us consider a magnet where a hard matrix of anisotropy K_h contains a soft inclusion with $K_1(\mathbf{r}) < K_h$. In an external field, magnetization reversal starts in the soft region (nucleation), which negatively affects the coercivity, but the exchange coupling to the hard phase counteracts the negative effect of the soft region. Starting from saturation, $\mathbf{M} = M_s\,\mathbf{e}_z$, nucleation involves a small magnetization component $\mathbf{m} = \mathbf{M} - M_s\,\mathbf{e}_z$ [59, 60]. The components m_x and m_y of $\mathbf{m}$ are approximately degenerate, so that we can restrict ourselves to the $m = |\mathbf{m}|$. The analysis starts from the micromagnetic free energy and yields [3, 32, 59]

$$-A\nabla^2 m + (K_1(\mathbf{r}) - \mu_o M_s H/2)m = 0 \tag{9.5}$$

Here, A is the exchange stiffness and the external field $\mathbf{H} = H\,\mathbf{e}_z$. In the approximation of Eq. 9.5, K_1 and H are understood to contain approximate shape and demagnetizing field contributions, respectively [59]. In the limit of very small soft inclusions, the exchange stiffness A dominates, so that $m = \text{const.}$ and $H_c = 2 < K_1(\mathbf{r}) > /\mu_o M_s$. In fact, $< K_1(\mathbf{r}) > = K_h$ in the present case, so that H_c is equal to the anisotropy field of the hard phase, $2K_h/\mu_o M_s$. In the limit of large soft inclusions, A has no effect, and the nucleation field $2K_s/\mu_o M_s$ is determined by the softest part of the soft region, $K_s = \min(K_1(\mathbf{r}))$.

9.3 Rare-Earth Free Permanent Magnets

The search for new intermetallic compounds by changing chemical composition and atomic structure has been a long-standing task in magnetism. Aside from using high-throughput experimental or computational methods to look for completely new ternary and quaternary alloys, there is continuing experimental and theoretical work on substitutional derivatives of existing materials. Table 9.2 shows magnetizations, Curie temperatures, and anisotropies for a variety of permanent magnet materials. In the following paragraphs, we will briefly discuss some classes of materials in the light of current research.

9.3.1 Iron-Rich Alloys

Permanent magnetism and hard magnetism are synonyms historically, and both reflect the difference between mechanically (and magnetically) soft iron and mechanically (and magnetically) hard steel ($Fe_{100-x}C_x$, $x \approx 4$) [3]. Steels magnets,

Table 9.2 Intrinsic properties of some compounds of interest in permanent magnetism

	$\mu_o M_s$ (RT)	T_c	K_1 (RT)	
Substance	T	K	MJ/m^3	Structure
YCo_5	1.06	987	5.2	Hexagonal ($CaCu_5$)
$SmCo_5$	1.07	1003	17.2	Hexagonal ($CaCu_5$)
$Pr_2Fe_{14}B$	1.41	565	5.6	Tetragonal ($Nd_2Fe_{14}B$)
$Nd_2Fe_{14}B$	1.61	585	4.9	Tetragonal ($Nd_2Fe_{14}B$)
$Sm(Fe_{11}Ti)$	1.14	584	4.9	Tetragonal ($ThMn_{12}$)
Sm_2Co_{17}	1.20	1190	3.3	Rhombohedral (Th_2Zn_{17})
$Sm_2Fe_{17}N_3$	1.54	749	8.9	Rhombohedral (Th_2Zn_{17})
γ-Fe_2O_3	0.47	863	–0.0046	Cubic (disordered spinel)
Fe_3O_4	0.60	858	–0.011	Cubic (ferrite)
$BaFe_{12}O_{19}$	0.48	723	0.33	Hexagonal (M ferrite)
$SrFe_{12}O_{19}$	0.46	733	0.35	Hexagonal (M ferrite)
$PbFe_{12}O_{19}$	0.40	724	0.22	Hexagonal (M ferrite)
CrO_2	0.56	390	0.025	Tetragonal (rutile)
Fe	2.15	1043	0.048	Cubic (bcc)
Co	1.76	1360	0.53	Hexagonal (hcp)
$Fe_{0.96}C_{0.04}$	2.0	(PT)	–0.2	Tetragonal
$Fe_{16}N_2$	2.7	(PT)	1.6	Tetragonal
Fe_3B	1.61	791	–0.32	Tetragonal
$Fe_{23}B_6$	1.70	698	0.01	Cubic (C_6Cr_{23})
$Fe_{0.65}Co_{0.35}$	2.43	1210	0.018	Cubic (bcc)
FeNi	1.60	(PT)	1.3	Tetragonal ($L1_0$)
FePd	1.37	760	1.8	Tetragonal ($L1_0$)
FePt	1.43	750	6.6	Tetragonal ($L1_0$)
CoPt	1.00	840	4.9	Tetragonal ($L1_0$)
Co_3Pt	1.38	1000	2.1	Hexagonal
MnAl	0.62	650	1.7	Tetragonal ($L1_0$)
MnBi	0.78	630	1.2	Hexagonal (NiAs)
Mn_2Ga	0.59	(PT)	2.35	Tetragonal ($D0_{22}$)
Mn_3Ga	0.23	(PT)	1.0	Tetragonal ($D0_{22}$)
Mn_3Ge	0.09	(PT)	0.91	Tetragonal ($D0_{22}$)

PT indicates interfering structural changes near the Curie temperature.

which have iconic horseshoe shapes due to their small coercivity, became the first commercial permanent magnets in the nineteenth century but are now obsolete. The high saturation magnetization of $Fe_{65}Co_{35}$ and its pronounced temperature stability remain valued in alnico permanent magnets. However, the definition of steel requires C as an alloying element, so $Fe_{65}Co_{35}$ is not a steel. The physics behind the mechanical and magnetic hardening of steel is that carbon causes a martensitic phase transition in bcc Fe, leading to a tetragonally distorted phase and improving both mechanical and magnetic hardnesses [1, 6, 67]. Honda steel, which contains

Co and C [5], has a coercivity of $\mu_o H_c = 0.020$ T, as contrasted to 0.004 T for ordinary carbon steel.

Steel magnets have attracted renewed attention [6, 27], especially under the keyword of tetragonally distorted Fe–Co [68, 69]. Experimental investigations of the magnetoelastic properties of these materials date back to the mid-twentieth century, showing that strain yields a strong magnetoelastic anisotropy in Fe–Co alloys [1]. Substantial anisotropy, $K_1 = 9.5$ MJ/m^3, and a magnetization of $\mu_o M_s = 1.9$ T have been predicted for strongly distorted Fe–Co with $c/a = 1.23$ [68], although such strains are difficult to sustain metallurgically. Experimental room temperature anisotropies per strained Fe or Co atom reach about 2.1 MJ/m^3, but this value does not account for the large amount of Pt in the system (about 75 vol.%) [69]. In carbon steel, K_1 is actually negative (Sect. 4.2), and both magnetoelastic and chemical effects contribute to the net anisotropy [6].

The behavior of interstitial N in Fe is similar to that of C [70]. An interesting noncubic high-magnetization materials is Fe_8N ($Fe_{16}N_2$ in hypercorrect unit-cell notation), where the nitrogen enhances the magnetization electronically and by expanding the Fe lattice [71]. The compound has a very high magnetization [71, 72], about 2.8 ± 0.4 T, but the actual value has remained subject to debate, and the structures are difficult to prepare. The room temperature K_1 of the material is about 1.6 MJ/m^3 [73].

9.3.2 Co-Rich Alloys

The hexagonal $CaCu_5$ compound YCo_5 has long been of interest in permanent magnetism, despite being overshadowed by the isostructural $SmCo_5$ in many regards [3, 12]. The alloy is a dense-packed derivative of the laves-phase YCo_2, one Y atom replacing two Co atoms per two formula units of YCo_2. While the alloys has the highest 3*d*-only anisotropy at room temperature (Table 9.2), it has been argued that the advantage of YCo_5 is at high temperatures [74], where the Sm anisotropy is small. However, neither Y nor Co are cheap raw materials.

Various transition-metal intermetallics, some of them including B, were investigated in the 1980s [75, 76]. Some of these compounds are derivatives of the Laves 1:2 (and $CaCu_5$ 1:5) phases, but some are not. Of some interest for future rare-earth-free permanent magnets are Zr_2Co_{11} and $HfCo_7$ [75, 77–87], whose structures are fairly dense-packed but not specifically related to the above 1:2 and 1:5 structures. They have magnetizations comparable to Sm–Co, for example, 1.09 T for $HfCo_7$, but lower anisotropies, of the order of 1–2 MJ/m^3. In all T_xCo_{100-x} alloys, the choice of the transition-metal atoms (T) is a key consideration. Specifically, the addition of early 4*d* and 5*d* atoms, such as W, tends to deteriorate the magnetization [82, 88]. Late 4*d* and 5*d* atoms are generally better regarding both magnetization and anisotropy, but they are often expensive, as are Ga and Hf.

There are several methods to produce Co alloys, such as mechanical alloying and melt spinning. To facilitate melt spinning, additives such as B [89] and Si [79] are used. These additions also affect the magnetic properties, intrinsically (by atomic substitution or quasi-interstitial occupancy at interruption sites) [81] or extrinsically (by creating pinning centers), but much work remains to be done to get a thorough understanding of these effects and to gauge their usefulness for permanent magnet development.

Vicalloy (Co–Fe–V) is a precipitation-hardened material reminiscent of steel but contains no carbon [2]. The material possesses a modest energy product but can easily be produced in the form of thin sheets (0.05 mm) and continues to be used for special applications. Compared to other classes of magnetic materials, some Co- and Fe-rich magnets are poorly understood, and there is ongoing exploratory research on systems like Fe–Co–Ti–B [90]. One aim is to combine shape and magnetocrystalline anisotropies with the high magnetizations and Curie temperatures of Co- and Fe-rich phases.

Many of the compounds considered in this section can also be prepared in nanoparticle form, by cluster deposition [91, 92]. This includes Co alloys such as YCo_5 [92, 93], $HfCo_7$ [84, 87], Zr_2Co_{11} [83, 87], W-Co [88], but also Fe_3Au, which does not exist as a bulk equilibrium phase [94]. To create a permanent magnet, these clusters must be compacted as far as possible while keeping them *c*-axis aligned. Some progress in this direction has been made by aligning the cluster during deposition [84], but further research is necessary. There are also chemical methods to produce Co nanoparticles and nanorods [95, 96], but compaction remains a challenge in these systems, too. Historically, lodex-type fine-particle magnets were used industrially until 1988 [9]. These magnets consist of elongated Fe–Co particles in a lead matrix.

9.3.3 $L1_0$-Ordered Magnets

$L1_0$-ordered alloys such as CoPt, FePt, and FePd [7–10, 97] have long been a part of permanent magnet research. The *c*/*a* ratio is normally close to 1, but the layered atomic environment is strongly anisotropic, both structurally and magnetically. However, Pt and Pd are very expensive, which has limited the use of these magnets. For example, recently produced Fe-Co-Pt thin-film magnets have a thickness of 20 nm and room temperature properties of up to $\mu_o H_c = 2.52$ T and $\mu_o M_s = 1.78$ T [42, 46]. In these structures, the compromise between magnetization and coercivity yields an impressive nominal energy product maximum of 510 kJ/m^3 for $Fe_{40}Co_{22}Pt_{38}$, but compaction into a bulk magnet remains a challenge. The substitution of Co for Fe deteriorates the magnetization, in striking contrast to elemental Fe. This is well understood in terms of the electronic structure of FePt, which does not need Co to become a strong ferromagnet [98, 99]. Note that perfectly ordered FePt has been predicted to be antiferromagnetic [100]. In fact, FePt probably needs some disorder or excess Fe to become ferromagnetic [27, 101], and there are also experimental indications in this direction, for example, the necessity of extra Fe for optimized performance [102, 103]. On the downside, the addition of Fe and/or Co ultimately yields the cubic $L1_2$ phase, whose anisotropy is very low.

A few other $L1_0$ alloys are much less expensive but difficult to process. MnAl was discovered in 1958 [104], and its $L1_0$-ordered (or τ) phase requires small C additions to become structurally stable [9]. MnAl exhibits appreciable intrinsic properties, namely, $\mu_o M_s = 0.75$ T, $K_1 = 1.7$ MJ/m^3 and $T_c = 650$ K [3, 10], and there are ongoing efforts to better understand this material and to develop it into commercial permanent magnets [105–107].

An interesting feature of MnAl is the *ferromagnetic* exchange between the close Mn–Mn nearest neighbors in the Mn planes of the $L1_0$ structure, which contradicts the general trend toward AFM order in half-filled bands. First-principle calculations yield a Mn–Mn intralayer exchange of $J = 502$ K, as contrasted to a Mn–Mn interlayer exchange of $J' = 73$ K, and an approximate Curie temperature of 718 K, as compared the experimental value of about 650 K [27, 108]. This finding also contradicts expectations from the popular Bethe–Slater–Néel curve, which predicts a decrease of J with decreasing Mn–Mn distance. This situation demonstrates that the Bethe–Slater–Néel curve is not rooted very deeply in the electronic structure of intermetallic compounds.

$L1_0$-ordered FeNi (tetrataenite) was originally discovered in meteorites, formed with cooling times much longer than one million years, but is now being explored from the viewpoint of permanent magnetism [109, 110]. An interesting feature is the relatively high coercivity of about 0.1 T [1 kOe] in a sample taken from the meteorite NWA 6259 [111]. This coercivity reflects the microstructure, where the three variants or "twins," namely <100>, <010>, and <001>, are likely to form boundaries that act as pinning sites. This creates pinning sites in a natural way and coercivities of about 120 mT ($\alpha = 0.059$) without any additional processing [111]. By contrast, hcp Co has intrinsic properties similar to FeNi, but it tends to form nearly perfect crystals with few pinning sites and low coercivity. A typical value is $\mu_o H_c = 1.2$ mT [1], corresponding to a discouraging Kronmüller factor $\alpha = 0.002$.

9.3.4 Manganese Alloys

Tripositive manganese has a moment of 5 μ_B per atom, as compared to 2.2 μ_B for Fe. If this moment could be exploited in industrial magnets, it would revolutionize technology far beyond magnetism and open the door for completely new technologies. Furthermore, Mn is a relatively cheap metal. Unfortunately, most manganese compounds are antiferromagnetic, which is typical for elements in the middle of the 3*d* series and easily understood in terms of band-filling arguments (Sects. 2.2 and 3.3). There are a few ferromagnetic Mn alloys with modest magnetizations of less than one Tesla, such as MnBi and the above-discussed MnAl [3, 9, 10, 105]. Other Mn compounds, such as Mn_2Ga and Mn_3Ga, are also being investigated in the context of permanent magnetism [112, 113], but their magnetizations are rather small, and Ga is expensive.

MnBi, which crystallizes in the hexagonal NiAs structure, has fascinated scientists for more than a century [114, 115]. The material exhibits structural phase transitions and a complicated interstitial behavior that affect the magnetism [115, 116] and complicate the magnet processing, which also includes the need to limit corrosion. Nevertheless, MnBi has long been considered a permanent magnet candidate [2], and there is ongoing research in various directions [117–119].

Manganese is also a constituent of many ternary Heusler compounds, such as Cu_2MnSn and Cu_2MnAl. The nonzero magnetization of these alloys aroused much interest around 1900, because the compounds exclusively consist of nonferromagnetic elements [1, 120]. In a broader sense, this group of materials also includes half-Heuslers (NiMnSb) and Heusler-like binary compounds (Mn_3Ga) [113, 121]. The magnetizations of these materials are usually small, more suitable for exchange bias (large ratio K_1/M_s) than for permanent magnet applications.

9.4 Nanoscale Permanent Magnetism

The limited range of compounds suitable for permanent magnets makes it necessary to explore and exploit magnetic nanostructuring. First, hard–soft nanostructuring can improve the energy product beyond that of the hard phase. Second, the realization of coercivity, remanence, and energy product in any permanent magnet material involves nanoscale effects. Third, magnets are often used above room temperature, which makes it important to understand thermal effects in nanostructures.

9.4.1 Geometrical and Optimization

In aligned hard–soft nanocomposites, the magnetically soft phase improves the hard magnetic performance of the main phase, sacrificing some anisotropy and coercivity but enhancing magnetization and energy product beyond that of the hard phase. The simple model of Fig. 9.5 illustrates the nanoscale interactions involved [3, 122, 123]. The model has only two magnetic degrees of freedom, namely, two rotation angles about a common axis. In an ideal aligned exchange-coupled hard–soft magnet (a), the magnetization remains nearly parallel in adjacent hard and soft grains. This quasicoherent regime is realized if the interatomic exchange dominates, that is, for small grains. The corresponding effective anisotropy is equal to the volume average of the anisotropy, $<K_1> = ½K_h$ in Fig. 9.5a, where K_h is the anisotropy of the hard phase. The corresponding nucleation field $2<K_1>/\mu_o<M_s>$ is reduced by a factor of about ½, that is, $\alpha \approx ½$. For large grain sizes, the soft phase switches first (Fig. 9.5b), and $\alpha \ll 1$, because the anisotropy K_s of the

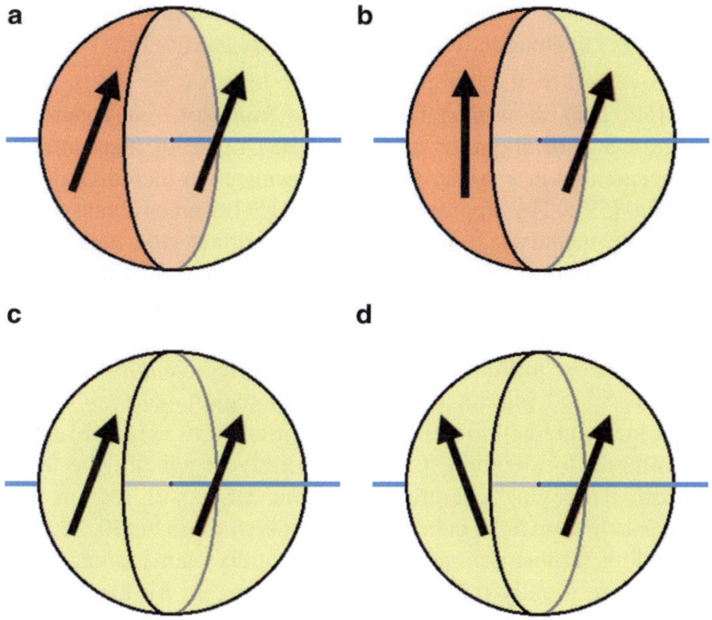

Fig. 9.5 Nanoscale magnetization reversal: (**a**) coherent rotation in two-phase particle, (**b**) localized nucleation in two-phase particle, (**c**) coherent rotation in single-phase particle, and (**d**) rudimentary "curling"

soft phase is approximately zero. The transition between (a) and (b) occurs when the radius of the sphere is about twice the Bloch wall width of the hard phase, $2\delta_B$, where $\delta_B = \pi\sqrt{A/K_h}$. In terms of Eq. 9.5, $\sqrt{A/K_h}$ describes the competition between exchange (A, measured in J/m) and anisotropy ($0 \leq K_1 \leq K_h$), the square root following from the dimensions of K_1 (J/m^3) and A (J/m).

Since the Bloch wall width δ_h of the hard phase scales as $1/\sqrt{K_h}$, it has been suggested to use semihard phases with rather small K_h to facilitate nanostructuring. The smaller K_h, the larger the range of the effective hard–soft coupling. However, the coercivity is proportional to $2K_h/\mu_o M_s$, so that anisotropy reductions are harmful in lowest order. An extreme limit is $K_h = 0$, which corresponds to $\delta_B = \infty$ and soft magnetism. A natural reduction in K_h and enhancement in δ_h occurs as the temperature approaches the Curie temperature, and nanostructuring can then be used to achieve a secondary coercivity improvement via α [86]. Physically, the exchange coupling may be two-phase like at room temperature, similar to Fig. 9.5b, but becomes single-phase like as the temperature increases, as in Fig. 9.5a.

Figure 9.5c, d shows the situation for soft-magnetic nanoparticles. In small particles (c), the magnetization remains uniform or coherent, whereas large particles undergo magnetization curling (d), in close analogy to compass needles located side by side. Curling or "vortex" states [60, 61, 124] are magnetostatically favorable but cost some exchange energy. Dimensional analysis of the exchange energy (J/m) and of the magnetostatic energy density $\mu_o M_s^2$ (J/m^3) indicates that the transition is governed by the "proper" exchange length $l_o = \sqrt{A/\mu_o M_s^2}$. Detailed calculations [59, 125] show that the transition from coherent rotation to curling occurs at $R_{coh} \approx 5\ l_o$ or about 10 nm for a broad range of materials. In Eq. 9.5, curling is incorporated in a crude way, by allowing H to include a demagnetizing field correction [59]. The approach of Eq. 9.15 becomes exact for very hard materials ($\kappa \gg 1$), where $K_h \gg \mu_o M_s^2$. As an important side note, the coherence radius R_{coh} is unrelated to the critical single-domain size. The latter is an equilibrium property that does not affect the hysteresis loop [3, 59].

Comparing the left and right subfigure columns in Fig. 9.6, we see a transition from a cooperative or "military" regime (a) with effective anisotropy to a noncooperative or "civilian" regime (b) where each region behaves differently and experiences a local interaction field. The transition from (a) to (b) is also known as nucleation-mode localization. In the cooperative state (a), the soft region is prevented from destroying coercivity, whereas large soft regions behave like (b) and reduce nucleation field, coercivity, and energy product [30, 32, 36]. Creating the corresponding nanostructures is experimentally demanding, and observed energy products are usually much smaller than expected for ideal nanostructures. However, this is not a failure of theory but a manifestation of Brown's paradox [60, 61], that is, an indicator of our limited knowledge about the real structure of the materials. Note that even advanced numerical simulation methods are not able to treat magnets having sizes larger than about 1 μm.

It is well-established that the soft phase of a two-phase system cannot be much larger than twice the Bloch wall width $\pi\sqrt{A/K_h}$ of the hard phase. In the worst case, a trivially small addition of the soft phase completely ruins the coercivity of the whole magnet by creating a harmful nucleus, and the energy product collapses. The question arises how this mechanism is affected by nanogeometry. After the investigation of the micromagnetism of layered hard–soft structures [30, 33–35], interest moved to other geometries, especially to three-dimensional nanostructures. The first calculations on spherical soft inclusions [31] were soon followed by several articles on a variety of geometries and limits, such as embedded soft cylinders [27, 44, 126], embedded soft cubes [59], disordered nanostructures [32, 126, 127], interacting soft regions [126, 127], granular nanocomposites [128], and core–shell structures [129]. These geometrical optimization or "rational design" calculations confirmed the involvement of $\sqrt{A/K_h}$ as a cornerstone of hard–soft exchange coupling.

Originally, it was believed that multilayers [30] and spherical inclusions [31, 32] behave similarly, in spite of the more pronounced micromagnetic localization behavior of one-dimensional structures [59]. In other words, the one-dimensional and three-dimensional calculations were treated on equal footing, both improving the energy product while not exhibiting fundamental differences. This view has partially been revised, both quantitatively and qualitatively.

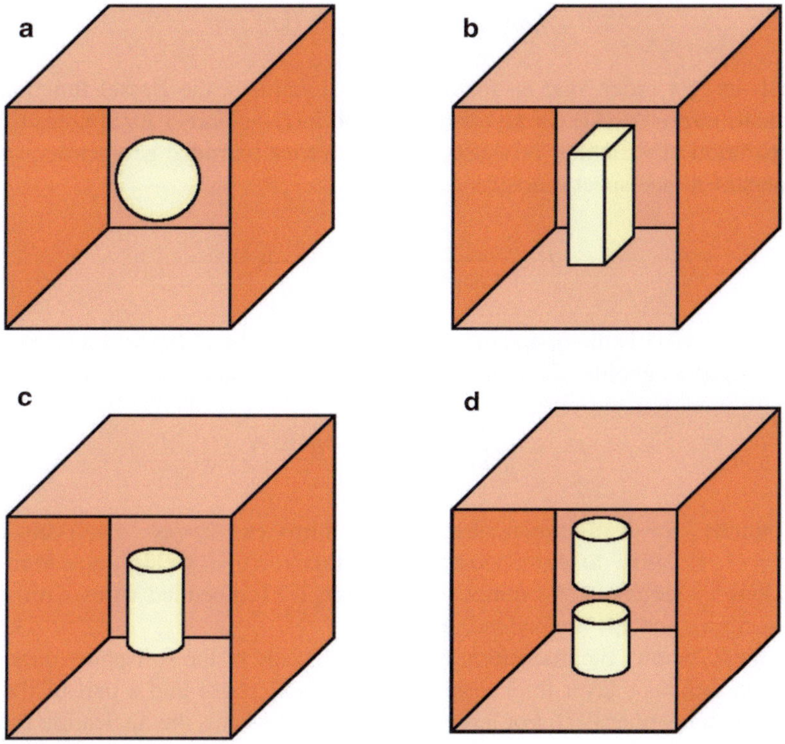

Fig. 9.6 Some soft-in-hard geometries: (**a**) embedded sphere, (**b**) embedded cuboid, (**c**) embedded cylinder, and (**d**) interacting soft regions. Dark and bright gray denote hard and soft regions, respectively, and the aspect ratios of the cuboids (**a**:**b**:**c**) and cylinders (L:R) can assume arbitrary values

The simplest nontrivial approach to the geometry of hard–soft composites is to embed the soft regions in a very hard matrix. Figure 9.6 shows some basic geometries [31, 32]. The aspect ratios of the embedded soft region are variable, so that (b) and (c) also include embedded layers and infinite cylinders, respectively. In the approximation of Eq. 9.5, these structures are described by $K_1(\mathbf{r}) = K_h$ in the hard matrix and by $K_1(\mathbf{r}) = 0$ in the soft inclusions. Furthermore, we assume that K_h is sufficiently high to ensure that the hard phase remains aligned, $\mathbf{M}(r) = M_h\ \mathbf{e}_z$. This means that we consider the switching of the soft phase only.

The calculation of the nucleation field, $H_N = H_c$ consists in the solution of Eq. 9.5, subject to the “clamped” boundary condition $m = 0$ at the hard–soft interface. Physically transparent analytical solutions in terms of Bessel functions $J_{d/2-1}$ exist for hyperspherical soft inclusions. This includes layers ($d = 1$), cylinders ($d = 2$), and spheres ($d = 3$), as well as extruded hyperspheres, such as cuboids, and for geometries with free soft surfaces [130]. In the hard magnetic limit considered in this section, the solutions of this equation are the Bessel functions $J_\alpha(x)$, so that

$$m(r) = r^{1-d/2} J_{d/2-1}(\mathrm{k}r) \tag{9.6}$$

For half-integer order, that is, for odd dimensionalities, the Bessel functions are analytical: m(r) ~ cos(κr) for layers (d = 1) and m(r) ~ sin(κr)/r for spheres (d = 3). By separation of variables, it is also possible to treat extruded geometries, such as cylinders of finite aspect ratios ξ = L/2R:

$$H_{\mathrm{N}} = \frac{2A}{\mu_o M_s R^2} \left({c_2}^2 + \pi^2/4\xi^2 \right) \tag{9.7}$$

where $c_2 = 2.4048$ is the first zero of the "two-dimensional" Bessel function $J_o(\kappa r)$. For rectangular cuboids of dimensions a, b, and c, the same procedure yields

$$H_{\mathrm{N}} = \frac{2A}{\mu_o M_s} \left(1/a^2 + 1/b^2 + 1/c^2 \right) \tag{9.8}$$

This includes the well-known limits of thin films (a = b = ∞, c = t) and cubes (a = b = c). Relative to embedded spheres, this method yields nucleation field reductions by factors of 3/4 (embedded cubes), 1/2 (embedded wires with square cross section), and 1/4 (embedded layers).

Figure 9.7 shows the nucleation field as a function of the soft-phase dimension (inclusion radius or layer thickness), where the soft phase and a part of the hard phase switch together [27]. For large soft regions (R → ∞), the nucleation mode m (**r**) is confined to the center of the soft inclusion, and one obtains $1/R^2$-type nucleation field expressions similar to Eqs. 9.6–9.8. The corresponding magnetization profiles are $m(z) = m_o \cos(\pi z/t)$ for platelike soft inclusions of thickness t, $m_o J_o(r/R)$ for cylindrical inclusions of radius R, and $2R\, m_o \sin(\pi r/2R)/\pi r$ for spherical inclusions of radius R. Explicitly, the ratio H_N/M_s is equal to $19.74\, {l_o}^2/R^2$ (spheres), $11.57{l_o}^2/R^2$ (cylinders), $19.74\, {l_o}^2/t^2$ (embedded plates), and $4.94\, {l_o}^2/t^2$ (soft layer with free surface).

For very small soft inclusions, the nucleation field approaches the anisotropy field H_A of the hard phase, but due to localization, the approach to the "plateau" at H_A is dimensionality dependent [27]. This is in close analogy to the quantum-mechanical delocalization of electrons in an inhomogeneous potential V(**r**) and to the behavior of impurity states in the band gaps of solids of different dimensionalities. Nucleation modes in one-dimensional systems (layers) localize most easily and exhibit a pronounced nucleation field drop, whereas three-dimensional (spherical) inclusions need a minimum size to be effective, namely, R = 4 nm in Fig. 9.7. Two-dimensional systems (cylinders) form a borderline case, with logarithmically weak localization corrections [56].

To optimize the permanent magnet performance, it is also important to maximize the volume fraction of the soft phase without having soft regions too far away from the hard phase. Spheres and cylinders cannot be packed densely in space, but

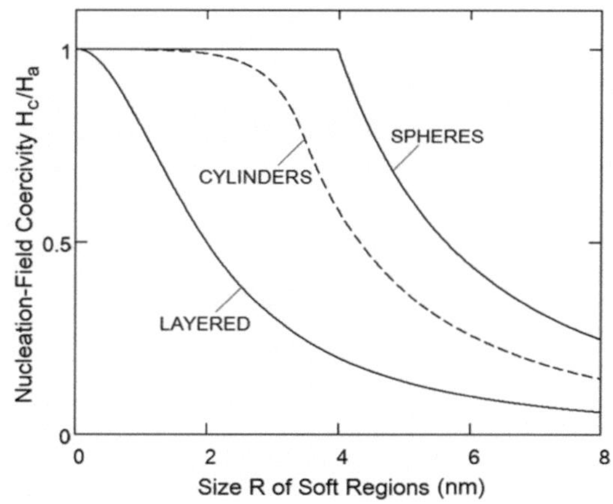

Fig. 9.7 Schematic soft-phase nucleation field as a function of the radius R of the soft phase. Note that 2R = t for the embedded layers (bottom curve). The assumed parameters are $l_o = 2$ nm, $K_1 = 5$ MJ/m^2 and $\mu_o M_h = 1.5$ T [27]

soft cubes can fill space nearly completely, surrounded by a thin "grain boundary" layer of hard material [119].

The volume fraction of the soft phase is limited by secondary nucleation-mode delocalization, meaning that neighboring soft regions couple and cooperatively reduce the coercivity. Figure 9.8 compares different delocalization and surface effects. The delocalization mechanism, where the nucleation-mode tunnels through the hard phase like an electron through a potential barrier (c–d), is unrelated to the geometrical connectivity and to the percolation behavior of the soft phase. Free soft-magnetic surfaces, as in Fig. 9.8b, are particularly harmful to the nucleation field, yielding a nucleation field factor of ¼ compared to embedded layers of the same thickness. The reason is that Fig. 9.8a, b is micromagnetically equivalent due to the free-surface boundary condition dm(x)/dx = 0. Both (a) and (b) yield an asymptotic $1/t^2$ behavior, but t is defined differently, yielding a factor 2 in t and a factor ¼ in the nucleation field.

The term "incomplete exchange coupling" is occasionally applied to separately switching hard and soft phases, corresponding to the tails in Fig. 9.8 and to two-phase hysteresis loops. However, this term is somewhat unfortunate. The hard–soft interface exchange is usually very strong, with the exception of antiferromagnetic exchange at the interface, and even moderately reduced exchange has little or no effect on the hysteresis loops [59]. Even for infinite exchange at the interface, one would obtain curves similar to Fig. 9.8.

9.4.2 Easy-Plane Micromagnetism

In ferromagnets with cubic crystal structure, it does not matter very much whether K_1 is positive or negative, because the sign of the "cubic" K_1 merely indicates the preference of sixfold or eightfold easy axes, respectively. However, the magnetic

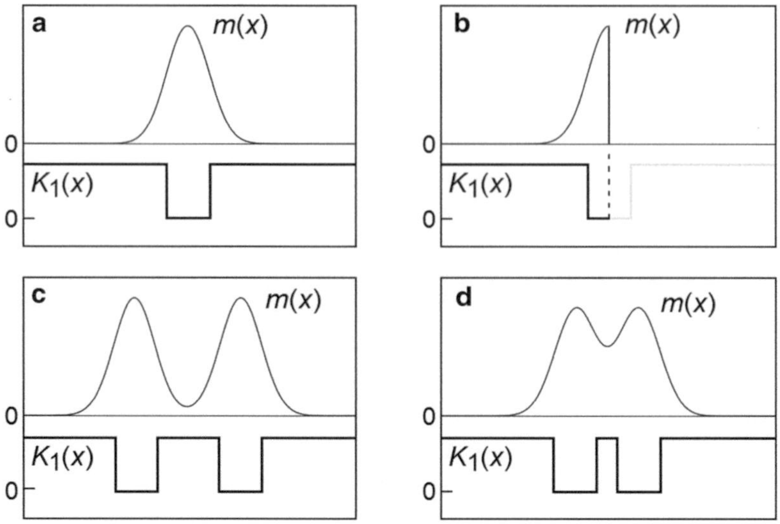

Fig. 9.8 Surface and interaction effects: (**a**) soft-in-hard inclusion, (**b**) surface, (**c**) weakly interacting soft regions, and (**d**) strongly interacting soft regions with emerging nucleation-mode delocalization. The modes of (**a**) and (**b**) are micromagnetically equivalent, meaning that the nucleation field $H_N \sim 1/t^2$ of an embedded layer of thickness t is four times as big compared to a layer having the same thickness but a free surface [130]

hardness of steel is associated with a tetragonal martensitic lattice distortion (Sect. 3.1).

It is intriguing to note that past research has not explained the coercivity of ordinary carbon steel. The volume magnetostriction constant of Fe is actually negative, as contrasted to that of $Fe_{65}Co_{35}$ [1] and therefore favors easy-plane anisotropy ($K_1 < 0$) (Fig. 9.9b. This corresponds to the unusual case of imaginary magnetic hardness. Complicating factors are that the strain configurations in free bulk and clamped thin-film configurations are very different, that the strain effect competes against the chemical or “crystal field” effect created by the carbon atoms, and that small Co additions change the sign of the magnetostriction constant. First-principle electronic structure calculations [6] indicate that strain and chemical effects are of comparable magnitude.

This section focuses on such *easy-plane magnets*, where K_1 is negative and κ is imaginary. In single crystals, H_c is exactly zero for negative K_1, because the magnetization is allowed to rotate freely. However, the situation changes in polycrystalline magnets. A simple example is two adjacent grains, where the net anisotropy is $|K_1|/2$ [131]. Figure 9.10 shows the situation for arbitrary angles γ between the c-axes of neighboring grains. The coupled system is no longer uniaxial for arbitrary $\mathbf{n}_1$ and $\mathbf{n}_2$ but must be described by the biaxial anisotropy energy density.

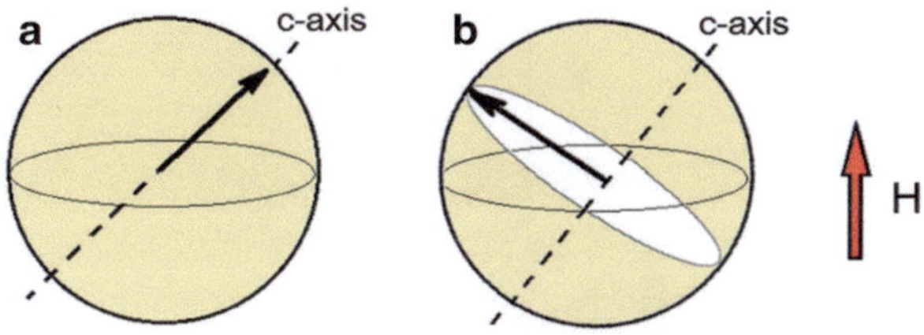

Fig. 9.9 Polycrystalline misalignment: (**a**) easy axis and (**b**) easy plane

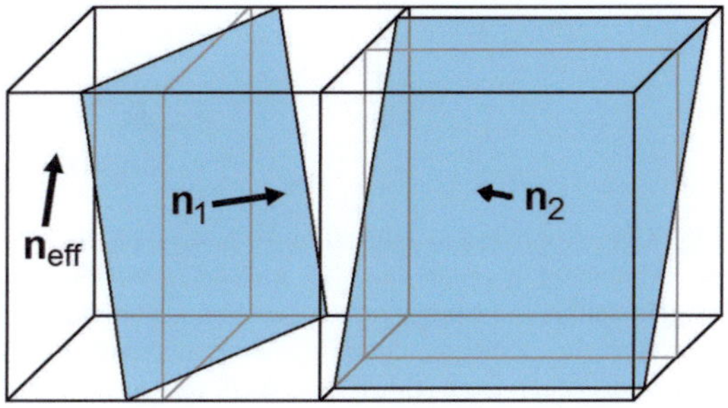

Fig. 9.10 Pair of coupled grain with arbitrary crystallite orientations. The effective easy-axis $\boldsymbol{n}_{\mathrm{eff}} \sim \boldsymbol{n}_1 \times \boldsymbol{n}_2$ is parallel to the direction where the two planes intersect

$$\eta = K_1 \sin^2\theta + K_1{}' \sin^2\theta \ \cos^2\phi \tag{9.9}$$

The new coordinate frame is given by $\boldsymbol{e}_x' \sim \boldsymbol{n}_1 + \boldsymbol{n}_2, \boldsymbol{e}_y' \sim \boldsymbol{n}_1 - \boldsymbol{n}_2$, and $\boldsymbol{e}_z' \sim \boldsymbol{n}_1 \times \boldsymbol{n}_2$, where $\boldsymbol{e}_z' = \boldsymbol{n}_{\mathrm{eff}}'$ is the net easy axis. The corresponding anisotropy energy densities are $\eta_x' = K_1 + K_1', \eta_y' = K_1 - K_1'$, and $\eta_z' = 0$, where $K_1 = ½|K|$ and $K_1' = ½|K| \cos\gamma$.

The effective anisotropy is equal to the energy difference between the y′ and z′ directions

$$K_{\mathrm{eff}} = \frac{|K|}{2}(1 - \cos\gamma) \tag{9.10}$$

The trivial limit $\gamma = 0$ (no grain boundary misalignment) means easy-plane anisotropy ($K_{\mathrm{eff}} = 0$), whereas $\gamma = 90°$ corresponds to $K_{\mathrm{eff}} = |K|/2$. For random grain orientation, three-dimensional averaging over γ yields $K_{\mathrm{eff}} = |K|/4$.

The effective anisotropy of Eq. 9.10 gives rise to domain wall pinning, meaning that the domain wall interacts with the effective local anisotropy K_{eff} near the grain boundaries. As in other inhomogeneous magnetic systems [59], the thickness of the interaction zone is of the order of $\delta_{\mathrm{eff}} = \pi(A/K_{\mathrm{eff}})^{1/2}$. The anisotropy constant of steel is of the order of 0.2 MJ/m^3 [6], corresponding to $\delta_{\mathrm{eff}} = 40$ nm. Grain sizes much larger than δ_{eff} mean that most of the magnet volume switches easily in a reverse magnetic field, which is detrimental to the coercivity and corresponds to very small Kronmüller factors α.

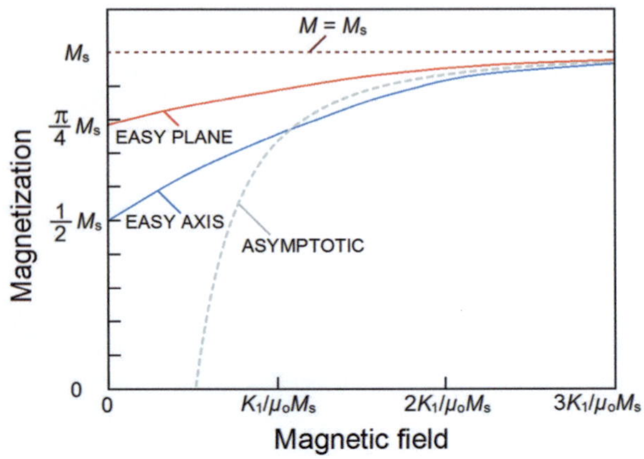

Fig. 9.11 Approach to saturation for noninteracting easy-axis and easy-plane ensembles. Isotropic permanent magnets such as melt-spun $Nd_2Fe_{14}B$ are of the easy-axis type, meaning the magnetization and energy product are reduced by factors of ½ and ¼, respectively. In the absence of nanoscale exchange coupling, as assumed in this figure, the magnetization of easy-plane ensembles drops from $\pi M_s/4$ to $-\pi M_s/4$ as one goes from the first of the hysteresis loop ($H > 0$) to the second quadrant ($H < 0$), so that $(BH)_{max} = 0$

A complication is that easy-plane and easy-axis anisotropies in polycrystalline magnets are difficult to distinguish experimentally. Let us consider large noninteracting grains. The anisotropy of polycrystalline magnets is often determined from approach to saturation in the hysteresis loop. Averaging over all c-axis directions reveals that easy-axis and easy-plane particle ensembles yield the same asymptotic behavior

$$M(H) = M_s\left(1 - 4K_1{}^2/15\mu_o M_s{}^2 H^2\right) \tag{9.11}$$

The behavior in small fields, that is, near remanence, is different, namely, $M(H) = M_s\,(1/2 + \mu_o M_s H/3\,K_1)$ for easy-axis anisotropy and $M(H) = M_s\,(\pi/4 + \mu_o M_s H/6\,K_1)$ for easy-plane anisotropy, but this region is less suitable for analysis than the high-field region, because hysteresis interferes. Figure 9.11 shows the corresponding curves.

9.4.3 *Curie Temperature Fitting*

Another fitting-related challenge is to determine the Curie temperature of nanocomposite materials. For example, melt-spun Co alloys tend to have some hcp Co background with a very high Curie temperature, and experiments are usually conducted in nonzero fields, which smoothens the Curie temperature singularity of the main phase [87]. To fit the experimental M–T data, one can use the implicit equation

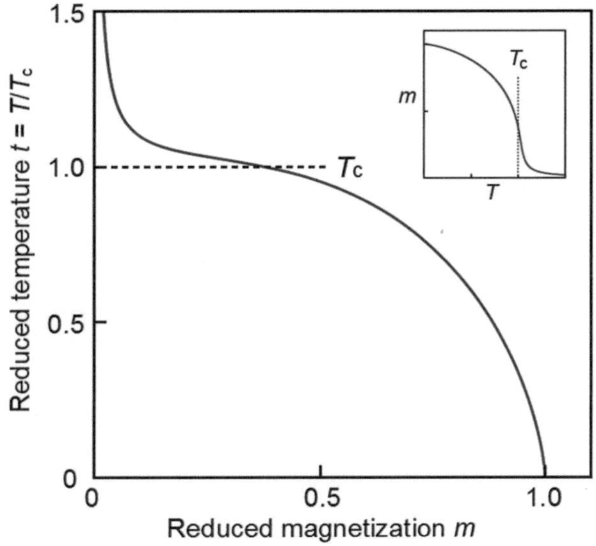

Fig. 9.12 Reduced temperature t as a function of the reduced magnetization m, where the inverse function *t(m)* is given by Eq. 9.12. The inset shows the original and more intuitive m(s) plot

$$t(m,h) = \left(1 - m^{1/\beta}\right)^{2/3} + h/m \tag{9.12}$$

Here, $m = (M - M_B)/(M_0 - M_B)$ is a normalized magnetization, M_B is the Co background contribution to the total magnetization, $t = T/T_c$, and M_0 is the main phase magnetization at $T = 0$. Finally, $h \approx H/H_{MF}$ is a magnetic field parameter, where H_{MF} denotes the molecular field. Analysis of low-temperature limit ($t \ll 1$), critical-point behavior ($t \approx 1$), and high-temperature limit ($t \gg 1$) shows that Eq. 9.12 reproduces Bloch's law $M = M_0(1 - \beta(T/T_c)^{3/2})$, the critical behavior $M \sim (1 - T/T_c)^\beta$ ($\beta \approx 0.30$ for the Heisenberg model), and the Curie–Weiss law $M \sim H/(T - T_c)$, respectively. Equation 9.12 is an interpolation formula rather than a rigid result, but it describes the critical region near T_c better than the mean-field approximation.

Since Eq. 9.12 is an inverse function, it is convenient to use the experimental T(M) curve instead of the more familiar M(T) curve. Figure 9.12 shows that the inverse function is counterclockwise rotated by 90° and then mirrored at the ordinate axis. The parameter β is essentially fixed, but h, T_c, M_0, and M_B need to be varied until Eq. 9.12 provides a good fit. Automated numerical least-square fitting using $T = T_c\, t(m, h)$ can be used, as can manual fitting until the desired accuracy is reached. In the latter case, it is convenient to start with estimates of M_0(low-temperature magnetization) and M_B (high-temperature limit), the latter being zero in the absence of a background. T_c is also estimated quite easily, whereas h may need a few tries. (The higher h, the smoother the curve.)

9.4.4 Thermal Excitations and Nanomagnetism

In a strict sense, micromagnetism is limited to zero temperature, so the question arises whether and how thermal excitations destroy the energy product. The leading finite-temperature effects have been known for a long time. First, the intrinsic temperature dependence of the micromagnetic parameters, especially of K_1, yields a direct temperature dependence of extrinsic properties, such as M_r and H_c. This is a big effect, but it is easily taken into account by using temperature-dependent intrinsic parameters, such as $M_s(T)$ and $K_1(T)$. Second, nonequilibrium thermal excitations yield waiting-time- and sweep-rate-dependent magnetic-viscosity corrections to M(H), affecting both remanence and coercivity. The early research on this aftereffect, including the physically correct explanation of the phenomenon as an Arrhenius effect, was reviewed by Becker and Döring in 1939 [132]. Third, there are minor corrections due to equilibrium excitations that correspond to nanoscale corrections to intrinsic properties [133].

These findings and their modern extensions have been reinvented multiple times, occasionally with improvements in details, and sometimes they have been incorrectly rejected. Such problems often arise if researchers from nearby fields enter nanomagnetism but are only partially aware of then specific scientific laws that govern the field.

The first thorough analysis of the nonlinear energy landscapes involved in magnetic viscosity

$$E_{\mathrm{a}} = E_{\mathrm{o}}(1 - H/\mathrm{H_c})^m \tag{9.13}$$

was performed by Néel in 1949 [134], and Brown used a Fokker–Planck equation to derive an exact solution for nanoparticles [135]. For this reason, it has become popular to refer to the 1930s approach as the Néel–Brown theory; a variant of this theory is that by Street and Woolley [136]. An expression for the magnetic-viscosity contribution to the temperature dependence of the coercivity was derived in the 1960s [137] and reinvented by Sharrock in 1994 [138]. A sound theoretical basis for Arrhenius-like activation mechanisms was established by Kramers in 1940 [139] and explicitly applied to the magnetic-viscosity problem by Skomski et al. [140]. A powerful numerical method is the elastic-band method, developed by Schrefl [141], which can be considered a computational extension of Kramer's theory. The basic features of the magnetic-viscosity research are [56, 59, 142]: (1) magnetic-viscosity effects are normally small corrections to the leading micromagnetic contributions, (2) the Arrhenius law $\Gamma = \Gamma_o \exp(-E_a/k_B T)$ implies an activation energy $E_a = 25 k_B T$ for a waiting time $1/\Gamma \approx 100$ s, (3) the energy barriers obey the above power law with m = 2 for symmetric energy landscapes and m = 3/2 for asymmetric energy landscapes, and (4) thermal activation does not alter the underlying magnetization reversal mechanism.

Over the years, there have been many attempts to "improve" our understanding of thermally activated magnetization processes by discarding the above rigid results. A naïve approach is to use $H_c = H_o \exp(E_a/k_BT)$, but putting $T = 0$ yields $H_c = \infty$, in striking contrast to experiment. In 1973, Egami [143] postulated a model where $E_a \sim 1/H$ so that $E_a = \infty$ for $H = 0$. In fact, this E_a is a zero-temperature domain wall propagation energy not associated with thermal activation; experiment shows that E_a remains finite even for $H = 0$, in agreement with Eq. 9.13.

Very recently, Hans-Benjamin Braun argued in an otherwise very interesting article [144] that magnetization reversal in long ellipsoids (nanowires) does not start by curling but by soliton–antisoliton nucleation, which is basically a new name for domain wall formation. Traditional micromagnetics includes domain wall formation, as well as related pinning-type effects that overcome the topological constraints emphasized by Braun, but only through spontaneous symmetry breaking that starts from a curling mode. In other words, traditional micromagnetism is dismissed through the fate of Buridan's ass. In this philosophical paradox, a donkey is placed precisely midway between two identical piles of hay—unable to choose between the two, the ungulate dies of hunger.

Braun discards the above exact findings as obsolete relicts from the "early days of zero-temperature micromagnetism," based on a linear theory that cannot be applied to magnetic nanostructures. His reasoning is intuitive: in very thin wires, thermal effects must be important, because there is no ferromagnetic long-range order in one-dimensional ferromagnets, which leads to the belief that thermal excitations directly create domain walls. He assumes that curling can be compared to a mountaineer who attempts to get into a neighboring valley via the least strenuous shallow path and decides only to find himself at the end of a basin surrounded by the highest peaks.

Brown's argument is wrong for four independent reasons. First, nobody seriously claims that curling occurs in very thin wires—it is limited to wires whose radius exceeds a coherence length of about 10 nm (Sect. 4.1). Second, as it is clear from Eq. 9.13, micromagnetic nucleation theory is not a linear theory. Figure 9.13 illustrates this point for an energy landscape with $m = 3/2$ [140]. Third, finite-temperature micromagnetism must converge to the correct $T = 0$ limit at low temperatures, and in micromagnetism, even T_c is a very low temperature. In this limit, thermal excitations lead to reversal paths very close to the path with the lowest energy barrier [59]. This is the main result of Kramer's escape rate theory [139, 145]. Fourth, the shallowness of the path toward the saddle point is included in Kramer's escape rate theory [139], where it leads to a relatively small activation entropy correction to Γ_o [140, 142, 145].

A fifth point, related to points three and four, concerns the relation between curling and direct domain formation. At zero temperature, magnetization reversal starts with curling, because all other mechanisms correspond to a more negative applied field and therefore to some energy difference ΔE. Curling is energetically favorable due to magnetostatic flux closure, which leads to the rough estimate $\Delta E = \mu_o M_s 2\pi R^2 \delta w/3$. Taking $\mu_o M_s = 1$ T, $\delta_w = 15$ nm, and $R = 10$ nm yields $\Delta E/k_B = 80{,}000$ K, or $\Delta E = 270\ k_B T_{RT}$. This energy is huge, albeit not unusual compared to magnetic-viscosity energies encountered in experimental systems,

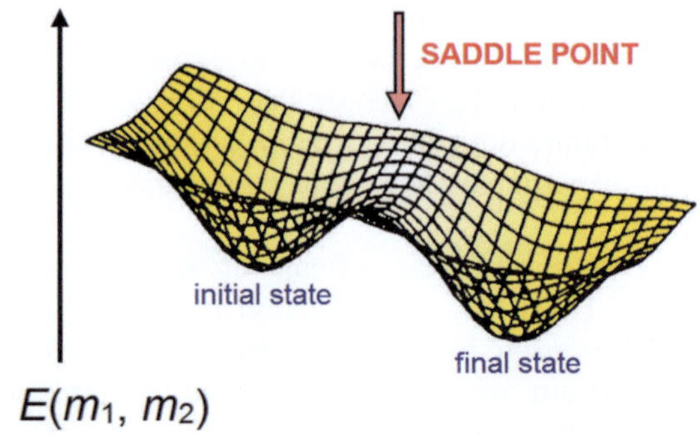

Fig. 9.13 Schematic magnetic energy landscape (after Ref. [140])

such as $Nd_2Fe_{14}B$ [62]. It means that direct domain wall formation has the character of a giant fluctuation [56, 142] and can safely be ignored.

The situation is different in ultrathin wires, such as monatomic chains, where thermal energies successfully compete against micromagnetic energies [142]. The difference becomes clear by considering the simpler case of equilibrium thermodynamics. It is well known that the spontaneous magnetization of one-dimensional magnets is zero for all temperatures $T > 0$, with additional problems due to quantum fluctuations at $T = 0$. This finding, which amounts to $T_c = 0$, includes nanowires, although experiment shows that nanowires have a T_c very close to the bulk value. This seeming contradiction has its roots in the different cross sections of monatomic and experimental wires [146].

Monatomic wires are best discussed in terms of the Ising model, $\mathcal{H} = -J\,\Sigma_i\, s_i\, s_{i+1}$, with atomic spins $s_i = \pm 1$. Since $J > 0$, the spins prefer to be parallel, but this does not lead to long-range order at $T > 0$. The easiest way to show this is to introduce bond variables $\tau_i = s_i s_{i+1}$, which can also assume the two values $\tau_i = \pm 1$ [147]. This procedure transforms the Hamiltonian into $\mathcal{H} = -J\ \Sigma_i\ \tau_i$, which is a sum of noninteracting or paramagnetic pseudospins τ_i, so that $T_c = 0$. Nevertheless, neighboring spins tend to yield parallel spin block of the type ↑↑↑↑ or ↓↓↓↓, and the average size ξ of these blocks (the correlation length) increases with increasing J and decreasing T. A single broken bond (↑↓) costs an energy of 2 J, corresponding to a Boltzmann probability $\exp(-2\ J/k_BT)$, and the average distance between broken bonds is therefore $\xi = a\ \exp(2\ J/k_BT)$, where a is the interatomic distance. Taking $J/k_B = 500$ K, $T = 300$ K, and $a = 0.25$ nm yields the fairly small correlation length $\xi = 5$ nm. For the Heisenberg model, this value is even smaller.

Let us next consider a Heisenberg square wire of cross section $L^2 = 5 \times 5\ nm^2$. The broken bond has now the character of a domain wall of energy $4(AK)^{1/2}L^2$. Taking $K = 0.4\ MJ/m^3$ and $A = 10$ pJ/m yields an energy of 15,000 K for a broken bond or $E/k_BT_{RT} = 50$ at room temperature. The corresponding correlation length, $\xi = a\ \exp(E/k_BT)$, is $1.3 \cdot 10^{12}$ m or about nine times the distance from the Earth to the Sun. This isn't long-range order in a strict sense, but it is long-range order from any practical viewpoint. The correlation length ξ strongly decreases with decreasing cross section, and for $L = 3$ nm, it is 64 million interatomic distances or about 1 cm.

9.5 Outlook

While it is impossible to precisely predict the future of permanent magnetism, it is in order to judge some trends. The search for new permanent magnet phases has been a cornerstone of research for many decades. It is unlikely that Nd–Fe–B will be replaced by non-rare-earth magnets in the foreseeable future, but at the lower end of the performance spectrum there are niches for alternative materials. However, simultaneously optimizing magnetization, Curie temperature, and anisotropy is a demanding challenge. High magnetic hardnesses ($\kappa > 1.2$) and high anisotropy fields are advantageous but need to be accompanied by a high magnetization. Proposals based on low magnetization, intrinsically or in nanostructures, ignore that the energy product $(BH)_{max} \leq \frac{1}{4}\mu_o M_s^2$ is the quintessential figure of merit in permanent magnetism. On the other hand, anisotropies of the order of 1–2 MJ/m^3 are a hardly "giant" or "surprising," and without a clear strategy concerning the microstructure, there is no hope of developing such materials into competitive permanent magnets.

Experimental, computational, and analytical methods all play an important role in the quest for new materials, and no method will probably be successful without the help of the others. One example is high-throughput simulations based on genetic algorithms. Considering up to four elements per system yields an astronomical number of quaternary phases, solid solutions, metastable nanostructures, and phase mixtures, and looking for a needle in a haystack is a euphemism for such a venture. As far as specific materials groups are concerned, one thrust is to search for derivatives of traditional Fe- and Co-based materials (Sects. 3.1 and 3.2). One example of a meaningful task is to search for materials that combine magnetocrystalline and shape anisotropies. The second thrust is to develop Mn-rich materials. Mn–Mn interactions are often antiferromagnetic due to band filling, but the strongly ferromagnetic exchange for very short Mn–Mn distances in MnAl is encouraging.

The exploration of new phases must be accompanied by micromagnetic optimization, with aim of maximizing the Kronmüller factor α. This research is material-specific and closely related to magnet processing. For example, bulk hcp Co exhibits quite good intrinsic properties (Table 9.2), but it has never been possible to develop a useful permanent from hcp Co. $L1_0$-ordered FeNi is much more forgiving micromagnetically, probably due to the coexistence of the three $L1_0$ variants.

Nanostructuring introduces an additional degree of freedom, although the processing of c-axis-aligned hard–soft nanostructure is very demanding and still in an exploratory stage. Note that isotropic two-phase nanostructures (esp. Nd–Fe–B plus Fe and/or Fe_3B) are much easier to produce but suffer from the strongly reduced magnetization of the hard phase, as shown in Fig. 9.12. Some rules to optimize the performance of aligned two-phase nanostructure have been outlined above. Soft-in-hard geometries are better than hard-in-soft geometries, and embedded soft spheres are better than sandwiched soft layers. However, in the latter case, it may be possible to exploit that the nanostructure of multilayers such as MnBi/FeCo resists demagnetization in the second quadrant [119]. This is actually an example of how magnetostatic and magnetocrystalline energies could be combined favorably. Concerning the choice of the hard phase, both a high magnetization and a high anisotropy are necessary. $SmCo_5$ is good in this regard, especially with respect to anisotropy [148], although it also exhibits processing-related challenges.

Thermal effects continue to be of interest in permanent magnetism, because many magnets operate above room temperature. One aim of current Nd–Fe–B research is to save Dy, and this includes the understanding of the positive role of Dy at temperatures around 150 °C [18]. Nanostructuring yields minor hysteresis loop corrections at nonzero temperatures but no substantial "low-dimensional" deterioration of the ferromagnetism. In fact, nanostructuring may actually be used to improve the Kronmüller factor as the temperature increases (Sect. 4.1).

Acknowledgment This chapter is partially based on original research supported by DOE BES (DE-FG02-04ER46152, Sect. 3), ARO (Nr. WF911NF-10-2-0099, Sect. 4), ARPA-E (PNNL/Maryland and Argonne/Delaware), DREaM (Ames), HCC, and NCMN. It has also benefitted from discussions and collaborations with B. Balamurugan, R. Choudhary, J. M. D. Coey, S. Constantinides, J. Cui, B. Das, A. Enders, G. C. Hadjipanayis, S. Hirosawa, Y. Jin, A. Kashyap, L.-Q. Ke, M. J. Kramer, L. H. Lewis, S.-H. Liou, J. P. Liu, Y. Liu, P. Kumar, P. Manchanda, R. W. McCallum, F. Pinkerton, T. Rana, S. G. Sankar, J. E. Shield, D. J. Sellmyer, S. Valloppilly, V. Sharma, I. Takeuchi, and W.-Y. Zhang.

Appendix: Units in Magnetism

It is generally recommended to use the international or SI system or transparent units differing by multiples of 10, such as Å = 100 pm. Some researchers, most notably in the USA and China, continue to us the cgs system, which was developed by Carl Friedrich Gauß around 1830. The British Association for the Advancement of Science officially endorsed and widely popularized the Gaussian system in 1874 but replaced it in 1889 by the MKS predecessor of the SI system.

In strict sense, today's Gaussian system is a "reduced" or dimensionless system as far as magnetism is concerned. The situation is similar to the atomic unit (a.u.) system, where all physical quantities are made dimensionless by division, using combinations of quantities such as Bohr's hydrogen radius $a_o = 0.529$ Å. Similar to "a.u.," "emu" is not a unit but a reminder that the moment is measured in a variant of the cgs system. The expression "emu/cm^3" is also such a reminder, albeit slightly differently structured by involving cm, which is a well-defined length unit. The Gaussian system exhibits some oddities that can never happen in a physically meaningful unit system. For example, multiplication of the magnetization by the dimensionless number 4π changes the units from emu/cm^3 to kG. In the SI, this problem does not occur, because the corresponding quantities are connected through the permeability of free space, $\mu_o = 4\pi \times 10^{-7}$ N/A^2. (N/A^2 can be written in a variety of equivalent SI forms, notably H/m, T · m/A, Wb/A · m and V · s/A · m.)

Note that electrostatic units (esu) are rarely used today, and few solid-state scientists can even recall the electron charge in esu units ($e = 4.803 \cdot 10^{10}$ esu).

As far as permanent magnetism is concerned, the only shortcoming of the SI system is that the magnetization is measured in A/m. This feature dates back to the nineteenth century, when scientists believed that the magnetization was caused by microscopic currents. We now know that this is incorrect: currents, or orbital moments, are largely quenched in materials like Fe and Co, where most of the magnetization is caused by the spin. Explaining the spin by local currents implies that the electron's charge distribution moves with a velocity larger than the velocity of light, which is not a meaningful physical concept. The role of μ_o in the conversion between A/m and T may be compared to the role of k_B in the conversion between temperature (K) and energy (J): a typical dust particle, of radius 1 μm and one millimeter above the ground, has a potential energy of about 10^{-16} J. There is nothing wrong with quoting this energy as a temperature, about 10^7 K, unless one believes that this temperature is actually the temperature of the dust particle.

The situation in permanent magnetism would be much easier if B, M, and H had the same unit (T). A seeming counterargument is that H and the flux density B are physically different and should therefore have different units, but the example of energy and torque, both measured in Nm, proves that different physical quantities do not need different units. $J = \mu_o H$ is sometimes used, but J also denotes exchange and the total angular momentum, which creates a messy situation in some contexts. Expressions such as $B_r = \mu_o M_r$ are common, but they obscure the situation as far as physics is concerned. A compromise, used in the present chapter, is to consider the magnetization $\mu_o M$ and the magnetic field $\mu_o H$, both measured in tesla (T). Here are some informal conversion rules for cgs and A/m aficionados: 1 T = 10 kG = 10 kOe, 1 T = 10/4π MA/m ≈ 800 kA/m, 1 emu/cm^3 = 1 kA/m, 1000 kA/m = 4π/10 T ≈ 1.25 T, 1 kA/m = 4π Oe ≈ 12.5 Oe, 1 MGOe = 100/4π kJ/m^3 ≈ 8 kJ/m^3, 1 kJ/m^3 = 4π/100 MGOe ≈ 0.125 MGOe, 1 kJ/m^3 = 1 kPa, 100 MGOe = 1 T^2, 1 kOe = 1000/4π kA/m ≈ 80 kA/m.

References

1. Bozorth, R.M.: Ferromagnetism. van Nostrand, Princeton (1951)
2. Chikazumi, S.: Physics of Magnetism. Wiley, New York (1964)
3. Skomski, R., Coey, J.M.D.: Permanent Magnetism. Institute of Physics, Bristol (1999)
4. Consider $\nabla \cdot (A \times H) = B \cdot H$, which follows from $B = \nabla \times A$ and $H = - \nabla \phi m$. Since $\int \nabla \cdot (A \times H)\, dV = \int (A \times H)\, dS$ and the respective A and H fields decay as 1/r2 and 1/r3 in infinity, the surface integral and therefore $\int B \cdot H\, dV$ are equal to zero
5. Yensen, T.D.: Development of magnetic material. Elec. J. **18**, 93–95 (1921)
6. Skomski, R., Sharma, V., Balamurugan, B., Shield, J.E., Kashyap, A., Sellmyer, D.J.: Anisotropy of doped transition-metal magnets. In: Kobe, S., McGuinness, P. (eds.) Proc. REPM'10, pp. 55–60. Jozef Stefan Institute, Ljubljana (2010)
7. Jellinghaus, W.: New alloys with high coercive force. Z. Tech. Physik **17**, 33–36 (1936)
8. Jin, S., Chin, G.Y.: Fe-Cr-Co magnets. IEEE Trans. Magn. **23**, 3187–3192 (1987)

9. Evetts, J.E. (ed.): Concise Encyclopedia of Magnetic and Superconducting Materials. Pergamon, Oxford (1992)
10. Klemmer, T., Hoydick, D., Okumura, H., Zhang, B., Soffa, W.A.: Magnetic hardening and coercivity in $L1_0$ ordered fepd ferromagnets. Scr. Met. Mater. **33**, 1793–1805 (1995)
11. Kooy, C., Enz, U.: Experimental and theoretical study of the domain configuration in thin layers of $BaFe_{12}O_{19}$. Philips Res. Rep. **15**, 7–29 (1960)
12. Strnat, K., Hoffer, G., Olson, J., Ostertag, W., Becker, J.J.: A family of new cobalt-base permanent magnet materials. J. Appl. Phys. **38**, 1001–1002 (1967)
13. Kumar, K.: $RETM_5$ and RE_2TM_{17} permanent magnets development. J. Appl. Phys. **63**, R13–57 (1988)
14. Sagawa, M., Fujimura, S., Yamamoto, H., Matsuura, Y.: Permanent magnet materials based on the rare earth-iron-boron tetragonal compounds. IEEE Trans. Magn. **20**, 1584–1589 (1984)
15. Sagawa, M., Hirosawa, S., Yamamoto, H., Fujimura, S., Matsuura, Y.: Nd-Fe-B permanent magnet materials. Jpn. J. Appl. Phys. **26**, 785–800 (1987)
16. Herbst, J.F.: $R_2Fe_{14}B$ materials: intrinsic properties and technological aspects. Rev. Mod. Phys. **63**, 819–898 (1991)
17. Coey, J.M.D., Sun, H.: Improved magnetic properties by treatment iron-based rare-earth intermetallic compounds in ammonia. J. Magn. Magn. Mater. **87**, L251–L254 (1990)
18. Moriya, H., Tsuchiura, H., Sakuma, A.: First-principles calculation of crystal field parameter near surfaces of Nd2Fe14B. J Appl Phys **105**, 07A740-1-3 (2009)
19. Sugimoto, S.: An overview of the Dy-saving project in Japan. In: Kobe, S., McGuinness, P. (eds.) Proc. REPM'10, pp. 103–105. Jozef Stefan Institute, Ljubljana (2010)
20. Sugimoto, S.: Current status and recent topics of rare-earth permanent magnets. J. Phys. D: Appl. Phys. **44**, 064001–1-11 (2011)
21. Skomski, R., Kashyap, A., Enders, A.: Is the magnetic anisotropy proportional to the orbital moment?. J. Appl. Phys. **109**, 07E143-1-3 (2011)
22. Tanaka, S., Moriya, H., Tsuchiura, H., Sakuma, A., Diviš, M., Novák, P.: First principles study on the local magnetic anisotropy near surfaces of Dy2Fe14B and Nd2Fe14B magnets. J. Appl. Phys. **109**, 07A702-1-3 (2011)
23. Nakamura, T., Yasui, A., Kotani, Y., Fukagawa, T., Nishiuchi, T., Iwai, H., Akiya, T., Ohkubo, T., Gohda,Y., Hono, K., Hirosawa, S.: Direct observation of ferromagnetism in grain boundary phase of Nd-Fe-B sintered magnet using soft x-ray magnetic circular dichroism. Appl. Phys. Lett. **105**, 202404–1-4 (2014)
24. Brown, D. N., Wu, Z., He, F., Miller, D. J., Herchenroeder, J.W.: Dysprosium-free melt-spun permanent magnets. J. Phys. Condens. Matter. **26**, 064202–1-8 (2014)
25. Coey, J.M.D.: Hard magnetic materials: Aperspective. IEEE Trans. Magn. **49**, 4671–4681 (2011)
26. Gutfleisch, O., Willard, M.A., Brück, E., Chen, C.H., Sankar, S.G., Liu, J.P.: Magnetic materials and devices for the 21st century: Stronger, lighter, and more energy efficient. Adv. Mater. **23**, 821–842 (2011)
27. Skomski, R., Manchanda, P., Kumar, P., Balamurugan, B., Kashyap, A., Sellmyer, D.J.: Predicting the future of permanent-magnet materials (invited). IEEE Trans. Magn. **49**, 3215–3220 (2013)
28. McCallum, R.W., Lewis, L.H., Skomski, R., Kramer, M.J., Anderson, I.E.: Practical aspects of modern and future permanent magnets. Ann. Rev. Mater. Res. **44**, 451–477 (2014)
29. Namai, A., Yoshikiyo, M., Yamada, K., Sakurai, Sh., Goto, T., Yoshida, T., Miyazaki, T., Nakajima, M., Suemoto, T., Tokoro, H., Ohkoshi, Sh.-I.: Hard magnetic ferrite with a gigantic coercivity and high frequency millimetre wave rotation. Nature Comm. **3**, 1035–1-6 (2012).
30. Kneller, E.F., Hawig, R.: The exchange-spring magnet: a new material principle for permanent magnets. IEEE Trans. Magn. **27**, 3588–3600 (1991)

31. Skomski, R.: Nucleation in Inhomogeneous permanent magnets. Phys. Stat. Sol. B **174**, K77–80 (1992)
32. Skomski, R., Coey, J.M.D.: Giant energy product in nanostructured two-phase magnets. Phys. Rev. B **48**, 15812–15816 (1993)
33. Goto, E., Hayashi, N., Miyashita, T., Nakagawa, K.: Magnetization and switching characteristics of composite thin magnetic films. J. Appl. Phys. **36**, 2951–2958 (1965)
34. Kronmüller, H.: Theory of nucleation fields in inhomogeneous ferromagnets. Phys. Stat. Sol. B. **144**, 385–396 (1987)
35. Nieber, S., Kronmüller, H.: Nucleation fields in periodic multilayers. Phys. Stat. Sol. B. **153**, 367–375 (1989)
36. Jones, N.: The pull of stronger magnets. Nature **472**, 22–23 (2011)
37. Fullerton, E.E., Jiang, S.J., Bader, S.D.: Hard/soft heterostructures: Model exchange-spring magnets. J. Magn. Magn. Mater. **200**, 392–404 (1999)
38. Jiang, J.S., Pearson, J.E., Liu, Z.Y., Kabius, B., Trasobares, S., Miller, D.J., Bader, S.D., Lee, D.R., Haskel, D., Srajer, G., Liu, J.P.: Improving exchange-spring nanocomposite permanent magnets. Appl. Phys. Lett. **85**, 5293–5295 (2004)
39. Zhang, J, Takahashi, Y.K., Gopalan, R., Hono K.: Sm(Co, Cu)5/Fe exchange spring multilayer films with high energy product. Appl. Phys. Lett. **86**, 122509–1-2 (2005)
40. Toga, Y., Moriya, H., Tsuchiura, H., Sakuma, A.: First principles study on interfacial electronic structures in exchange-spring magnets. J. Phys. Conf. Ser. **266**, 012046–1-5 (2011)
41. Neu, V., Sawatzki, S., Kopte, M., Mickel, C., Schultz, L.: Fully epitaxial, exchange coupled SmCo/Fe multilayers with energy densities above 400 kJ/m^3. IEEE Trans. Magn. **48**, 3599–3602 (2012)
42. Sahota, P. K., Liu, Y., Skomski, R., Manchanda, P., Zhang, R., Fangohr, H., Franchin, M., Hadjipanayis, G. C., Kashyap, A., Sellmyer D. J.: Ultrahard magnetic nanostructures. J. Appl. Phys. **111**, 07E345-1-3 (2012)
43. Poudyal N., Liu, J. P.: Advances in nanostructured permanent magnets research. J. Phys. D. Appl. Phys. **46**, 043001–1-23 (2013)
44. Jiang, J. S., Bader, S. D: Rational design of the exchange-spring permanent magnet. J. Phys. Condens. Matter. **26**, 064214–1-9 (2014)
45. Liu, P., Luo, C.P., Liu, Y., Sellmyer, D.J.: High energy products in rapidly annealed nanoscale Fe/Pt multilayers. Appl. Phys. Lett. **72**, 483–485 (1998)
46. Liu, Y., George, T. A., Ralph Skomski., Sellmyer, D. J.: Aligned and exchange-coupled FePt-based films. Appl. Phys. Lett. **99**, 172504–1-3 (2011)
47. Roy, D., Anil Kumar, P. S.: Enhancement of (BH)max in a hard-soft-ferrite nanocomposite using exchange spring mechanism. J. Appl. Phys. **106**, 073902–1-4 (2009)
48. Cui, W.-B., Takahashi, Y.K., Hono, K.: $Nd_2Fe_{14}B$/FeCo anisotropic nanocomposite films with a large maximum energy product. Adv. Mater. **24**, 6530–6535 (2012)
49. Skomski, R.: Aligned two-phase magnets: permanent magnetism of the future? J. Appl. Phys. **76**, 7059–7064 (1994)
50. Coehoorn, R., de Mooij, D. B., Duchateau, J. P. W. B., Buschow, K. H. J.: Novel permanent magnetic materials made by rapid quenching. J. Physique. **49**, C-8, 669–670 (1988)
51. Schneider, J., Eckert, D., Müller, K.-H., Handstein, A., Mühlbach, H., Sassik, H., Kirchmayr, H.R.: Magnetization processes in $Nd_4Fe_{77}B_{19}$ permanent magnetic materials. Mater. Lett. **9**, 201–203 (1990)
52. Manaf, A., Buckley, R.A., Davies, H.A.: New nanocrystalline high-remanence Nd-Fe-B alloys by rapid solidification. J. Magn. Magn. Mater. **128**, 302–306 (1993)
53. Skomski, R.: Spin-glass permanent magnets. J. Magn. Magn. Mater. **157–158**, 713–714 (1996)
54. Osborn, J.A.: Demagnetizing factors of the general ellipsoid. Phys. Rev. **67**, 351–357 (1945)
55. Note that the total magnetostatic self-energy, or dipole-dipole energy, –½μo∫ M · H dV = ½μo D M2 V, is the sum of Ea = ½μo D (1 – D) M2 V and Ei = ½μo D2 M2 V
56. Skomski, R.: Simple Models of Magnetism. Oxford University Press, Oxford (2008)

57. Bloch, F., Gentile, G.: Zur anisotropie der magnetisierung ferromagnetischer einkristalle. Z. Phys. **70**, 395–408 (1931)
58. Skomski, R., Kashyap, A., Solanki, A., Enders, A., Sellmyer, D. J.: Magnetic anisotropy in itinerant magnets. J. Appl. Phys. **107**, 09A735-1-3 (2010)
59. Skomski, R.: Nanomagnetics. J. Phys. Condens. Matter. **15**, R841–R896 (2003)
60. Brown, W.F.: Micromagnetics. Wiley, New York (1963)
61. Aharoni, A.: Theoretical search for domain nucleation. Rev. Mod. Phys. **34**, 227–238 (1962)
62. Givord, D., Rossignol, M.F.: Coercivity. In: Coey, J.M.D. (ed.) Rare-Earth Iron Permanent Magnets, pp. 218–285. University Press, Oxford (1996)
63. Kronmüller, H., Yang, J. B., Goll, D.: Micromagnetic analysis of the hardening mechanisms of nanocrystalline MnBi and nanopatterned FePt intermetallic compounds. J. Phys. Condens. Matter. **26**, 064210–1-7 (2014)
64. McCurrie, R.A.: Ferromagnetic materials—structure and properties. Academic, London (1994)
65. Skomski, R., Liu, Y., Shield, J. E., Hadjipanayis, G. C., Sellmyer, D. J.: Permanent magnetism of dense-packed nanostructures. J. Appl. Phys. **107**, 09A739-1-3 (2010)
66. Zhou, L., Miller, M.K., Lu, P., Ke, L., Skomski, R., Dillon, H., Xing, Q., Palasyuk, A., McCartney, M.R., Smith, D.J., Constantinides, S., McCallum, R.W., Anderson, I.E., Antropov, V., Kramer, M.J.: Architecture and magnetism of alnico. Acta Mater. **74**, 224–233 (2014)
67. Fast, J.D.: Gases in Metals. Macmillan, London (1976)
68. Burkert, T., Nordström, L., Eriksson, O., Heinonen, O.: Giant magnetic anisotropy in tetragonal FeCo alloys. Phys. Rev. Lett. **93**, 027203–1-4 (2004)
69. Andersson, G., Burkert, T., Warnicke, P., Björck, M., Sanyal, B., Chacon, C., Zlotea, C., Nordström, L., Nordblad, P., Eriksson, O.: Perpendicular magnetocrystalline anisotropy in tetragonally distorted Fe-Co alloys. Phys. Rev. Lett. **96**, 037205–1-4 (2006)
70. Jack, K.W.: The iron—nitrogen system: The crystal structures of ε-phase iron nitrides. Acta Crystallogr. **5**, 404–411 (1952)
71. Coey, J.M.D., O'Donnell, K., Qinian, Q., Touchais, E., Jack, K.H.: The magnetization of α "Fe16N2". J. Phys. Condens. Matter. **6**, L23–L28 (1994)
72. Kim, T.K., Takahashi, M.: New magnetic material having ultrahigh magnetic moment. Appl. Phys. Lett. **20**, 492–494 (1972)
73. Takahashi, H., Igarashi, M., Kaneko, A., Miyajima, H., Sugita, Y.: Perpendicular uniaxial magnetic anisotropy of $Fe_{16}N_2$(001) single crystal films grown by molecular beam epitaxy. IEEE Trans. Magn. **35**, 2982–2984 (1999)
74. Al-Omari, I.A., Skomski, R., Thomas, R.A., Leslie-Pelecky, D., Sellmyer, D.J.: High-temperature magnetic properties of mechanically alloyed $SmCo_5$ and YCo_5 magnets. IEEE Trans. Magn. **37**, 2534–2536 (2001)
75. Buschow, K.H.J.: Differences in magnetic properties between amorphous and crystalline alloys. J. Appl. Phys. **53**, 7713–7716 (1982)
76. Buschow, K.H.J.: New developments in hard magnetic materials. Rep. Prog. Phys. **54**, 1123–1213 (1991)
77. Demczyk, B.G., Cheng, S.F.: Structures of Zr_2Co_{11} and $HfCo_7$ intermetallic compounds. J. Appl. Crystallogr. **24**, 1023–1026 (1991)
78. Ivanova, G.V., Shchegoleva, N.N., Gabay, A.M.: Crystal structure of Zr_2Co_{11} hard magnetic compound. J. Alloys Comp. **432**, 135–141 (2007)
79. Das, B., Balamurugan, B., Kumar, P., Skomski, R., Shah, V.T., Shield, J.E., Kashyap, A., Sellmyer, D.J.: $HfCo_7$-based rare-earth-free permanent-magnet alloys. IEEE Trans. Magn. **49**, 3330–3333 (2013)
80. Zhao, X., Nguyen, M. C., Zhang, W. Y., Wang, C. Z., Kramer, M. J., Sellmyer, D. J., Li, X. Z., Zhang, F., Ke, L. Q., Antropov, V. P., Ho, K. M.: Exploring the structural complexity of intermetallic compounds by an adaptive genetic algorithm. Phys. Rev. Lett. **112**, 045502–1-5 (2014)

81. Zhao, X., Ke, L. Q., Nguyen, M. C., Wang, C.-Zh., Ho, K.-M.: Structures and magnetic properties of Co-Zr-B magnets studied by first-principles calculations. J. Appl. Phys. **117**, 243902–1-6 (2015)
82. Kumar, P., Kashyap, A., Balamurugan, B., Shield, J. E., Sellmyer, D. J., Skomski, R.: Permanent magnetism of intermetallic compounds between light and heavy transition-metal elements. J. Phys. Condens. Matter. **26**, 064209–1-8 (2014)
83. Das, B., Balamurugan, B., Zhang, W. Y., Skomski, R., Krage, E. S., Valloppilly, S. R., Shield, J. E., Sellmyer, D. J.: Magnetism of less common cobalt-rich alloys. Proc. REPM'12, Nagasaki, pp. 427–430 (2012)
84. Balamurugan, B., Das, B., Shah, V. R., Skomski, R., Li, X. Z., Sellmyer, D. J.: Assembly of uniaxially aligned rare-earth-free nanomagnets. Appl. Phys. Lett. **101**, 122407–1-5 (2012)
85. Balamurugan, B., Das, B., Skomski, R., Zhang, W.-Y., Sellmyer, D.J.: Novel nanostructured rare-earth-free magnetic materials with high energy products. Adv. Mater. **25**, 6090–6093 (2013)
86. Balamurugan, B., Mukherjee, P., Skomski, R., Manchanda, P., Das, B., Sellmyer, D. J.: Magnetic nanostructuring and overcoming Brown's paradox to realize extraordinary high-temperature energy products. Sci. Rep. **4**, 6265–1-6 (2014)
87. Balamurugan, B., Das, B., Zhang, W. -Y., Skomski, R., Sellmyer, D. J.: Hf-Co and Zr-Co alloys for rare-earth-free permanent magnets. J. Phys. Condens. Matter **26**, 064204–1-8 (2014)
88. Golkar, F., Kramer, M. J., Zhang, Y., McCallum, R. W., Skomski, R., Sellmyer, D. J., Shield, J. E.: Structure and magnetic properties of Co-W clusters produced by inert gas condensation. J. Appl. Phys. **111**, 07B524-1-3 (2012)
89. Zhang, W.-Y., Li, X.Z., Valloppilly, S., Skomski, R., Sellmyer, D.J.: Effect of annealing on nanostructure and magnetic properties of Zr_2Co_{11} material. Mater. Sci. Eng. **B186**, 64–67 (2014)
90. Zhang W. -Y., et al.: in preparation (2015)
91. Sellmyer, D. J., Balamurugan, B., Das, B., Mukherjee, P., Skomski, R., Hadjipanayis, G. C.: Novel structures and physics of nanomagnets (invited). J. Appl. Phys. **117**, 172609–1-6 (2015)
92. Balasubramanian, B., Skomski, R., Li, X.-Z., Valloppilly, S.R., Shield, J.E., Hadjipanayis, G. C., Sellmyer, D.J.: Cluster synthesis and direct ordering of rare-earth transition-metal nanomagnets. Nano Lett. **11**, 1747–1752 (2011)
93. Balamurugan, B., Skomski, R., Li, X. Z., Shah, V. R., Hadjipanayis, G. C., Shield, J. E., Sellmyer, D. J.: Magnetism of cluster-deposited Y-Co nanoparticles. J. Appl. Phys. **109**, 07A707-1-3 (2011)
94. Mukherjee, P., Manchanda, P., Kumar, P., Zhou, L., Kramer, M.J., Kashyap, A., Skomski, R., Sellmyer, D.J., Shield, J.E.: Size-induced chemical and magnetic ordering in individual Fe-Au nanoparticles. ACS Nano **8**, 8113–8120 (2014)
95. Harris, V. G., Chen, Y., Yang, A., Yoon, S., Chen, Z., Geiler, A. L., Gao, J., Chinnasam, C. N., Lewis, L. H., Vittoria, C., Carpenter, E. E., Carroll, K. J., Goswami, R., Willard, M. A., Kurihara, L., Gjoka, M., Kalogirou, O.: High coercivity cobalt carbide nanoparticles processed via polyol reaction: a new permanent magnet material. J. Phys. D. Appl. Phys. **43**, 165003–1-7 (2010)
96. Gandha, K., Elkins, K., Poudyal, N., Liu, X., Liu, J. P.: High energy product developed from cobalt nanowires. Sci. Rep. **4**, 5345–1-5 (2014)
97. Skomski, R.: Phase formation in L10 magnets. J. Appl. Phys. **101**, 09N517-1-3 (2007)
98. McHenry, M.E., Ramalingum, B., Willoughby, S., MacLaren, J., Sankar, S.G.: First principles calculations of the electronic structure of $Fe_{1-x}Co_xPt$. IEEE Trans. Magn. **37**, 1277–1279 (2001)
99. Manchanda, P., Skomski, R., Shield, J. E., Constantinides, S., Kashyap, A.: Intrinsic magnetic properties of L10-based Mn-Al and Fe-Co-Pt Alloys. Proc. REPM'12, pp. 115–118 (2012)

100. Brown, G., Kraczek, B., Janotti, A., Schulthess, T. C., Stocks, G. M., Johnson, D. D.: Competition between ferromagnetism and antiferromagnetism in FePt. Phys. Rev. B. **68**, 052405–1-4 (2003)
101. Skomski, R., Kashyap, A., Zhou, J.: Atomic and micromagnetic aspects of $L1_0$ magnetism. Scr. Mater. **53**, 391–396 (2005)
102. Cheng Lai, Y., Chang, Y.H., Chen, G.-J., Chiu, K.-F., Chen, Y.-C.: Abnormal enhancement of ordered phase in sputter-deposited (Fe1-xCox)59Pt41 thin films. Mater. Trans. **47**, 2086–2091 (2006)
103. Choudhary, R., Kumar, P., Manchanda, P., Liu, Y., Kashyap, A., Sellmyer, D. J., Skomski, R.: Atomic magnetic properties of Pt-Lean FePt and CoPt derivatives. Proc. REPM'14, Annapolis, p. 289–291 (2014)
104. Kono, H.: On the ferromagnetic phase in Mn-Al System. J. Phys. Soc. Jpn. **13**, 1444–1451 (1958)
105. Jiménez-Villacorta, F., Marion, J. L., Sepehrifar, T., Daniil, M., Willard, M. A., Lewis, L. H.: Exchange anisotropy in the nanostructured MnAl system. Appl. Phys. Lett. **100**, 112408–1-4 (2012)
106. Chaturvedi, A., Yaqub, R., Baker, I.: A comparison of τ-MnAl particulates produced via different routes. J. Phys. Condens. Matter. **26**, 064201–1-7 (2014)
107. Pasko, A., LoBue, M., Fazakas, E., Varga, L. K., Mazaleyrat, F.: Spark plasma sintering of Mn-Al-C hard magnets. J. Phys. Condens. Matter **26,** 064203–1-7 (2014)
108. Manchanda, P., Kumar, P., Kashyap, A., Lucis, M.J., Shield, J.E., Mubarok, A., Goldstein, J., Constantinides, S., Barmak, K., Lewis, L.-H., Sellmyer, D.J., Skomski, R.: Intrinsic properties of Fe-substituted $L1_0$ magnets. IEEE Trans. Magn. **49**(10), 5194–5198 (2013)
109. Lewis, L. H., Barmak, K., Goldstein, J. G., Pinkerton, F., Skomski, R.: Towards stabilization of L10-type FeNi compounds for permanent magnet applications. Proc. REPM'12, Nagasaki, p. 102–105 (2012)
110. Lewis, L. H., Mubarok, A., Poirier, E., Bordeaux, N., Manchanda, P., Kashyap, A., Skomski, R., Goldstein, J., Pinkerton, F. E., Mishra, R. K., Kubic R. C., Jr., Barmak, K.: Inspired by nature: investigating tetrataenite for permanent magnet applications. J. Phys. Condens. Matter **26**, 064213–1-10 (2014)
111. Lewis, L. H., Pinkerton, F. E., Bordeaux, N., Mubarok, A., Poirier, E., Goldstein, J. I., Skomski, R. Barmak, K.: De magnete et meteorite: Cosmically motivated materials. IEEE Magn. Lett. **5**, 5500104–1-4 (2014)
112. Coey, J.M.: Magnetism and Magnetic Materials. University Press, Cambridge (2010)
113. Coey, J. M. D.: New permanent magnets; manganese compounds. J. Phys. Condens. Matter **26**, 064211–1-6 (2014)
114. Heusler, F.: Über manganbronze und über die synthese magnetisierbarer legierungen aus unmagnetischen metallen. Z. Angew. Chem. **17**, 260–264 (1904)
115. Goodenough, J.B.: Magnetism and the Chemical Bond. Wiley, New York (1963)
116. Kharel, P., Skomski, R., Lukashev, P., Sabirianov, R., Sellmyer, D. J.: Spin correlations and Kondo effect in a strong ferromagnet. Phys. Rev. B. **84**, 014431–1-5 (2011)
117. Kang, K., Lewis, L. H., Moodenbaugh, A. R.: Alignment and analyses of MnBi/Bi nanostructures. Appl. Phys. Lett. **87**, 062505–1-3 (2005)
118. Cui, J., Choi, J. P., Li, G., Polikarpov, E., Darsell, J., Overman, N., Olszta, M., Schreiber, D., Bowden, M., Droubay, T., Kramer, M. J., Zarkevich, N. A., Wang, L. L., Johnson, D. D., Marinescu, M., Takeuchi, I., Huang, Q. Z., Wu, H., Reeve, H., Vuong, N. V., Liu, J. P.: Thermal stability of MnBi magnetic materials. J. Phys. Condens. Matter **26**, 064212–1-10 (2014)
119. Skomski, R., Manchanda, P., Takeuchi, I., Cui, J.: Geometry dependence of magnetization reversal in nanocomposite alloys. J. Metals. **66**, 1144–1150 (2014)
120. Heusler, F.: Über magnetische manganlegierungen. Verhandl. Deut. Physik. Ges. **5**, 219–223 (1903)
121. Kharel, P., Huh, Y., Al-Aqtash, N., Shah, V. R., Sabirianov, R. F., Skomski, R., Sellmyer, D. J.: Structural and magnetic transitions in cubic Mn3Ga. J. Phys. Condens. Matter **26**, 126001–1-8 (2014)

122. Liu, J.P., Skomski, R., Liu, Y., Sellmyer, D.J.: Temperature dependence of magnetic hysteresis of RCo_x:Co nanocomposites (R = Pr and Sm). J. Appl. Phys. **87**, 6740–6742 (2000)
123. Lyubina, J., Müller, K.-H., Wolf, M., Hannemann, U.: A two-particle exchange interaction model. J. Magn. Magn. Mater. **322**, 2948–2955 (2010)
124. Skomski, R., Liu, J.-P., Sellmyer, D.J.: Quasicoherent nucleation mode in two-phase nanomagnets. Phys. Rev. B **60**, 7359–7365 (1999)
125. Aharoni, A.: Introduction to the Theory of Ferromagnetism. University Press, Oxford (1996)
126. Skomski, R.: Micromagnetic localization. J. Appl. Phys. **83**, 6503–6505 (1998)
127. Skomski, R., Coey, J.M.D.: Exchange coupling and energy product in random two-phase aligned magnets. IEEE Trans. Magn. **30**(2), 607–609 (1994)
128. Schrefl, T., Fidler, J.: Micromagnetic simulation of magnetizability of nanocomposite Nd–Fe–B magnets. J. Appl. Phys. **83**, 6262–6264 (1998)
129. Szlaferek, A.: Model exchange-spring nanocomposite magnetic grains. Phys. Stat. Sol. B. **241**, 1312–1315 (2004)
130. Skomski, R.: Optimum hard-soft geometries: science, wishful thinking, and technology. (invited). Proc. REPM'14, Annapolis, p. 129–132 (2014)
131. Skomski, R., Hadjipanayis, G. C., Sellmyer, D. J.: Graded permanent magnets. J. Appl. Phys. **105**, 07A733-1-3 (2009)
132. Becker, R., Döring, W.: Ferromagnetismus. Springer, Berlin (1939)
133. Skomski, R., Kumar, P., Hadjipanayis, G.C., Sellmyer, D.J.: Finite-temperature micromagnetism. IEEE Trans. Magn. **49**(7), 3229–3232 (2013)
134. Néel, L.: Théorie du trainage magnétique des ferromagnétiques en grains fins avec applications aux terres cuites. Ann. Geophys. **5**, 99–136 (1949)
135. Brown, W.F.: Thermal fluctuations of a single-domain particle. Phys. Rev. **130**, 1677–1686 (1963)
136. Street, R., Woolley, J.C.: A study of magnetic viscosity. Proc. Phys. Soc. A **62**, 562–572 (1949)
137. Kneller, E.: Theorie der magnetisierungskurve kleiner kristalle. In: Wijn, H.P.J. (ed.) Handbuch der Physik XIII/2: Ferromagnetismus, pp. 438–544. Springer, Berlin (1966)
138. Sharrock, M.P.: Time dependence of switching fields in magnetic recording media. J. Appl. Phys. **76**, 6413–6418 (1994)
139. Kramers, H.A.: Brownian motion in a field of force and the diffusion model of chemical reactions. Physica **7**, 284–304 (1940)
140. Skomski, R., Kirby, R.D., Sellmyer, D.J.: Activation entropy, activation energy, and magnetic viscosity. J. Appl. Phys. **85**, 5069–5071 (1999)
141. Dittrich, R., Schrefl, T., Kirschner, M., Suess, D., Hrkac, G., Dorfbauer, F., Ertl, O., Fidler, J.: Thermally induced vortex nucleation in permalloy elements. IEEE Trans. Magn. **41**(10), 3592–3594 (2005)
142. Skomski, R.: Role of thermodynamic fluctuations in magnetic recording (invited). J. Appl. Phys. **101,** 09B104-1-6 (2007)
143. Egami, T.: Theory of intrinsic magnetic after-effect i. thermally activated process. Phys. Stat. Sol. A. **19**, 747–758 (1973)
144. Braun, H.-B.: Topological effects in nanomagnetism: from superparamagnetism to chiral quantum solitons. Adv. Phys. **61**, 1–116 (2012)
145. Hänggi, P., Talkner, P., Borkovec, M.: Reaction-rate theory: fifty years after Kramers. Rev. Mod. Phys. **62**, 251–341 (1990)
146. Shen, J., Skomski, R., Klaua, M., Jenniches, H., Manoharan, S.S., Kirschner, J.: Magnetism in one dimension: Fe on Cu(111). Phys. Rev. B **56**, 2340–2343 (1997)
147. This procedure works in one dimension only, because the τ_i cannot be defined unambiguously in two and three dimensions
148. Cui, B. Z., Gabay, A. M., Li, W. F., Marinescu, M., Liu, J. F., Hadjipanayis, G. C.: Anisotropic $SmCo_5$ nanoflakes by surfactant-assisted high energy ball milling. J. Appl. Phys. **107**, 09A721-1-3 (2010)

Chapter 10
Bulk Metallic Glasses and Glassy/Crystalline Materials

Dmitri V. Louzguine-Luzgin

10.1 Introduction

Natural glasses are formed in various materials, for example, oxides, and polymers, while commercial metallic alloys have a crystalline structure either after slow or rapid cooling on casting. Metallic glassy alloys from the melt were first produced in Au–Si system [1] by using a rapid solidification technique at a very high cooling rate of 10^6 K/s. Pd–Cu–Si and Pd–Ni–P system alloys were first macroscopic metallic glassy articles produced in the shape of 1–2 mm diameter rods [2]. Larger-size Pd–Ni–P samples were obtained later after flux treatment which helps to suppress heterogeneous nucleation of crystals [3].

However, these noble-metal based alloys did not attract significant attention of the materials research community until the breakthrough achieved in the end of the past century. At that time various large-scale bulk metallic glassy (BMG) alloys (also called bulk metallic glasses with the same abbreviation) arbitrarily defined as three-dimensional massive glassy (amorphous) objects with a size of not less than 1 mm in each spatial dimension (10 mm by other definition) were produced [4, 5] and at present attract significant attention of the scientific community. The high glass-forming ability (GFA) of some alloys allowed formation of bulk metallic glasses up to about 10^2 mm in size (the dimension limiting cooling rate) (Fig. 10.1) by using various mold casting and water cooling processes [6, 7]. As a result BMG alloys were obtained in a variety of alloy systems, including Rare-Earth (RE) metals-, Mg-, Zr-, Ti-, Fe-, Co-, Pd-, Pt-, Au-, Ag-, Cu-, Ni- and Ca-based

A Chapter for "Novel Functional Magnetic Materials: Fundamentals and Applications", Springer book.

D.V. Louzguine-Luzgin (✉)
WPI Advanced Institute for Materials Research, Tohoku University,
2-1-1 Katahira, Aoba-Ku, Sendai 980-8577, Japan
e-mail: dml@wpi-aimr.tohoku.ac.jp

A. Zhukov (ed.), *Novel Functional Magnetic Materials*, Springer Series in Materials Science 231, DOI 10.1007/978-3-319-26106-5_10

Fig. 10.1 Optical image of a Zr–Cu-Fe–Al BMG sample

system alloys [8, 9]. Large samples of hard magnetic Nd-based alloys [10] and soft ferromagnetic Fe-based [11, 12] and Co-based [13] BMGs samples were also produced.

10.2 Formation of a Glassy Phase from the Melt

The transition from liquid to glassy state taking place at the glass transition temperature (T_g) is observed universally in various types of liquids, including molecular liquids, ionic liquids, metallic liquids, oxides, and chalcogenides [14, 15]. The glass-transition phenomenon is characterized by the change in specific heat capacity (C_p) and in the volumetric thermal expansion coefficient as one can observe in Fig. 10.2. Such changes on glass-transition were clearly demonstrated in many works and the transition takes place in a temperature interval [16, 17]. An arbitrary glass-transition point is defined as a temperature at which viscosity of a liquid reaches 10^{12} Pa·s on cooling. For some materials this value belongs to the glass-transition region defined by C_p measurement but it is not the case for other materials [18].

Metallic glassy alloys with low glass-forming ability can be produced by condensation from a vapor phase [19, 20] or by a solid state reaction using mechanical attrition [21], for example, ball milling [22] though such materials often contain residual nanocrystalline particles. Glassy or amorphous powder samples produced by ball milling or gas atomization technique [23]' can be consolidated into bulk form using hot pressing [24], spark plasma sintering (SPS) [25] or microwave furnace sintering [26] processes to obtain bulk samples owing to good viscous flow formability in the supercooled liquid region [27]. Electrodeposition is another method for producing glassy alloys from a solution in terms of glassy coatings [28, 29] while the structural inhomogeneity may be an issue.

Good metallic glass formers like BMGs possess three common features summarized by Inoue [30]: (1) these alloys belong to multicomponent systems, (2) the constituent elements have significant atomic size ratios exceeding 1.12 (or less than 0.89), and (3) most of the alloying elements in such alloys have a large and negative

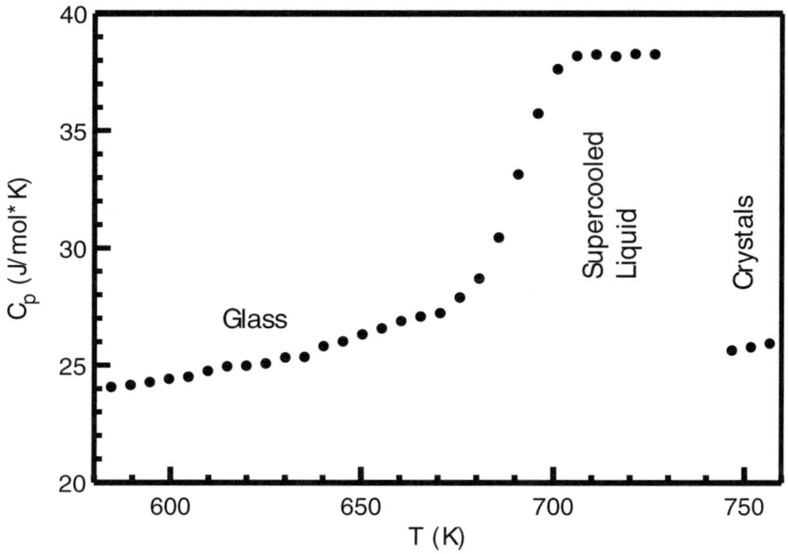

Fig. 10.2 Variation in C_p of a $Zr_{55}Cu_{30}Al_{10}Ni_5$ glassy alloy with temperature. The samples of were heated at a heating rate of 0.083 K/s, while the annealing (waiting) time between the steps was maintained at 60 s. Data taken from Ref. [17]

mixing enthalpy with each other. The first principle is responsible for the formation of dense packed structures leading, together with large difference in electronegativity [31], to low-temperature ternary and quaternary eutectics and elemental "confusion" on solidification [32]. Good correlation is found between the alloy system complexity (the number of alloying elements) and critical diameter of the glassy sample [33]. The role of the second principle (atomic size ratio) has been rationalized through minimum atomic concentration necessary to destabilize terminal solid solution phases [34, 35]. The third principle (large negative mixing enthalpy) is responsible for the formation of low-temperature eutectics and dense packing of atoms of different kinds.

These principles represent indispensable conditions required in order to form BMG alloys leading to good GFA [1, 33] and relatively high thermal stability against crystallization [36]. However, it was found that the higher GFA of the Ge–Ni–Nd alloy compared to the Si–Ni-Nd alloy cannot be explained on the basis of the widely used parameters, geometrical and chemical factors, viscosity and diffusion data [37] while the electronic structure characteristics [38, 39], for example, electronegativity difference [31], should be taken into consideration.

Even pure metals, for example, Ni, Fe can be made amorphous at high enough cooling rate estimated at 10^8–10^{10} K/s [40–44] which are, however, not stable at room temperature unless separated into the nanometer scale spheres [45]. Alloying with other metals improves their GFA. Although, binary BMG alloys exist [46–48], their GFA is low and critical thickness for the glassy samples does not exceed 2 mm (except for the flux-treated Pd–Si alloys). Pd–Si binary BMG alloys with the diameter ranging from 7 to 8 mm were prepared by a combination of fluxing and water cooling [49]. An addition of a certain third element drastically enhances the glass-forming ability of binary alloys [33]. The role of minor additions in the formation of metallic glasses is discussed in Ref. [50].

The specific volume of a liquid phase decreases faster while its density increases faster with temperature than that of a competing crystal (Fig. 10.3a). By analogy with the well-known Kauzmann's entropy crisis [51] one can suggest that a liquid metal should not have a volume lower (or a higher density), at a given temperature, than its crystalline counterpart provided that it contracts upon solidification and there are no changes in the chemical bond character. Face-centered cubic and hexagonal close packed lattices are the densest packing structures for crystalline pure metals, and thus, a liquid or a glassy monoatomic phase should not become denser than them. Quantum mechanics [52] as well as classical [53] molecular dynamics simulations of Ni and Fe confirmed vitrification of the liquid phase (Fig. 10.3b) at temperatures very close to the temperatures of equal volume for liquid and crystalline phases (Fig. 10.3a) [54].

Different criteria used to correlate with the observed GFA include the reduced glass transition temperature, $T_{rg} = T_g/T_l$ [55] where T_g is the glass-transition temperature and T_l is the liquidus temperature though overall validity of this criterion has been questioned [56, 57]; the width of the supercooled liquid region ($\Delta T_x = T_x - T_g$) [58] where T_x is the onset crystallization temperature; $\gamma = T_x/(T_g + T_l)$ parameter [59], which combines both ΔT_x and T_g/T_l criteria into a single parameter. One should also mention a topological criterion λ [60], the thermal conductivity of a molten alloy λ_l [61], δ [62], electronegativity difference between the constituent elements [31, 63], σ [64] and many other criteria [65, 66]. The addition of Zr or Sc substituting for Y reduces the effective ΔEN values among the constituent elements and increases ΔT_x of an Al–Y–Ni–Co alloy [67].

However, certain recently studied binary alloys with a large difference in the electronegativity between the constituent elements, large negative mixing enthalpy in the liquid state, large atomic size difference did not form metallic glasses even by rapid solidification [68]. It was found that formation of densely packed intermetallic compounds with complex structure in the alloy systems neither favors nor disfavors glass-formation. Glass-formation is hampered in metalloid-rich alloys likely owing to unfavorably large or small atomic size ratios strongly deviating from the unity and possible covalent bonding leading to large fragility of the melt. No clear correlation with critical thickness for glass-formation was found for the difference in the atomic size, the entropy of mixing, the difference in electronegativity of the elements and valence electron concentration. This emphasizes the importance of the electronic structure and fragility of the melt. For example, intensification of a covalent bonding between metallic atoms and P was found in the Pd–Cu–Ni–P melts cooled in-situ close to the glass-transition region [54, 69] which was responsible for the changes in the structure of a liquid [70], and thus, fragile behavior of this melt. Fragile liquids are generally predisposed to have lower GFA compared to strong liquids [56] as can be illustrated by using an Angell viscosity plot [71]. Even at the

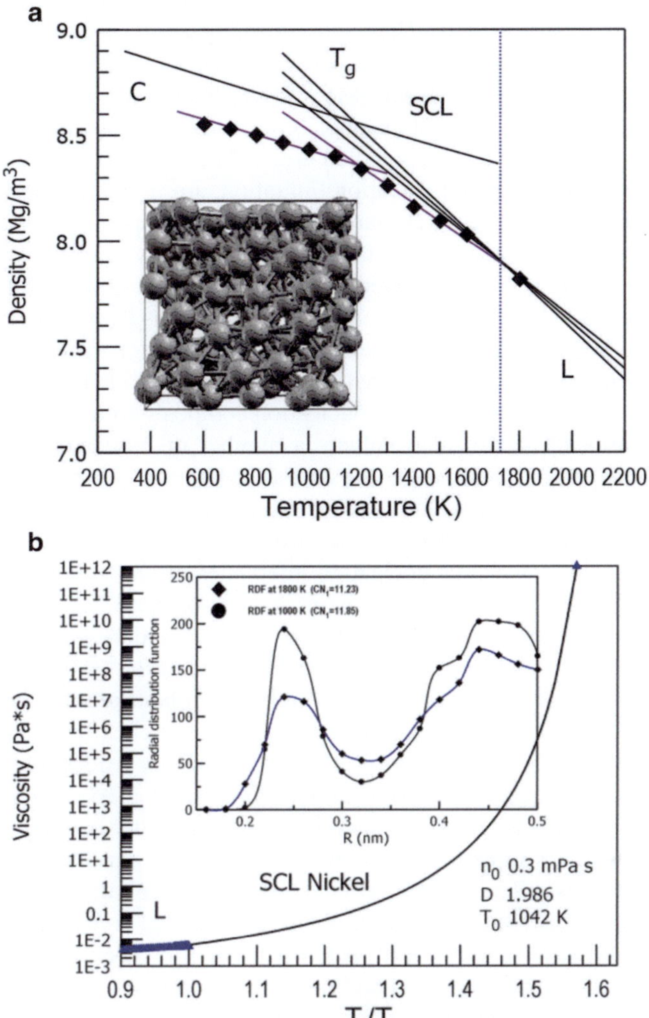

Fig. 10.3 (**a**) *Solid lines*—density versus temperature diagram for Ni according to the literature data. *Diamonds* represent the results of ab-initio MD simulation. Several plots for liquid Ni represent data from different literature sources. The *inset*—atomic structure of Ni cell obtained in an MD simulation at 1000 K. The *dotted line* represents the liquidus temperature. (**b**) The viscosity of liquid Ni as a function of temperature fitted considering T_g of 1100 K. The *inset*—RDFs for liquid and glassy Ni at 1800 and 1000 K, respectively. Reprinted from Ref. [54] with permission of Elsevier

same values of T_{rg} fragile liquid is going to have lower viscosity on cooling in the entire range between T_l and T_g, while low viscosity of the melt facilitates both nucleation and growth rate of crystals [72].

At the same time, Fe–Cr–Mo–C–B–RE (RE—rare earth metals) alloys are distinct from the usual BMGs. Their glass-forming ability appears to be limited by the rate of crystal growth, and even rapidly solidified samples contain pre-existing nuclei of the χ-$Fe_{36}Cr_{12}Mo_{10}$ phase, which are formed because the additions of Y and Tm elements cannot be redistributed fast enough to trigger eutectic crystallization of the supercooled liquid. The lack of incubation period and growth limitation rather than nucleation limitation, are in contrast to the behavior expected for BMGs, in general. Destabilization of the competing crystalline phases (to be formed by eutectic crystallization) and low growth rate of the χ-$Fe_{36}Cr_{12}Mo_{10}$ phase are considered to be the dominant reasons for the significantly improved GFA of these bulk metallic glassy alloys by RE metals [73].

The factors influencing GFA are separated as intrinsic (characteristic of a glass) and extrinsic (depending upon external conditions) factors [74]. By using intrinsic factors we assume that homogeneous nucleation competes with glass formation. As has been mentioned above they include: a number of fundamental and derived thermal parameters; physical properties such as heat capacity, thermal conductivity and diffusivity, thermal expansion coefficients and a topological contribution from efficient atomic packing in the atomic structure. Extrinsic factors are usually operative when heterogeneous nucleation intervenes during solidification. Important extrinsic factors include: crystalline inclusions or dissolved impurities in the melt; poor mould surface finish or cleanliness; turbulence during solidification; and the degree of liquid metal superheat. The critical size of the BMG samples can be significantly limited by extrinsic factors which should be taken into consideration. Fluxing using B_2O_3 actively suppresses heterogeneous nucleation and increases the critical diameter of Pd–Ni–P [3], Pd–Cu–P [75] and Fe-based samples [76]. Similarly Fe–(Co, Cr, Mo, Ga, Sb)–P–B–C bulk ferromagnetic glasses were prepared in the form of 4-mm diameter rods by flux treatment and water cooling [77].

Although the glass-transition phenomenon in metallic glasses has been studied extensively there is still no common conclusion on its nature [78–80]. In some works glassy phase is treated as a frozen liquid, and thus, glass-transition is a kinetic phenomenon and no thermodynamic phase transformation takes place. On the other hand glass-transition may be a second-order transformation as follows from the shape of the curves for the thermodynamic parameters which exhibit a continuity at the glass-transition temperature while their derivatives like thermal expansion coefficient or heat capacity exhibit a discontinuity (in a certain approximation) at the glass-transition temperature. Moreover, although equilibrium liquids are homogeneous on the laboratory timescale there are reasons to believe that deeply supercooled liquids and glasses, which inherited their structure, are rather heterogeneous in terms of their dynamics [81]. Such heterogeneities in glasses are widely discussed at present and make the picture of glass-transition phenomenon even more complicated [82].

10.3 Metallic Glassy Structure

Metallic glassy structure is usually studied by using X-ray diffractometry including synchrotron radiation beam experiments, transmission electron microscopy (TEM) and high-resolution TEM investigations (HRTEM). For example, one can see formation of a disordered structure without crystallinity and halo patterns in both selected-area electron diffraction (SAED) and nanobeam diffraction (NBD) images (Fig. 10.4).

The structure of various glassy alloys has been studied by diffraction experiments and computer modeling [83–91]. For example, the structure of the $Cu_{60}Zr_{30}Ti_{10}$ alloy was studied by synchrotron-radiation X-ray diffraction [92] (Fig. 10.5) and it was suggested that the medium-range order in this alloy maintains up to about 2 nm distance (R). It is also shown that the interatomic distances correspond to those of oC68 $Cu_{10}Zr_7$ compound.

One of the first models of glassy and liquid structure was dense random packing [93]. However, owing to small density difference between bulk metallic glasses and the corresponding crystalline structures it become clear that the metallic glasses consist of the efficiently packed clusters rather than dense random packing of atoms [94]. Cluster-based models of glassy structure become particularly popular recently [95, 96] and importance of an efficient atomic packing for the formation of metallic glasses is emphasized [97]. Atomic clusters about 5–10 nm size were directly observed using scanning tunneling microscopy [98]. As for Fe-based alloys, the structure models created for $Fe_{80}B_{20}$ and $Fe_{70}Nb_{10}B_{20}$ indicated that prismatical clusters observed in some Fe–boride compounds form around B atoms [99, 100]. One should also mention that some of bulk glassy alloys, especially

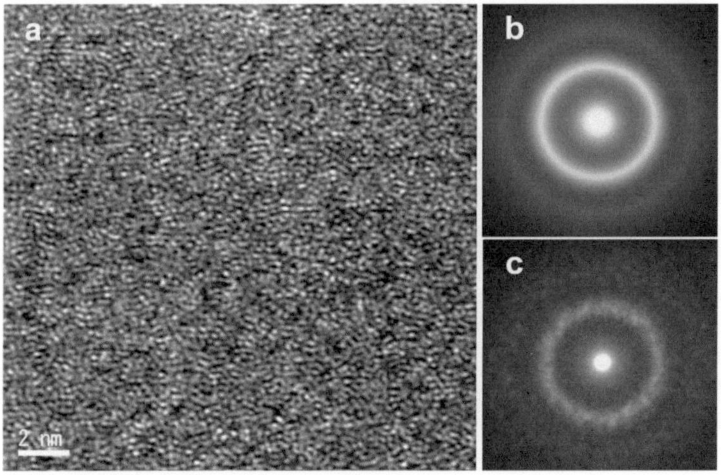

Fig. 10.4 (**a**) Typical HRTEM image of a Zr-based bulk metallic glassy sample, (**b**) typical SAED pattern and (**c**) NBD pattern with a probe size of about 1 nm

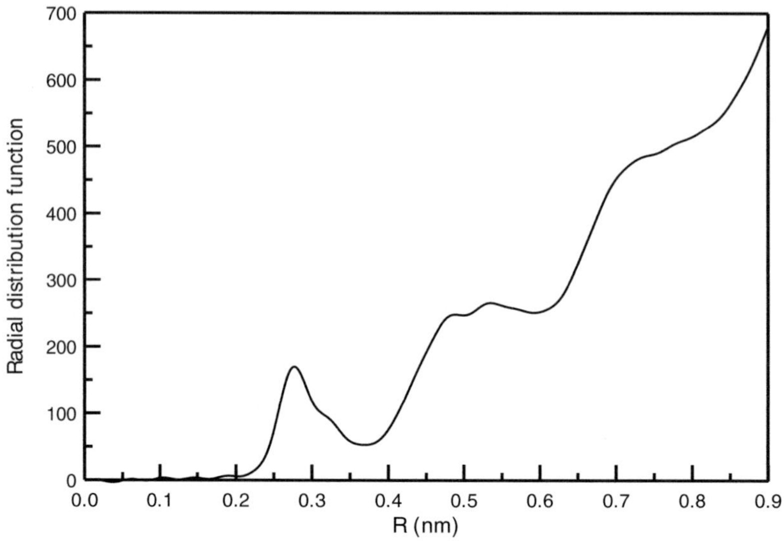

Fig. 10.5 Radial distribution function of $Cu_{60}Zr_{30}Ti_{10}$ alloy. Data taken from Ref. [92] with permission of Elsevier

Pd–Si [101], Pd–Ni–P [102–104], Pd–Cu–Si [105] and Ni–Pd–P [106] alloys contain clear medium-range order (MRO) zones.

Quasicrystalline-type short-range order was reported in Zr-based metallic glasses. It has been found that reduced supercooling (undercooling) before crystallization from the melt was found to be the smallest for quasicrystals, larger for crystal approximants and the largest for crystal phases [107]. However, these icosahedra are distorted as it has been recently demonstrated by direct observation using sub-nanometer beam electron diffraction [108].

Quite an unusual structure is observed on the surface of the as-cast $Au_{49}Cu_{26.9}Ag_{5.5}Pd_{2.3}Si_{16.3}$ BMG rods which have a thin (micrometer-scale) crystalline surface layer which consists of the Au-based solid solution grains embedded in the amorphous phase (with a small part of an intermetallic phase), as illustrated in Fig. 10.6a, b, responsible for nice yellow color of the samples surface (Fig. 10.7) [109]. Internal part of the ingot is grey metallic owing to relatively high Si content.

At the same time, ribbon samples had no such a layer owing to significantly higher cooling rate (about 10^6 K/s) compared to bulk samples of 2–3 mm in size, which were reported [110] to have a cooling rate of about 10^3 K/s. The above-mentioned observation is in line with the finding that such a layer exists in the melt close to the liquidus temperature [111], and thus surface Au solid solution can grow upon solidification of the ingot.

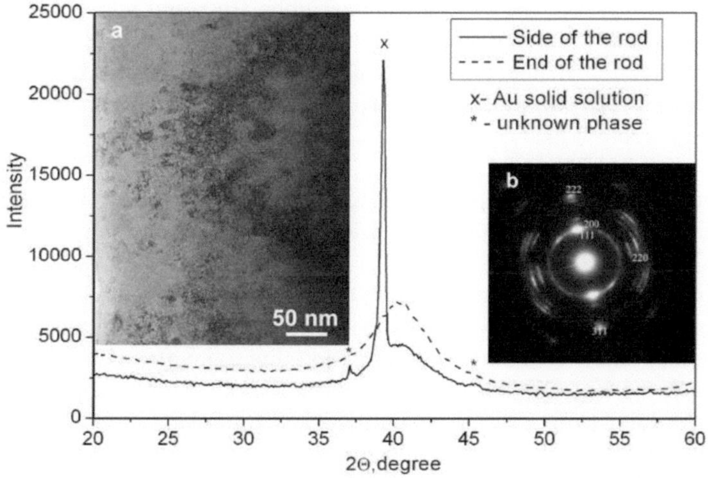

Fig. 10.6 XRD patterns of the as cast Au-based BMG rod. "*End of the rod*" indicates upper surface of the ingot after cutting. (**a**) Bright field TEM image of the side surface layer exhibiting dendrites of a crystalline Au-based solid solution phase. (**b**) Represents selected-area diffraction pattern (SAED) obtained from this region. Reprinted from Ref. [109] with permission of the American Institute of Physics

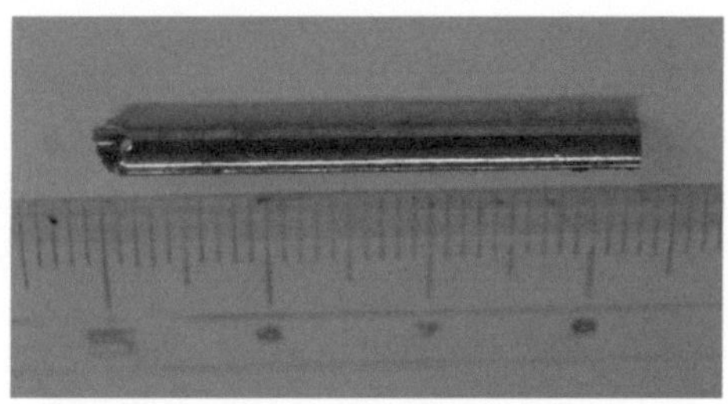

Fig. 10.7 Optical image of the as-cast $Au_{49}Cu_{26.9}Ag_{5.5}Pd_{2.3}Si_{16.3}$ ingot and the ruler scaled in centimeters

10.4 Phase Transformations on Heating and Under Mechanical Exposure

10.4.1 Structural Relaxation and Rejuvenation

Glasses vitrified at a higher cooling rate have higher T_g and a larger specific volume. They increase density on heating below T_g. For example, $Cu_{55}Zr_{30}Ti_{10}Pd_5$ BMG exhibits a large amount of the excess volume of about 0.8 % [112]. This process is called structural relaxation and leads to increase in hardness and Young's Modulus and often causes embrittlement of the samples though hypoeutectic alloys are less sensitive to this process [113]. In Mg- and Ge-based glasses this process also leads to the formation of clusters close in their structure to those of crystalline

Mg and Ge [114, 115]. The volume changes upon structural relaxation on heating have been studied by density measurements [116], dilatometry [117] and by X-ray radiation diffraction of the $Pd_{40}Cu_{30}Ni_{10}P_{20}$ [118], $Zr_{55}Cu_{30}Ni_5Al_{10}$ [119, 120], La-based BMG [121] as well as the $Cu_{55}Hf_{25}Ti_{15}Pd_5$ and $Cu_{55}Zr_{30}Ti_{10}Ni_5$ glassy alloys [122].

The mechanical relaxation processes in metallic glasses are investigated by using dynamic mechanical spectroscopy. At least two relaxations are observed: the α relaxation related to the glass transition and the β relaxation below the glass transition [123].

The process of structural rejuvenation (process opposite to relaxation) in BMGs in terms of structural disordering was induced by anelastic deformation at room and elevated temperature by intensive deformation. For example, the heat of structural relaxation of the $Zr_{50}Cu_{40}Al_{10}$ metallic glass deformed by high pressure torsion increased on increasing rotation revolutions and on raising the deformation temperature when it is below that for starting structural relaxation on subsequent heating [124]. Rejuvenation of a glassy structure was also observed as a result of cyclic cryogenic treatment [125].

10.4.2 Phase Separation Prior to Crystallization

Some metallic glassy alloys exhibit phase separation [126] owing to solid- or liquid-state immiscibility among the constituent elements. A complete review paper is published recently [127]. For example, the solid-state immiscibility between Zr and Y caused phase separation in the Zr–Y–Al–Ni system during heating upon the glass transition to the Y-rich and Zr-rich glassy phases [128]. Zr-rich/La-rich glassy phases were obtained in the melt-spun Zr–La–Al–Ni–Cu alloy [129]. A globular phase in the $Zr_{30}Nd_{30}Al_{15}Ni_{25}$ alloy is Nd-rich while the surrounding matrix phase is Zr-rich one [130]. Phase separation was also observed in the Gd–Zr–Al–Ni metallic glass [131]. The $Zr_{35}Gd_{20}Al_{23}Ni_{22}$ glassy alloy of 1 mm diameter undergoes metastable liquid phase separation and forms Zr-rich and Gd-rich amorphous phases. Similar behavior was found in the Y–Ti–Al–Co [132] and Ni–Nb–Y alloy [133]. Phase separation on a submicron level was also observed in the supercooled liquid region of $Cu_{35}Zr_{45}Ag_{20}$ just before crystallization can be expected according to a slightly negative mixing enthalpy in Cu–Ag pair. It has been also observed in the $Cu_{43}Zr_{43}Al_7Ag_7$ [134] and $Ag_{20}Cu_{48}Zr_{32}$ [135] glassy alloys on a nanoscale. This process may lead to the formation of a composite material with useful mechanical properties [136].

The $Cu_{36}Zr_{48}Al_8Ag_8$ alloy is found to exhibit phase separation rather by binodal mechanism in the glassy phase prior to crystallization [137]. There is an apparent decrease in viscosity observed upon crystallization that is connected with the densification of the sample, but the intrinsic viscosity does not rise significantly until the initial crystallization process is almost complete. During the process the partly crystallized supercooled liquid still behaves like a fluid until a large fraction of the crystalline phase (~90 %) is formed. Nanocrystallization of Ag-rich particles takes place in these regions after separation. Intensive crystallization of the studied alloy starts at higher temperature. The intrinsic viscosity does not rise significantly until the initial crystallization process is almost complete, and thus, the crystal-liquid mixture exhibits quite a low viscosity.

Formation of a mesostructure (micrometer to submicron scale level) was found in some of the bulk metallic glass-forming Zr–Cu–Al and Zr–Cu–Fe–Al alloys [138] though it is not found in similar alloys prepared at different conditions. The reason for its formation is not so clear so far.

10.4.3 *Crystallization of Glassy Alloys*

10.4.3.1 General Features

Metallic glasses crystallize on heating [139]. In general, glassy alloys transform to the supercooled liquid region on heating before crystallization at high enough heating rate. Some alloys do not transform to a supercooled liquid before devitrification and crystallize directly from a glassy phase. Marginal glass-formers have pre-existing nuclei or even nanoparticles in the amorphous matrix which start to grow on heating. Recently observed crystal growth control over crystallization in a BMG is rather unusual for bulk glass formers [140].

Glassy alloys crystallize by the following reactions: polymorphous (a product phase has the same composition as the glassy phase), primary (a product phase has a composition different from that of the glassy phase) and eutectic (two or more phases nucleate and grow conjointly).

All spontaneous thermodynamic processes at constant temperature and pressure lead to a decrease of Gibbs free energy. Below liquidus temperature the liquid and glassy phases have higher free energy than the corresponding crystalline phase. Nucleation of the daughter phase can be homogeneous and heterogeneous. In the case of heterogeneous nucleation the energy barrier against nucleation can be significantly reduced enhancing nucleation and reducing the necessary supercooling. If crystallization occurs by *nucleation and growth mechanism* (there are no pre-existing nuclei), then nucleation and growth rates have different temperature dependences. The number of critical-size nuclei per unit volume is proportional to $e^{-\Delta G_c/RT}$ while the frequency of atomic transfer through the interface is proportional to $e^{-Q_N/RT}$. Thus, the nucleation rate I is:

$$I = I_0 e^{-Q_N/RT} \times e^{-\Delta G_c/RT}, \tag{10.1}$$

where $Q_N \sim 1/\Delta G^2$ activation energy for transfer of atoms across the surface of a nucleus and ΔG_c (as a function of ΔT-supercooling) is a free energy required to form a critical nucleus. The growth rate (u) is proportional to the frequencies of the transition from liquid/glass ν_{l-s} to solid and back ν_{s-l}. Thus, in case of interface controlled growth:

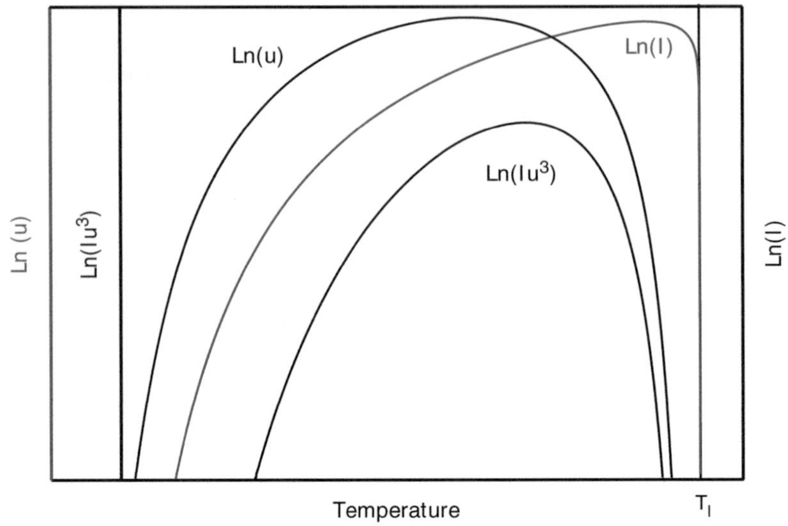

Fig. 10.8 Natural logarithm of the nucleation and growth rates as a function of supercooling. Iu^3 is proportional to the reaction constant *K*. *Left corner* of the temperature axis corresponds to room temperature

$$u = u_0 e^{-Q_g/RT} \times \left(1 - e^{-\Delta G/RT}\right), \tag{10.2}$$

where Q_g is the activation energy for growth; ΔG(ΔT-undercooling) is the Gibbs free energy difference, that is the driving force. The decencies of nucleation and growth rates on the supercooling are schematically shown in Fig. 10.8.

The kinetics of the crystallization process was analyzed substantially [141]. If I and u are time independent and the reaction is interface-controlled then kinetics of the formation of spherical crystals in the glassy alloys can be analyzed by Kolmogorov [142]—Johnson-Mehl [143]—Avrami [144] general exponential equation for the fraction transformed *x*:

$$x = 1 - \exp\left(-\pi/3 \times Iu^3 t^n\right) \tag{10.3}$$

where n is the Avrami exponent.

The kinetic reaction constant and the nucleation rate in several metallic glassy alloys are higher in the initial crystallization stage compared to the steady state. Such a behavior may be a good illustration of the fact that metallic glasses have heterogeneous nucleation sites where nucleation is enhanced and the energy barrier lowered [145]. These sites are pre-existed nuclei which together with possible surface-induced crystallization lead to fast nucleation in the beginning of the process, and thus, larger fraction transformed than expected in the case of purely homogeneous nucleation. These sites are being employed and saturate with time, after which predominantly homogeneous nucleation takes place. As found by optical microscopy, TEM [146, 147] atomic force [148] and scanning tunneling microscopy [149] the surface-induced crystallization takes place in the glassy samples at the initial state.

The time–temperature-transformation diagrams [150] created in the isothermal mode or under continuous heating are useful for comparison of the thermal stabilities of different glasses against devitrification as well as for the selection of the heat treatment regimes. Such diagrams have been created for different metallic glasses [151–154]. Comparison of the long-term thermal stabilities of different metallic glasses has been done using continuous heating transformation (CHT) diagrams [155]. CHT diagrams also can be recalculated from the isothermal ones [156]. A good correspondence is found at least in the mid-temperature range. Recent investigations also showed that a large-size BMG sample crystallizes similarly to a glassy alloy sample produced by melt spinning [157].

10.4.3.2 Nanocrystallization

Metallic glassy alloys crystallize at a large supercooling from the liquidus temperature when the growth rate is relatively low while the nucleation rate is high [158]. In many alloys this process leads to the formation of a nanostructure. Nanostructured materials can be defined as the substances having a very small grain (particle) size typically ranging from 1 to 100 nm. The nanocrystalline materials attract a constantly high interest of the scientists working in different areas of physics, chemistry and materials science [159]. Nanostructured alloys are readily obtained on primary devitrification of glasses with a long-range diffusion controlled growth [160].

Low crystal growth rate of the χ-$Fe_{36}Cr_{12}Mo_{10}$ phase is found that $Fe_{48}Cr_{15}Mo_{14}C_{15}B_6RE_2$ (RE = Y or Tm) BMG alloys (Figs. 10.9 and 10.10) because Y and Tm elements cannot be redistributed fast enough to trigger eutectic crystallization of the supercooled liquid. The low growth rate of the χ-$Fe_{36}Cr_{12}Mo_{10}$ phase, important for the GFA of $Fe_{48}Cr_{15}Mo_{14}C_{15}B_6RE_2$ may be related to large, inhomogeneous internal strains in the nanocrystals. These strains, related to the large volume difference between the glassy and crystalline phases, are partly reduced but redistributed by the incorporation of slow-diffusing Y element in the crystalline phase. Slow diffusion of RE metals at the crystallization temperature does not allow solute partitioning to complete and trigger eutectic crystallization observed in the RE metals free alloys.

Not only pure metals and limited solid solutions but some intermetallic compound particles can also have a nanoscale size. For example, the devitrification of nanocrystal-forming Ti-based alloys, for example, the $Ti_{50}Ni_{20}Cu_{23}Sn_7$ alloy begins from the primary precipitation of a nanoscale equiaxed, almost spherical particles of cF96 Ti_2Ni solid solution phase [161, 162]. Formation of such a nanoscale cF96 phase having a large cubic unit cell was observed in the Zr- and Hf-based alloys [163, 164]. Extremely small size and low growth rate of cF96 crystals were observed in the $Hf_{55}Co_{25}Al_{20}$ glassy alloy [165].

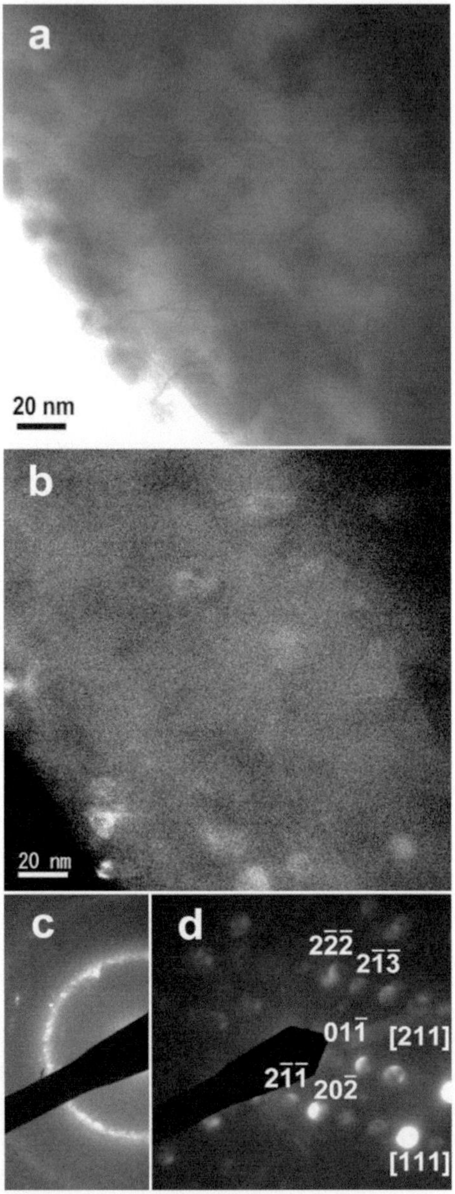

Fig. 10.9 Bright- (**a**) and dark-field (**b**) images, selected-area electron diffraction (**c**) and nanobeam diffraction (**d**) patterns of the $Fe_{48}Cr_{15}Mo_{14}C_{15}B_6Y_2$ alloy annealed at 898 K for 900 s. Two somewhat overlapping NBD patterns are indexed according to the χ-$Fe_{36}Cr_{12}Mo_{10}$ phase (zone axis [211] and [111]). Reprinted from [73] with permission of Elsevier

Flash annealing was applied to obtain nanostructures, and thus, to improve the magnetic properties of the $Fe_{78}Si_9B_{13}$ and $Co_{71.5}Fe_{2.5}Mn_2Mo_1Si_9B_{14}$ glasses [166].

Phase transformations were also observed, recently, upon mechanical cyclic loading of the $Zr_{62.5}Cu_{22.5}Fe_5Al_{10}$ bulk metallic glassy samples in the elastic region at room temperature. It is found that kinetically frozen anelastic deformation accumulates at room temperature and causes crystallization of a metallic glassy

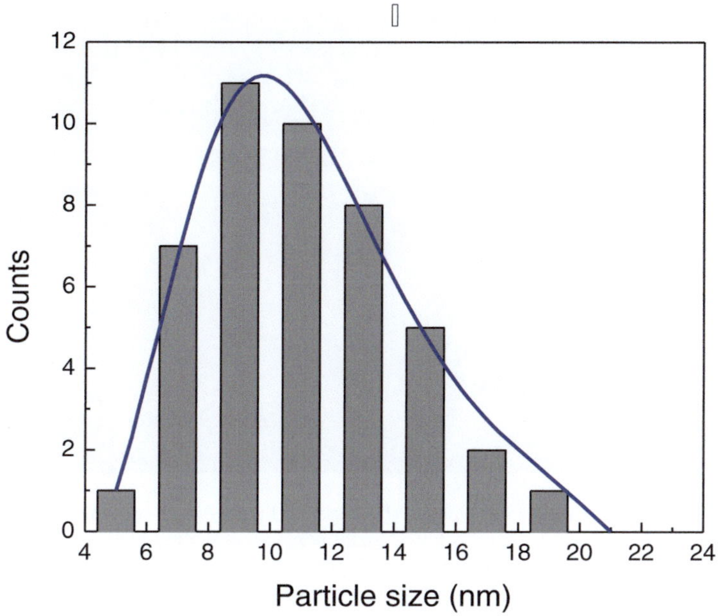

Fig. 10.10 Particle size distribution and a log-normal function fit obtained using the data of Fig. 10.9a. Reprinted from [73] with permission of Elsevier

phase forming either precursors of a metastable crystalline phase [167] or clear crystalline nanoparticles [168] depending on the sample size and the applied load. Such a nanocrystallization was treated as a consequence of anelastic effects connected with operation of the localized areas of viscoelastic deformation [169], taking place even in the deformation regime which is characterized by the linear character of the stress–strain curve.

10.4.3.3 Formation of Quasicrystals

An icosahedral phase—a quasicrystal having a long-range quasiperiodic translational order and an icosahedral orientational order, but with no three-dimensional translational periodicity was initially discovered in Al–Mn alloys [170, 171] and in the other binary Al–TM alloys [172]. Later many other icosahedral quasicrystals, even thermodynamically stable, have been found [173]. Their classification was put forward based on the centering largest atoms in the constituent quasicrystals [174]. Quasicrystals can be quasiperiodic in one, two or three dimensions. The symmetry and stability of an icosahedral structure can be described by a six-dimensional space group. Crystal approximants are similar to quasicrystals but they have periodic structures with relatively large unit cells compared to ordinary crystals which exhibit translational periodicity [175].

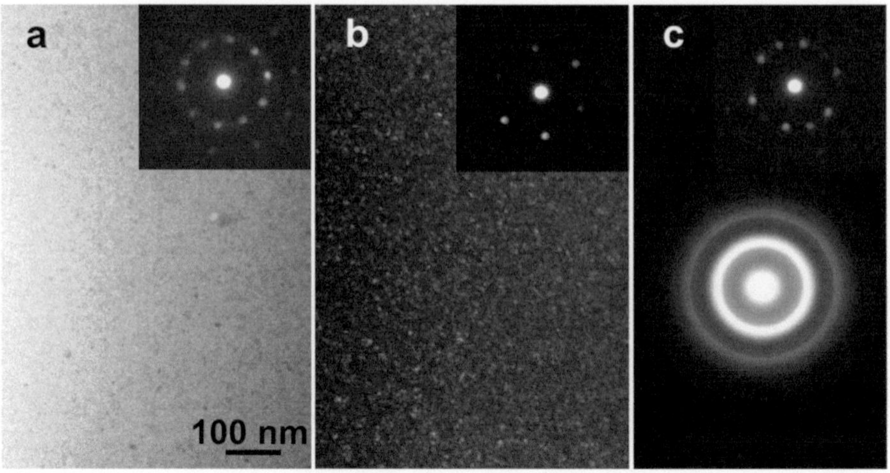

Fig. 10.11 TEM: (**a**) Bright-field image, (**b**) dark-field image and (**c**) selected-area electron diffraction pattern obtained from a Cu-based metallic glass containing nanocrystals of the icosahedral phase. Dark-field image was taken with the sharp rings in (**c**). Nanobeam diffraction patterns of five, three and twofold symmetries are inserted in (**a**, **b**, **c**), respectively. Reprinted from [183] with permission of Elsevier

The nanoscale icosahedral phase was obtained in the devitrified Zr–Al–Ni–Cu–Ag [176], Zr–Al–Ni–Cu–Pd [173, 177] and various other alloys system alloys. Later the nanoscale icosahedral phase has been produced in the NM-free Zr–Cu–Ti–Ni [178] and other glassy alloys. A nanoscale icosahedral quasicrystalline phase has been also produced upon heating glassy Hf-based alloys [179, 180]. In the rapidly solidified Ti–Zr–Hf–Ni system alloys the nanoscale icosahedral phase is formed in the composition ranges close to the cI2 β solid solution phase and complex cF96 phase formation ranges [181]. The $Zr_{65}Al_{7.5}Ni_{10}Pd_{17.5}$ bulk glassy alloy is presumed to contain pre-existing icosahedral nuclei [182].

Formation of the nanoscale (3–10 nm in size) Cu-based icosahedral phase was observed in the Cu–Zr–Ti–NM (NM-noble metal) alloys containing Pd [183] and Au [184] after heat treatment of the glassy phase (Fig. 10.11). The energy-dispersive X-ray analysis and 3D atom probe analysis showed that the icosahedral phase is a Cu-based one [185]. The grain boundaries between residual glassy and icosahedral phases are quite diffuse. The largest particles of the icosahedral phase after the completion of the primary devitrification process reach the size of about 10 nm. The icosahedral phase was formed only in the $Cu_{55}Zr_{30}Ti_{10}Pd_5$ alloy while $Cu_{55}Zr_{40}Pd_5$ alloy did not form the icosahedral phase. $Cu_{55}Zr_{40}Pd_5$ alloy exhibited formation of the equilibrium $(Cu,Pd)_{10}Zr_7$ phase from the supercooled liquid likely by polymorphous mechanism indicating that the Cu–Zr–Ti–Pd icosahedral phase is different from the Zr–Cu–Pd one [186].

10.5 Mechanical Properties and Deformation Behavior at Room Temperature

10.5.1 Bulk Metallic Glasses

Owing to the absence of a crystalline lattice and dislocations, a unique deformation mechanism is realized in bulk glassy alloys [187, 188], which thus, exhibit high strength (~2 GPa for Cu-, Ti-, Zr-based, ~3 GPa for Ni-based, ~4 GPa for Fe-based, ~5 GPa for Co-based alloys), high hardness, good wear resistance [189] and large elastic deformation [190]. Co–Fe–Ta–B glassy alloy exhibited ultrahigh fracture strength exceeding 5 GPa [191] while tensile strength of Co–Ta–B alloys approaches 6 GPa [192]. Fe–(Co,Ni)–(Zr,Nb,Ta)–(Mo,W)–B system bulk glassy alloys exhibit a high compressive strength of 3.8 GPa and good corrosion resistance [193].

Early works of Spaepen [194] and Argon [195] described the concept of the shear transformation in small volumes as the underlying microscopic mechanism of plasticity in glasses. At relatively low homologous temperature the inhomogeneous plastic flow of glassy alloys occurs by propagation of shear bands [196] which are 10–20 nm thick and make steps on the surface up to several micrometers in height [197]. As the localized shear deformation is a dominant plastic-deformation mode at room temperature, tensile ductility of metallic glasses is not found except for few special cases in thin sections of hypoeutectic alloys tested at a relatively high strain rate [198–200]. Tensile deformation behavior of Zr-based glassy thin foils has been also studied recently in-situ in TEM and they were found to be more ductile than larger samples [201, 202]. Hypoeutectic alloys [203] were found to be much more ductile compared to hypereutectic ones owing to a larger excess volume [113].

The relatively large dimensions of bulk metallic glasses with good glass-forming ability permitted fracture toughness tests. The $Zr_{61}Ti_2Cu_{25}Al_{12}$ alloy was reported to have nearly highest toughness among monolithic BMGs. The sample showed fatigue pre-cracked fracture toughness, K_{max}, in excess of 100 MPa$\sqrt{m}$ [204]. The correlations of the toughness were made with the Poisson's ratio (ν), the product of shear modulus and molar volume (μV_m), and the glass transition temperature (T_g). Also, invention of the $Pd_{79}Ag_{3.5}P_6Si_{9.5}Ge_2$ BMG alloy with high fracture toughness [205] increased scientific interest to these fascinating materials.

Novel fracture behavior of Pd-based metallic glasses was reported recently [206]. Although the majority of BMG alloys are brittle even under compression, some of bulk metallic glasses exhibit significantly higher compressive plasticity [207, 208] compared to the others. Such an improved plasticity can be related to Poisson's ratio ν [209], in-situ nano-crystallization [210] or glassy phase separation [211]. Also, ductility of the $Pd_{40}Ni_{40}Si_4P_{16}$ BMG is drastically improved by application of B_2O_3 fluxing technique [212]. Further investigations on this subject were taken recently and embrittlement of Mg- and La-based alloys is caused by the inclusions of oxides (oxygen is insoluble in solid Mg and La, while it dissolves very well in Zr and Ti) acting as stress concentrators [213].

It was possible to induce residual stresses in a bulk metallic glassy sample by shot peening method which improved the mechanical performance, in particular plasticity. The effects of shot peening are noteworthy, and the compressive plasticity drastically increased [214].

While the strain-rate sensitivity (SRS) of polycrystalline alloys is generally positive, for bulk metallic glasses (BMGs) it was reported to be negative for the $Zr_{57}Ti_5Cu_{20}Ni_8Al_{10}$ alloy at strain rates ranging from 10^{-4} to 3×10^3 s^{-1} [215]. Also a negative value was obtained for a similar composition the $Zr_{52.5}Ti_5Cu_{17.9}Ni_{14.6}Al_{10}$ and its composite, at strain rates of 3.3×10^{-4} and 3.7×10^{-3} s^{-1} [216]. The $Zr_{65}Cu_{20}Fe_5Al_{10}$ BMG alloys exhibited negative strain rate sensitivity and the parameter ($m = -0.0026$) has been calculated from the true stress–strain curves. A different behavior observed on nanoindentation could be mainly attributed to the stress state which is closer to three-axial compression [217, 218] when tested by indentation. Moreover, there is a strong size effect on the deformation behavior upon indentation in which the deformed volume is significantly smaller while local plastic deformation is larger than in compression and metallic glasses are proven to deform homogeneously on nanoscale by [201, 219].

Two deformation stages are observed in the bulk metallic glassy samples [220]. Figure 10.12a represents analysis of the continuous loading stress–plastic strain curve of the cylindrical $Zr_{62.5}Cu_{22.5}Fe_5Al_{10}$ glassy alloy in its plastic deformation range. The stress drop spectrum changes its shape after about 2 % of deformation. The frequency distribution of the magnitude of stress drops (Fig. 10.12b) calculated from the continuous strain–stress curve in (Fig. 10.12a) in two ranges, namely, from 0 to 2.5 % and from 2.5 to 4 % of plastic deformation clearly indicates different deformation behaviors at these two ranges: formation of new shear bands and propagation of the existing shear bands at the first stage versus the formation of a major shear band throughout the sample and its propagation at the second stage. Such a behavior is also illustrated by the changes in the stress drop and strain per stress drop plots (Fig. 10.12b). A video file was recorded during its deformation (see the snapshots in Fig. 10.12a, inset). The snapshots in the inset in Fig. 10.12a clearly demonstrate the formation and propagation of a major shear band in the second part of the deformation process. The formation of a dominant shear band in different samples takes place at different strain values and is considered to be a stochastic process. A similar behavior was observed in the case of the $Pd_{40}Ni_{40}Si_4P_{16}$ BMG and $Ti_{43.6}Zr_{7.9}Cu_{40.4}Ni_{7.1}Co_1$ BMG samples containing 1000 ppm B.

The two-stage deformation process is schematically illustrated in Fig. 10.13. These BMG samples initially show multiple shear bands and microscopically relatively homogeneous deformation at nearly constant flow stress and the formation of a dominant shear band at the late deformation stage which leads to a significant decrease in the apparent flow stress owing to the decrease in the efficient sample cross-section. The formation of a dominant shear band and its subsequent critical propagation is likely a stochastic process [221] depending on the stress field distribution.

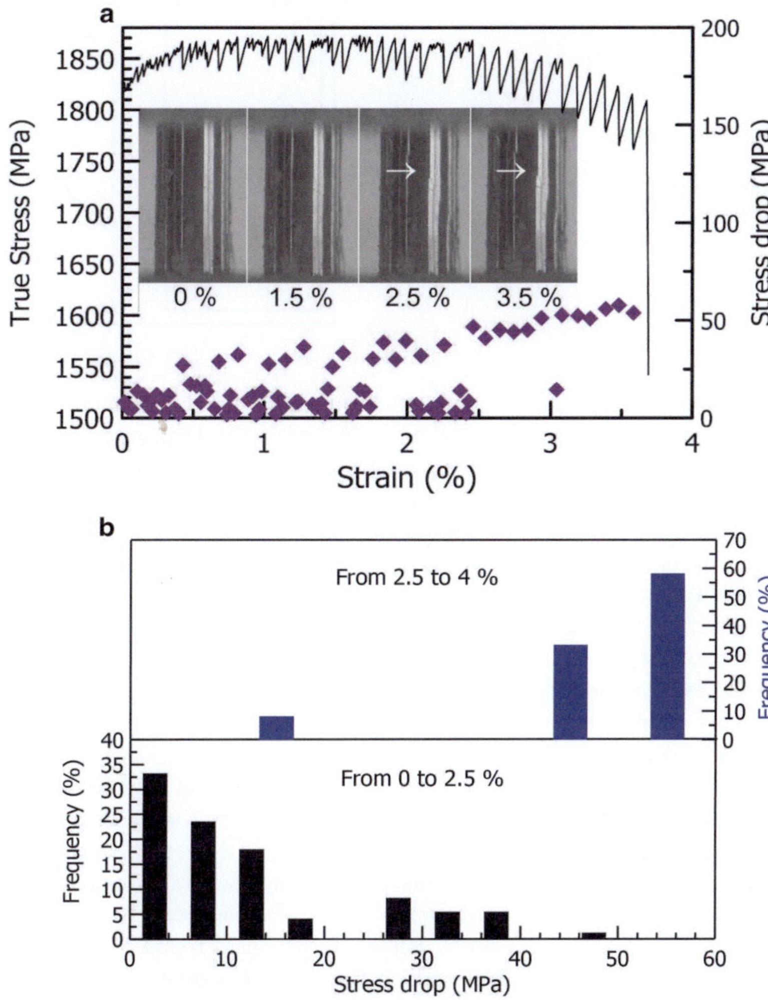

Fig. 10.12 (**a**) Stress and stress drops (*diamonds*) as a function of plastic strain for the cylindrical $Zr_{62.5}Cu_{22.5}Fe_5Al_{10}$ sample of 2:1 aspect ratio upon the compression test. The *insets* in (**a**) illustrate a sequence of the video file snapshots corresponding to (0 %) the initial sample before deformation, (1.5 %) the sample deformed plastically to 1.5 %, (2.5 %) the sample deformed plastically to 2.5 % and (3.5 %) the sample before fracture: deformed plastically to 3.5 %. (**b**) The frequency distribution of the magnitude of stress drops calculated from the continuous strain–stress curve in (**a**) in two ranges, namely, from 0 to 2.5 % and from 2.5 to about 4 % in plastic deformation. Reprinted from [220] with permission of Elsevier

BMG's deformation starts from the formation of a shear transformation zone (STZ) in a certain area of the sample with a higher local stress near a stress concentrator (flaw or surface notch) leading to its structural disordering [222, 223]. Then several STZs deform, unite and form a front, which propagates

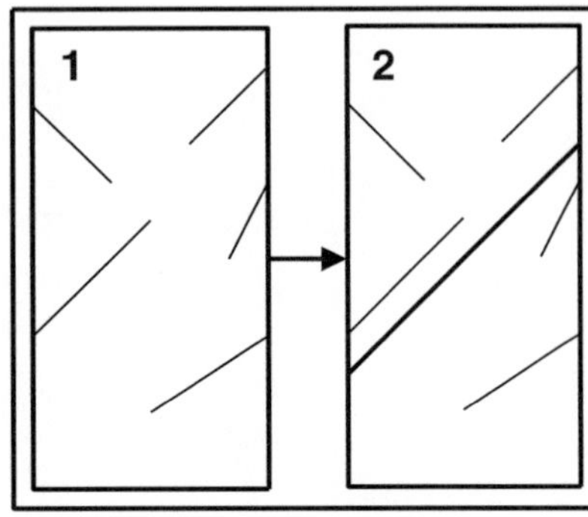

Fig. 10.13 Two-stage deformation mode: *1*—multiple shear bands mostly not crossing the entire sample, *2*—dominant shear band from one side to another leading to stick-slip flow behavior. Reprinted from [220] with permission of Elsevier

as a shear band. At this stage the strain values in the band and the shear offset values are rather small leading to small stress drops. A certain band can gradually reach the other surface of the sample, leaving either a relatively flat or complex-shaped curved two-dimensional interface between two pieces of the sample (see Fig. 10.13) further exhibiting the slip-stick process [224] and concentration of the subsequent deformation activity in this band corresponding to stage 2 in Fig. 10.13. This moment coincides with the change in the shape of the stress–strain curve in Fig. 10.12. At the first stage, the formation of new cold shear bands with relatively small shear offsets prevails. At the second stage, deformation proceeds mainly through the existing dominant shear band(s). These two types of the deformation regimes may be responsible for the observed bimodal distribution of the shear offsets observed earlier [225].

Shear softening may be a result of local heating [226] but may be stress driven [227] when friction in the liquid layer may be the heat source in a shear band [228]. At a high enough strain rate within the shear band the central shear zone heats up due to friction forces [228] and its viscosity can drop several orders in magnitude further increasing the shear speed in the band. The band can be stopped likely owing to the decrease in axial stress waiting for the next round of the increase in stress. At the critical shear offset sudden fracture takes place owing to the low viscosity reduced by the acting friction force heat release [228].

The onset of the shear band propagation has been observed approximately 350 MPa below the ultimate strength by thermography, which serves as a fore-warning of the final fracture of Zr-based BMGs during tensile testing [229]. The instant melting of BMG specimens at the moment of the failure was captured by thermography while temperature rise in the shear band was much lower.

However, the two stage deformation observed in the sample having 2:1 aspect ratio may not be the case in the samples of significantly larger or smaller size or different geometry. The samples with 1:2 size ratio can be deformed to large strain values without formation of a dominant shear band [230]. Figure 10.14 represents the continuous loading stress–plastic strain curve of the cylindrical $Zr_{61}Cu_{27}Fe_2Al_{10}$ glassy alloy. The true stress values (S) were calculated from nominal stress assuming homogeneous deformation and constant volume of the

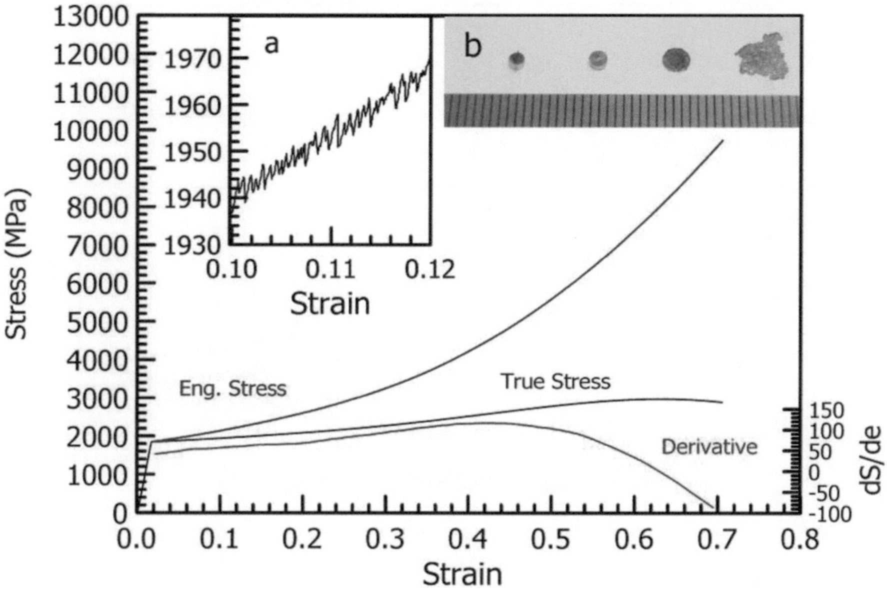

Fig. 10.14 The engineering and calculated compressive true stress–strain curves of the sample having 1:2 aspect ratio tested at room temperature at 5×10^{-4} s^{-1} (*left axis*) and the derivative dS/de (right axis). The *inset* (**a**) a close-up of the curve between 0.10 and 0.12 strain. (**b**) An optical image of the samples before compression and deformed for 0.24, 0.68 and 0.80 strain, respectively, from *left* to *right*. Reprinted from [230] with permission of Elsevier

deforming sample throughout the test. As one can see from the inset Fig. 10.14b the samples deformed for 0.24 and 0.68 strain (engineering strain is used if not specified for better visibility) were intact while the sample deformed to 0.80 was destroyed to pieces.

The XRD patterns (Fig. 10.15) obtained from the samples deformed to 0.68 and 0.80 show no traces of crystallinity.

As can be seen from shear offsets on the lateral surface in Fig. 10.16a, b the sample showed formation of multiple shear bands and no clear cracks are seen inside the sample in the cross-section in Fig. 10.16c.

At the lateral surface of the samples deformed to 0.68 beginning of the formation of cracks is seen. The length of the shear steps (L) on the lateral surface per its area (A) at first approximation is supposed to correlate with the fraction of shear bands (shear surfaces) per sample volume. The thermal properties of the samples were tested by DSC and the resulted values (glass-transition temperature (T_g), crystallization temperature (T_x), exothermic peak temperature (T_p), as well as enthalpies of relaxation (H_r), overshoot (H_o) and crystallization (H_c)) are also shown in Fig. 10.17. ΔH_r and ΔH_o change significantly while the other values are less affected. The structural relaxation process starts at about 500 K.

Owing to the geometrical constraints the $Zr_{61}Cu_{27}Fe_2Al_{10}$ BMG samples of 1:2 aspect ratio exhibited nearly homogeneous deformation (on a macroscopic scale).

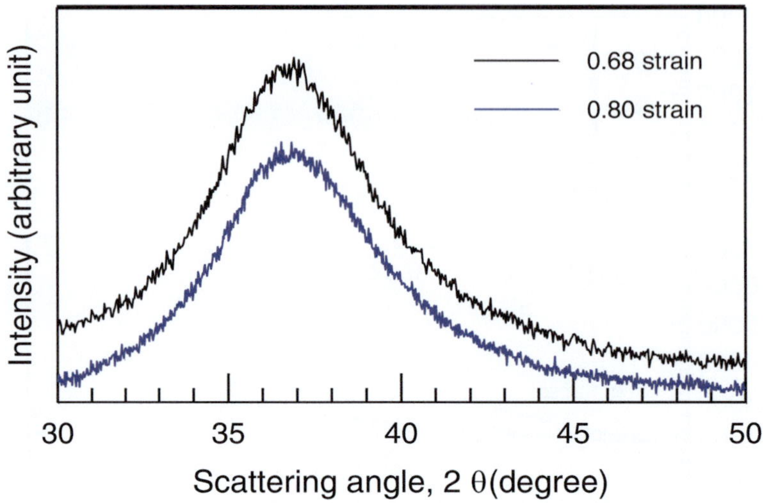

Fig. 10.15 XRD pattern obtained from the sample deformed to 0.68 and 0.80 strain: main broad maximum scanned at low enough scanning rate to reduce the noise. Reprinted from [230] with permission of Elsevier

Such deformation scheme allowed achieving large values of relatively homogeneous deformation. Similarly plastic deformation of crystalline alloys is localized in slip bands which, however, usually have larger number density per surface area than the shear bands in metallic glasses. Nevertheless, even in crystalline materials shear banding starts to dominate over dislocation slip and twinning after a certain amount of strain or at high strain rate [231]. The formation of multiple shear bands is observed in the present BMG samples and no clear cracks are seen up to about 0.5 of plastic strain. Even at 0.68 of plastic strain (1.14 true strain) the sample is still intact. The number of the shear bands increases with strain from 0.24 to 0.68. The volume fraction of the shear bands is only 9×10^{-10}. This is negligibly small value to cause significant changes in the thermal properties of materials if the rest of the sample is not affected.

According to the XRD data obtained with relatively high signal to noise ratio the samples remain glassy after severe plastic deformation and even failure of the sample. Even after fracture of the sample deformed to 0.8 engineering strain (1.61 true strain) the enthalpy of crystallization is not affected. The reduced values of T_x and T_p after deformation (Fig. 10.17b) may be connected with ease of heterogeneous nucleation at the shear steps on the surface or inside the shear bands.

The $Zr_{61}Cu_{27}Fe_2Al_{10}$ BMG sample of 1:2 aspect ratio deforming rather homogeneously without a dominant shear band, nevertheless, also exhibited negative strain rate sensitivity with $m = -0.002$. The Vickers Microhardness measured in the direction of applied stress increases after deformation (Fig. 10.17c). This may be connected with large fraction of the elastic stresses in the glassy matrix induced by so-called not well developed shear bands, the shear bands which "shear front"

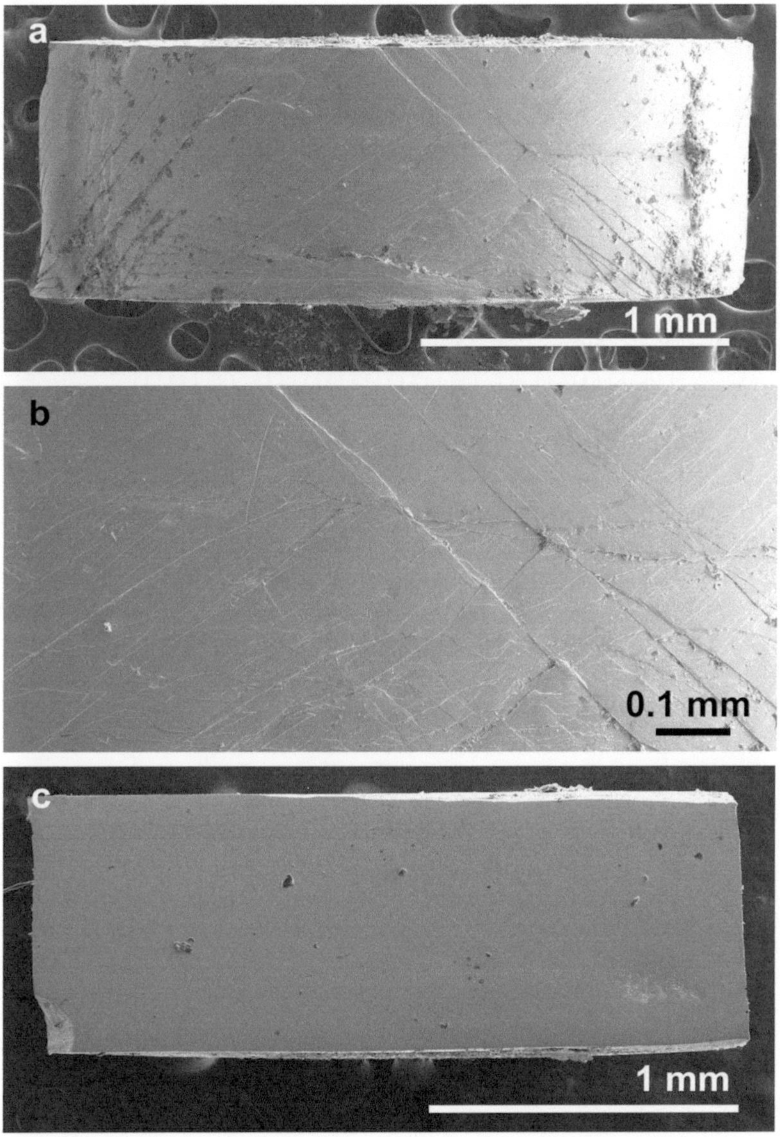

Fig. 10.16 (**a**, **b**) Lateral surface and (**c**) cross section of the sample deformed at $5 \times 10^{-4}\ s^{-1}$ for 0.24 strain at different magnification. Reprinted from [230] with permission of Elsevier

[232] did not cross the entire sample. According to the DSC results in Fig. 10.17b the glass-transition and crystallization temperatures decreased after strain to 0.24 and 0.68 while the heat of relaxation increased significantly consistent with larger hardness owing to stresses. The stored plastic strain energy at large deformation was also found in a Pd-based alloy [233]. This may indicate that intensive plastic

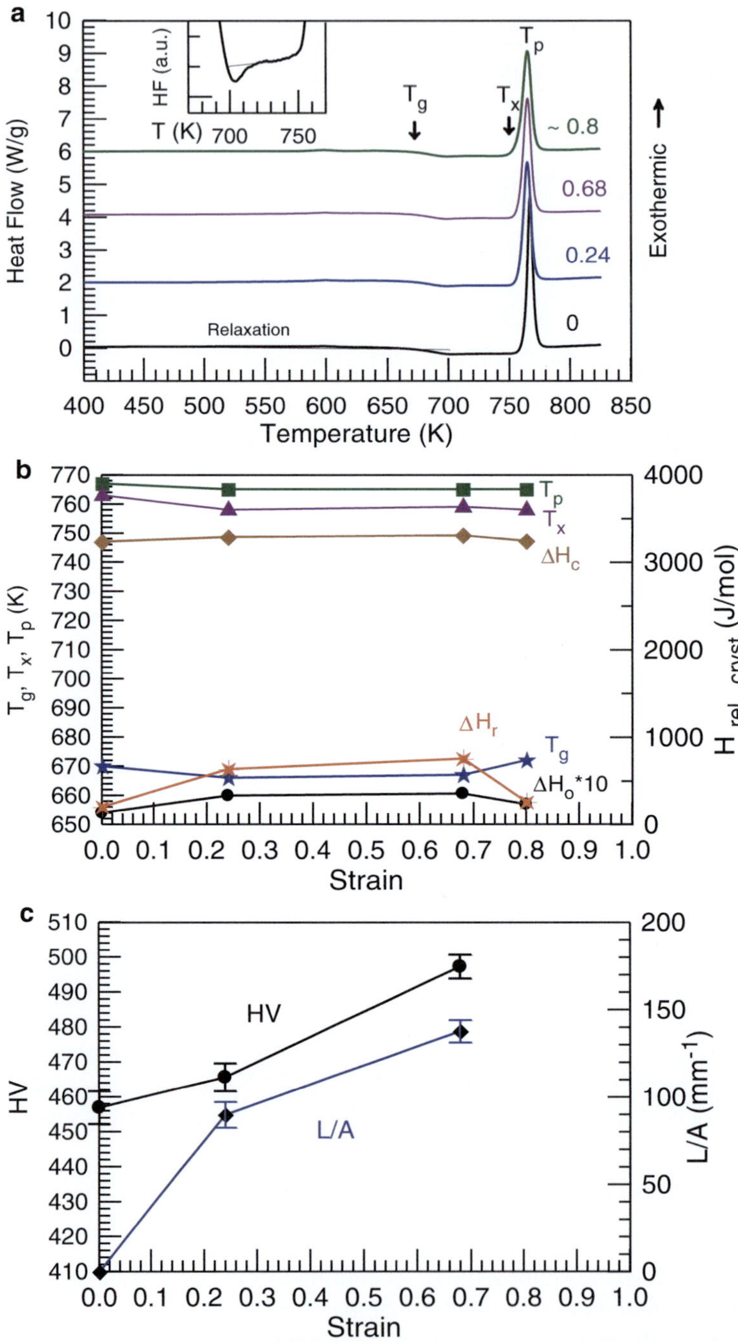

Fig. 10.17 (**a**) DSC traces of the studied alloys at 0.67 K/s, *inset*—illustration of an overshoot of heat flow after the completion of glass-transition. (**b**) The characteristic temperatures and (**c**) HV and L/A values from Table 10.1 as a function of strain. Reprinted from [230] with permission of Elsevier

deformation with a large number of the shear bands brought the bulk metallic glass to a higher energy state which is released only after the failure of the sample. On the other hand the heat of the overshoot (ΔH_o) after glass-transition, ordinary indicative of the degree of structural relaxation [234] also increased.

Intensive plastic deformation of a bulk metallic glass with a large number of the shear bands before intensive crack initiation is supposed to have a twofold effect on its properties. In local areas a higher energy state is reached (similar to glassy phase rejuvenation) which is released after failure of the sample. The plastically (within the shear bands) and elastically (at the edges of not well developed shear bands) deformed areas release a significant amount of extra heat upon structural relaxation while the decreased T_x may indicate ease of crystallization connected with the preferential states for heterogeneous nucleation in the locally deformed areas. On the other hand another part of the glassy matrix may suffer from the opposite effect of structural relaxation leading to an overall lower T_g and larger heat overshoot of the DSC signal after glass-transition indicative of more relaxed glass.

Fatigue properties of BMG alloys have been also studied and fatigue-endurance limits of some Zr-based alloys are comparable with those of the high-strength crystalline alloys [235, 236]. In-situ observations of the mechanical-damage behavior of BMGs during both fatigue and tensile testing with video resolved thermography showed that the relationship between temperature evolutions and stress–strain diagrams during high-cycle fatigue indicate three effects: the thermoelastic effect, the inelastic effect, and the heat-transfer effect [229]. The specimen temperature was observed to oscillate regularly corresponding to the stress oscillation each fatigue cycle. A thermoelastic-degradation behavior was observed at the center of the specimen. This behavior was attributed to the free-volume accumulation inside the specimen during fatigue. However, no shear bands were observed before failure during fatigue experiments, which indicated that the crack-initiation mechanisms of BMGs during fatigue testing could be surface-voids and defects based, and different from the shear-band mechanisms during tensile testing.

Deformation behavior of the Zr-based BMG alloys was also tested at the boiling liquid nitrogen temperature [237–239]. The sample tested in liquid nitrogen exhibits notably higher compression strength compared to the sample tested at room temperature. Also no clear serrated flow typical for bulk glassy samples tested at room temperature is observed in case of the samples tested in liquid nitrogen [239]. The mechanical behavior and the kinetics of shear deformation in bulk metallic glasses were also investigated at room and liquid nitrogen temperature using the acoustic emission technique. It was demonstrated that the intensive acoustic emission reflecting the activity of strongly localized shear bands at room temperature vanishes at the transition from serrated to non-serrated plastic flow at low temperature. The disappearance of the acoustic emission signals clearly suggests that the shear band propagation velocity significantly decreases at low temperature, and sliding along the principle shear band is observed at the machine-driven rate [240].

On heating BMGs first show non-Newtonian flow near T_g and slightly above it, and then Newtonian flow in the supercooled liquid region [241, 242]. However, the boundaries of all these temperature regions (inhomogeneous deformation, non-Newtonian flow and Newtonian flow) are strain rate dependent, i.e., transition from one region to another can take place with decrease in the strain rate near T_g.

10.5.2 Glassy-Crystalline Alloys

For the samples with a large fraction of stress concentrators such as porous BMGs [243] or dual-phase crystal/glassy materials [244] the deformation behavior can change. A different type of the stress–strain curve was observed in a dual-phase $Ti_{42.3}Zr_{7.7}Cu_{41.7}Ni_{7.3}Co_1$ sample with 1000 ppm of B containing the crystalline phase embedded in the glassy matrix (Fig. 10.18). This sample also exhibited the formation of multiple shear offsets on the lateral surface of the sample (Fig. 10.18 inset). Strain hardening and no detectable serrated flow were observed at the initial stage of deformation while weak stress drops of 1–3.5 MPa are seen after 4.5 % of plastic deformation. Fracture occurred when strain hardening ($d\sigma/d\varepsilon$) (stress σ and strain ε) mechanisms in the crystalline phase were exhausted. Deformation of such a dual-phase material likely takes place by a series of smaller sample displacements in the shear bands though the shear deformation mechanism is still active in this case (Fig. 10.15a inset). The ductile crystal/glassy composite sample contains

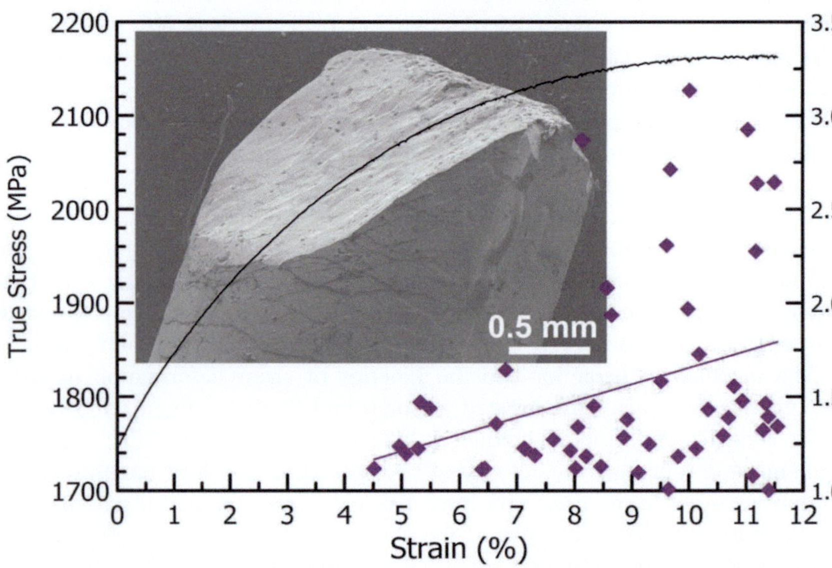

Fig. 10.18 Stress and stress drops as a function of plastic strain for the cylindrical $Ti_{42.3}Zr_{7.7}Cu_{41.7}Ni_{7.3}Co_1$ crystal-glassy sample with 1000 ppm of B of 2:1 aspect ratio upon the compression test. The *inset* represents a lateral surface of the sample tested to fracture. The *diamond* symbols denote the stress drop values exceeding 1 MPa. Reprinted from [220] with permission of Elsevier

multiple shear bands, exhibits very weak stress drops of the stress–strain curve and fractures when the strain hardening mechanisms in the crystalline phase are exhausted.

The achievements related to ductile BMGs samples with in situ dendritic crystals in the La–Al–Cu,Ni [245] (hcp α-La phase as the dendritic phase) and Zr–Ti–Nb–Cu–Be (in which the dendritic phase is a body-centered cubic (Zr, Ti, Nb) solid solution) [246, 247] as well as $Ti_{46}Zr_{20}V_{12}Cu_5Be_{17}$ [248] alloys inspired optimistic viewpoints on applications of bulk metallic glasses. There materials were created using the strategy of microstructural toughening and ductility enhancement in metallic glasses. The two basic principles are: (1) introduction of the inhomogeneities in a metallic glass matrix to initiate local shear band propagation from the inhomogeneity; and (2) matching of microstructural length scales to the characteristic length scale for plastic shielding of an opening crack tip in order to avoid crack development. These principles are applicable to other ductile phase reinforced metallic glass systems provided several criteria are met: the new alloy system must be a highly processable metallic glass in which a dendritic phase nucleates and grows while the remaining liquid is vitrified on subsequent cooling.

Moreover, although Zr–Ti–Nb–Cu–Be βZr-BMG dual phase alloys show necking owing to stress softening of the glassy phase, it was reported that Cu–Zr–Al–Co [249, 250] and Cu–Zr–Al composites [251] show strain hardening owing to transformation-induced plasticity upon martensitic transformation. 1 mm diameter rods of $Ti_{50}Cu_{43}Ni_7$ and $Ti_{50}Cu_{41}Ni_9$ glassy/crystal alloys developed recently also exhibited tensile plasticity [252].

Although, most of these ductile and tough materials so far are based on the expensive elements like Zr, Pt and Pd, the success achieved, nevertheless, allows to suggest that BMG alloys and the composites containing glassy phase are promising structural and functional materials of the present century.

10.6 Magnetic Properties

Ferromagnetic alloys exhibit either hard or soft magnetism depending on their coercivity. Fe- and Co-based bulk metallic glassy alloys exhibit good soft magnetic properties [253–255] and have a coercivity below 10^3 A/m while Nd-based alloys show hard magnetic properties with a coercivity above about 10^4 A/m.

Coercive force (H_c) of soft magnetic materials changes with grain size (D) with maximum at about 10^2 nm as schematically shown in Fig. 10.19 and the coarse grain and nanostructured alloys are known to have low H_c. Suggested structural units of glassy alloys: atomic clusters having sub nanometer size makes them the ideal soft magnetic materials. For long time soft magnetic glassy alloys were limited to marginal glass-formers. However, as soon as a ferromagnetic Fe-(Al, Ga)-metalloid bulk glassy alloy was produced in 1995 [256], Fe-based metallic glasses have attracted significant interest in the scientific community owing to their

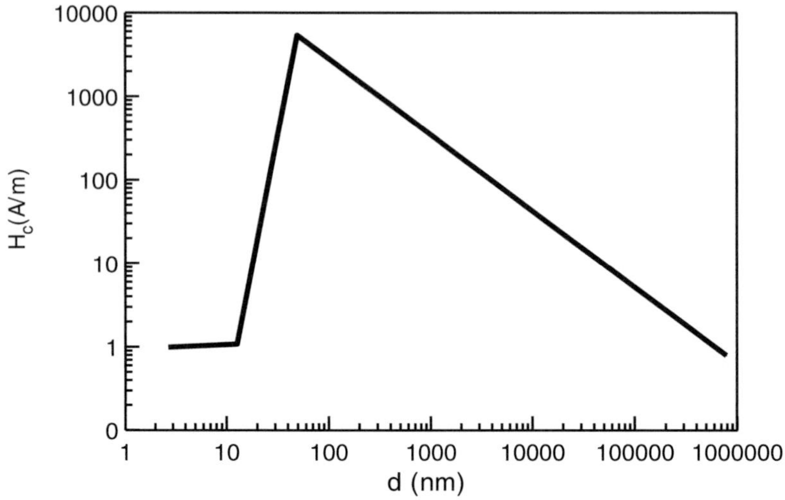

Fig. 10.19 Schematic representation of the changes in H_c of a soft magnetic materials as a function of the grain size (d)

good soft magnetic properties [257]. Figure 10.20 shows, for example, the magnetization curves for the as-cast $Fe_{56}Co_7Ni_7Nb_{10}B_{20}$ bulk rod samples [258].

The (Fe, Ni, Co)$_{70}$$Mo_5P_{10}C_{10}B_5$ bulk metallic glasses with critical sample diameter exceeding 4 mm showed a good thermoplastic formability and good soft magnetic and mechanical properties. These glasses exhibit a wide supercooled liquid region of up to 89 K and low viscosity of about 10^7 Pa s in the supercooled liquid state [259].

The bulk glassy alloys having Curie temperature T_c of 580–610 K were developed in the Fe–(Al, Ga)-(P, C, B) and Fe–(Al–Ga)-(P, C, B, Si) systems [260, 261]. These bulk glassy alloys also exhibited good soft magnetic properties. For example, a ring-shaped sample of the $Fe_{70}Al_5Ga_2P_{9.65}C_{5.75}B_{4.6}Si_3$ glassy alloy with a thickness of 1 mm, an outer diameter of 10 mm and an inner diameter of 6 mm formed by the copper mold casting method exhibits high saturation magnetization of 1.2 T, low coercivity H_c of 2.2 A/m and rather low saturated magnetostriction (λ_s) of 21×10^{-6}. The maximum permeability (μ_{max}) is 110,000. The H_c and μ_{max} values are superior to those (3.7 A/m and 27,000) for the melt-spun samples.

The remarkable improvement of the soft magnetic properties has been demonstrated to result from the significant difference in magnetic domain structure. The domain walls are arranged along the circumference direction for the cast-ring alloy and radial direction for the ribbon ring sheet. The difference in the domain wall structure was reported to originate from the difference in the residual stress during the preparations of the ring sample and the melt-spun ribbon.

In the case of the (Fe, Co, Ni)$_{70}$$Zr_{10}B_{20}$ glassy alloys [262] H_c decreases gradually from 6 to 3 A/m with increasing Fe content. The I_s increases from 0.3 to 0.9 T with

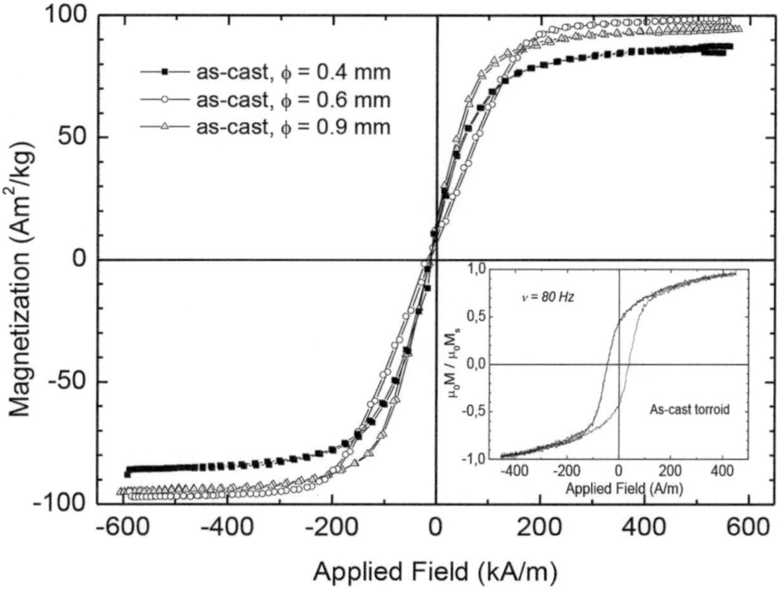

Fig. 10.20 Magnetization curves for as-cast rod samples with the nominal composition $Fe_{56}Co_7Ni_7Nb_{10}B_{20}$. *Inset*: AC MHL curve for an as-cast toroid with the same composition; the reduced magnetization (μ_0M/μ_0M_s) is given instead of absolute magnetization values Reprinted from [258] with permission of IEEE Magnetics Society

increasing Fe content, while λ_s equals zero in the Co-rich composition range and increases monotonously to 15×10^{-6} with increasing Fe content. μ_e reaches the maximum of about 20000 in the case of the Fe- and Co-rich composition ranges. These glassy alloys exhibit good soft magnetic properties including I_s up to 0.9 T, H_c of 3–6 A/m, λ_s of 12–15 $\times 10^{-6}$ and μ_e of 20000 in the Fe-rich range and I_s of 0.5 T, H_c of 6 A/m, nearly zero λ_s and μ_e of 20,000 in the Co-rich range.

Compared to other bulk metallic glasses (BMGs), (Fe–Co)-based alloys are also particularly attractive for engineering applications due to their combination of ultrahigh strength (the highest reported in the literature), superior wear resistance, good glass forming ability and rather low cost [191]. The $Co_{43}Fe_{20}Ta_{5.5}B_{31.5}$ glassy alloy exhibited extremely low value of coercive force as low as 0.25 A/m together with the maximum permeability of 550,000. However, the saturation magnetization was only 0.49 T.

The Fe–(Co, Ni)–(Zr, Nb, Ta)–(Mo, W)–B system glassy alloys exhibit a large saturation magnetization of 0.74–0.96 T, low coercivity of 1.1–3.2 A/m, high permeability exceeding 1.2×10^4 at 1 kHz, and low magnetostriction of about 12×10^{-6} [193].

The $Fe_{76}Si_9B_{10}P_5$ alloy exhibited saturation magnetization J_s of 1.51 T, H_c of 0.8 A/m and a critical diameter of D_{cr} 2.5 mm [263]. Minor addition of Cu improved plasticity of this alloy [264]. Recently developed Fe-metalloid $Fe_{80}P_{11}C_9$ BMG alloy showed good soft-magnetic properties and high strength of

3.2 GPa [265]. The partially devitrified alloy also exhibited good soft-magnetic properties including magnetization of 1.49 T and coercivity of 4 A/m. Magnetic properties of the $(Fe_{1-x}Ni_x)_{72}B_{20}Si_4Nb_4$ ($x = 0.0$–0.5) bulk metallic glasses were studied recently and found that the magnetization decreases from 1.15 to 0.69 T with increasing Ni content from $x = 0.0$ to $x = 0.5$; the Curie temperature reaches its maximum of 598 K for composition $x = 0.1$, and is then followed by a quick decrease with higher Ni content [266].

The $(Fe_{1-x}Co_x)_{76}Si_9B_{10}P_5$ ($x = 0 \sim 0.4$) ferromagnetic bulk glassy alloys exhibited a high saturation magnetization of 1.49 T and low coercive force of 1.2 A/m. These BMGs demonstrate excellent combination of high GFA, good soft-magnetic properties as well as high strength [267].

Formation of a nanostructure within glassy phase improves soft magnetic properties. The saturation magnetization (M_s) and permeability (μ) of the $Fe_{76.5}C_6Si_{3.3}B_{5.5}P_{8.7}$ BMG alloy were enhanced from 1.35 T and 3500 in as-cast state to 1.57 T and 9890, respectively, after annealing at 873 K for 30 s, while coercivity was below 20 A/m [268]. After annealing MRO zones of 1–3 nm size were observed in the glassy matrix. Further the effect of minor Co and Ni alloying on soft magnetic properties of the $Fe_{75.5}C_{7.0}Si_{3.3}B_{5.5}P_{8.7}$ bulk metallic glass was investigated. The saturation magnetization continuously decreases with increasing Co or Ni content, while the Curie temperature and the permeability increase. The Co-bearing alloys show smaller coercivity and larger permeability than the Ni-bearing alloys [269].

The $Fe_{82}(Zr,Hf,Nb)_7B_{10}Cu_1$ alloys exhibit good soft magnetic properties especially in high-frequency range by [270]. The Fe–M–B (M = Zr, Hf, or Nb) alloys also show low core losses [271]. Soft magnetic properties of the (Fe, Co)–RE–B glassy alloys with large thickness were also studied [272].

Y addition was confirmed to improve GFA of the $Fe_{72}B_{24}Nb_4$ alloy. The maximum critical glass forming diameter in the Fe–B–Nb–Y system alloys was 7 mm. Saturation magnetization and coercivity of the as-cast $Fe_{0.72}B_{0.24}Nb_{0.04})_{95.5}Y_{4.5}$ glassy ring were found to be 0.8 T and 0.8 A/m, respectively, in the relaxed state after annealing 821 K (T_g-50 K) for 300 s [273]. 0.8 A/m is an extremely low value of coercivity.

The $Fe_{76+x}Si_{9-x}B_{10}P_5$ bulk metallic glasses showed the saturation magnetization above 1.6 T, low coercive force of 1.6–1.9 A/m, high effective permeability of 16,500–17,200 and low core losses [274]. Large increase of coercivity from 5 A/m up to 4.4 kA/m was reported for the $Fe_{75}Si_{11}B_{10}Nb_3Sn_1$ glassy samples after the first stage of crystallization. This outstanding change was ascribed to the generation of metastable nanocrystallites that disappear at higher temperature [275]. The properties of some alloys are summarized in Table 10.1. These values are also plotted in the diagram in Fig. 10.21 together with the values for well known alloys.

New Fe-based bulk glassy alloys exhibiting a high J_s of 1.51 T reported recently require further investigation as new promising magnetic materials owing to the high Fe content more than 3/4 and absence of other metallic elements, except for Fe. After successful attempt of simultaneous achievement of high Fe content and high GFA in Fe-based BMGs further research activities are expected in this field. Fe, C,

Table 10.1 Composition, magnetic properties (B_s, H_c) of typical ferromagnetic BMGs and their critical diameter (Dcr)

Composition	B_s (T)	H_c (A/m)	D_{cr} (mm)	Ref.
$Fe_{76}Si_9B_{10}P_5$	1.51	0.8	2.5	[263]
$(Fe_{0.75}Si_{0.1}B_{0.15})_{96}Nb_4$	1.47	2.9	1.5	[263]
$Fe_{73}Al_5Ga_2P_{11}C_5B_4$	1.29	6.3	1	[263]
$Fe_{77}Ga_3P_{9.5}C_4B_4Si_{2.5}$	1.36	4.25	2.5	[276]
$Fe_{72}Al_5Ga_2P_{10}C_6B_4Si_1$	1.14	2.8	2	[263]
$Fe_{76}Mo_2Ga_2P_{10}C_4B_4Si_2$	1.32	2.9	2	[263]
$Fe_{30}Co_{30}Ni_{15}Si_8B_{17}$	0.92	3.4	1.2	[263]
$Co_{43}Fe_{20}Ta_{5.5}B_{31.5}$	0.49	0.25	3	[191]
$Fe_{74}Nb_6Y_3B_{17}$	0.81	15	2	[263]
$Fe_{56}Co_7Ni_7Zr_8Ti\ _2B_{20}$	0.82	1.9	1	[277]
Fe–(Co,Ni)–(Zr,Nb,Ta)–(Mo,W)–B	0.74–0.96	1.1–3.2	2–6	[193]
$Fe_{70}Mo_5P_{10}C_{10}B_5$	0.93	2.36	3	[259]
$Fe_{65}Ni_5Mo_5P_{10}C_{10}B_5$	0.88	3.2	3	[259]
$Fe_{55}Ni_{15}Mo_5P_{10}C_{10}B_5$	0.74	1.47	3	[259]
$Fe_{55}Co_{10}Ni_5Mo_5P_{10}C_{10}B_5$	0.85	2.56	4	[259]
$Fe_{74}Nb_6Y_3B_{17}$	0.806	15	2	[278]
$Fe_{76}Si_9B_{10}P_5$	1.51	0.8	2.5	[279]
$(Fe_{0.75}B_{0.15}Si_{0.1})_{96}Nb_4$	1.2	3.7	1.5	[280]
$[(Fe_{0.8}Co_{0.1}Ni_{0.1})_{0.75}B_{0.2}Si_{0.05}]_{96}Nb_4$	1.1	3	2.5	[280]
$[(Fe_{0.6}Co_{0.1}Ni_{0.3})_{0.75}B_{0.2}Si_{0.05}]_{96}Nb_4$	0.8	2.5	3	[280]
$[(Fe_{0.6}Co_{0.2}Ni_{0.2})_{0.75}B_{0.2}Si_{0.05}]_{96}Nb_4$	0.86	2.5	4	[280]

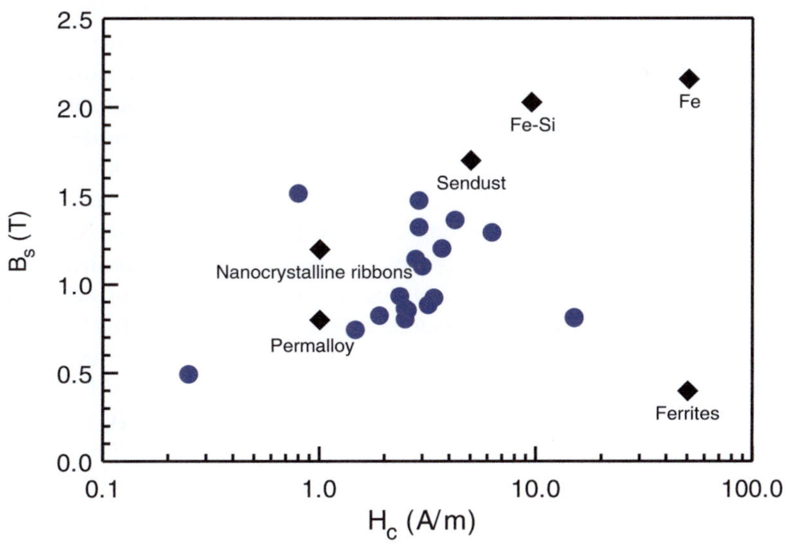

Fig. 10.21 Magnetic properties of the BMGs listed in Table 10.1 (*blue circles*) and typical average values of classical soft magnetic alloys (*black diamonds*)

Si, and P, are the constituent elements in cast-iron produced in a blast furnace. As the low-priced Fe–Si and Fe–P ferroalloys are in mass production there is no restriction in availability of such materials which offers advantages of lower material cost for industry.

As it was shown a Fe-base metallic glass with good soft magnetic properties is a promising material for mirror actuation. Metallic glasses provide a high value of elastic strain limit and fracture toughness resulting from its amorphous structure without structural defects like dislocations and grain boundaries. These attractive functional properties are claimed for micro-electro-mechanical systems.

Moreover, large-scale consolidated powder cores are made from metallic glassy powders. This technique allows overcoming limitation connected with small glass-forming ability of some metallic glassy alloys. Recent work shows possibility to form magnetic glassy-ceramics composites with reduced core losses [281]. The addition of the SiC particulates was effective in improving the high frequency magnetic properties. This approach is very promising for creation net shape products the shape and dimension of which require minimum or no further adaption.

Apart from the massive glassy samples metallic glassy nanowires and nanoglasses attract increasing attention at present. Here there is a field for further employment of magnetic materials. Arrays of magnetic nanowires attracted considerable interest from the viewpoint of perpendicular magnetic recording. The nanowires are potentially capable of producing recording densities in excess of several tens of Gbit per square inch. It has been also reported (Nielsch et al. 2001) [282] that the magnetic nanowires of Fe, Co and Ni show significantly enhanced magnetic coercivity than that of their bulk counterparts. Such an approach can be applied to metallic glassy nanowires as well.

Hard magnetic alloys with a large coercive force exceeding 10 kA/m can be used as permanent magnets. High magnetic induction is retained because of a strong resistance to demagnetization, for example, as a result of high anisotropy. Permanent magnets must have large coercive force, high saturation magnetization and the energy product (BH_{max}). Ferromagnetic $Nd_{90-x}Fe_xAl_{10}$ bulk amorphous alloys exhibit a large coercive force of about 280 kA/m [283]. The bulk amorphous $Nd_{70}Fe_{20}Al_{10}$ alloy shows ferromagnetism with the Curie temperature (T_c) of about 600 K. The remanence (B_r) and coercive force for the bulk $Nd_{60}Fe_{30}Al_{10}$ alloy are 0.122 T and 277 kA/m, respectively. The hard magnetic properties for the bulk amorphous alloys are presumably due to the development of homogeneous ferromagnetic clusters with large random magnetic anisotropy [284]. Hard magnetic alloys can also be produced by crystallization of the glassy phase, for example, on annealing $Nd_{4.5}Fe_{77}B_{18.5}$ rapidly solidified alloys [285]. High remanence (B_r) of 0.8 T is obtained. The maximum energy product $(BH)_{max}$ is 97 kJ/m^3 and $H_c = 240$ kJ/m.

10.7 Corrosion Resistance

Various metallic glassy alloys have a high corrosion resistance [286, 287]. Anodic polarization curves of Ti-based BMG samples exhibit spontaneous passivation behavior and lower passive current densities [288] than their competitors before pitting corrosion by increasing potential indicating that these glassy rod samples have spontaneously passivated regions. The $(Ti_{0.45}Zr_{0.1}Pd_{0.1}Cu_{0.31}Sn_{0.4})_{98}Nb_2$ glassy alloy exhibits passivation with the lower passive current densities of approximately 10^{-3} A · m^{-2} in PBS and 10^{-2} A · m^{-2} in HBSS, which are lower than those of the competitors (pure Ti metal, Ti–6Al–4 V alloy and the basic BMG alloy). Also the current density of the $Ti_{44.1}Zr_{9.8}Pd_{9.8}Cu_{30.38}Sn_{3.92}Nb_2$ bulk glassy is below 10^{-2}A/m^2 in lactic acid, 10^{-3}A/m^2 in PBS and 10^{-2}A/m^2 in HBSS, respectively, and the passivated region exceeds the range of the body potential by [289, 290].

Fe-based BMGs also have high corrosion resistance. The corrosion rates of the $Fe_{50-x}Cr_{16}Mo_{16}C_{18}B_x$ glassy alloys were in the range of 10^{-3}–10^{-2} mm year^{-1} in 1, 6 and 12 N HCl solutions. These bulk glassy alloys are spontaneously passivated in 1 and 6 N HCl solutions and do not exhibit pitting corrosion in 12 N HCl solution up to the potential of 1.0 V (Ag/AgCl). The high corrosion resistance takes place owing to the formation of chromium-rich passive films during immersion in HCl solutions [291].

10.8 Biocompatibility

The above-mentioned Ti–Zr–Pd–Cu–Sn–Nb BMGs [288] can be applicable as biomaterials. Porous bulk metallic glassy samples produced by hydrogenation treatment [292], infiltration [293] mixing the melt with hydrated B_2O_3 [294] or by glassy powder sintering [295] can also be used as bio-implants or damping materials owing to their reduced Young's Modulus compared to monolithic BMGs.

Biodegradable BMGs were developed in Mg–Zn–Ca [296–298], and Ca–Mg–Zn ($Ca_{65}Mg_{15}Zn_{20}$) [299] systems. Alloying with Yb improved the ductility of Mg–Zn–Ca BMGs [300]. Also in comparison with that of Yb-free control, an in vitro cell culture study confirms an improved biocompatibility of these Mg-based BMGs alloyed with Yb. $Ca_{48}Zn_{30}Mg_{14}Yb_8$ BMG exhibited a low degradation rate without observable hydrogen evolution in Hank's solution [301]. $Ca_{65}Li_{6.46}Mg_{5.54}Zn_{23}$, $Ca_{65}Li_{7.54}Mg_{6.46}Zn_{21}$, $Ca_{65}Li_{9.96}Mg_{8.54}Zn_{16.5}$, and $Ca_{65}Li_{14.54}Mg_{12.46}Zn_8$ are very light and have good thermoplastic formability close to room temperature but degrade rapidly [302]. The $Sr_{40}Mg_{20}Zn_{15}Yb_{20}Cu_5$ BMG showed relatively slowest degradation rate [303]. Zn-based metallic glasses like $Zn_{38}Ca_{32}Mg_{12}Yb_{18}$ were also invented [304].

10.9 Applications

Bulk metallic glasses have important applications [305, 306] owing to high glass-forming ability, good casting ability, good formability in the supercooled liquid region as well as good mechanical and chemical properties. These materials are used in sportive goods, watches, electromagnetic wave shields, optical devices, power inductors, mini transformers, micro-geared motor parts, pressure sensors, Coriolis flow meters, coating materials, in medical instruments, etc. [307].

Glassy alloys exhibit high hydrogen solubility and significant resistance to embrittlement. Such materials were found to be promising for future applications as separators for the fuel cells owing to hydrogen permeation characteristics of Zr–Hf–Ni [308], Ni–Nb–Zr [309] and other glassy membranes [310].

Soft magnetic powder core with a higher saturation magnetization of 1.3 T were developed in the Fe–Nb–B–Si and Fe–Nb–Cr–P–B–Si systems [311]. Fe–Ni–Cr–Mo–B–Si glassy alloy powders produced by water atomization have also been commercialized with the commercial name AMO-beads owing to the high glass-forming ability for the Fe-based alloy. The AMO-beads have much longer endurance times compared to those made of steel [312]. Good mechanical properties, such as high Vickers Hardness of 900, high fracture strength of 3000 MPa and large elastic strain of 0.02 together with high corrosion resistance and a smooth outer surface make AMO-beads a good tool for the shot peening treatment.

Dense Fe-based glassy-alloy-coated layer in the Fe–Cr–Mo–C–B system has been produced on various metallic alloy substrates using the high-velocity powder-spray coating layer technique [313]. The Fe–Cr–Mo–C–B coating layer exhibits better corrosion resistance than that of SUS304 and higher Vickers hardness than that of a hard chromium plating plate [314].

Bulk metallic glasses have a smooth surface and soften on heating above T_g which allows creation of micron and nanoscale patterns on their surfaces by molding (Fig. 10.22) [315–318]. After cooling such a material becomes hard again and can be used as a stamp for other materials.

Metallic nanowires [319, 320] may have analytical applications ranging from interconnects to sensors [321]. Micrometer long multi-segment nanowires have been used as barcodes for biological multiplexing [322]. The micrometer long nickel nanowires could be internalized by cells allowing the manipulation of living cells through magnetic field [323]. Metallic glasses are suitable materials for scanner micro-mirrors. In order to achieve a large scanning angle without mechanical failure during actuation, the micro-mirror structure was fabricated using Fe-based metallic glass [324]. High values of mechanical strength and elastic strain limit are desired for the torsion bar for providing high performance of the mirror, including the large tilting angle and good stability.

Recently developed Au-based metallic nano-glasses with a large surface area produced from a BMG forming alloy by [325] open up a new application area of such a material as catalyst. This material showed a good catalytic activity for the following reaction of dimethylphenylsilane ($PhMe_2Si$-H) with water in the

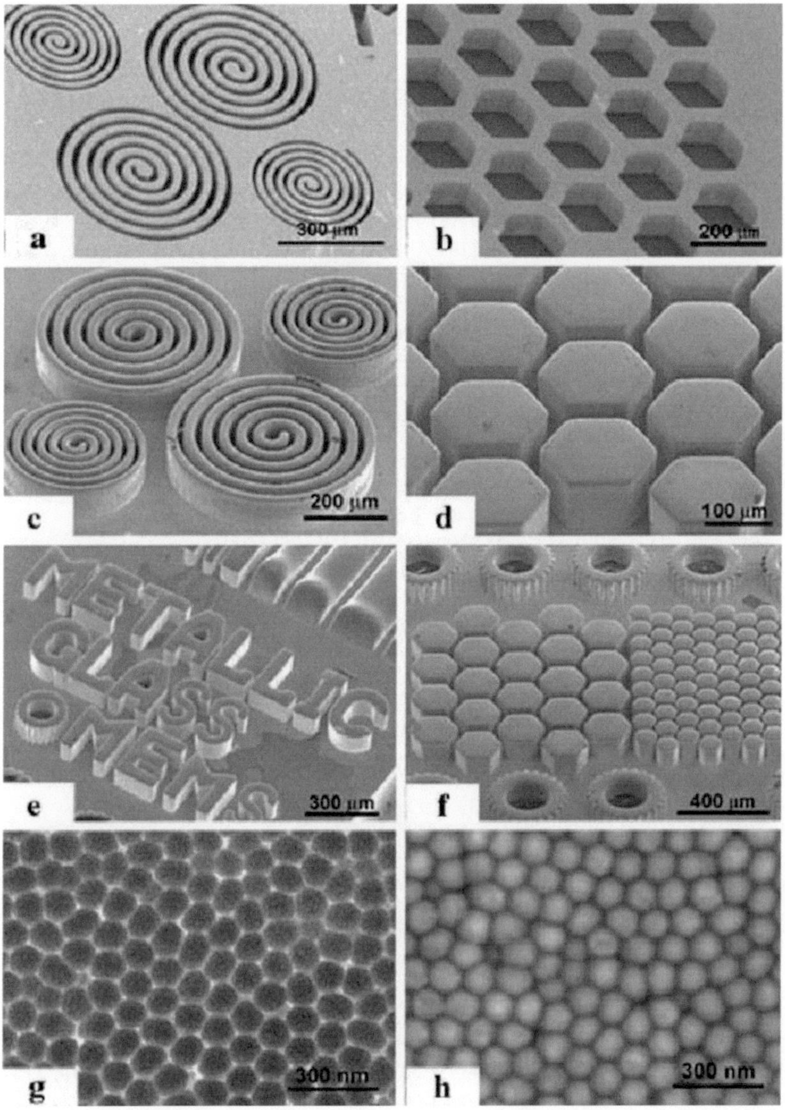

Fig. 10.22 SEM microphotographs of (**a**) and (**b**) are Si molds, (**c–f**) are the patterns fabricated with Si molds, (**g**) is porous alumina, and (**h**) is the corresponding nano rods fabricated with the porous alumina (**g**). All the above structures are obtained using molding with the $Pd_{40}Ni_{40}Si_4P_{16}$ glassy ribbons. Reprinted from [318] with permission of Springer

presence of Au-based metallic nanoglass. The reaction proceeded at room temperature for 24 h and the desired dimethylphenylsilanol compound was obtained. As similar granular structure was obtained in the case of the $Pd_{78}Si_{22}$ nanoglass other catalysts can also be prepared.

References

1. Klement, W., Willens, R.H., Duwez, P.: Nature **187**, 869 (1960)
2. Chen, H.S.: Acta Metall. **22**, 1505 (1974)
3. Kui, H.W., Greer, A.L., Turnbull, D.: Appl. Phys. Lett. **45**, 716 (1982)
4. Inoue, A., Zhang, T., Masumoto, T.: Mater. Trans. JIM **30**, 965 (1989)
5. Molokanov, V.V., Chebotnikov, V.N., Key Engineering Materials: **40–41**, 319, (1991)
6. Inoue, A.: Mater. Trans. JIM **36**, 866 (1995)
7. Johnson, W.L.: MRS Bull. **24**, 42 (1999)
8. Suryanarayana, C., Inoue, A.: Bulk Metallic Glasses, 17th edn. Taylor and Francis, New York (2011)
9. D. V. Louzguine-Luzgin, A. Inoue, Bulk Metallic Glasses. Formation, Structure, Properties, and Applications. in: K.H.J. Buschow (ed) Handbook of Magnetic Materials, Elsevier, New York, pp. 131-171 (2013).
10. Inoue, A., Zhang, T.: Mater. Sci. Eng. A **226–228**, 393 (1997)
11. Inoue, A., Gook, J.S.: Fe-based ferromagnetic glassy alloys with wide supercooled liquid region. Mater. Trans. JIM **36**, 1180–1183 (1995)
12. Ponnambalam, V., Poon, S.J., Shiflet, G.J.: J. Mater. Res. **19**, 1320 (2004)
13. Amiya, K., Inoue, A.: Rev. Adv. Mater. Sci. **18**, 27 (2008)
14. Debenedetti, P.G., Stillinger, F.H.: Nature **410**, 259–267 (2001)
15. Ojovan, M.I.: Entropy **10**, 334–364 (2008)
16. Lu, I.-R., Gorler, G.P., Fecht, H.J., Willnecker, R.: J. Non-Cryst. Solids **312–314**, 547 (2002)
17. Louzguine-Luzgin, D.V., Seki, I., Yamamoto, T., Kawaji, H., Suryanarayana, C., Inoue, A.: Intermetallics **23**, 177–181 (2012)
18. Martinez, L.M., Angell, C.A.: Nature **410**, 663–667 (2001)
19. Leung, P.K., Wright, J.G.: Philos. Mag. **30**, 185–194 (1974)
20. Gleiter, H.: In: N. Hansen, T. Leffers, H. Lilholt (eds.) Proceedings of the Second Riso International Symposium on Metallurgy and Materials Science, Risø National Laboratory (Risø), Roskilde, Denmark, p. 15 (1981)
21. Fecht, H.J.: Nanostruct. Mater. **6**, 33 (1995)
22. Weeber, A.W., Bakker, H.: Physica B **153**, 93 (1988)
23. Inoue, A.: Prog. Mater. Sci. **43**, 365 (1998)
24. Inoue, A., Kimura, H.M.: J. Metast. Nanocr. Mater. **9**, 41 (2001)
25. Xie, G., Louzguine-Luzgin, D.V., Kimura, H.M., Inoue, A.: Appl. Phys. Lett. **90**, 241902 (2007)
26. Louzguine-Luzgin, D.V., Xie, G.Q., Li, S., Inoue, A., Yoshikawa, N., Sato, M.: Philos. Mag. Lett. **89**, 86–94 (2009)
27. Li, N., Chen, Y., Jiang, M.Q., Li, D.J., He, J.J., Wu, Y., Liu, L.: Acta Mater. **61**, 1921–1931 (2013)
28. Yamasaki, T., Schlossmacher, P., Ehrlich, K., Ogino, Y.: Nanostruct. Mater. **10**, 375 (1998)
29. Ahmad, J., Asami, K., Takeuchi, A., Louzguine, D.V., Inoue, A.: Mater. Trans. **44**, 1942 (2003)
30. Inoue, A.: Acta Mater. **48**, 279 (2000)
31. Louzguine, D.V., Inoue, A.: Appl. Phys. Lett. **79**, 3410–3412 (2001)
32. Greer, A.L.: Nature **366**, 303–304 (1993)
33. Louzguine-Luzgin, D.V., Miracle, D.B., Louzguina-Luzgina, L.V., Inoue, A.: J. Appl. Phys. **108**, 103511 (2010)
34. Egami, T., Waseda, Y.: J. Non-Cryst. Solids **64**, 113–134 (1984)
35. Egami, T.: J. Non-Cryst. Solids **205–207**, 575–582 (1996)
36. Louzguine, D.V., Inoue, A.: Scr. Mater **47**, 887–891 (2002)
37. Louzguine, D.V., Louzguina, L.V., Inoue, A.: Appl. Phys. Lett. **80**, 1556 (2002)
38. Hasegawa, M., Sato, H., Takeuchi, T., Soda, K., Mizutani, U.: J. Alloys Comp. **483**, 638 (2009)

39. Oreshkin, A.I., Mantsevich, V.N., Savinov, S.V., Oreshkin, S.I., Panov, V.I., Maslova, N.S., Louzguine-Luzgin, D.V.: Appl. Phys. Lett. **101**, 181601 (2012)
40. Tamura, K., Endo, H.: Phys. Lett. A **29**, 52–53 (1969)
41. Davies, H.A.: Metallic glass formation. In: Luborsky, F.E. (ed.) Amorphous Metallic Alloys, pp. 8–25. Butterworths, London (1983)
42. Louzguine-Luzgin, D.V., Belosludov, R., Saito, M., Kawazoe, Y., Inoue, A.: J. Appl. Phys. **104**, 123529 (2008)
43. Belousov, O.K.: Russ Metall **6**, 488–497 (2009)
44. Zhong, L., Wang, J., Sheng, H., Zhang, Z., Mao, S.X.: Nature **512**, 177–182 (2014)
45. Kim, Y.W., Lin, H.M., Kelly, T.F.: Acta Metall. **37**, 247–255 (1989)
46. Inoue, A., Zhang, W., Zhang, T., Kurosaka, K.: Acta Mater. **49**, 2645 (2001)
47. Xu, D., Lohwongwatana, B., Duan, G., Johnson, W.L., Garland, C.: Acta Mater. **52**, 2621 (2004)
48. Wang, D., Li, Y., Sun, B.B., Sui, M.L., Lu, K., Ma, E.: Appl. Phys. Lett. **84**, 4029 (2004)
49. Yao, K.F., Chen, N.: Sci. China, Ser. G **51**, 414 (2008)
50. Wang, W.H.: Prog. Mater. Sci. **52**, 540 (2007)
51. Kauzmann, W.: Chem. Rev. **43**, 219–256 (1948)
52. Louzguine-Luzgin, D.V., Belosludov, R., Saito, M., Kawazoe, Y., Inoue, A.: J. Appl. Phys. **104**, 123529 (2008)
53. Evteev, A.V., Kosilov, A.T., Levchenko, E.V.: J. Exp. Theor. Phys. **99**, 522 (2004)
54. Louzguine-Luzgin, D.V.: J. Alloys Compd. **586**, S2–S8 (2014)
55. Turnbull, D., Cohen, M.H.: J. Chem. Phys. **34**, 120 (1961)
56. Senkov, O.N.: Phys. Rev. B **76**, 104202 (2007)
57. Louzguine-Luzgin, D.V., Belosludov, R., Saito, M., Kawazoe, Y., Inoue, A.: J. Appl. Phys. **104**, 123529 (2008)
58. Inoue, A., Zhang, T., Masumoto, T.J.: J. Non-Cryst. Sol. **156–158**, 473 (1993)
59. Lu, Z.P., Liu, C.T.: Acta Mater. **50**, 3501 (2002)
60. Egami, T., Waseda, Y.: J. Non-Cryst. Sol. **64**, 113 (1984)
61. Louzguine-Luzgin, D.V., Setyawan, A.D., Kato, H., Inoue, A.: Philos. Mag. **87**, 1845 (2007)
62. Louzguine-Luzgin, D.V., Inoue, A.: J. Mater. Res. **22**, 1378–1383 (2007)
63. Ma, C.S., Zhang, J., Chang, X.C., Hou, W.L., Wang, J.Q.: Philos. Mag. Lett. **88**, 917–924 (2008)
64. Park, E.S., Kim, D.H., Kim, W.T.: Appl. Phys. Lett. **86**, 061907 (2005)
65. Senkov, O.N., Miracle, D.B., Mullens, H.M.: J. Appl. Phys. **97**, 103502 (2005)
66. Suryanarayana, C., Inoue, A.: Bulk Metallic Glasses, p. 565. Tailor and Francis CRC, New York (2010)
67. Louzguine-Luzgin, D.V., Inoue, A., Botta, W.J.: Appl. Phys. Lett. **88**, 011911 (2006)
68. Louzguine-Luzgin, D.V., Chen, N., Churymov, A.Y., Louzguina-Luzgina, L.V., Polkin, V.I., Battezzati, L., Yavari, A.R.: J. Mater. Sci. **50**, 1783–1793 (2015)
69. Louzguine-Luzgin, D.V., Belosludov, R., Yavari, A.R., Georgarakis, K., Vaughan, G., Kawazoe, Y., Egami, T., Inoue, A.: J. Appl. Phys. **110**, 043519 (2011)
70. Ojovan, M.I.: J. Non-Cryst. Solids **382**, 79–86 (2013)
71. Angell, C.A.: Science **267**, 1924 (1995)
72. Uhlmann, D.R.: J Non-Cryst Sol. **7**, 337–348 (1972)
73. Louzguine-Luzgin, D.V., Bazlov, A.I., Ketov, S.V., Greer, A.L., Inoue, A.: Acta Mater. **82**, 396–402 (2015)
74. Louzguine-Luzgin, D.V., Miracle, D.B., Inoue, A.: Adv. Eng. Mater. **10**, 1008 (2008)
75. He, Y., Shen, T.D., Schwarz, R.B.: Metall. Mater. Trans. A **29**, 1795–1804 (1998)
76. Stoica, M., Hajlaoui, K., Lemoulec, A., Yavari, A.R.: Philos. Mag. Lett. **86**, 267–275 (2006)
77. Shen, T.D., Schwarz, R.B.: Appl. Phys. Lett. **75**, 49 (1999)
78. Cohen, M.H., Grest, G.S.: Phys. Rev. B **20**, 1077 (1979)
79. Van Den Beukel, A., Sietsma, J.: Acta Metall. Mater. **38**, 383 (1990)

80. Angell, C. A.: Glass Transition. In Encyclopedia of Materials: Science and Technology, Vol. 4, Pergammon, Elsevier. **4**, 3365 (2001)
81. Ediger, M.D.: Annu. Rev. Phys. Chem. **51**, 99–128 (2000)
82. Wang, Z., Sun, B.A., Bai, H.Y., Wang, W.H.: Nat. Commun. **5**, 5823 (2014)
83. Waseda, Y.: The Structure of Non-crystalline Solids. McCraw-Hill, New York (1980)
84. K. Suzuki in Proceedings of the 4th International Conference on Rapidly Quenched Metals, edited by Masumoto, T. and Suzuki, K., Japan Institute of Metals, Sendai, 1, 309, (1982).
85. Waseda, Y., Chen, H.S.: Phys. Stat. Solidi **49**, 387 (1978)
86. Matsubara, E., Waseda, Y.: Mater. Trans. JIM **36**, 883 (1995)
87. Kramer, M.J., Ott, R.T., Sordelet, D.J.: J. Mater. Res. **22**, 382 (2007)
88. Hirata, A., Hirotsu, Y., Ohkubo, T., Tanaka, N., Nieh, T.G.: Intermetallics **14**, 903 (2006)
89. Haruyama, O., Sugiyama, K., Sakurai, M., Waseda, Y.: J. Non-Cryst. Solids **353**, 3053 (2007)
90. Yavari, A.R.: Nature **439**, 405 (2006)
91. Antonowicz, J., Pietnoczka, A., Drobiazg, T., Almyras, G.A., Papageorgiou, D.G., Evangelakis, G.A.: Philos. Mag. **92**, 1865–1875 (2012)
92. Louzguine-Luzgin, D.V., Antonowicz, J., Georgarakis, K., Vaughan, G., Yavari, A.R., Inoue, A.: J. Alloys Compd. **466**, 106–110 (2008)
93. Bernal, J.D.: Nature **185**, 68–70 (1960)
94. Miracle, D.B.: Nat. Mater. **3**, 697–702 (2004)
95. Miracle, D.B.: Acta Mater. **54**, 4317 (2006)
96. Sheng, H.W., Luo, W.K., Alamgir, F.M., Bai, J.M., Ma, E.: Nature **439**, 419 (2006)
97. Cheng, Y.Q., Ma, E.: Prog. Mater. Sci. **56**, 379–473 (2011)
98. Oreshkin, A.I., Maslova, N.S., Mantsevich, V.N., Oreshkin, S.I., Savinov, S.V., Panov, V.I., Louzguine-Luzgin, D.V.: JETP Lett. **94**, 58–62 (2011)
99. Waseda, Y., Chen, H.S.: Solid State Commun. **27**, 809 (1978)
100. Hirata, A., Hirotsu, Y., Ohkubo, T., et al.: Phys. Rev. B **74**, 214206 (2006)
101. Ohkubo, T., Hirotsu, Y.: Phys. Rev. B **67**, 094201 (2003)
102. Hirotsu, Y., Ohkubo, T., Matsushita, M.: Microsc. Res. Tech. **40**, 284 (1998)
103. Haruyama, O., Sugiyama, K., Sakurai, M., Waseda, Y.: J. Non-Cryst. Solids **353**, 3053 (2007)
104. Hirata, A., Hirotsu, Y., Ohkubo, T., Tanaka, N., Nieh, T.G.: Intermetallics **14**, 903 (2006)
105. Hirotsu, Y., Uehara, M., Ueno, M.: J. Appl. Phys. **59**, 3081 (1986)
106. Louzguine-Luzgin, D.V., Zeng, Y.H., Setyawan, A.D., Nishiyama, N., Kato, H., Saida, J., Inoue, A.: J. Mater. Res. **22**, 1087 (2007)
107. Kelton, K.F.: J. Non-Cryst. Solids **334–335**, 253 (2004)
108. Hirata, A., Kang, L.J., Fujita, T., Klumov, B., Matsue, K., Kotani, M., Yavari, A.R., Chen, M. W.: Science **341**, 376–379 (2013)
109. Ketov, S.V., Chen, N., Caron, A., Inoue, A., Louzguine-Luzgin, D.V.: Appl. Phys. Lett. **101**, 241905 (2012)
110. Louzguine-Luzgin, D.V., Setyawan, A.D., Kato, H., Inoue, A.: Philos. Mag. **87**, 1845–1854 (2007)
111. Mechler, S., Yahel, E., Pershan, P.S., Meron, M., Lin, B.: App. Phys. Lett. **98**, 251915 (2011)
112. Louzguine-Luzgin, D.V., Yavari, A.R., Fukuhara, M., Ota, K., Xie, G., Vaughan, G., Inoue, A.: J. Alloys Compd. **431**, 136 (2007)
113. Aljerf, M., Georgarakis, K., Yavari, A.R.: Acta Mater. **59**, 3817–3824 (2011)
114. Louzguine, D.V., Louzguina, L.V., Inoue, A.: Philos. Mag. **83**, 203 (2003)
115. Louzguine, D.V., Saito, M., Waseda, Y., Inoue, A.: J. Phys. Soc. Jpn. **68**, 2298 (1999)
116. Ishii, A., Hori, F., Iwase, A., Fukumoto, Y., Yokoyama, Y., Konno, T.J.: Mater. Trans. **49**, 1975–1978 (2008)
117. Schermeyer, D., Neuhäuser, H.: Mater. Sci. Eng. A **226–228**, 846 (1997)
118. Mattern, N., Hermann, H., Roth, S., Sakowski, J., Macht, M.P., Jovari, P., Jiang, J.: Appl. Phys. Lett. **82**, 2589 (2003)
119. Yavari, A.R., Nikolov, N., Nishiyama, N., Zhang, T., Inoue, A., Uriarte, J.L., Heunen, G.: Mater. Sci. Eng. A **375–377**, 709 (2004)

120. Yavari, A.R., Le Moulec, A., Inoue, A., Nishiyama, N., Lupu, N., Matsubara, E., Botta, W.J., Vaughan, G., Di Michiel, M., Kvick, A.: Acta Mater. **53**, 1611 (2005)
121. Jiang, Q.K., Chang, Z.Y., Wang, X.D., Jiang, J.Z.: Metall. Mater. Trans. A **41A**, 1634 (2010)
122. Louzguine, D.V., Yavari, A.R., Ota, K., Vaughan, G., Inoue, A.: J. Non-Cryst. Solids **351**, 1639 (2005)
123. Qiao, J., Pelletier, J.-M., Casalini, R.: J. Phys. Chem. B **117**, 13658–13666 (2013)
124. Meng, F.Q., Tsuchiya, K., Yokoyama, Y.: Mater. Trans. **55**, 220–222 (2014)
125. Ketov S. V., Sun Y. H., Nachum S., Lu Z., Checchi A., Beraldin A. R., Bai H. Y., Wang W. H., Louzguine-Luzgin D. V., Carpenter M. A. & Greer A. L., Nature, **524**, 200–203 (2015).
126. Cahn, J.W.: Acta Metall. **9**, 795 (1961)
127. Kim, D.H., Kim, W.T., Park, E.S., Mattern, N., Eckert, J.: Prog. Mater. Sci. **58**(8), 1103–1172 (2013)
128. Inoue, A., Chen, S., Masumoto, T.: Mater. Sci. Eng. **A179/A180**, 346 (1994)
129. Kündig, A., Ohnuma, M., Ping, D., Ohkubo, T., Hono, K.: Acta Mater. **52**, 2441 (2004)
130. Wada, T., Louzguine-Luzgin, D.V., Inoue, A.: Scr. Mater. **57**, 901 (2007)
131. Sohn, S.W., Yook, W., Kim, W.T., Kim, D.H.: Intermetallics **23**, 57–62 (2012)
132. Park, B.J., Chang, H.J., Kim, D.H., Kim, W.T.: Appl. Phys. Lett. **85**, 6353–6355 (2004)
133. Mattern, N., Kühn, U., Gebert, A., Gemming, T., Zinkevich, M., Wendrock, H., Schultz, L.: Scr. Mater. **53**, 271–274 (2005)
134. Oh, J.C., Ohkubo, T., Kim, Y.C., Fleury, E., Hono, K.: Scr. Mater. **53**, 165 (2005)
135. Kundig, A.A., Ohnuma, M., Ohkubo, T., Abe, T., Hono, K.: Scr. Mater. **55**, 449 (2006)
136. Chang, H.J., Yook, W., Park, E.S., Kyeong, J.S., Kim, D.H.: Acta Mater. **58**, 2483–2491 (2010)
137. Louzguine-Luzgin, D.V., Wada, T., Kato, H., Perepezko, J., Inoue, A.: Intermetallics **18**, 1235–1239 (2010)
138. Caron, A., Zhang, Q.S., Minkow, A., Zadorozhnyy, V.Y., Fukuhara, M., Fecht, H.J., Louzguine-Luzgin, D.V., Inoue, A.: Mater. Sci. Eng. A **555**, 57–62 (2012)
139. Greer, A.L.: Mater. Sci. Eng. A **179**, 41–45 (1994)
140. Louzguine-Luzgin, D.V., Bazlov, A.I., Ketov, S.V., Greer, A.L., Inoue, A.: Acta Mater. **82**, 396–402 (2015)
141. Perepezko, J.H., Hebert, R.J., Tong, W.S.: Intermetallics **10**, 1079 (2002)
142. Kolmogorov, A.N.: Isz. Akad. Nauk. USSR Ser. Matem. **3**, 355 (1937)
143. Johnson, M.W.A., Mehl, K.F.: Trans. Am. Inst. Mining Met. Eng. **135**, 416 (1939)
144. Avrami, M.: J. Chem. Phys. **9**, 177 (1941)
145. Louzguine-Luzgin, D.V.: J. Alloys Compd. **586**, 216–219 (2014)
146. Koster, U., Punge-Witteler, B., Steinbrink, G.: Key Eng. Mater. **40-41**, 53 (1990)
147. Kelton, K.F., Spaepen, F.: Acta Metall. **33**, 455 (1985)
148. Watanabe, Y., Nakamura, Y.: J. Mater. Res. **7**, 2126 (1992)
149. Oreshkin, A.I., Mantsevich, V.N., Savinov, S.V., Oreshkin, S.I., Panov, V.I., Yavari, A.R., Miracle, D.B., Louzguine-Luzgin, D.V.: Acta Mater. **61**, 5216–5222 (2013)
150. Christian, J.W.: The Theory of Transformations in Metals and Alloys, p. 369. Pergamon, Oxford (1975)
151. Peker, A., Johnson, W.L.: Mater. Sci. Eng. A **179–180**, 173 (1994)
152. Nishiyama, N., Inoue, A.: Acta Mater. **47**, 1487 (1999)
153. Kim, J.H., Kim, S.G., Inoue, A.: Acta Mater. **49**, 615 (2001)
154. Schroers, J., Johnson, W.L.: Mater. Sci. Eng. A **375–377**, 781 (2004)
155. Louzguine, D.V., Inoue, A.: Scr. Mater. **47**, 887 (2002)
156. Louzguine-Luzgin, D.V., Inoue, A.: Physica B. Condens. Matter. **358**, 174 (2005)
157. Louzguine-Luzgin, D.V., Xie, G., Li, S., Zhang, Q.S., Zhang, W., Suryanarayana, C., Inoue, A.: J. Mater. Res. **24**, 1886–1895 (2009)
158. Greer, A.L.: Science **267**, 1947 (1995)
159. Gleiter, H.: Prog. Mater. Sci. **33**, 223 (1989)
160. Yavari, A.R., Negri, D.: Nanostruct. Mater. **8**, 969 (1997)

161. Louzguine, D.V., Inoue, A.: Scr. Mater. **43**, 371 (2000)
162. He, G., Eckert, J., Loser, W.: Acta Mater. **51**, 1621 (2003)
163. Altounian, Z., Batalla, E., Strom-Olsen, J.O., Walter, J.L.: J. Appl. Phys. **61**, 149 (1987)
164. Xing, L.Q., Eckert, J., Loser, W., Schultz, L., Herlach, D.M.: Phil. Mag. A. **79**, 1095 (1999)
165. Louzguine, D.V., Kato, H., Kim, H.S., Inoue, A.: J. Alloys Compd. **359**, 198 (2003)
166. Kulik, T., Matyja, H.: Mater. Sci. Eng. A **133**, 232–235 (1991)
167. Churyumov, A.Y., Bazlov, A.I., Zadorozhnyy, V.Y., Solonin, A.N., Caron, A., Louzguine-Luzgin, D.V.: Mater. Sci. Eng. A **550**, 358–362 (2012)
168. Caron, A., Kawashima, A., Fecht, H.-J., Louzguine-Luzguin, D.V., Inoue, A.: Appl. Phys. Lett. **99**, 171907 (2011)
169. Louzguine-Luzgin, D.V., Seki, I., Wada, T., Inoue, A.: Metall. Mater. Trans. A **43**, 2642–2648 (2012)
170. Shechtman, D., Blech, I.A., Gratias, D., Cahn, J.W.: Phys. Rev. Lett. **53**, 1951 (1984)
171. Levine, D., Steinhardt, R.J.: Phys. Rev. Lett. **53**, 2477 (1984)
172. Inoue, A., Kimura, H.M., Masumoto, T.: J. Mater. Sci. **22**, 1758 (1987)
173. Louzguine-Luzgin, D.V., Inoue, A.: Ann Rev. Mater. Res. **38**(403–423) (2008)
174. Ranganathan, S., Inoue, A.: Acta Mater. **54**, 3647 (2006)
175. Srivastava, A.K., Ranganathan, S.: Acta Mater. **44**, 2935 (1996)
176. Chen, M.W., Zhang, T., Inoue, A., Sakai, A., Sakurai, T.: Appl. Phys. Lett. **75**, 1697 (1999)
177. Inoue, A., Zhang, T., Saida, J., Matsushita, M., Chen, M.W., Sakurai, T.: Mater. Trans. JIM **40**, 1137 (1999)
178. Louzguine, D.V., Inoue, A.: Appl. Phys. Lett. **78**, 1841 (2001)
179. Louzguine, D.V., Ko, M.S., Inoue, A.: Appl. Phys. Lett. **76**, 3424 (2000)
180. Li, C., Saida, J., Matsushita, M., Inoue, A.: Appl. Phys. Lett. **77**, 528 (2000)
181. Chen, N., Louzguine, D.V., Ranganathan, S., Inoue, A.: Acta Mater. **53**, 759 (2005)
182. Louzguine-Luzgin, D.V., Zeng, Y., Setyawan, A.D.H., Nishiyama, N., Kato, H., Saida, J., Inoue, A.: J. Mater. Res. **22**, 1087–1092 (2007)
183. Louzguine, D.V., Inoue, A.: Scr. Mater. **48**, 1325 (2003)
184. Louzguine, D.V., Inoue, A.: J. Alloys Compd. **361**, 153 (2003)
185. Louzguine-Luzgin, D.V., Inoue, A., Nagahama, D., Hono, K.: Appl. Phys. Lett. **87**, 211918 (2005)
186. Louzguine-Luzgin, D.V., Churyumov, A.Y.: Mater. Charact. **96**, 6–12 (2014)
187. Argon, A.S.: J. Phys. Chem. Solids **43**, 945 (1982)
188. Liu, F.X., Liaw, P.K., Wang, G.Y., Chiang, C.L., Smith, D.A., Rack, P.D., Chu, J.P., Buchanan, R.A.: Intermetallics **14**, 1014 (2006)
189. Togashi, N., Ishida, M., Nishiyama, N., Inoue, A.: Rev. Adv. Mater. Sci. **18**, 93 (2008)
190. Yavari, A.R., Lewandowski, J.J., Eckert, J.: MRS Bull. **32**, 635 (2007)
191. Inoue, A., Shen, B., Koshiba, H., Kato, H., Yavari, A.R.: Nat. Mater. **2**, 661–663 (2003)
192. Wang, J.F., Li, R., Hua, N.B., Zhang, T.: J. Mater. Res. **26**, 2072–2079 (2011)
193. Inoue, A., Zhang, T., Takeuchi, A.: Appl. Phys. Lett. **71**, 464 (1997)
194. Spaepen, F.: Acta Metall. **25**, 407 (1977)
195. Argon, A.: Acta Metall. **27**, 47 (1979)
196. Conner, R.D., Li, Y., Nix, W.D., Johnson, W.L.: Acta Mater. **52**, 2429–2434 (2004)
197. Donovan, P.E., Stobbs, W.M.: Acta Metall. **29**, 1419–1436 (1981)
198. Yokoyama, Y., Fujita, K., Yavari, A.R., Inoue, A.: Philos. Mag. Lett. **89**, 322 (2009)
199. Song, K.K., Pauly, S.Y., Zhang, S., Scudino, P., Gargarella, K., Surreddi, B., Kühn, U., Eckert, J.: Intermetallics **19**, 1394 (2011)
200. Yokoyama, Y., Tokunaga, H., Yavari, A.R., Kawamata, T., Yamasaki, T., Fujita, K., Sugiyama, K., Liaw, P.K., Inoue, A.: Metall. Mater. Trans. A **42**, 1468–1475 (2011)
201. Guo, H., Yan, P.F., Wang, Y.B., Tan, J., Zhang, Z.F., Sui, M.L., Ma, E.: Nat. Mater. **6**, 735 (2007)
202. Louzguine-Luzgin, D.V., Yavari, A.R., Xie, G.Q., Madge, S., Li, S., Saida, J., Greer, A.L., Inoue, A.: Phil. Mag. Lett. **90**, 139 (2010)

203. Liu, Y.H., Wang, G., Wang, R.J., Zhao, D.Q., Pan, M.X., Wang, W.H.: Science **315**, 1385–1388 (2007)
204. He, Q., Cheng, Y.-Q., Ma, E., Xu, J.: Acta Mater. **59**, 202–215 (2011)
205. Demetriou, M.D., Launey, M.E., Garrett, G., Schramm, P.J., Hofmann, D.C., Johnson, W.L., Ritchie, R.O.: Nat. Mater. **10**, 123 (2011)
206. Chen, N., Louzguine-Luzgin, D.V., Xie, G.Q., Inoue, A.: Appl. Phys. Lett. **94**, 131906 (2009)
207. Greer, A.L., Ma, E.: MRS Bull. **32**, 611 (2007)
208. Schuh, C.A., Hufnagel, T.C., Ramamurty, U.: Acta Mater. **55**, 4067 (2007)
209. Lewandowski, J.J., Wang, W.H., Greer, A.L.: Philos. Mag. Lett. **85**, 77 (2005)
210. Hajlaoui, K., Yavari, A.R., Doisneau, B., Lemoulec, A., Botta, W.J., Vaughan, G., Greer, A. L., Inoue, A., Zhang, W., Kvick, A.: Scr. Mater. **54**, 1829 (2006)
211. Kim, K.B., Das, J., Baier, F., Tang, M.B., Wang, W.H., Eckert, J.: Appl. Phys. Lett. **88**, 051911 (2006)
212. Chen, N., Pan, D., Louzguine-Luzgin, D.V., Xie, G.Q., Chen, M.W., Inoue, A.: Scr. Mater. **62**, 17–20 (2010)
213. Madge, S.V., Louzguine-Luzgin, D.V., Lewandowski, J.J., Greer, A.L.: Acta Mater. **60**, 4800–4809 (2012)
214. Zhang, Y., Wang, W.H., Greer, A.L.: Nat. Mater. **5**, 857 (2006)
215. Hufnagel, T.C., Jiao, T., Li, Y., Xing, L.Q., Ramesh, K.T.: J. Mater. Res. **17**, 1441 (2002)
216. Dalla Torre, F.H., Dubach, A., Siegrist, M., Löffler, J.F.: Appl. Phys. Lett. **89**, 091918 (2006)
217. Sanditov, D.S., Mantatov, V.V., Sangadiev, S.S.: Glas. Phys. Chem. **30**, 4150419 (2004)
218. Srikant, G., Chollacoop, N., Ramamurty, U.: Acta Mater. **54**, 5171 (2006)
219. Guo, H., Yan, P.F., Wang, Y.B., Tan, J., Zhang, Z.F., Sui, M.L., Ma, E.: Nat. Mater. **6**, 735 (2007)
220. Louzguine-Luzgin, D.V., Zadorozhnyy, V.Y., Chen, N., Ketov, S.V.: J. Non-Cryst. Solids **396–397**, 20–24 (2014)
221. Louzguine-Luzgin, D.V., Packwood, D.M., Xie, G., Churyumov, A.Y.: J. Alloys Compd. **561**, 241 (2013)
222. Greer, A.L., Cheng, Y.Q., Ma, E.: Mater. Sci. Eng. R **74**, 71 (2013)
223. Liu, Z.Y., Yang, Y., Liu, C.T.: Acta Mater. **61**, 5928 (2013)
224. Klaumünzer, D., Maaß, R., Löffler, J.F.: J. Mater. Res. **26**, 1453 (2011)
225. Perepezko, J.H., Imhoff, S.D., Chen, M.W., Gonzalez, S., Inoue, A.: J. Alloys Compd. **536**, S55 (2012)
226. Lewandowski, J.J., Greer, A.L.: Nat. Mater. **5**, 15–18 (2006)
227. Guan, P., Chen, M.W., Egami, T.: Phys. Rev. Lett. **104**, 205701 (2010)
228. Ketov, S., Louzguine-Luzgin, D.V.: Sci. Rep. **3**, 2798 (2013)
229. Yang, B., Liaw, P.K., Wang, G., Morrison, M., Liu, C.T., Buchanan, R.A., Yokoyama, Y.: Intermetallics **12**, 1265–1274 (2004)
230. Louzguine-Luzgin, D.V., Ketov, S.V., Wang, Z., Miyama, M.J., Tsarkov, A.A., Churyumov, A.Y.: Mater. Sci. Eng. A **616**, 288–296 (2014)
231. Jia, N., Eisenlohr, P., Roters, F., Raabe, D., Zhao, X.: Acta Mater. **60**, 3415–3434 (2012)
232. Schuh, C.A., Hufnagel, T.C., Ramamurty, U.: Acta Mater. **55**, 4067–4109 (2007)
233. Mear, F.O., Lenk, B., Zhang, Y., Greer, A.L.: Scr. Mater. **59**, 1243–1246 (2008)
234. Louzguine-Luzgin, D.V., Seki, I., Yamamoto, T., Kawaji, H., Suryanarayana, C., Inoue, A.: Intermetallics **23**, 177–181 (2012)
235. Zhang, Z.F., Eckert, J., Schultz, L.: Metall. Mater. Trans. A. **35**, 3489–3498 (2004)
236. Wang, G.Y., Liaw, P.K., Peter, W.H., Yang, B., Yokoyama, Y., Benson, M.L., Green, B.A., Kirkham, M.J., White, S.A., Saleh, T.A., McDaniels, R.L., Steward, R.V., Buchanan, R.A., Liu, C.T., Brook, C.R.: Intermetallics **12**, 885–892 (2004)
237. Kawashima, A., Zeng, Y., Fukuhara, M., Kurishita, H., Nishiyama, N., Miki, H., Inoue, A.: Mater. Sci. Eng. A **498**, 475–481 (2008)
238. Tabachnikova, E.D., Podol'ski, A.V., Bengus, V.Z., Smirnov, S.N., Luzgin, D.V., Inoue, A.: Low Temp. Phys. **34**, 675–677 (2008)

239. Louzguine-Luzgin, D.V., Vinogradov, A., Li, S., Kawashima, A., Xie, G., Yavari, A.R., Inoue, A.: Metall. Mater. Trans. A. **42**, 1504–1510 (2011)
240. Vinogradov, A., Lazarev, A., Louzguine-Luzgin, D.V., Yokoyama, Y., Li, S., Yavari, A.R., Inoue, A.: Acta Mater. **58**, 6736 (2010)
241. Kawamura, Y., Nakamura, T., Inoue, A., Masumoto, T.: Mater. Trans. JIM **40**, 794 (1999)
242. Schuh, C.A., Hufnagel, T.C., Ramamurty, U.: Acta Mater. **55**, 4067 (2007)
243. Inoue, A., Wada, T., Louzguine-Luzgin, D.V.: Mater. Sci. Eng. A **471**, 144 (2007)
244. Zadorozhnyy, V.Y., Inoue, A., Louzguine-Luzgin, D.V.: Intermetallics **31**, 173 (2012)
245. Lee, M.L., Li, Y., Schuh, C.A.: Acta Mater. **52**, 4121–4131 (2004)
246. Hofmann, D.C., Suh, J.Y., Wiest, A., Lind, M.L., Demetriou, M.D., Johnson, W.L.: Proc. Natl. Acad. Sci. U. S. A. **105**, 20136 (2008)
247. Hofmann, D.C., Suh, J.Y., Wiest, A., Duan, G., Lind, M.L., Demetriou, M.D., Johnson, W.L.: Nature **451**, 1085 (2008)
248. Qiao, J.W., Zhang, T., Yang, F.Q., Liaw, P.K., Pauly, S., Xu, B.S.: Sci Rep **3**, 2816 (2011)
249. Wu, Y., Xiao, Y., Chen, G., Liu, C.T., Lu, Z.: Adv. Mater. **22**, 2270 (2010)
250. Hofmann, D.C.: Science **329**, 1294 (2010)
251. Pauly, S., Gorantla, S., Wang, G., Kühn, U., Eckert, J.: Nat. Mater. **9**, 473 (2010)
252. Gargarella, P., Pauly, S., Songa, K.K., Hu, J., Barekar, N.S., Khoshkhoo, M.S., Teresiak, A., Wendrock, H., Kühn, U., Ruffing, C., Kerscher, E., Eckert, J.: Acta Mater. **61**, 151–162 (2013)
253. Makino, A., Inoue, A., Masumoto, T.: Mater. Trans. JIM **36**, 924 (1995)
254. Amiya, K., Urata, A., Nishiyama, N., Inoue, A.: J. Appl. Phys. **101**, 112 (2007)
255. Inoue, A., Shen, B.L.: Mater. Sci. Eng. A **375–377**, 302 (2004)
256. Inoue, A., Shinohara, Y., Gook, J.S.: Mater. Trans. JIM **36**, 1427 (1995)
257. Inoue, A., Makino, A., Mizushima, T.: J. Magn. Magn. Mater. **215–216**, 246 (2000)
258. Chiriac, H., Lupu, N.: IEEE Trans. Magn. **41**, 3289–3291 (2005)
259. Zhang, W., Fang, C.F., Li, Y.H.: Scr. Mater. **69**, 77–80 (2013)
260. Mizushima, T., Ikarashi, K., Yoshida, M., Makino, A., Inoue, A.: Mater. Trans. JIM **40**, 1019 (1999)
261. Bitoh, T., Makino, A., Inoue, A.: Mater. Trans. **45**, 1219 (2004)
262. Inoue, A., Zhang, T., Itoi, T.: Mater. Trans. JIM **38**, 359 (1997)
263. Makino, A., Kubota, T., Chang, C., Makabe, M., Inoue, A.: J. Mater. Res. **23**, 1339 (2008)
264. Makino, A., Li, X., Yubuta, K., Chang, C., Kubota, T., Inoue, A.: Scr. Mater. **60**, 277–280 (2009)
265. Wang, J.F., Li, R., Hua, N.B., Huang, L., Zhang, T.: Scr. Mater. **65**, 536–539 (2011)
266. Yang, W., Liu, H., Xue, L., Li, J., Dun, C., Zhang, J., Zhao, Y., Shen, B.: J. Magn. Magn. Mater. **35**, 172–176 (2013)
267. Li, X., Zhang, Y., Kato, H., Makino, A., Inoue, A.: Key Eng. Mater. **508**, 112–116 (2012)
268. Jung, H.Y., Yi, S.: Intermetallics **49**, 18–22 (2014)
269. Jung, H.Y., Stoica, M., Yi, S., Kim, D.H., Eckert, J.: J. Magn. Magn. Mater. **364**, 80–84 (2014)
270. Moon, Y.M., Kim, K.S., Yu, S.C., Rao, K.V.: J. Magn. Magn. Mater. **177**, 968 (1998)
271. Suzuki, K., Makino, A., Inoue, A., Masumoto, T.: J. Appl. Phys. **74**, 3316 (1993)
272. Zhang, W., Inoue, A.: Mater. Trans. **429**, 1835 (2001)
273. Lee, S., Kato, H., Kubota, T., Yubuta, K., Makino, A., Inoue, A.: Mater. Trans. **49**, 506 (2008)
274. Zhang, J.H., Chang, C.T., Wang, A., Shen, B.L.: J. Non-Cryst. Solids **358**, 1443–1446 (2012)
275. Cremaschi, V., Arcondo, B., Sirkin, H., Vazquez, M., Asenjo, A., Garcia, J.M., Abrosimova, G., Aronin, A.: J. Mater. Res. **15**, 1936 (2000)
276. Shen, B., Inoue, A.: Mater. Trans. **43**, 1235 (2002)
277. Makino, A., Inoue, A., Mizushima, T.: Mater. Trans. **41**, 1471 (2000)
278. Song, D.S., Kim, J.H., Fleury, E., Kim, W.T., Kim, D.H.: J. Alloys Compd. **389**, 159 (2005)
279. Makino, A., Kubota, T., Chang, C.T., Makabe, M., Inoue, A., Magn, J.: Magn. Mater. **320**, 2499 (2008)

280. Shen, B.L., Chang, C.T., Inoue, A.: Intermetallics **15**, 9–16 (2007)
281. Xie, G., Kimura, H.M., Louzguine-Luzgin, D.V., Men, H., Inoue, A.: Intermetallics **20**, 76 (2012)
282. Nielsch, K., Wehrspohn, R.B., Fischer, S.F., Kronmiller, H., Kirsehner, J., Gosele, U.: Mater. Res. Soc. Symp. Proc. **9**(636) (2001)
283. Inoue, A., Zhang, T.: Mater. Sci. Eng. A **226–228**, 393 (1997)
284. Inoue, A., Zhang, T., Zhang, W., Takeuchi, A.: Mater. Trans. JIM **37**, 99 (1996)
285. Coehoorn, R., Mooij, D. B., Duchateau, J. P. W. B, Buschow, K. J. H: J. Phys. C. **8**, 669, (1988)
286. Pang, S., Zhang, T., Asami, K., Inoue, A.: Mater. Sci. Eng. A. **375**, 368 (2004)
287. Peter, W.H., Buchanan, R.A., Liu, C.T., Liaw, P.K., Morrison, M.L., Horton, J.A., Carmichael, C.A., Wright, J.L.: Intermetallics **10**, 1157 (2002)
288. Oak, J.J., Louzguine-Luzgin, D.V., Inoue, A.: Mater. Sci. Eng. C **29**, 322 (2009)
289. Eisenbarth, E., Velten, D., Müller, M., Thull, R., Breme, J.: Biomaterials **25**, 5705 (2004)
290. Steinemann, S.G., Winter, G.D., Leray, J.L., de Goot, K. (eds.): Evaluation of Biomaterials, 33rd edn. Wiley, Chichester, NY (1980)
291. Pang, S.J., Zhang, T., Asami, K., Inoue, A.: Corros. Sci. **44**, 1847 (2002)
292. Wada, T., Inoue, A., Greer Appl, A.L.: Phys. Lett. **86**, 251907 (2005)
293. Brothers, A.H., Dunand, D.C.: Appl. Phys. Lett. **84**, 1108 (2004)
294. Schroers, J., Veazey, C., Johnson, W.L.: Appl. Phys. Lett. **82**, 370 (2003)
295. Xie, G., Zhang, W., Louzguine-Luzgin, D.V., Kimura, H.M., Inoue, A.: Scr. Mater. **55**, 687 (2006)
296. Zberg, B., Uggowitzer, P.J., Loffler, J.F.: Nat. Mater. **8**, 887–891 (2009)
297. Gu, X.N., Zheng, Y.F., Zhong, S.P., Xi, T.F., Wang, J.Q., Wang, W.H.: Biomaterials **31**, 1093–1103 (2010)
298. Wessels, V., Le Mené, G., Fischerauer, S.F., Kraus, T., Weinberg, A.-M., Uggowitzer, P.J., Löffler, J.F.: Adv. Eng. Mater. **14**, B357–B364 (2012)
299. Wang, Y.B., Xie, X.H., Li, H.F., Wang, X.L., Zhao, M.Z., Zhang, E.W., Bai, Y.J., Zheng, Y. F., Qin, L.: Acta Biomater. **7**, 3196–3208 (2011)
300. Yu, H.-J., Wang, J.-Q., Shi, X.-T., Louzguine-Luzgin, D.V., Wu, H.-K., Perepezko, J.H.: Adv. Funct. Mater. **23**, 4793–4800 (2013)
301. Jiao, W., Li, H.F., Zhao, K., Bai, H.Y., Wang, Y.B., Zheng, Y.F., Wang, W.H.: J. Non-Cryst. Solids **357**, 3830–3840 (2011)
302. Li, J.F., Zhao, D.Q., Zhang, M.L., Wang, W.H.: Appl. Phys. Lett. **93**, 17190 (2008)
303. Li, H.F., Zhao, K., Wang, Y.B., Zheng, Y.F., Wang, W.H.: J. Biomed. Mater. Res. B Appl. Biomater. **100B**, 368–378 (2012)
304. Jiao, W., Zhao, K., Xi, X.K., Zhao, D.Q., Pan, M.X., Wang, W.H.: J. Non-Cryst. Solids **356**, 1867–1870 (2010)
305. Ashby, M.F., Greer, A.L.: Scr. Mater. **54**, 321 (2006)
306. Nishiyama, N., Amiya, K., Inoue, A.: Mater. Sci. Eng. A. **449**, 79 (2007)
307. Nishiyama, N., Amiya, K., Inoue, A.: J. Non-Cryst. Solids **353**, 3615 (2007)
308. Hara, S., Hatakeyama, N., Itoh, N., Kimura, H.-M., Inoue, A.: J. Membr. Sci. **211**, 149 (2003)
309. Yamaura, S., Shimpo, Y., Okouchi, H., Nishida, M., Kajita, O., Kimura, H.M., Inoue, A.: Mater. Trans. **44**, 1885 (2003)
310. Jayalakshmi, S., Vasantha, V.S., Fleury, E., Gupta, M.: Appl. Energy **90**, 94 (2012)
311. Matsumoto, H., Urata, A., Yamada, Y., Inoue, A.: IEEE Trans. Magn. **46**, 373 (2010)
312. Okumura, K., Kajita, J., Kurosaki, J., Kimura, H.M., Inoue, A.: 10th Int'l Conf on Shot Peening, Tokyo, Japan, Sept 15–18 2008
313. Kobayashi, A., Yano, S., Kimura, H.M., Inoue, A.: Surf. Coat. Technol. **202**, 2513 (2008)
314. Kobayashi, A., Yano, S., Kimura, H.M., Inoue, A.: Mater. Sci. Eng. **B148**, 110 (2008)
315. Saotome, Y., Miwa, S., Zhang, T., Inoue, A.: J. Mater. Process. Technol. **113**, 64–69 (2001)
316. Saotome, Y., Noguchi, T., Zhang, T., Inoue, A.: J Mater. Sci. Eng. A. **375–377**, 389 (2004)

317. Saotome, Y., Imai, K., Shioda, S., Shimizu, S., Zhang, T., Inoue, A.: Intermetallics **10**, 1241 (2002)
318. Chen, N., Yang, H.A., Caron, A., Chen, P.C., Lin, Y.C., Louzguine-Luzgin, D.V., Yao, K.F., Esashi, M., Inoue, A.: J. Mater. Sci. **46**, 2091–2096 (2011)
319. Nakayama, K.S., Yokoyama, Y., Ono, T., Chen, M.W., Akiyama, K., Sakurai, T., Inoue, A.: Adv. Mater. **22**, 863 (2010)
320. Carmo, M., Sekol, R.C., Ding, S., Kumar, G., Schroers, J., Taylor, A.D.: ACS Nano **5**, 2979 (2011)
321. Walter, E.C., Penner, R.M., Liu, H., Ng, K.H., Zach, M.P., Favier, F.: Surf. Interface Anal. **34**, 409–412 (2002)
322. Keating, C.D., Natan, M.J.: Adv. Mater. **15**, 451 (2003)
323. Tanase, M., Felton, E.J., Gray, D.S., Hultgren, A., Chen, C.S., Reich, D.H.: Lab Chip **5**, 598 (2005)
324. Lee, J.-W., Lin, Y.C., Chen, N., Louzguine, D.V., Esashi, M., Gessner, T.: Jpn. J. Appl. Phys. **50**, 087301 (2011)
325. Chen, N., Frank, R., Asao, N., Louzguine-Luzgin, D.V., Sharma, P., Wang, J.Q., Xie, G.Q., Ishikawa, Y., Hatakeyama, N., Lin, Y.C., Esashi, M., Yamamoto, Y., Inoue, A.: Acta Mater. **59**, 6433 (2011)

ERRATUM TO

Novel Functional Magnetic Materials

Arcady Zhukov

A. Zhukov (ed.), *Novel Functional Magnetic Materials*, Springer Series
in Materials Science 231, DOI 10.1007/978-3-319-26106-5

DOI 10.1007/978-3-319-26106-5

The spelling of the editor name **Arkady Zhukov** was incorrect. It has been updated as **Arcady Zhukov** throughout this book.

The original version of the book was revised because Arcady Zhukov's name was misspelled. An erratum explaining this can be found at http://dx.doi.org/10.1007/978-3-319-26106-5

A. Zhukov (ed.), *Novel Functional Magnetic Materials*, Springer Series
in Materials Science 231, DOI 10.1007/978-3-319-26106-5_11

Index

A. Zhukov (ed.), *Novel Functional Magnetic Materials*, Springer Series in Materials Science 231, DOI 10.1007/978-3-319-26106-5

N

O

P

MIX
Papier aus verantwortungsvollen Quellen
Paper from responsible sources
FSC® C105338

If you have any concerns about our products,
you can contact us on
ProductSafety@springernature.com

In case Publisher is established outside the EU,
the EU authorized representative is:
Springer Nature Customer Service Center GmbH
Europaplatz 3, 69115 Heidelberg, Germany

Printed by Libri Plureos GmbH
in Hamburg, Germany